Geological Hazards

Geological Hazards

Their assessment, avoidance and mitigation

F. G. Bell

Department of Geology and Applied Geology, University of Natal, Durban, South Africa

E & FN SPON
An Imprint of Routledge

London and New York

First published in 1999 by E & FN Spon, an imprint of Routledge
11 New Fetter Lane, London EC4P 4EE

Simultaneously published in the USA and Canada
by Routledge
29 West 35th Street, New York, NY 10001

Typeset in Garamond by
J&L Composition Ltd, Filey, North Yorkshire
Printed and bound in Great Britain by
Biddles Ltd, Guildford and King's Lynn

British Library Cataloguing in Publication Data
A catalogue record for this book is available
from the British Library

Library of Congress Cataloging in Publication Data
Bell, F. G. (Frederic Gladstone)
  Geological hazards : their assessment, avoidance, and
mitigation/F.G. Bell.
    p.   cm.
  Includes bibliographical references and index.
  ISBN 0–419–16970–9 (hb)
  1. Natural disasters. 2. Geodynamics. 1. Title.
GB5014.B45 1999
363.34'9—dc21   98–39190
                                                        CIP

ISBN 0-419-16970-9

Contents

Chapter 1

Geohazards: assessment and planning

1.1 Introduction

It has been estimated that natural hazards cost the global economy over $50 000 million per year. Two-thirds of this sum is accounted for by damage, and the remainder represents the cost of predicting, preventing and mitigating disasters. Man-made hazards such as groundwater pollution and subsidence and damage caused by expansive soils add to this figure.

A natural hazard has been defined by UNESCO (see Varnes, 1984) as the probability of occurrence within a specified period of time and within a given area of a potentially damaging phenomenon. However, geological hazards are not all natural; some are influenced by or brought about by man. Hazards pose a threat to man, his property and the environment and so they are of more significance when they occur in highly populated areas. Moreover, as the global population rises so the significance of hazard is likely to rise. For instance, according to the United Nations Disaster Relief Coordinator (UNDRO, 1991), the number of people affected by natural disasters increases by some 6% annually. In developed societies, hazards can cause great damage to property with associated high costs. In developing areas, loss of life and injury are often of far more consequence. This is not to say that developing countries do not suffer heavy economic losses due to natural disasters. They do. For example, Mora (1995) indicated that the Limón–Telire earthquake of 1991 cost Costa Rica between 5 and 8.5% of that country's gross national product that year. Indeed, estimates made by the Comisión Económica Para América Latina (CEPAL, 1990) suggested that between 1960 and 1985 the countries of Central America lost on average 2.7% of their gross national product annually as a result of natural disasters. This compares with 1% for a developed country like the United States over the same period. Furthermore, of the three million people killed by natural disasters between 1960 and 1990 (excluding the former Communist bloc) only 0.7% were killed in developed countries, over three-quarters were killed in third world countries, and the rest were in countries with average gross national product. Unfortunately, however, development in many of these countries is placing increasing numbers of people and property at risk as a consequence of unplanned occupancy and the use of marginal and high-risk zones. In particular, overcultivation and deforestation aggravate the problem. However, the associated problems of soil erosion and excessive run-off have not been confined to the developing world; such problems occurred, often with serious effects, in the United States in the 1930s. As geological hazards are widespread and fairly common events they obviously give cause for concern. Many such geohazards impose constraints on development or exert a substantial influence on society. Hence, it is important that geological hazards are understood in order that their occurrence and behaviour can be predicted and that measures can be taken to reduce their impact.

Geohazards generally are not one-off events but tend to occur again and again. The frequency of a hazard event can be regarded as the number of events of a given magnitude in a particular period of time. As such, a recurrence interval for such an event can be determined in terms of the average length of time between events of a certain size. Generally, however, the most catastrophic events are highly infrequent, so their impact does have the greatest overall significance. Similarly, frequent events are normally too small to have significant aggregate effect.

A hazard involves a degree of risk, the elements at risk being life, property, possessions and the environment. Risk involves quantification of the probability that a hazard will be harmful, and the tolerable degree of risk depends upon what is being risked, life being much more important than property. The risk to society can be regarded as the magnitude of a hazard multiplied by the probability of its occurrence. If there are no mitigation measures, no warning systems and no evacuation plans for an area that is subjected to a recurring hazard, then such an area has the highest vulnerability. Vulnerability, V, was defined by UNESCO as the degree of loss to a given element or set of elements at risk resulting from the occurrence of a natural phenomenon of a given magnitude. It is expressed on a scale from 0 (no damage) to 1 (total loss). The factors that usually influence loss are population distribution, social infrastructure, and socio-cultural and socio-political differentiation and diversity.

The effects of a disaster may be lessened by reduction of vulnerability. Short-term forecasts a few days ahead of an event may be possible and will complement relief and rehabilitation planning. In addition, it is possible to reduce the risk of disaster by a combination of preventative and mitigative measures (see below). To do this successfully, the patterns of behaviour of the geological phenomena posing the hazards need to be understood and the areas at risk identified. The level of potential risk may be decreased and the consequence of disastrous events mitigated by introducing regulatory measures or other inducements into the physical planning process. The impact of disasters may be reduced further by incorporating into building codes and other regulations appropriate measures so that structures will withstand or accommodate potentially devastating events.

UNESCO went on to define specific risk, elements at risk and total risk.

Specific risk, R, refers to the expected degree of loss associated with a particular hazard. It may be expressed as the product of hazard, H, and vulnerability ($R = H \times V$). Elements at risk, E, refers to the population, property and economic activity at risk in a given area. Lastly, total risk, R_T, refers to the expected number of lives lost, of injuries, property damage and disruption to economic activity brought about by a given hazard. It can be regarded as the product of specific risk and the elements at risk ($R_T = R \times E$ or $R_T = E \times H \times V$). The net impact of a hazard can be looked upon in terms of the benefits derived from inhabiting a zone in which a hazard occurs minus the costs involved in mitigation measures. Mitigation measures may be structural or technological on the one hand or regulatory on the other (Figure 1.1).

The response to geohazards by risk evaluation leading to land-use planning and appropriate engineering design of structures is lacking in many parts of the world, yet most geohazards are amenable to avoidance or prevention

Figure 1.1 Non-structural and structural methods of disaster mitigation (after Alexander, 1993).

Non-structural methods: (a) Short-term: Emergency plans:	Civil and military forces	Coordinator(s) police and fireman Red Cross and charities volunteer groups medical service
Evacuation plans: Prediction of impact: Warning processes:	Routes and reception centres for the general public, for vulnerable groups: the very young, elderly, sick or handicapped, Monitoring equipment, forecasting methods and models, General and specialized warning (e.g. ethnic)	

(b) Long-term
    Building codes and construction norms
    Hazard micro-zonation:
 	
    Land-use control:
 	
    Probabilistic risk analysis
    Insurance
    Taxation
    Education and training
	Selected risks
	All risks
	Regulations, prohibitions, moratoria, compulsory purchase	

Structural methods:
    Retrofitting of existing structures
    Reinforcement of new structures:
 	
    Safety features:
 	
    Engineering phenomenology
    Probabilistic prediction of impact strength
	Design features
	overdesign
	Structural safeguards
	fail-safe design	

measures. In addition, the causes of geological hazards are often reasonably well understood, so most can be identified and even predicted with a greater or lesser degree of accuracy. Nonetheless, geological hazards can represent obstacles that impede economic development, especially in underdeveloped countries. Consequently, measures should be included in planning processes in order to avoid severe problems attributable to hazards that lead to economic and social disruption. However, in underdeveloped and developing countries the capacity may not exist to effect planning processes or to establish mitigating measures, even though in some instances economic losses due to natural disasters could be reduced with only small investment.

Geological hazards vary in their nature and can be complex. One type of hazard, for example, an earthquake, can be responsible for the generation of others such as liquefaction of sandy soils, landslides or tsunamis. Certain hazards such as earthquakes and landslides are rapid-onset hazards and so give rise to sudden impacts. Others such as soil erosion and subsidence due to the abstraction of groundwater may take place gradually over an appreciable period of time. Furthermore, the effects of natural geohazards may be difficult to separate from the impact of man. For example, although soil erosion is a natural process it frequently has been accentuated by the interference of man. In fact, modification of nature by man usually increases the frequency and severity of natural geohazards, and at the same time these increase the threats to human occupancy. Yet other geohazards are primarily attributable to the activity of man such as pollution of groundwater and subsidence due to mining.

As the world population increases the risks attributable to natural geohazards increase. Nevertheless, people continue to live in areas of known geohazard despite the known levels of risk. The reasons for this are many and various but in doing so a demand for protection arises in developed countries. In poor countries, people remain in hazard areas because there is nowhere else to go, no other place to offer them a living. They are therefore obliged to accept the risk of hazard and its consequences. In such situations a sense of resignation, of fatalism, frequently develops and may be manipulated by unscrupulous political and economic interests. The extent to which protection measures can be adopted depends not just upon the type of geohazard and the technology available to combat the hazard but also on the ability and willingness to pay for protective measures. Permanent evacuation of areas of high hazard, even if feasible, involves compensation of the people involved. Again this can only be accomplished in societies that are rich enough to pay and where the authorities are willing to do so.

Many losses attributable to geohazards may be compensated for by way of government grants or loans. Such aid is frequently linked to a defined hazard zone such as the 100-year flood, which in the United States defines the area that is subject to flood damage compensation. If a community wishes to qualify for federal aid, then it must join the Federal Flood Insurance Program. However, financial assistance in the form of government payouts does not help to reduce the hazard. Furthermore, the likelihood of financial aid in an emergency may encourage development in hazard-prone areas, which is contrary to the principles that should be embodied in planning in relation to hazard.

1.2 Risk assessment

Risk, as mentioned, should take account of the magnitude of the hazard and the probability of its occurrence. Risk arises out of uncertainty due to insufficient information being available about a hazard and to incomplete understanding of the mechanisms involved. The uncertainties prevent accurate predictions of hazard occurrence. Risk analysis involves identifying the degree of risk, then estimating and evaluating it. An objective assessment of risk is obtained from a statistical assessment of instrumental data and/or data gathered from past events (Wu *et al.*, 1996). The risk to society can be quantified in terms of the number of deaths attributable to a particular hazard in a given period of time and the resultant damage to property. The costs involved can then be compared with the costs of hazard mitigation. Unfortunately, not all risks and benefits are readily amenable to measurement and assessment in financial terms. Even so, risk assessment is an important objective in decision making by urban planners because it involves the vulnerability of people and the urban infrastructure on the basis of probability of occurrence of an event. The aims of risk analysis are to improve the planning process, as well as reducing vulnerability and mitigating damage.

The occurrence of a given risk in a particular period of time can be expressed in terms of probability as follows:

$$R = \Sigma P(E)Ct \tag{1.1}$$

where R is the level of risk per unit time; E is events expressed in terms of probability, P; C refers to the consequences of the events; and t is a unit of time. Risk analysis should include the magnitude of the hazard as well as its probability. The magnitude of the hazard will be related to the size of the population at risk. The assessment should mention all the assumptions and conditions on which it is based and, if these are not constant, the conclusions should vary according to the degree of uncertainty, a range of values representing the probability distribution. The confidence limits of the predictions should be provided in the assessment.

Jefferies *et al.* (1996) referred to the use of the Bayesian approach in the assessment of risk probability. The Bayesian approach differs from the traditional use of statistics in assessing probability. In the conventional statistical assessment, probability expresses the relative frequency of occurrence of an event based upon data collected from various sources. In the Bayesian approach, probability expresses current judgement of likely events and does not require occurrence data to estimate probabilities, although, if available, such data can be used. In other words, in the Bayesian approach probabilities quantify judgement. A number of computer programs are available to carry out such analyses and are referred to by Jefferies *et al.*. They used the Bayesian approach to illustrate how the risk of contamination of groundwater from a landfill could be assessed.

The acceptable level of risk is inversely proportional to the number of people exposed to a hazard. Acceptable risk, however, is complicated by cultural factors and political attitudes. It is the latter two factors that are likely to determine public policy in relation to hazard mitigation and regulatory control.

Although it can be assumed that there is a point where the cost involved in the reduction of risk equals the savings earned by the reduction of risk, society tends to set arbitrary tolerance levels based largely on the perception of risk and the priorities for their management. Some hazards are far more emotive than others, notably the disposal of radioactive waste, and therefore may attract more funds from politically sensitive authorities to deal with associated problems. In essence, something is safe if its risks are judged to be acceptable, but there is little consensus as to what is acceptable.

The Health and Safety Executive (1988) in Britain defined a concept of tolerability. This it regarded as a willingness to live with a risk in order to obtain certain benefits in the confidence that the risk was being properly controlled. To tolerate a risk does not mean that it can be regarded as negligible or that it can be ignored. A risk needs to be kept under review and further reduced if and when necessary. The use of tolerability to guide public policy involves both technical and social considerations.

A number of problems are inherent in risk management. For instance, the degree of risk does not increase linearly with the length of exposure to a hazard. Moreover, with time the response to risk can change so that mitigation and risk reduction can change. As Alexander (1993) pointed out, the dichotomy between actual and perceived risk does not help attempts to reduce risk and promotes a conflict of objectives. Nonetheless, risk management has to attempt to determine the level of risk that is acceptable. This has been referred to as risk balancing. Public and private resources may then be allocated to meet a level of safety that is acceptable to the public. Cost–benefit analyses can be made use of in order to develop a rational economic means for risk reduction expenditure.

Risk management requires a value system against which decisions are made, that is, the risk basis. In order to effect risk control certain actions need to be taken. Risk assessment and control are made either in terms of monetary value or in terms of potential loss of life. According to Jefferies *et al.* (1996), despite efforts to assign a monetary value to life in order to develop a single value system, it is desirable to use separate health and money value systems in terms of law. Risk management decisions that do not involve health risk can use monetary value for assessment of costs and benefits. If a health risk is used as a value system, then the average incremental mortality rate of a population affected by a hazard may be used as the value criterion. This mortality rate is the expected additional deaths per year divided by the total affected population. Other health risk measures can be used, such as average loss of life expectancy.

As far as risk is concerned, people more readily accept risks that are associated with voluntary activities than those that may be imposed upon them. In addition, people are more conscious of risk immediately after a disastrous hazard and then call for greater protection than during periods of quiescence. Mora (1995), for instance, mentioned that after the seismic events of 1990–1992 in Costa Rica there was a greater awareness of geohazards and their consequences. Sadly, however, within a short while the interest had declined.

1.3 Remote sensing, aerial photographs and hazards

Remote sensing can provide data for the study of earthquake and volcanic activity, floods, marine inundation and tsunamis, and soil erosion and desertification. Remote sensing involves the identification and analysis of phenomena on the Earth's surface by using devices that are usually borne by aircraft or spacecraft (Sabins, 1996). Most techniques used in remote sensing depend on recording energy from some part of the electromagnetic spectrum, ranging from gamma rays through the visible spectrum to radar. The scanning equipment used measures both emitted and reflected radiation, and the employment of suitable detectors and filters permits the measurement of certain spectral bands. Signals from several bands of the spectrum can be recorded simultaneously by multi-spectral scanners. Two of the principal systems of remote sensing are infrared linescan (IRLS) and side-looking airborne radar (SLAR).

IRLS depends on the fact that all objects emit electromagnetic radiation generated by the thermal activity of their component atoms. It involves scanning a succession of parallel lines across the track of an aircraft. The radiation is picked up by a detector, which converts it to electrical signals, which in turn are transformed into visible light via a cathode ray tube, thereby enabling a record to be made on film. The data can be processed in colour, as well as black and white. Warm areas give rise to light tones and cool areas to dark tones. Relatively cold areas are depicted as purple and relatively hot areas as red on a colour print.

In SLAR, short pulses of energy in a selected part of the radar waveband are transmitted sideways to the ground from antennae on both sides of an aircraft. The pulses of energy strike the ground along successive range lines and are reflected back at time intervals related to the height of the aircraft above the ground. The reflected pulses are transformed into black and white photographs with the aid of a cathode ray tube. As SLAR is not affected by cloud cover, imagery can be obtained at any time.

The capacity to detect surface features and landforms from imagery obtained by multi-spectral scanners on satellites is facilitated by energy reflected from the ground surface being recorded within four specific wavebands, namely, visible green, visible red, and two invisible infrared bands. The images are reproduced on photographic paper and are available for the four spectral bands plus two false-colour composites.

Aerial photographs are generally taken from an aeroplane that is flying at an altitude of between 800 and 9000 m, the height being governed by the amount of detail that is required. Four main types of film are used in normal aerial photography, namely, black and white, infrared monochrome, true colour and false colour. Black and white film is used for normal interpretation purposes, the other types being used for special purposes. Examination of consecutive pairs of aerial photographs with a stereoscope allows observation of a three-dimensional image of the ground surface, which means that height can be determined and contours drawn. However, the topographic relief obtained in this way is exaggerated so that slopes appear steeper than they really are, and in mountainous areas it is difficult to distinguish between steep and very steep slopes. Photomosaics represent a combination of aerial photographs, which if controlled, that is, based on a number of geodetically surveyed points, can be regarded as having similar accuracy to topographic maps.

Heat emission from volcanoes can be detected using infrared sensors (Rothery *et al.*, 1988). Admittedly not all volcanic eruptions are preceded by changes in ground temperature, but the slow rise of hot magma, notably andesitic magma, can give rise to a detectable thermal anomaly. Hence, a map of thermal imagery can be produced from satellite, aerial or ground-based infrared sensors. Weather satellites can be used to monitor the patterns of emission and direction of volcanic ash and dust. The Earth Observation System (EOS) will carry ultraviolet, visible, infrared and microwave sensors and will monitor volcanic eruptions, as well as other hazards. The Orbiting Volcano Observatory (OVO) will assess gas emissions, thermal anomalies and changes in volcanic landforms.

Aerial photographs and satellite imagery can be used for mapping landslides, and monitoring can be undertaken if surveys are repeated. Excess groundwater, which is one of the principal causes of landslides, can be assessed by infrared imagery, water-bearing soils or rocks being cooler and therefore appearing as dark tones or purple/blue colours.

Imagery and aerial photographs can be used to record flooding, by either rivers or marine inundation. The extent of the flooding can be monitored, as can the drainage of flood waters (Rasid and Pramanik, 1990). The characteristics of a flood plain also can be mapped from imagery and aerial photographs.

Areas undergoing soil erosion can be detected by a series of aerial photographs taken at different times. Gully enlargement may be noted, as may be excessive sediment carried by streams. Aerial photographs and satellite images, taken over a period of time, can also be used to monitor the spread of desertification. Changes can be noted in the type of vegetation, land use, soil, soil moisture, surface water and migrating sand dunes.

1.4 Hazard maps

Any spacial aspect of hazard can be mapped, providing there is sufficient information on its distribution. Hence, when risk assessment is made over a large area, the results can be expressed in the form of hazard and risk maps. An ideal hazard map should provide information relating to the spacial and temporal probabilities of the hazard mapped (Wu *et al.*, 1996). In addition to data gathered from surveys, hazard maps are frequently compiled from historical data related to past hazards that have occurred in the area concerned. The concept is based on the view that past hazards provide some guide to the nature of future hazards. Hazard-zoning maps usually provide some indication of the degree of risk involved with a particular geological hazard (Seeley and West, 1990). The hazard is often expressed in qualitative terms as high, medium or low. These terms must be adequately described so that their meaning is understood. The variation in intensity of a hazard from one location to another can be depicted by risk mapping. The latter attempts to quantify the hazard in terms of potential victims or damage (see Figure 3.9). Risk mapping therefore attempts to estimate the location, probability and relative severity of future hazardous events so that potential losses can be estimated, mitigated or avoided. Specific risk-zoning maps divide a region into zones indicating exposure to a specific hazard.

Hazard maps dealing with particular hazards have been described by several authors. In particular, the principles of landslide hazard mapping were discussed by Varnes (1984). Hansen (1984) also discussed mapping of landslide hazard and risk. Seismic hazard mapping has been discussed by Karnik and Algermissen (1979). In fact, seismic phenomena are among the most extensively mapped geological hazards. The early hazard maps were based on the known locations of epicentres (see Figure 3.10) or on seismic intensity ratings (see Figure 3.11). In the United States, such maps have given way to hazard zoning based upon maximum acceleration in a given number of years (see Figure 3.12). The latter can be related to earthquake-resistant design. Vulnerability maps, for example, for the recognition of areas that offer potential risk to groundwater supply if developed, were described by Fobe and Goosens (1990).

A more sophisticated development of the concept and use of hazard maps has been provided by Soule (1980). He outlined a method of mapping areas prone to geological hazards by using map units based primarily on the nature of the potential hazards associated with them. The resultant maps, together with their explanation, are combined with a land-use/geological hazard area matrix, which provides some idea of the problems that may arise in the area represented by the individual map. For instance, the matrix indicates the effects of any changes in slope or the mechanical properties of rocks or soils, and attempts to evaluate the severity of hazard for various land uses. As an illustration of this method, Soule used a landslide hazard map of the Crested Butte–Gunnison area, Colorado (Figure 1.2). This map attempts to show which factors within individual map units have the most significance as far as potential

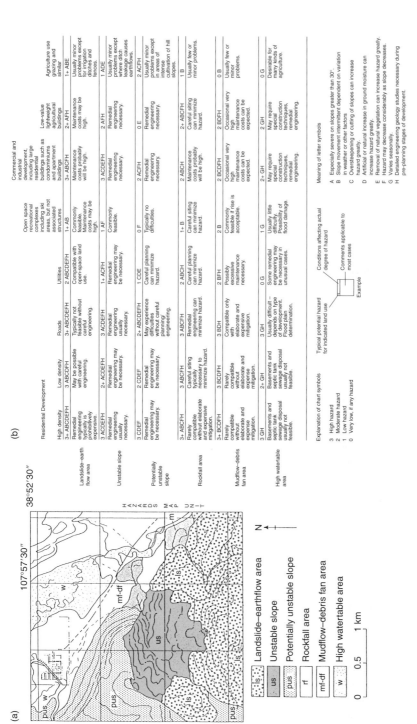

Figure 1.2 (a) An example of engineering geological hazard mapping in the Crested Butte–Gunnison area, Colorado (after Soule, 1980). (b) The matrix is formatted to indicate to the map user that several geology-related factors should be considered when contemplating the land use in a given type of mapped area. This matrix can also serve to recommend additional types of engineering geological study that may be needed for a site. Thus the map can be used to model or anticipate the kinds of problem that a land-use planner or land developer may have to overcome before a particular activity is permitted or undertaken.

hazard is concerned. The accompanying matrix outlines the problems likely to be encountered as a result of human activity.

However, hazard maps, like other maps, do have disadvantages. For instance, they are highly generalized and represent a static view of reality. They need to be updated periodically as new data become available. The hazard map of Mount St Helens had to be completely redrawn after the eruption of 1980. As noted above, catastrophic events occur infrequently, so information may be completely lacking in an area where such an event subsequently occurs.

Micro-zonation has been used to depict the spacial variation of risk in relation to particular areas. Figure 1.2 may be regarded as an example of micro-zonation, as may Figure 3.13. The latter distinguishes three zones of ground response to seismic events. Ideally, land use should adjust to the recommendations suggested by such micro-zoning so that the impacts of the hazard event(s) depicted have a reduced or minimum influence on people, buildings and infrastructure. Foster (1980) distinguished single-hazard–single-purpose, single-hazard–multiple-purpose and multiple-hazard–multiple-purpose maps. The first type of map is obviously appropriate where one type of hazard occurs and the effects are relatively straightforward. Single-hazard–multiple-purpose maps are prepared where the hazard is likely to affect more than one activity and shows the varying intensity of and response to the hazard (see Figure 3.13). The most useful map as far as planners are concerned is the multiple-hazard–multiple-purpose map. Such a map can be used for risk analysis and assessment of the spatial variation in potential loss and damage. Both the hazard and the consequences should be quantified and the relative significance of each hazard compared.

One means by which the power, potential and flexibility of mapping may be increased is by the use of geographical information systems (GIS), which represent a form of technology that is capable of storing, retrieving, editing, analysing, comparing and displaying spatial information. Star and Estes (1990) indicated that a GIS consists of four fundamental elements, namely, data acquisition, data processing and management, data manipulation and analysis, and the generation of products. Data can be obtained from various sources, including remote-sensing imagery, aerial photography, aero-magnetometry, gravimetry and various types of map, as well as the field and the laboratory, and historical data. These data are recorded in a systematic manner in a computer database. The manipulation and analysis of the data allow it to be combined in various ways to evaluate what will happen in certain situations. The output from the analysis can be the generation of hazard maps (see Chapter 4).

Mejía-Navarro and Gracia (1996) referred to several attempts to use GIS for geological hazard and vulnerability assessment, especially in relation to the assessment of landslide, seismic and fluvial hazards. They then went on to describe a decision support system for planning purposes that evaluates a number of variables by the use of GIS. This integrated computer support system, termed Integrated Planning Decision Support System (IPDSS), was designed to assist urban planning by organizing, analysing and evaluating existing or needed spatial data for land-use planning. The system incorporates GIS software that allows comprehensive modelling capabilities for geological hazards, vulnerability and risk assessment. The IPDSS uses data on topography, aspect, solid and superficial geology, structural geology, geomorphology, soil, land cover and use, hydrology and floods, and historical data on hazards. As a consequence, it has been able to delineate areas of high risk from those where future urban development could take place safely, and it is capable of producing hazard-susceptibility maps.

1.5 Land-use planning and hazards

Land-use planning represents an attempt to reduce the number of conflicts and adverse environmental impacts in relation to both society and nature (Bell *et al.*, 1987). In the first instance, land-use planning involves the collection and evaluation of relevant data from which plans can be formulated. The policies that result depend on the economic, sociological and political influences in addition to the perception of the problem. Geology and geotechnical engineering need to be involved in the first steps of planning decisions to provide basic information in order to develop acceptable conditions for people to live under. In the context of geological hazards, in particular, the role of geologists should be to provide planners and engineers with sufficient information so that, ideally, they can develop the environment in harmony with nature. The most convenient method of providing the necessary geological information is usually, at least partially, in the form of a map or maps (Siddle *et al.*, 1987).

Hence, one of the aspects of planning that intimately involves geology is the control or prevention of geological hazards that militate against the interests of man. The development of planning policies for dealing with hazards, in particular, requires an assessment of the severity, extent and frequency of the hazard in order to evaluate the degree of risk. Therefore land-use planning for the prevention and mitigation of geological hazards should be based on criteria establishing the nature and degree of the risks present and their potential impact. Both the probable intensity and frequency of the hazard(s), and the susceptibility (or probability) of damage to human activities in the face of such hazards, are integral components of risk assessment. Vulnerability analyses comprising risk identification and evaluation should be carried out in order to make rational decisions on how best the effects of potentially disastrous events can be reduced or overcome through systems of permanent controls on land development. The geological data needed in planning and decision making, as summarized by Hays and Shearer (1981), are given in Table 1.1. Just as important is the quality of information transmission and how readily it can be understood by politicians and planners, since this influences their reactions.

Once the risk has been assessed methods whereby the risk can be reduced have to be investigated and evaluated in terms of public costs and benefits. The

Table 1.1 Data required to reduce losses from geological hazards (after Hays and Shearer, 1981).

Reduction decisions	Technical information needed about the hazards from earthquakes, floods, ground failures and volcanic eruptions
Avoidance	Where has the hazard occurred in the past? Where is it occurring now? Where is it predicted to occur in the future? What is the frequency of occurrence?
Land-use zoning	Where has the hazard occured in the past? Where is it occurring now? Where is it predicted to occur in the future? What is the frequency of occurrence? What is the physical cause? What are the physical effects of the hazard? How do the physical effects vary within an area? What zoning within the area will lead to reduced losses to certain types of construction?
Engineering design	Where has the hazard occurred in the past? Where is it occurring now? Where is it predicted to occur in the future? What is the frequency of occurrence? What is the physical cause? What are the physical effects of the hazard? How do the physical effects vary within an area? What engineering design methods and techniques will improve the capability of the site and the structure to withstand the physical effects of a hazard in accordance with the level of acceptable risk?
Distribution of losses	Where has the hazard occurred in the past? Where is it occurring now? Where is it predicted to occur in the future? What is the frequency of occurrence? What is the physical cause? What are the physical effects of the hazard? How do the physical effects vary within an area? What zoning has been implemented in the area? What engineering design methods and techniques will improve the capability of the site and the structure to withstand the physical effects of a hazard in accordance with the level of acceptable risk? What annual loss is expected in the area? What is the maximum probable annual loss?

risks associated with geological hazards may be reduced, for instance, by control measures carried out against the hazard-producing agent; by pre-disaster community preparedness, monitoring, surveillance and warning systems that allow evacuation; by restrictions on development of land; and by the use of appropriate building codes together with structural reinforcement of property.

In addition, the character of the ground conditions can affect both the viability and implementation of planning proposals. The incorporation of geological information into planning processes should mean that proposals can be formulated that do not conflict with the ground conditions present.

Obviously, societies make demands upon land, which frequently mean that it is degraded. One of the major causes of the degradation of land is waste disposal from various sources such as industry, mining and domestic. Not only has this led to the degradation of land but it frequently has also been responsible for contamination of ground and pollution of groundwater. Such sites require restoration. In fact, an appreciable contribution to the general economy can be made by bringing this derelict land back into worthwhile use. Eventually such land must be restored. Furthermore, derelict land has a blighting effect on the surrounding area, which makes its restoration highly desirable. Whatever the ultimate use to which the land is put after restoration, it is imperative that it should fit the needs of the surrounding area and be compatible with other forms of land use that occur in the neighbourhood. Accordingly, planning of the eventual land use must take account of overall plans for the area and must endeavour to include the ecological integration of the restored area into the surrounding landscape. Restoring a site that represents an intrusion in the landscape to a condition that is well integrated into its surroundings upgrades the character of the environment far beyond the confines of the site.

References

Alexander, D. (1993) *Natural Disasters.* University College London Press, London.

Bell, F. G., Cripps, J. C., Culshaw, M. G. and O'Hara, M. (1987) Aspects of geology in planning. In *Planning and Engineering Geology, Engineering Geology Special Publication No 4,* Culshaw, M. G., Bell, F. G., Cripps, J. C. and O'Hara, M. (eds), Geological Society, London, 1–38.

Bullard, R. M. (1976) *Volcanoes of the Earth.* University of Texas Press, Austin.

CEPAL (1990) Efectos económicos y sociales de los desatres naturales en América Latina. *Taller Regional de Capacitación para Desatres,* PNUD/UNDRO.

Fobe, B. and Goosens, M. (1990). The groundwater vulnerability map for the Flemish region: its principles and uses. *Engineering Geology,* 29, 355–363.

Foster, H. D. (1980) *Disaster Planning: The Preservation of Life and Property.* Springer, New York.

Hansen, A. (1984) Landslide hazard analysis. In *Slope Instability,* Brunsden, D. and Prior, D. B. (eds), Wiley–Interscience, Chichester, 523–602.

Hays, W. W. and Shearer, C. F. (1981) Suggestions for improving decision making to face geologic and hydrologic hazards. In *Facing Geologic and Hydrologic Hazards – Earth Science Considerations,* Hays, W. W. (ed.), US Geological Survey, Professional Paper 1240–B, B103–B108.

Health and Safety Executive. (1988) *The Tolerability of Risk from Nuclear Power Stations.* HMSO, London.

Jefferies, M., Hall, D., Hinchliff, J. and Aiken, M. (1996) Risk assessment: where are we, and where are we going? In *Engineering Geology of Waste Disposal, Engineering*

Geology Special Publication No 11, Bentley, S. P. (ed.), Geological Society, London, 341–359.

Karnik, V. and Algermissen, S. T. (1979) Seismic zoning. In *The Assessment and Mitigation of Earthquake Risk,* UNESCO, Paris, 171–184.

Mejía-Navarro, M. and Garcia, L. A. (1996) Natural hazard and risk assessment using decision support systems, application: Glenwood Springs, Colorado. *Environmental and Engineering Geoscience,* 2, 299–324.

Mora, S. (1995) The impact of natural hazards on the socio-economic development in Costa Rica. *Environmental and Engineering Geoscience,* 1, 291–298.

Rasid, H. and Pramanik, M. A. H. (1990) Visual interpretation of satellite imagery for monitoring floods in Bangladesh. *Environmental Management,* 14, 815–821.

Rothery, D. A., Francis, P. W. and Wood, C. A. (1988) Volcano monitoring using short wavelength infrared data from satellites. *Journal Geophysical Research,* 93, 7993–8008.

Sabins, F. F. (1996) *Remote Sensing: Principles and Interpretation,* third edition. Freeman, New York.

Seeley, M. W. and West, D. O. (1990) Approach to geological hazard zoning for regional planning, Inyo National Forest, California and Nevada. *Bulletin Association Engineering Geologists,* 27, 23–36.

Siddle, H. J., Payne, H. R. and Flynn, M. J. (1987) Planning and development control in an area susceptible to landslides. In *Planning and Engineering Geology, Engineering Geology Special Publication No 4,* Culshaw, M. G., Bell, F. G., Cripps, J. C. and O'Hara, M. (eds), Geological Society, London, 247–254.

Soule, J. M. (1980) Engineering geology mapping and potential geologic hazards in Colorado. *Bulletin International Association Engineering Geology,* 21, 121–131.

Star, J. and Estes, J. (1990) *Geographical Information Systems.* Prentice-Hall, Englewood Cliffs, NJ.

UNDRO (1991) *Mitigating Natural Disasters: Phenomena and Options. A Manual for Policy Makers and Planners.* United Nations, New York.

Varnes, D. J. (1984) *Landslide Hazard Zonation: A Review of Principles and Practice.* Natural Hazards 3, UNESCO, Paris.

Wu, T. H., Tang, W. H. and Einstein, H. H. (1996) Landslide hazard and risk assessment. In *Landslides Investigation and Mitigation,* Special Report 247, Turner, A.K. and Schuster, R.L. (eds), Transport Research Board, National Research Council, Washington, DC, 106–117.

Chapter 2

Volcanic activity

2.1 Introduction

Volcanic zones are associated with the boundaries of the crustal plates (Figure 2.1). The type of plate boundary offers some indication of the type of volcano that is likely to develop. Plates can be largely continental, oceanic, or both oceanic and continental. Oceanic crust is composed of basaltic material, whereas continental crust varies from granitic to basaltic in composition. At destructive plate margins oceanic plates are overridden by continental plates. The descent of the oceanic plate, together with any associated sediments, into zones of higher temperature leads to melting and the formation of magmas. Such magmas vary in composition, but some may be comparatively rich in silica. The latter form andesitic magmas and are often responsible for violent eruptions. By contrast, at constructive plate margins, where plates are diverging, the associated volcanic activity is a consequence of magma formation in the upper mantle. The magma is of basaltic composition, which is less viscous than andesitic magma. Hence there is relatively little explosive activity, and associated lava flows are more mobile. However, certain volcanoes, for example, those of the Hawaiian Islands, are located in the centres of plates. Obviously, these volcanoes are totally unrelated to plate boundaries. They owe their origins to hot spots in the Earth's mantle, which have 'burned' holes through the overlying plates as the plates moved over them.

Volcanic activity is a surface manifestation of a disordered state within the Earth's interior, which has led to the melting of rock material and the consequent formation of a magma. This magma travels to the surface, where it is extravasated from either a fissure or a central vent. In some cases, instead of flowing from the volcano as a lava, the magma is exploded into the air by the rapid escape of the gases within it. The fragments produced by explosive activity are collectively known as pyroclasts.

Eruptions from volcanoes are spasmodic rather than continuous. Between eruptions, activity may still be witnessed in the form of steam and vapours issuing from small vents named fumaroles or solfataras. But in some volcanoes even this form of surface manifestation ceases and such a dormant state may

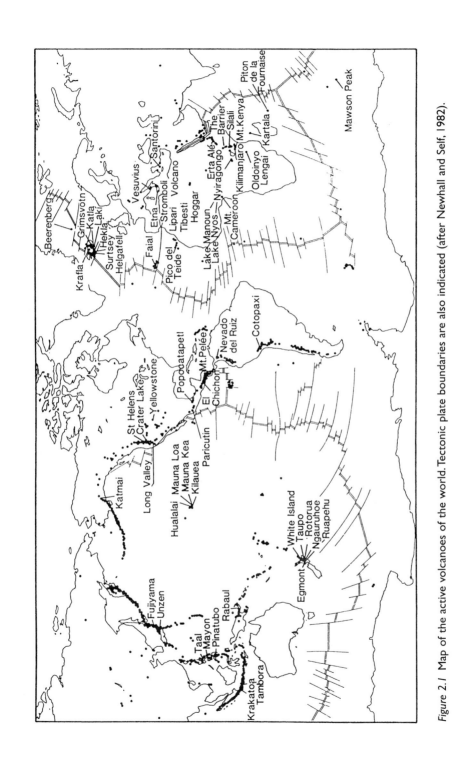

Figure 2.1 Map of the active volcanoes of the world. Tectonic plate boundaries are also indicated (after Newhall and Self, 1982).

continue for centuries. To all intents and purposes these volcanoes appear extinct. For example, this was believed about Vesuvius prior to the fateful eruption of AD 79. In old age, the activity of a volcano becomes limited to emissions of gases from fumaroles, and hot water from geysers and hot springs.

Eruptions can be classified as effusive or explosive according to the violence with which they occur. Low-viscosity magmas escape as lava flows from effusive volcanoes, whereas if a magma has a high viscosity the eruption is explosive. Volcanic eruptions also can be classified on the basis of their size, that is, on the amount of ejectamenta that is erupted (Table 2.1). They can be further classified according to the type and violence of the eruption, the volcanic products (including lahars) and their duration (Table 2.2). The length of time over which eruption occurs is also significant and should be considered when comparing volcanic events.

Most volcanic material is of basaltic composition. As mentioned above, it is believed that basaltic magmas are generated within the upper mantle, where temperatures approach the fusion point of the material from which they are derived. Their development appears to be associated with huge rifts and thrusts that fissure the crust and extend into the mantle. However, the small, local magma chambers that supply individual volcanoes need not be located at such depths. Indeed, it has been estimated that the chamber beneath Monte Somma is situated at not more than 3 km below the surface. By contrast, continuous tremors starting at a depth of 60 km a few months before an eruption of Kilauea were regarded as having been caused by magma streaming into a conduit leading to a chamber immediately beneath the crater.

According to Rittmann (1962), basalt magmas generated in the mantle, presumably in the low-velocity layer, cannot reach the Earth's surface under their own power but do so via open fissures, which act as channels of escape. These fissures are formed by the convection currents in the mantle. The hydrostatic pressure of a magma that penetrates a fissure helps to widen it and escaping gases from the magma help to clear the channel by blast action. Furthermore, the reduction of pressure on the magma as it ascends causes a lowering of its viscosity, which means that it can penetrate more easily along the fissure. The potential eruptive energy of the magma is governed by the quantity of volatile constituents it contains. Indeed, basalt magmas formed in the mantle could not reach the continental surface if they lacked volatiles, since their density would be higher than that of the upper regions of the crust. They would consequently obtain a state of hydrostatic equilibrium some kilometres below the surface. The release of original gaseous constituents in the magma lowers its density and thereby aids its ascent.

2.2 Volcanic form and structure

The form and structure that a volcano adopts depends upon the type of magma feeder channel, the character of the material emitted and the number of

Table 2.1 Volcanic eruption scales.

Volcanic explosivity index (VEI)[a]	Volcanic intensity[b]	Tsuya scale[c]	Eruption rate (kg s^{-1})	Volume of ejecta (m^3)	Eruption column height (km)	Thermal power output (log kW)	Duration (hours of continuous blast)
0	V	I	10^2–10^3	$<10^4$	0.8–1.5	5–6	<1
1	VI	II–III	10^3–10^4	10^4–10^6	1.5–2.8	6–7	<1
2	VII	IV	10^4–10^5	10^6–10^7	2.8–5.5	7–8	1–6
3	VIII	V	10^5–10^6	10^7–10^8	5.5–10.5	8–9	1–12
4	IX	VI	10^6–10^7	10^8–10^9	10.5–17.0	9–10	1–>12
5	X	VII	10^7–10^8	10^9–10^{10}	17.0–28.0	10–11	1–>12
6	XI	VIII	10^8–10^9	10^{10}–10^{11}	28.0–47.0	11–12	6–>12
7	XII	IX	$>10^9$	10^{11}–10^{12}	>47.0	>12	>12
8	—	—	—	$>10^{12}$	—	—	>12

Notes:
[a] Newhall and Self 1982.
[b] Fedetov 1985.
[c] Tsuya 1955.

Table 2.2 Other determinants of volcanic explosivity index (VEI) (after Newhall and Self, 1982).

VEI	Explosivity	Qualitative description	Classification	Tropospheric injection	Stratospheric injection
0	non-explosive	effusive	Hawaiian	negligible	none
1	small	gentle	Hawaiian Strombolian	minor	none
2	moderate	explosive	Strombolian	moderate	none
3	moderate–large	severe	Strombolian, Vulcanian	great	possible
4	large	violent	Vulcanian, Plinian	great	definite
5	very large	cataclysmic	Plinian Ultraplinian	great	significant
6	very large	paroxysmal	Ultraplinian	great	significant
7	very large	colossal	Ultraplinian	great	significant
8	very large	terrific	Ultraplinian	great	significant

eruptions that occur. As far as the feeder channel is concerned this may be either a central vent or a fissure, which give rise to radically different forms. The composition and viscosity of a magma influence its eruption. For instance, acid magmas are more viscous than basic and so gas cannot escape as readily from them. As a consequence the more acid, viscous magmas are generally associated with explosive activity, and the volcanoes they give rise to may be built mainly of pyroclasts. Alternatively, fluid basalt magmas construct volcanoes that consist of piles of lavas with very little pyroclastic material. Volcanoes built largely of pyroclasts grow in height much more rapidly than do those formed of lavas. As an illustration, in 1538 Monte Nuovo, on the edge of the Bay of Naples, grew to a height of 134 m in a single day. It would take anything up to a million years for lava volcanoes like those of Hawaii to grow to a similar height. The number of eruptions that take place from the same vent allows the recognition of monogenetic (single eruption) and polygenetic (multiple eruption) volcanoes. Monogenetic central-vent volcanoes are always small and have a simple structure, the eruptive centre moving on after the eruption; Paracutin in Mexico provides an example. Monogenetic volcanoes do not occur alone but are clustered in fields. Polygenetic central-vent volcanoes are much larger and more complicated. The influence of the original topography upon their form is obscured. Displacement of the vent frequently occurs in polygenetic types. Fissure volcanoes are always monogenetic.

Some initial volcanic perforations only emit gas, but the explosive force of the escaping gas may be sufficient to produce explosion vents (Figure 2.2). These are usually small in size and are surrounded by angular pyroclastic material formed from the country rock. If the explosive activity is weaker, then the country rocks are broken and pulverized in place rather than thrown from the vent. Accordingly, breccia-filled vents are formed.

Pyroclastic volcanoes are formed when viscous magma is explosively erupted. They are often monogenetic and are generally found in groups, their deposits interdigitating with one another. They are small when compared with some shield and composite volcanoes. Some of the cones formed earlier in this category are often destroyed or buried by later outbreaks. In pumice cones, banks of ash often alternate with layers of pumice. Cinder cones are not as common as those of pumice and ash. They are formed by explosive eruptions of the Strombolian or Vulcanian type. Some of these cones are symmetrical with an almost circular ground plan, the diameter of which may measure several kilometres. They may reach several hundred metres in height. Parasitic cones may arise from the sides of these volcanoes. For example, in 1943, steam began to rise from a field in Mexico and within eight months a cinder cone 410 m in height, named Paracutin, was formed. It emitted 8×10^7 m^3 of pyroclastic material, at a maximum rate of 6×10^6 m^3 a day, activity continuing until 1952.

Fissure pyroclastic volcanoes arising from basic and intermediate magmas are not common. By contrast, sheets of ignimbrite of rhyolitic composition

Figure 2.2 Ubehebe, an explosion vent in Death Valley, California (Newhall and Self, 1982).

have often been erupted from fissures. When enormous quantities of material are ejected from fissures associated with large volcanoes, because their magma chambers are emptied, they may collapse to form huge volcano-tectonic sinks.

Mixed or composite volcanoes consist of accumulations of lavas and pyroclastic material. They have an explosive index in excess of 10 (according to Rittmann (1962) the explosive index (E) is the percentage of fragmentary material in the total material erupted). They are the most common type of volcano.

Strato-volcanoes are polygenetic and consist of alternating layers of lava flows and pyroclasts. The simplest form of strato-volcano is cone-shaped with concave slopes and a crater at the summit from which eruptions take place. However, as it grows in height the pressure exerted by the magma against the conduit walls increases and eventually the sides are ruptured by radial fissures, from which new eruptions take place. Cinder cones form around the uppermost centres, while lava wells from the lower. The shape of strato-volcanoes may be changed by a number of factors; for example, migration of their vents is not uncommon, so they may exhibit two or more summit craters. Significant changes in form are also brought about by violent explosions, which may blow part of the volcano or even the uppermost portion of the magma chamber away. When the latter occurs the central part of the volcanic structure collapses because of loss of support. In this way huge summit craters are formed in which new volcanoes may subsequently develop.

Most calderas measure several kilometres across and are thought to be formed by collapse of the superstructure of a volcano into the magma chamber below (Figure 2.3), since this accounts for the small proportion of pyroclastic deposits

Figure 2.3 Crater Lake, Oregon.

that surround the crater. If they had been formed as a result of tremendous explosions, then fragmentary material would be commonplace.

Rittmann (1962) used the term 'lava volcano' to include those volcanic structures that had an explosive index of less than 10; in fact it is usually between 2 and 3. Monogenetic central-vent volcanoes rarely occur as independent structures and are generally developed as parasitic types on the flanks of larger lava volcanoes. The former type of volcano invariably consists of a small cone surrounding the vent from which a lava stream has issued. Polygenetic lava volcanoes are represented by shield volcanoes such as those of the Hawaiian Islands. These volcanoes are built by successive outpourings of basaltic lava from their lava lakes, which occur at the summit in steep-sided craters. When emitted, the lava spreads in all directions and because of its fluidity covers large areas. As a consequence, the slopes of these volcanoes are very shallow, usually between 4 and 6°.

A volcanic dome is a mass of rock that has been formed when viscous lava is extruded from a volcanic vent. The lava is too viscous to flow more than a few tens or hundreds of metres. Domes are usually steep-sided; for example, the Puy de Sarcoui in Auvergne reaches a height of 150 m with a base only 400 m wide. Domes may develop in the summit of a volcano, on its flanks or at vents along a fissure at or beyond the base of a volcano. They may grow rapidly, reaching their maximum size in less than a year. In the centre of a dome movement is primarily upward, which frequently causes the sides to become unstable. As a result, the height of domes is generally limited to a few hundred metres. At times, viscous lava may block the vent of a volcano so that subsequently rising lava accumulates beneath the obstruction and exerts an increasing pressure on the sides of the volcano until they are eventually fissured. The lava is then intruded forcibly through the cracks and emerges to form streams, which flow down the flanks of the dome. Cryptodomes (i.e. hidden domes) occur as a result of magmatic intrusion at shallow depth, the hydrostatic pressure of the magma causing the roof material to swell upwards. Eruption of pyroclasts and/or lava occurs if the roof fissures. The unroofing of a cryptodome on the north flank of Mount St Helens in 1980 gave rise to a destructive explosive eruption.

The term 'flood basalt' was introduced by Tyrrell (1937) to describe large areas, usually at least 130 000 km^2, that are covered by basaltic lava flows (Figure 2.4). These vast outpourings have tended to build up plateaux; for instance, the Deccan plateau extends over 640 000 km2 and at Bombay reaches a thickness of approximately 3000 m. The individual lava flows that comprise these plateau basalt areas are relatively thin, varying between 5 and 13 m in thickness. In fact, some are less than 1 m thick. They form the vast majority of the sequence, pyroclastic material being of very minor importance. That the lavas were erupted intermittently is shown by their upper parts, which are stained red by weathering. Where weathering has proceeded further, red earth or bole has been developed. It is likely that flood basalts were erupted by both fissures and central-vent volcanoes. For instance, Anderson and Dunham (1966) wrote that the distribution of lava types on Skye suggests that they were extruded by several fissures, which were related to a central volcano. The sev-

Figure 2.4 Lava flows, overlying chalk, that form part of the Antrim Plateau, Northern Ireland.

eral groups into which the lavas of Skye have been divided all thin away from centres, which have been regarded as the sites of their feeders. Consequently, the flood basalts are believed to have been built up by flows from several fissures operating at different times and in different areas, the lavas from which met and overlapped to form a succession that does not exceed 1200 m. The most notable flood basalt eruption that occurred in historic times took place at Laki in Iceland in 1783 (Thorarinsson, 1970). Torrents of lava were emitted from a fissure approximately 32 km long and overwhelmed 560 km^2. As the volcanic energy declined the fissure was choked, but eventually pent-up gases broke through to the surface at innumerable points and small cones, ranging up to 30 m or so high, were constructed.

2.3 Types of central eruption

The Hawaiian type of central eruption is characterized in the earliest stages by the effusion of mobile lava flows from the places where rifts intersect, the volcano growing with each emission (Figure 2.5). When eruptive action was taking place under the sea, steam-blast activity caused a much higher proportion of pyroclasts to be formed. Above sea level, Hawaiian volcanoes are typified by quiet emissions of lavas. This is due to their low viscosity, which permits the ready escape of gas. Low viscosity also accounts for the fact that certain flows at outbreak have been observed to travel at speeds of up to 55 km h^{-1}. Lava flows are emitted both from a summit crater and from rift zones on the flanks

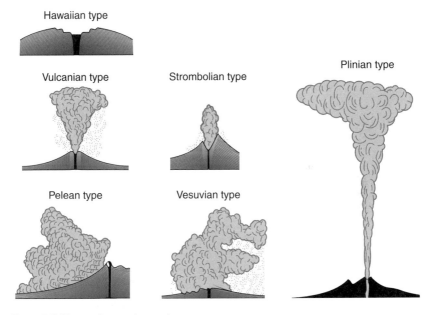

Figure 2.5 Types of central eruption.

of the volcano. Eruptive activity may follow a pattern similar to that associated with Mauna Loa (Bullard, 1976). There, after the initial earthquakes, bulging of the volcano is caused by up-welling magma. Fissures open across the floor of the summit crater, from which lavas, under initial-release gas pressures, shoot into the air to form a 'curtain of fire'. After a day or so the fissures are sealed with freshly congealed lava and emission becomes localized in the largest vents. A major lava flow is erupted from Mauna Loa on average once in every three or four years.

If the lava emitted by a volcano is somewhat more viscous than the basaltic types that flow from the Hawaiian volcanoes then, since gas cannot escape as readily, moderate explosions ensue. Clots of lava are thrown into the air to form cinders and bombs, and so pyroclasts accompany the extrusion of lavas. Stromboli typifies this kind of central eruption, hence the designation Strombolian type (see Figure 2.5). This volcano has erupted with amazing consistency throughout historical times. After the ejection of lavas some of the molten material congeals in the vent. At the next eruption, gas breaks through this thin crust and throws blocks of it, together with fragments of lava, into the air. Most of this material falls back into the crater. The steam and gases that are emitted form clouds.

Vulcano gives its name to the Vulcanian type of central eruption (see Figure 2.5). In this type the lava is somewhat more viscous than that of the Strombolian type and so after an eruption, the lava remaining in the conduit quickly solidifies. This means that gases accumulate beneath the obstruction until they have gathered enough strength to blow it into the air. The fragmented crust, together with exploded lava and gas, give rise to large clouds. The lavas that are emitted tend to fill the crater and spill over the sides. After a time they congeal and the process begins again.

The Vesuvian type of central eruption is yet more violent and may be preceded by eruptions of the Strombolian and Vulcanian type (see Figure 2.5). Once again the conduit of the volcano becomes sealed with a plug of solidified lava so that great quantities of gas become pent up in the magma below. This may cause lava to escape through fissures or parasitic vents on the slopes of the volcano, so the conduit may be emptied down to a considerable depth. The release of pressure so created allows gas, held in solution in the remaining magma, to escape. It does so with great force, thereby clearing the conduit of its obstruction, throwing its fragmented remains and exploded lava far into the air, and forming great clouds of ash. Vesuvian eruptions are followed by periods of long quiescence; for example, since AD 79 Vesuvius itself has broken into violent eruption only ten times.

The Plinian type of central eruption is the most violent form of the Vesuvian type (see Figure 2.5). Such an eruption from Vesuvius was responsible for the destruction of Pompeii and Herculaneum in AD 79. This tremendous eruption was recorded by the Roman historian Pliny the Elder, hence the name. The Plinian type of eruption is usually preceded by the Vesuvian and is characterized by an extremely violent upsurge of gas shooting from the neck of the volcano to heights of 12 km or more, where it levels out to form a canopy (Carey and Sigurdsson, 1989). The amount of ash involved in such eruptions is relatively small and consists chiefly of material eroded from the conduit wall by the tremendous gas activity. The last Plinian eruption by Vesuvius occurred in 1944. It lasted for a few hours but had been preceded by nine days of Vesuvian paroxysms. These increased in gas activity and periodically fountains of lava were thrown into the air, sometimes reaching heights of 1000 m above the summit. Indeed, the preceding Vesuvian eruptions almost exhausted the lava supply. The eruption of 1944 ended with a few days of pseudo-Vulcanian eruptions.

In the Pelean type of central eruption the magma concerned is extremely viscous, which leads to catastrophic explosions (Figure 2.5). Prior to the eruption of Mont Pelée in 1902 this type of volcanic activity had been unknown. Once again the neck of the volcano is blocked with congealed lava so that the magma beneath becomes very highly charged with gas. The pressures that are thereby developed reach a point at which they have sufficient strength to fissure the weakest zone in the flanks of the volcano. The lava that is extruded through the fissure is violently exploded by the escape of gas that is held in solution. In this way a cloud of very hot gas and ash is formed. This rushes with tremendous speed down the side of the volcano, destroying everything in its path. Such glowing clouds are referred to as *nuées ardentes*. In the case of the Mont Pelée eruption of 1902, Vulcanian activity began in the early spring but the crater filled up and was sealed by the growth of a dome within the conduit. Hot mud continued to pour down the Rivière Blanche from a notch in the rim, and it was from this point of weakness that all the *nuées ardentes* were later to arise. On 8 May the first, and easily the most violent, *nuée* rushed down the slope to envelop the port of St Pierre within seconds, killing all but two of its inhabitants. Because of its high density the cloud hugged the ground and because of its intense heat, when it reached the sea, the sea boiled. *Nuée* activity continued and the dome blocking the conduit was forced slowly upwards until it reached about 300 m above the summit. However, it was soon reduced to a large heap of rubble by weathering.

The dense lower part of a *nuée ardente* tends to follow the valleys, filling them with pyroclastic debris. These deposit a hot sandy and/or gravelly mass several metres thick, which may remain hot for months or years. A cloud of steam and fine ash, escaping from the debris, forms the upper part of the *nuée ardente*. The heavy base surge *nuées ardentes*, which occurred at Mont Pelée, may have moved at up to 100 km h^{-1} for distances of up to 10 km. A larger form of *nuée*, referred to as an ash flow or pyroclastic flow, can travel at 200 km h^{-1} for distances of up to 25 km. They are also caused by the expansion of trapped gases that are released from hot pyroclastic debris. The temperature of *nuées ardentes* may be around 900 °C where eruption occurs, falling to 400 °C at their furthest extent of flow. Ash flows may extend up to 10 km^3 in volume.

2.4 Volcanic products: volatiles

When a magma is erupted it separates at low pressures into incandescent lava and a gaseous phase. If the magma is viscous (the viscosity is to a large extent governed by the silica content), then separation is accompanied by explosive activity. On the other hand, volatiles escape quietly from very fluid magmas. The composition of gases emitted varies from one volcano to another, and from one eruption of the same volcano to another.

Steam accounts for over 90% of the gases emitted during a volcanic eruption. Other gases present include carbon dioxide, carbon monoxide, sulphur dioxide, sulphur trioxide, hydrogen sulphide, hydrogen chloride and hydrogen fluoride. Small quantities of methane, ammonia, nitrogen, hydrogen thiocyanate, carbonyl sulphide, silicon tetrafluoride, ferric chloride, aluminium chloride, ammonium chloride and argon have also been noted in volcanic gases. It has often been found that hydrogen chloride is, next to steam, the major gas emitted during an eruption but that in the later stages the sulphurous gases take over this role. These gases, depending on their concentrations in the air, may be toxic to many animals and plants.

Water tends to move towards the top of a body of magma, where the temperatures and pressures are lower. For example, in a chamber extending from 7 to 14 km depth the saturation water content would occur at the top, whereas at the bottom it would be 2%. This relationship offers an explanation of a common sequence in volcanic eruptions, that is, a highly explosive opening phase with a resultant vigorous gas column carrying pumice to a great height, followed by the formation of ash flows, and lastly by the comparatively quiet effusion of lava flows.

At high pressure, gas is held in solution but as it falls gas is released by the magma. The amount and rate at which it escapes determines the explosiveness of the eruption. An explosive eruption occurs when, because of its high viscosity, magma cannot readily allow the escape of gas. It is only secondarily related to the amount of gas a magma holds. Nevertheless, because obsidians and basaltic glasses contain approximately 0.1% original water, it can be inferred therefrom that explosive eruption is impossible when a magma possesses this amount of dissolved water or less.

Lavas frequently contain bubble-like holes or vesicles, which have been left behind by gas action. Even the escape of small quantities of gas is sufficient to cause frothing of the surface of a lava, since pumice can be produced by the exsolution of less than 0.1% of dissolved water.

Solfataras and fumaroles occur in groups and are referred to as solfatara or fumarole fields. They are commonly found in the craters of dormant volcanoes or are associated with volcanoes that have reached the age of senility in their life cycle. Superheated steam may issue continuously from the fissures of larger solfataras and at irregular intervals from the smaller ones. Fenner (1923) estimated that water vapour constitutes 99% of the exhalations from fumaroles in the Valley of Ten Thousand Smokes. The steam commonly contains carbon dioxide and hydrogen sulphide. Atmospheric oxygen reacts with the hydrogen sulphide to form water and free sulphur, the latter being deposited around the steam holes. Sulphuric acid is also formed by oxidation of hydrogen sulphide and this, together with superheated steam, frequently causes complete decomposition of the rocks in the immediate neighbourhood, leaching out their bases and replacing them with sulphates. Boiling mud pits may be formed on the floor of a crater, where steam bubbles through fine dust and ash.

The composition of fumarolic gases depends not only upon their initial composition in the magma but also upon their temperature; the length of time since they began to form, the more insoluble gases being more abundant in the early emanations; the place where the gases are emitted, whether from the eruptive vent or from a lava flow; the extent of mixing with air or meteoric water; and upon reactions with air, water and country rock. Cool fumaroles, the gas temperatures of which exceed the boiling point of water by only a few degrees, are more frequent in their occurrence than are solfataras. The water vapour emitted generally contains some carbon dioxide but no hydrogen sulphide. In the coolest fumaroles, the gases are often only a few degrees warmer than air temperature and have either been cooled during their ascent or arise from evaporating thermal water. By contrast, the temperature of steam may reach 900 °C in very hot fumaroles, which occur on active volcanoes. These emissions always contain hydrogen chloride and volatile chlorides, particularly sodium chloride and ferric chloride, together with the usual constituents of solfataric gases.

Hot springs are found in all volcanic districts, even some of those where the volcanoes are extinct. They originate from hot steam and gases given off by masses of intruded magma that are in the last stages of crystallization. On their passage to the surface the steam and gases often encounter groundwater, which is thereby heated and forms part of the hot springs. Many hot springs contain carbon dioxide and hydrogen sulphide together with dissolved salts. Indeed, some very hot springs found in active volcanic districts may contain dissolved silica, which on cooling is deposited to form sinter terraces. If the hot springs contain dissolved calcium carbonate, this is precipitated to form travertine terraces. Geysers are hot springs from which a column of hot water is explosively discharged at intervals, the water spout in some cases rising over 100 m. The periodicity of their eruptions varies from a matter of minutes to many days and changes with time. Geysers are generally short-lived. Again, boiling mud pits occur in association with hot springs.

2.5 Volcanic products: pyroclasts

The term 'pyroclast' is applied collectively to material that has been fragmented by explosive volcanic action. 'Tephra' is a synonym for the phrase 'pyroclastic material'. Pyroclasts may consist of fragments of lava exploded on eruption, of fragments of pre-existing solidified lava or pyroclasts, or of fragments of country rock that, in both latter instances, have been blown from the neck of the volcano. These three types have been distinguished as essential, accessory and accidental ejectamenta, respectively.

The size of pyroclasts varies enormously. It is dependent upon the viscosity of the magma, the violence of the explosive activity, the amount of gas coming out of solution during the flight of the pyroclast, and the height to which it is thrown. The largest blocks thrown into the air may weight over 100 tonnes, whereas the smallest consist of very fine ash, which may take years to fall back to the Earth's surface. The largest pyroclasts are referred to as volcanic bombs. These consist of clots of lava or fragments of wall rock. They may fall in significant quantities within a 5 km radius of the eruption. The bombs may destroy structures on which they fall, and incandescent bombs may ignite homes, crops or woodlands.

The term 'lapilli' is applied to pyroclastic material that has a diameter ranging from approximately 10 to 50 mm (Figure 2.6). Cinder or scoria is irregularly shaped material of lapilli size. It is usually glassy and fairly to highly vesicular and represents the ejected froth of a magma. Lapilli can be ejected during an explosive eruption over a radius of several kilometres from a volcano.

The finest pyroclastic material is called ash (Figure 2.7). Much more ash is produced on eruption of acidic than basaltic magma. This is because acidic material is more viscous, so gas cannot escape as readily from it as it can from basaltic lava. For example, ash forms less than 1% of those parts of the Hawaiian

Figure 2.6 Lapilli from near the Crater Lake caldera, Oregon.

Figure 2.7 Ash deposit from Tarawera, North Island, New Zealand.

shield volcanoes exposed above sea level. By contrast, Monte Nuovo, near Naples, is formed mostly of ash and cinders of trachytic composition.

Beds of ash commonly show lateral variation as well as vertical. In other words, with increasing distance from the vent the ash becomes finer and, in the second case, because the heavier material falls first, ashes frequently exhibit graded bedding, coarser material occurring at the base of a bed, it becoming finer towards the top. Reverse grading may occur as a consequence of an increase in the violence of eruption or changes in wind velocity. The spatial distribution of ash is very much influenced by wind direction, and deposits on the lee side of a volcano may be much more extensive than on the windward side; indeed, they may be virtually absent from this side.

Because the fall of ash can be widespread, it can cover or bury houses and farmland, cause roofs to collapse, ruin crops, and block streams and sewers. For example, 10 mm of ash may add 19 kg to the weight of each square metre of roof, the density of ash varying from 0.5 to 1.0 Mg m^{-3}, depending on its wetness. Roofs need to have a high pitch (e.g. 20°) so that the ash will slide off. Deposits of ash may suffer very rapid erosion by streams which may become blocked in the process. They are frequently metastable and on wetting are subject to hydrocompaction.

Extremely fine particles can be thrown into the stratosphere and circle the Earth for months or years. This can affect solar radiation by absorbing, reflecting or scattering it and may obscure the Sun for several hours. It also can lead to small reductions in average temperatures (i.e. 0.3–1.0 °C). Repeated eruptions of this sort can cause respiratory problems in people and animals, and can lead to silicosis. Crops can be damaged or destroyed by fine ash-fall. Long, dry periods coupled with windy conditions can increase the ash content in the air, which can also be responsible for respiratory ailments.

Rocks that consist of fragments of volcanic ejectamenta set in a fine-grained groundmass are referred to as agglomerate or volcanic breccia, depending on whether the fragments are rounded or angular, respectively.

After pyroclastic material has fallen back to the surface it eventually becomes indurated. It is then described as tuff. According to the material of which the tuff is composed, distinction can be drawn between ash tuff, pumiceous tuff and tuff breccia. Tuffs are usually well bedded, and the deposits of individual eruptions may be separated by thin bands of fossil soil or old erosion surfaces. Chaotic tuffs are formed from the deposits of glowing clouds or mud streams. Glowing clouds give rise to chaotic tuffs in which blocks of all dimensions are present along with very fine ash. Lenses of breccia, pumice and volcanic sand are found in chaotic tuffs that are formed by mud flows, and they indicate that some amount of incomplete sorting has occurred during flow. Pyroclastic deposits that accumulate beneath the sea are often mixed with a varying amount of sediment and are referred to as tuffites. They are generally well sorted and well bedded.

When clouds or showers of intensely heated, incandescent lava spray fall to the ground, they weld together. Because the particles become intimately fused with each other they attain a largely pseudo-viscous state, especially in the deeper parts of the deposit. The resultant massive rock frequently exhibits columnar jointing. The term 'ignimbrite' is used to describe these rocks (Cook, 1966). If ignimbrites are deposited on a steep slope, then they begin to flow, hence they resemble lava flows. The considerable mobility of some pyroclastic flows, which allows them to move over distances that may be measured in tens of kilometres, has been explained by the process of fluidization. Fluidization involves the rapid escape of gas in which pyroclastic material has become suspended. Ignimbrites are associated with *nuées ardentes*.

2.6 Volcanic products: lava flows

Lavas are emitted from volcanoes at temperatures only slightly above their freezing points (MacDonald, 1972). During the course of their flow the temperature falls from within outwards until solidification takes place somewhere between 600 and 900 °C, depending upon their chemical composition and gas content. Basic lavas solidify at a higher temperature than do acidic ones.

Generally, flow within a laval stream is laminar. The rate of flow of a lava is determined by the gradient of the slope down which it moves and by its viscosity, which in turn is governed by its composition, temperature and volatile content. Also, lava flows fastest near its source and becomes progressively slower at increasing distance from the source since it cools on contact with the ground and atmosphere. It has long been realized that the greater the silica content of a lava, the greater is its viscosity. Thus basic lavas flow much faster and further than do acid lavas. Indeed, the former type has been known to travel at speeds of over 80 km h^{-1}. Tazieff (1977) recorded that extremely fluid basalt was erupted from Mount Nyiragongo in Zaire in 1977 and reached velocities of between 30 and 100 km h^{-1}, the lava spreading over an area of 20 km^2 in less than an hour. Basalt lavas may extend more than 50 km from their sources, whereas andesitic flows rarely move over more than 20 km. Dacite and rhyolite lavas, because of their normally low discharge rate, rarely form long flows. They typically produce domes and short, thick flows.

Many lava flows consist of several flow units, which represent separate sheets of liquid that were poured over one another during a single eruption. According to MacDonald (1967), the basaltic lava flows on the slopes of Hawaiian volcanoes range in thickness from a few hundred millimetres to about 20 m.

The upper surface of a recently solidified lava flow develops a hummocky, ropy pahoehoe; rough, fragmental, clinkery, spiny aa; or blocky structure. The reasons for the formation of these different structures are not fully understood but certainly the physical properties of the lava and the amount of disturbance it has to undergo must play an important part. The pahoehoe is the most fundamental type; however, some way downslope from the vent it may give way to aa or block lava. In other cases aa, or block lava may be traceable into the vent. It would appear that the change from pahoehoe to aa takes place as a result of increasing viscosity or stirring of the lava. Increasing viscosity occurs due to loss of volatiles, cooling and progressive crystallization, while a lava flow may be stirred by an increase in the gradient of the slope down which it is travelling. Moreover, if strong fountaining occurs in a lava while it is in the vent, this increases stirring and it may either issue as an aa flow or the likelihood of it changing to aa is accordingly increased. Melts that give rise to block lavas are more viscous than those that form aa; they are typically andesitic, although many are basaltic.

Pahoehoe is typified by a smooth, billowy or rolling, and locally ropy surface (Figure 2.8). Such surfaces are developed by dragging, folding and twisting of the still plastic crust of the flow due to the movement of the liquid lava beneath. Although the hummocky surface may give the appearance of smoothness it is usually interrupted by small sharp projections, which mark the places where bubbles have burst into the air. Lava that has oozed through the crust may be drawn into threads oriented in the direction of flow movement. The ridges of ropy lavas are commonly curved, the convex sides pointing in the downstream direction of flow.

The skin of an active pahoehoe flow is very tough and flexible and impedes the escape of gas from the lava. As a consequence, it is sometimes found that newly consolidated flows have a skin a millimetre or so in thickness that overlies a vesicular layer, at times the vesicles becoming so abundant that they merge. The skin tends to flake off, revealing the vesicular surface beneath.

Figure 2.8 Ropey lava (pahoehoe), Craters of the Moon, Idaho.

Sometimes the crust of the lava may be broken into a series of slabs by the movement of the still liquid material beneath.

With the exception of flows of flood eruptions, all large pahoehoe flows consist of several units. Large flows are fed by a complex of internal streams beneath the crust, each stream being surrounded by less mobile lava. When the supply of lava is exhausted, the stream of liquid may drain out of the tunnel through which it has been flowing (Figure 2.9).

The edge of a large, slow-moving pahoehoe flow does not generally advance as a single unit but rather by the extension of one toe of lava after another. These toes may reach a metre across and rarely advance more than 2 m before they become immobile. When this happens, the crust of the flow front cracks and another toe emerges. Some toes are hollow due to being inflated with gas.

Aa lava flows are characterized by very rough fragmental surfaces (Figure 2.10). The fragments are commonly referred to as clinker and have numerous sharp, jagged spines. Most fragments are less than 150 mm across, but some may be twice this dimension. Clinker is not readily compacted. This is illustrated by highly permeable clinker horizons in Hawaiian lavas several million years old and buried at depths of 500 to 1000 m. The spinose nature of the clinker is partly due to the still plastic surface being pulled apart when lava breaks into fragments. However, the crusts of some spinose aa flows are unbro-

Figure 2.9 Collapsed tunnel in lava flow near Flagstaff, Arizona.

Figure 2.10 An aa lava flow, Craters of the Moon, Idaho.

ken, and spinose protuberances have been observed growing on clinker-free aa surfaces. The massive part of aa flows is usually much less vesicular than pahoehoe, the vesicularity frequently being less than 30%.

Aa flows are fed by streams that lie approximately in the centre of each flow. These streams are usually a few metres in width. Levees, a metre or so in height, may be constructed along the sides of the streams as a result of numerous overflows congealing. These streams are only rarely roofed over by solidified lava and if they are, it is only for a short distance. The stream retains a close connection with the interior of the flow beneath the clinker. After an eruption much of the lava drains out of the stream and leaves behind a channel, which may be several metres deep.

A steep bank of clinker several metres high develops at the margin of a slowly moving aa flow. Movement takes place by a part of the front slowly bulging forward until it becomes unstable and separates from the parent body. The process is repeated over and over again and so a talus of clinkery deposits builds up along the foot of the lava front. These deposits are eventually buried under the flow. At the margin of a more active flow the process is just the same, except that bulges grow more quickly and they form an almost continuous line along the front. Clinker on the surface of the flow is carried forward and deposited over the front. Much of this is broken in the process. The actual motion of the flow is simply brought about by its interior, still a viscous liquid, spreading. Because this pasty material overrides the talus deposits of clinker, a congealed flow has a tripartite structure, that is, a massive inner layer sandwiched between two layers of clinker.

Although the term 'block lava' refers to lavas with fragmented surfaces, it is usually restricted to those flows in which the fragments are much more regular in shape than are those of aa flows. Most of the flows of orogenic regions are of block lava. The individual fragments of block lava are polyhedral in shape and they have smooth surfaces, although some may develop spines.

Block lava flows have very uneven surfaces. What is more, they generally exhibit a series of fairly regularly arranged ridges at right angles to the direction of flow. The ridges may be several metres high and are also covered with blocky rubble. Although a central massive layer is usually present, the fragmental material may form the whole of the flow at times .

It appears that the movement of block lava is similar to that of aa flows. However, individual parts of the lava front probably move forwards more quickly and there is a tendency for discrete layers to push forwards, which sometimes results in several layers shearing over one another. Such action is responsible for the production of considerable quantities of crushed fragmental material. Since this material is still plastic it may weld together. Block lavas may also move by sliding over the underlying surface.

Vesicles in block lava are not as common as in aa lava. It has been found that both the blocks and a massive portion of the flow generally contain more glassy material than do the corresponding parts of aa flows.

2.7 Mudflows or lahars

A lahar is a flowing slurry of volcanic debris and water that originates on a volcano. Lahars in which 50% of the volcanic debris is of sand size or smaller are termed mudflows. Beverage and Cuthbertson (1964) regarded lahars as water–sediment mixtures in which the volcanic debris accounts for 40 to 80% by weight, and which may move by turbulent flow. As the proportion of debris increases, laminar flow replaces turbulent flow.

Mudflows or lahars (Figure 2.11) are regarded as primary when they result directly from eruption and secondary if there are other causes. Volcanic lahars are generally formed by the spontaneous release of a large amount of water (Verstappen, 1992). For example, deposits of loose or fragmented volcanic ejectamenta, frequently in the form of ash or lapilli but possibly including large boulders, may become saturated by torrential rain and/or melted snow associated with an eruption. Lahars may also be generated by rainfall several months or years after the last eruption. Alternatively, crater lake lahars, as for example occur periodically in Indonesia, result from the discharge of huge volumes of water from craters during the course of eruption, giving rise to highly destructive hot lahars. For instance, in 1919 it was estimated that 3.85×10^7 m^3 of water was expelled from the crater of Mount Kelut and mixed with pyroclastic material to form lahars that killed over 5100 people and destroyed over 9000 homes (Suryo and Clarke, 1985). Streams may be impounded by pyroclastic material or lava, the dam eventually being breached and releasing copious quantities of water for mudflow formation; or more simply, pyroclastic flows may

Figure 2.11 Mudflows (lahars) on the flanks of Mount St Helens, Washington.

mix with river water to form a lahar. Lahars can occur at any time before, during or after an eruption, and their properties vary depending on whether the material is hot or cold. Both are very poorly sorted but tend to fine upwards. Cold lahars may be regarded as similar to debris flows. Such debris flows are usually triggered by torrential rainstorms, earthquakes or slope failures. Hot lahars are accompanied by the release of large quantities of steam and lesser amounts of volcanic gases. They are usually more destructive than cold lahars. Hot pyroclastic material may accumulate in the ravines around the summit of a volcano after *nuées ardentes*. Subsequent heavy rain results in lahars forming from the pyroclastic material. The latter still may be very hot, several hundred degrees Celsius, so that steam is formed, resulting in a boiling mixture of mud and water. The release of steam lessens the frictional drag, thereby allowing the lahar to travel further. Lahars can be long (over 100 km), and one mudflow can overlie another, having followed the same route as the previous one. They may be several kilometres in width. The rate of flow may reach 50 km h^{-1} on relatively shallow slopes with viscous material, or twice that down steep slopes with low-viscosity debris.

A lahar often moves in a series of waves or pulses. At the break of slope, lahars tend to spread out to form fans, which may interfere with natural drainage channels. Further downslope, beyond the usual extent of coarse lahar flow, the streams frequently carry large quantities of sediment. This may cause the rivers they flow into to aggrade their beds, which can cause flooding. Estuaries may silt up.

2.8 Volcanic hazard and prediction

The obvious first step in any attempt to predict volcanic eruption is to determine whether the volcano concerned is active or extinct. If a volcano has erupted during historic times it should be regarded as active. However, many volcanoes that have long periods of repose will not have erupted in historic times. Nonetheless, they are likely to erupt at some future date. It has been estimated that the active life span of most volcanoes is probably between one and two million years (Baker, 1979).

The term 'extinct' is commonly applied to volcanoes that have not erupted during historic times, but catastrophic eruptions of volcanoes believed to have been extinct have occurred. Witness the violent eruptions of Mount Lamington, Papua (now Papua New Guinea), 1951, and Bezymianny, USSR, 1956. Indeed Crandell *et al.* (1984) maintained that one so-called extinct volcano comes to life every decade. Consequently, they suggested that the use of the term 'extinct' should be discontinued. In fact, they suggested the use of the terms 'live' and 'dead'[*] based on the probability that a volcano will or will not erupt. Such a

---

[*] The same argument could be used against the term 'dead' as it could against the term 'extinct'.

probability can be assessed from the prehistoric record of a volcano, which involves geological investigation of the volcanic materials associated with a volcano, as well as consideration of any historical evidence. The probability can be expressed in terms of the ratio between the time that has elapsed since the last eruption and the length of the longest repose period of the volcano. The smaller this ratio, the greater is the probability of a future eruption. Live volcanoes should be kept under surveillance, especially those that present short-term hazards, that is, volcanic events that are likely to occur within the lifetime of a person.

Monitoring the activity of a volcano, and mapping and interpreting previous eruptions aid the delineation of volcanic hazard zones (Crandell *et al.*, 1984). This includes recording the direction and lateral distance reached by lethal and non-lethal ejectamenta, the build-up of potential lahar material, the path taken by lavas flows and their maximum extension, and the distances and direction of *nuées ardentes*. Aerial photographs and satellite imagery can be used to record lahar, lava flow and ash plume development.

Determination of the expected recurrence interval of particular types of eruption, the distribution of their deposits, the magnitude of events and the recognition of short-term cycles or patterns of volcanic activity are all of value as far as prediction is concerned. It must be admitted, however, that no volcano has revealed a recognizable cycle of activity that could be used to predict the time of an eruption within a decade. Nonetheless, individual volcanoes require mapping in order that their evolution can be reconstructed. The geological data so gathered are used in conjunction with historical records, where available, to postulate future events If it is assumed that future volcanic activity will be similar to previously observed activity, then it is possible to make certain general predictions about hazard.

Tazieff (1979) pointed out that no volcanic catastrophes have occurred at the very start of an eruption. Consequently, this affords a certain length of time to take protective measures. Even so, because less than one out of several hundred eruptions proves dangerous for a neighbouring population, evacuation, presuming that an accurate prediction could be made, would not take place before an eruption became alarming. Nevertheless, it is still important to predict whether or not a developing eruption will culminate in a dangerous climax and, if it does, when and how. According to Tazieff, it is more difficult to predict the evolution of a developing eruption than to predict the initial outbreak.

2.8.1 Methods of prediction

Prediction studies are based on the detection of reliable premonitory symptoms of eruptions (Gorshkov, 1971). Such forerunners may be geophysical or geochemical in nature (Mori *et al.*, 1989). Geophysical observations, principally tiltmetry, seismography and thermometry, provide the basis for forecasting volcanic eruptions (Anon., 1971). Unfortunately, however, because of the cost involved and the problem of accessibility intensive monitoring only takes place at a small number of volcanoes (about 12), so these are the ones for which full-scale forecasts can be made. Even in these cases it is not possible to determine the size and time of an eruption. If the prediction of volcanic eruptions is to be successful, then it must provide sufficient time for safety measures to be put into effect.

Data from satellites can be used to monitor volcanic activity, especially in developing countries (Rothery, 1992). Three aspects of volcanism, namely, the movement of eruption plumes, changes in thermal radiation of the ground and output of gases, can be measured using remote sensing data so that in some instances prior warning of eruptions may be obtained.

A volcanic eruption involves the transfer towards the surface of millions of tonnes of magma. This leads to the volcano concerned undergoing changes in elevation, notably uplift (Figure 2.12). For instance, the summit of Kilauea, Hawaii, has been known to rise by almost a metre in the months preceding an eruption. Changes in elevation significantly larger than this have been associated with other volcanoes. For example, the side of Mount St Helens bulged at up to 2 m per day, being finally elevated through 100 m. The uplift is usually measured by a network of geodimeters and tiltmeters set up around a vol-

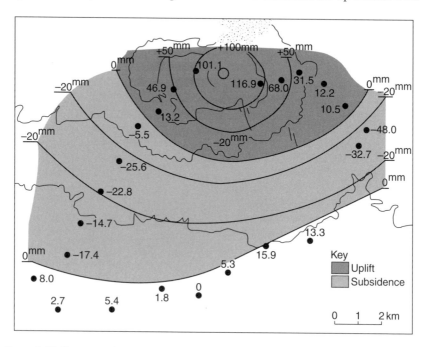

Figure 2.12 Elevation changes (in millimetres) at Mount Asama, Japan, during the period 1934–50 (from Decker and Kinoshita, 1971). The upper cone and the crater area have risen and are surrounded by a zone of subsidence (from Decher and Kinoshita, 1971).

cano. Gravity meters can be used to detect any vertical swelling. Tide gauges are used to measure changes in the level of the lake water in Rabaul caldera in Papua.

Such uplift also means that rocks are fractured, so volcanic eruptions are generally preceded by seismic activity (Figure 2.13). However, this is not always the case. For instance, this did not happen at Heimaey, Iceland, in 1973. Conversely, earthquake swarms need not be followed by eruptions: the tremors that were felt on Guadeloupe in 1976 were not followed by an eruption. Where earthquake swarms do occur the number of tremors increases as the time of eruption approaches. For example, tremors average six per day on Kilauea but at the beginning of 1955 these increased markedly, 600 being detected on 26 February. Two days later an eruption occurred. However, tremors may continue for a few days to a year or more. A network of seismic stations is set up to monitor the tremors and from the data obtained, the position and depth of origin of the tremors can be ascertained. The earthquakes are called A-type if they have hypocentres 1 to 10 km deep, B-type if the foci are more than 10 km deep and are unclassified if they result from explosions in or just beneath the crater. They may reach magnitude 5. The harmonic tremor is a characteristic form of volcano-seismicity. It is a narrow band of nearly continuous seismic vibrations dominated by a single frequency and is associated with the rise of magma or volcanically heated fluids.

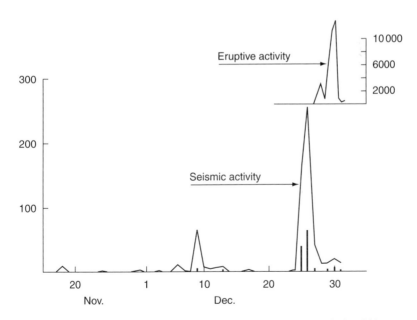

Figure 2.13 Seismic activity and daily numbers of earthquakes recorded at Mihara-yama (O-sima), November–December 1964 (after Shimozuro, 1971).

Infrared techniques have been used in the prediction of volcanic eruptions since, due to the rising magma, the volcano area usually becomes hotter than its surroundings. Thermal maps of volcanoes can be produced quickly by ground-based surveys using infrared telescopes (Francis, 1979). However, consistent monitoring is necessary in order to distinguish between real and apparent thermal anomalies. Aerial surveys provide better data but are too expensive to be used for routine monitoring.

According to Baker (1979), basaltic magmas of low viscosity prove an exception if their ascent is more rapid than the rate at which heat is conducted from them. Indeed, he suggested thermal anomalies are more likely to be produced by rising andesitic or rhyolitic magmas. A more or less contrary view has been advanced by Tazieff (1979). He maintained that thermal techniques have not proved satisfactory when monitoring explosive andesitic and dacitic volcanoes, since the ascent of highly viscous acid magmas is presumably too slow to give rise to easily detectable temperature changes. Perhaps both workers are correct in that the rise of magma can be too rapid or too slow to produce thermal anomalies that can be rapidly monitored.

Be that as it may, detectable anomalies are more likely to develop when heat is transferred by circulating groundwater rather than by conduction from a magma. The temperatures of hot springs may rise, as may the water in crater lakes, with the approach of a volcanic eruption. Hot groundwater also gives rise to the appearance of new fumaroles or to an increase in temperatures at the existing fumaroles. Again caution must be exercised since fumarole temperatures vary due to other factors, such as the amount of rain that has fallen. Steam gauges are used to monitor changes in gas temperatures, pH values and amounts of suspended mineral matter. The eruption of Taal, in the Philippines, in 1965 was predicted because of the rise in temperature of the water in the crater lake. This allowed evacuation to take place.

An increase in temperature also leads to demagnetization of rock as the magnetic minerals are heated above their Curie points. This can be monitored by magnetic surveying. Changes also occur in the gravitational and electrical properties of rocks.

Geophysical methods, however, cannot detect the climax of an eruption, because the magma is already very close to the surface and therefore rock fracturing, volcano inflation and increased heat transfer are not significant enough to record. As pointed out above, it is more important to forecast the climax, and its character, rather than the outbreak of an eruption.

The evolution of an eruption involves changes in matter and energy. The most significant variations take place in the gas phase and therefore it seems appropriate to gather as much information relating to this phase as possible, especially when this phase is the active agent of the eruptive phenomenon. Gas sensors can be stationed on the ground or carried in aircraft and record changes in the composition of gases. Unfortunately, reliable data are extremely scarce and the interpretation of what is available is in its infancy. Gas emitted from

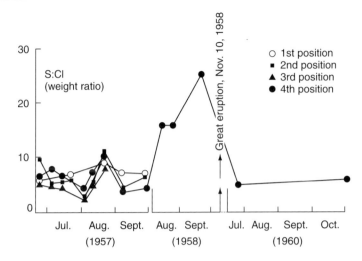

Figure 2.14 Changes in the S:Cl ratio in relation to explosions at Mount Asama (after Noguchi and Kamiya, 1963).

fumaroles on the flanks of a volcano may contain increases in HCl, HF and SO_2, or the ratios of, for example, S to Cl may change (Figure 2.14). These may be related to an impending eruption, because the proximity of magma tends to emit more sulphur, chlorine and fluorine. Moreover, increases in the dissolution of acid volcanic gases in the hydrothermal system often result in small decreases in the pH value. In addition, lake water chemistry may alter; for example, at Ruapehu volcano, New Zealand, the ratio of magnesium to chlorine increased by 25 to 33% before an eruption.

Monitoring rainfall and storm build-up is necessary for lahar prediction. Telemetered rain gauges may be used and are activated when a certain amount of rain falls in a given time.

2.8.2 Assessment of volcanic hazard and risk

When volcanic activity occurs in areas of high population density, it poses various kinds of hazard to the people living in the vicinity. It is impossible to restrict people from all hazardous areas around volcanoes, especially those that are active only intermittently. Hence, it is important to recognize the various types of hazard that may occur in order to prevent or mitigate disasters but still allow people to live on the fertile lower slopes of volcanoes.

Ten percent of the population of the world live on or near potentially active volcanoes, at least 91 of which are in high-risk areas (42 in southeast Asia and the western Pacific, 42 in the Americas and 7 in Europe and Africa). An eruption, or the precarious conditions it creates, may continue for months, and therefore volcanic emergencies are often long-lasting in comparison with other sudden-impact natural disasters. Nonetheless, most dangerous volcanic phenomena happen very quickly. For instance, the time interval between the beginning of an eruption and the appearance of the first nuées ardentes may be only a matter of hours. Fortunately, such events are usually preceded by visible signs of eruption.

Volcanic eruptions and other manifestations of volcanic activity are variable in type, magnitude, duration and significance as hazard. Volcanic hazard has been defined by Crandell *et al.* (1984) as a potentially damaging or destructive volcanic event. Ideally, in order to assess the significance of volcanic hazard it is helpful to estimate the time or period when a volcanic event is likely to occur, the frequency of such events, and the extent and frequency of areas likely to be affected. However, adequate data are seldom available to make such an assessment.

In any assessment of risk due to volcanic activity the number of lives at stake, the capital value of the property and the productive capacity of the area concerned have to be taken into account. Evacuation from danger areas is possible if enough time is available. However, the vulnerability of property is frequently close to 100% in the case of most violent volcanic eruptions. Hazard must also be taken into account in such an assessment (Tilling, 1989). It is a complex function of the probability of eruptions of various intensities at a given volcano and of the location of the site in question with respect to the volcano (Bondal and Robin, 1989; Forgione *et al.*, 1989). Hazard is the most difficult of factors to estimate, mainly because violent eruptions are rare events about which there are insufficient observational data for effective analysis. For example, in the case of many volcanoes, large eruptions occur at intervals of hundreds or thousands of years. Thorough stratigraphic study and dating of the deposits will help to provide the evidence needed to calculate the risk factors of such volcanoes. Because in the foreseeable future man is unlikely to influence the degree of hazard, then the reduction of risk can only be achieved by reducing the exposure of life and property to volcanic hazards. This can be assessed by balancing the loss of income resulting from non-exploitation of a particular area against the risk of loss in the event of an eruption.

Booth (1979) divided volcanic hazards into six categories, namely, premonitory earthquakes, pyroclast falls, pyroclast flows and surges, lava flows, structural collapse, and associated hazards. Each type represents a specific phase of activity during a major eruptive cycle of a polygenetic volcano and may occur singly or in combination with other types. Damage resulting from volcanoseismic activity is rare. However, intensities on the Mercalli scale varying from 6 to 9 have been recorded over limited areas.

Pyroclastic fall deposits may consist of bombs, scoria, lapilli, pumice, dense lithic material, crystals and/or any combination of these. There are, on average, about 60 pyroclast or tephra falls per century that are of social importance. In violent eruptions, intense falls of ash interrupt human activities and cause serious damage. The size of the area affected by pyroclastic falls depends on the amount of material ejected and the height to which it is thrown, as well as the wind speed and direction. They can affect areas up to several tens of kilometres from a volcano within a few hours from the commencement of an eruption. For example, the ash cloud associated with the Mount St Helens eruption in 1980 travelled 400 km downwind in the first six hours after the eruption. During the eruption, darkness occurred within 390 km of the volcano. Ashfall during and after the eruption contained 3 to 7% free silica, which if inhaled over long periods could cause silicosis of the lungs. The principal hazards to property resulting from pyroclastic falls are burial, impact damage and fire if the material has a high temperature. The latter hazard is most dangerous within a few kilometres of the vent. The weight of the material collected on roofs may cause them to collapse.

If violent Strombolian eruptions occur near a centre of population, then the rain of incandescent pyroclasts will cause fires, and buildings with flat roofs can collapse under the weight of ash they collect. Permanent damage to vegetation can be brought about by toxic gases, and they make evacuation more difficult. Plinian eruptions are several orders of magnitude greater than those of Strombolian type; for instance, Booth (1973) recorded that Plinian eruptions on Tenerife, Canary Islands, have covered 4200 km^2 with 1 m or more of ash. Welded air-fall ash can occur during a phase of high emission during a Plinian eruption. Although this action is generally concentrated near the parent crater, such deposits have occasionally been found up to 6 or 7 km distant.

Laterally directed blasts are among the most destructive of volcanic phenomena (Figure 2.15). They travel at high speeds; for example, the lateral blast associated with the Mount St Helens eruption in 1980 initially travelled at 600 km h^{-1}, slowing to 100 km h^{-1} some 25 km from the volcano. They can occur with little or no warning in a period of a few minutes and can affect hundreds of square kilometres. The material carried by lateral blasts can vary from cold to temperatures high enough to scorch vegetation and start fires. Such blasts kill virtually all life by impact, abrasion, burial and heat. A high concentration of ash contaminates the air. Crandell et al. (1984) indicated that laterally directed blasts are the result of sudden decompression of magmatic gases or explosion of a high-pressure hydrothermal system. Release of gases may be caused by volatile pressures exceeding the weight of the overlying rocks, so giving rise to an explosion, which may hurl rock fragments on ballistic trajectories or generate pyroclastic flows or surges, or combinations of these.

Pyroclastic flows are hot, dry masses of clastic volcanic material that move over the ground surface. Most pyroclastic flows consist of a dense basal flow, the pyroclastic flow proper, one or more pyroclastic surges and clouds of ash. Two major types of pyroclastic flow can be recognized. Pumiceous pyroclastic flows are concentrated mixtures of hot to incandescent pumice, mainly of ash and lapilli size. Ashflow tuffs and ignimbrites are associated with these flows. Individual flows vary in length from less than 1 up to 200 km, covering areas of up to $20\,000 \text{ km}^2$ with volumes from less than 0.001 to over 1000 km^3.

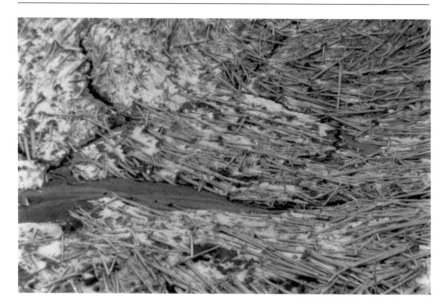

Figure 2.15 Snapped-off trees caused by the blast from Mount St Helens during the eruption of 18 August 1980.

Pyroclastic flows formed primarily of scoriaceous or lithic volcanic debris are
known as hot avalanches, glowing avalanches, *nuées ardentes* or block and ash-
flows. They generally affect a narrow sector of a volcano, perhaps only a single
valley. Pyroclastic flows may be classified according to the ratio of gas to water
they contain on the one hand and temperature on the other (Table 2.3).
Maximum temperatures of pyroclastic flow material soon after it has been
deposited range from 350 to 550 °C. Hence they are hot enough to kill anything in their path. Because of their high mobility (up to 160 km h^{-1} on the
steeper slopes of volcanoes) they constitute a great potential danger to many
populated areas. Other hazards associated with pyroclastic flows, apart from
incineration, include burial, impact damage and asphyxiation.

The collapse of lava domes and coulees on the steep slopes of volcanoes can
give rise to lava debris flows or pyroclast flows, especially if vesiculation occurs
in the freshly exposed hot lava interior. For example, Mount Merapi, Indonesia,
has produced numerous lava domes of 5×10^6 m^3 or more, which periodically
collapse, spilling down the flanks of the volcano. The lava fragments release
gas and give rise to *nuées ardentes* of the avalanche type (Suryo and Clarke, 1985).

A pyroclastic surge is a turbulent, low density cloud of gases and rock debris
that hugs the ground over which it moves (Sparks, 1976). Hot pyroclastic
surges can originate by explosive disruption of volcanic domes caused by rapidly
escaping gases under high pressure or by collapse of the flank of a dome. They
can also be caused by lateral explosive blast. Hot pyroclastic surges can occur
together with pyroclastic flows. Generally, surges are confined to a narrow valley

Table 2.3 Classification of volcanoclastic flows for risk assessment (after Booth 1979).

	Gas concentration in flow (air, volcanic gas or water vapour)			Water concentration in flow	
	High gas/solid ratio Hot Pyroclast surges	Low gas/solid ratio Hot Pyroclast flows	Cold	High solid/water ratio	Low solid/water ratio
		Pumice or scoria / Non-vesiculated			
Base surges	Nuées ardentes	Pumice slurry flows / Ash flows / Scoria flows — Lava debris flow	Rock avalanches	Mudflows	Torrents
				Contain as little as 20% water	Contain as much as 80 to 90% water
Base surge deposit	Nuée ardente/ Surge deposit	Ignimbrite (Welded tuffs)		Stratified sedimentary deposits (Commonly with cross bedding)	

of a volcano, but they may reach speeds of up to 300 km h^{-1}, so escape from them is virtually impossible. They give rise to similar hazards to pyroclastic flows. Cold pyroclastic surges are produced by phreatic and phreato-magmatic explosions. Vertical explosions can give rise to a primary surge, which moves away from the volcano in all directions. Subsequently, secondary surges may be formed when volcanic material falls to the ground. The high speed of primary surges is attributable to the explosive force, while that of secondary surges is due to the kinetic energy gained during falling and the speed gathered as they descend the slopes of the volcano. Surges decelerate rapidly and tend not to travel more than 10 km from their source. Fortunately, pyroclastic flows and surges tend to affect limited areas, and approximately twenty occur every 100 years.

If the movement of water within a volcano is restricted and a heat source is available, water can exist at temperatures that approach or exceed boiling point. A phreatic explosion takes place when the temperature of the water system rises, leading to the vapour pressure exceeding the load of overburden. Such phreatic explosions are generally associated with strato-volcanoes since they often contain alternating layers of relatively permeable and impermeable rocks, which inhibit the flow of groundwater. These explosions may eject pyroclastic material. Phreatic explosions may precede magmatic eruptions. For example, phreatic activity occurred prior to the eruptions of Mont Pelée in 1902 and Mount St Helens in 1980. Phreatic magmatic eruptions involve varying amounts of magma, as well as steam.

The distance that a lahar travels depends on its volume and water content on the one hand and the gradient of the slope of the volcano down which it moves on the other. Some may travel more than 100 km. The speed of a lahar is also influenced by the water–sediment ratio and its volume, as well as the gradient and shape of the channel it moves along. For example, speeds of up to 165 km h^{-1} have been claimed for some lahars generated by the Mount St Helens eruption of 1980 (Janda *et al.*, 1981). However, the average speed over several tens of kilometres was generally less than 25 km h^{-1}. Because of their high bulk density, lahars can destroy structures in their path and block highways. They can reduce the channel capacity of a river and so cause flooding, as well as adding to the sediment load of a river. Hence lahars may prove as destructive as pyroclastic flows over limited areas. Destructive lahars average 50 per century.

Lava effusions of social consequence average 60 per century. Fortunately, because their rate of flow is usually sufficiently slow and along courses that are predetermined by topography, they rarely pose a serious threat to life. The arrival of a lava flow along its course can be predicted if the rate of lava emission and movement can be determined. Damage to property, however, may be complete, destruction occurring by burning, crushing or burial of structures in its path. Burial of land by lava flows commonly means that its previous use is terminated. Lava flood eruptions are the most serious and may cover large areas with immense volumes of lava; for example, the Laki, Iceland, eruption of 1783 produced 12.3 km^3 of lava, which spread over 560 km^2, the flows reaching a maximum distance of 65 km from the fissure (Thorarinsson, 1970).

The likelihood of a given location being inundated with lava at a given time can be estimated from information relating to the periodicity of eruptions in time and space, the distribution of rift zones on the flanks of a volcano, the topographic constraints on the directions of flow of lavas, and the rate of covering of the volcano by lava. The length of a lava flow is dependent upon the rate of eruption, the viscosity of the lava and the topography of the area involved. Given the rate of eruption it may be possible to estimate the length of flow. Each new eruption of lava alters the topography of the slopes of a volcano to a certain extent and therefore flow paths may change. What is more, prolonged eruptions of lava may eventually surmount obstacles that lie in their path and that act as temporary dams. This then may mean that the lava invades areas that were formerly considered safe.

The formation of a caldera and landslip scars due to the structural collapse of a large volcano is a rare event (0.5–1 per century). They are frequently caused by magma reservoirs being evacuated during violent Plinian eruptions. Because calderas develop near the summits of volcanoes and subside progressively as evacuation of their magma chambers takes place, caldera collapses do not offer such a threat to life and property as does sector collapse. Sector collapse involves subsidence of a large area of a volcano. It takes place over a comparatively short period of time and may involve volumes of up to tens of cubic kilometres. Collapses that give rise to landslides may be triggered by volcanic explosions or associated earthquakes. For instance, a collapse on the north of Mount St Helens in 1980 was caused by an earthquake (M_L about 5). According to Voight et al. (1981), the volume of material involved in the landslide was around 2.8 km^3 and covered an area of 60 km^2. It moved about 25 km from the volcano.

Hazards associated with volcanic activity include destructive floods caused by sudden melting of the snow and ice that cap high volcanoes, or by heavy downfalls of rain (vast quantities of steam may be given off during an eruption), or the rapid collapse of a crater lake. Far more dangerous are the tsunamis (see Chapter 6) generated by violent explosive eruptions and sector collapse. Tsunamis may decimate coastal areas. Dense poisonous gases normally offer a greater threat to livestock than to man. Nonetheless, the large amounts of carbon dioxide released at Lake Manoun in Cameroon in 1984 and neighbouring Lake Nyos in 1986 led to the deaths of 37 and 1887 people, respectively (Kling et al., 1987). According to Le Guern et al. (1982), 149 people were killed by emission of gas, notably carbon dioxide and hydrogen sulphide, on the Dieng Plateau, Indonesia, in 1979. The gases moved downslope from vents and accumulated in low-lying areas. A hazard zone map was prepared for the area after this catastrophe and was based on known sites of gas emission and topography (Figure 2.16). In addition, gases can be injurious to the person, mainly because of the effects of acid compounds on the eyes, skin and respiratory system. They

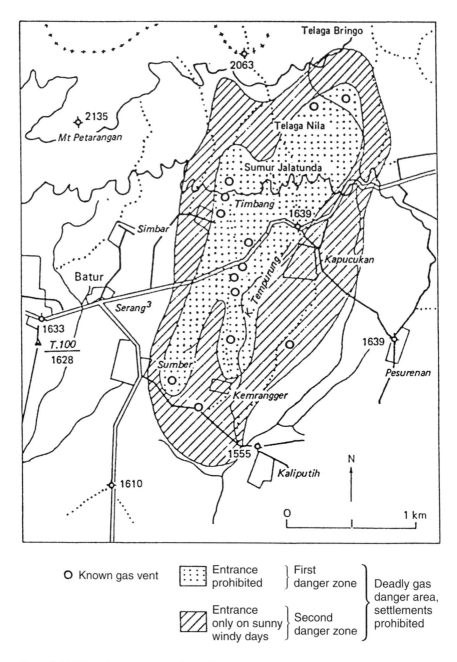

Figure 2.16 Hazard zonation map for volcanic gases for an area in the Dieng Mountains of Java (from Crandell et al., 1984).

can kill crops. Acid rain can form as a result of rain mixing with aerosols and gases adhering to tephra. Such rain can cause severe damage to natural vegetation and crops, and skin irritation to people. Air blasts, shock waves and counter-blasts are relatively minor hazards, although they can break windows several tens of kilometres away from major eruptions.

According to Booth (1979), four categories of hazard have been distinguished in Italy:

1   Very high-frequency events with mean recurrence intervals (MRI) of less than two years. The area affected by such events is usually less than 1 km^2.
2   High-frequency events with MRI values of 2 to 200 years. In this category damage may extend up to 10 km^2.
3   Low-frequency events with MRI values of 200 to 2000 years. Areal damage may cover 1000 km^2.
4   Very low-frequency events are associated with the most destructive eruptions and have MRI values in excess of 2000 years. The area affected may be greater than 10 000 km^2.

Hazard zoning involves mapping deposits that have formed during particular phases of volcanic activity and their extrapolation to identify areas that would be likely to suffer a similar fate at some future time. The zone limits on such maps normally assume that future volcanic activity will be similar to that recorded in the past. Unfortunately, this is not always the case: witness the 1980 eruption of Mount St Helens. Volcanic hazard maps produced by the Volcanological Survey of Indonesia define three zones: namely, the forbidden zone, the first danger zone and the second danger zone (Suryo and Clarke, 1985). The forbidden zone is meant to be abandoned permanently since it is affected by *nuées ardentes* (Figure 2.17). The first danger zone is not affected by *nuées ardentes* but may be affected by bombs. Lastly, the second danger zone is that likely to be affected by lahars. This zone is subdivided into an abandoned zone, from which there is no escape from lahars, and an alert zone, where people are warned and from where evacuation may be necessary. Preliminary volcanic hazard maps are prepared where fewer data are available. These distinguish danger zones that must be evacuated immediately when increased activity occurs and alert zones (Figure 2.18).

Volcanic risk maps indicate the specified maximum extents of particular hazards such as lava and pyroclastic flow paths, expected ashfall depths and the areal extent of lithic missile fall-out (Figure 2.19). They are needed by local and national government so that appropriate land uses, building codes and civil defence responses can be incorporated into planning procedures. It has been suggested by Fournier d'Albe (1979) that events with an MRI of less than 5000 years should be taken into account in the production of maps of volcanic hazard zoning. He further suggested that data on any events that have taken place in the last 50 000 years are probably significant. He proposed that two types

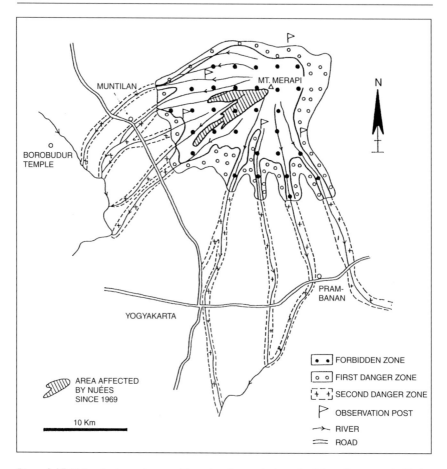

Figure 2.17 Volcanic hazard map, Merapi volcano, Indonesia (after Suryo and Clarke, 1985).

of map would be useful for economic and social planning. One type would indicate areas liable to suffer total destruction by lava flows, *nuées ardentes* and/or lahars. The other would show areas likely to be affected temporarily by damaging but not destructive phenomena, such as heavy falls of ash, toxic emissions, pollution of surface or underground waters, etc.

Losses caused by volcanic eruptions can be reduced by a combination of prediction, preparedness and land-use control. Emergency measures include alerting the public to the hazard, followed by evacuation. Unfortunately, false predictions and unnecessary evacuations do occur. Appropriate structural measures for hazard reduction are not numerous, but they include building steeply pitched reinforced roofs that are unlikely to be damaged by ash-fall, and con-

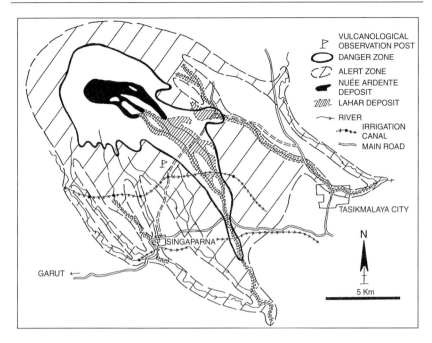

Figure 2.18 Preliminary hazard map of Galunggung volcano, Indonesia, showing extent of nuée ardente and lahar in 1982 (after Suryo and Clarke, 1985).

structing walls and channels to deflect lava flows. Hazard zoning, insurance, local taxation and evacuation plans are appropriate non-structural measures to put into effect. Risk management depends on identifying hazard zones and forecasting eruptions. The levels of risk must be defined and linked to appropriate social responses. Volcanic zoning is constrained by the fact that the authorities and public are reluctant to make costly adjustments to a hazard that may have a mean recurrence interval of 1000 years or more. Data on recurrence intervals are thus critical to zonation. In the long term, appropriate controls should be placed on land use and the location of settlements.

2.8.3 Dealing with volcanic activity

Obviously, it is impossible to control violent volcanic phenomena such as *nuées ardentes* and pyroclastic fallout. Their only mitigation is by recognition of their potential extent and so making possible either temporary or permanent evacuation of areas likely to be badly affected. The threat of lava flows can be dealt with, with varying degrees of success, by diverting, disrupting or stopping them. For example, during the eruption from Kirkefell on the island of Heimaey, Iceland, in 1973, ash from the volcano was bulldozed to form a wall

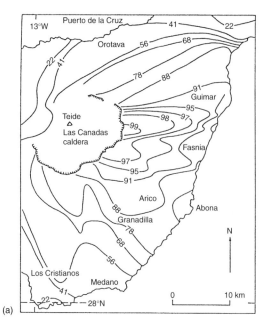

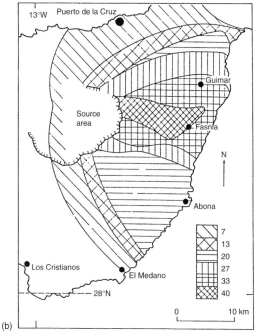

Figure 2.19 (a) Southeastern Tenerife, showing the extent and percentage probability of burial by more than 1 m of air-fall ash if a Plinian-type eruption occurs; figures are based on deposits formed during twenty-seven eruptions. (b) Frequency with which specific areas of Tenerife have been affected by air-fall deposits during the last 50 000 years (after Booth, 1979).

in order to divert lava flows away from the town of Vestmannaeyjar. Masonry and earth walls have been used in Japan and Hawaii for the same purpose. They should be constructed at an angle to the direction of flow so that it can be diverted rather than dammed. In the latter case, a lava flow would eventually spill over the dam. Topographic expression is therefore important; it should permit diversion so that no serious damage results. Lavas frequently flow along pre-existing channels across which such diversionary barriers can be built. Another diversionary technique is to dam the summit crater of a volcano at the usual exit and to breach it somewhere else so that the lavas will then flow in a different direction, to where they do little or no harm.

Consolidation, check and diversion check dams have also been constructed to deal with lahars. Dams can be designed to retain the downslope movement of larger particulate material but to allow water and suspended sediment to pass on. Such dams may be constructed of permeable gabions or of concrete with weep-holes. Suryo and Clarke (1985) suggested that the afforestation of the upper slopes of intermittently active volcanoes may inhibit the development of areas of potentially unstable soil, as well as preventing excessive run-off, and thereby reduce the likelihood of lahar formation in the event of an eruption. Bombing was used successfully in 1935 and 1942 to disrupt lava flowing from Mauna Loa, Hawaii, away from the town of Hilo. The technique might also be used to breach a summit wall in order to release lava in a harm-less direction.

Water has been sprayed on to advancing lavas to cool them and thereby cause them to solidify and stop (Figure 2.20). This technique eventually may have

Figure 2.20 Spraying water on to lava in an attempt to halt its advance, Heimaey, Iceland, 1973.

proved successful on Heimaey in 1973. However, it obviously is not known whether the lavas would have advanced much further without such treatment.

Crater lakes with a known history of volcanic eruptions have been drained, the classic example being provided by Mount Kelut. Following the eruption of 1919 a series of siphon tunnels were constructed to lower the surface of the lake and thereby reduce the potential lahar hazard.

References

Anderson, F.W. and Dunham, C.K. (1966) *The Geology of Northern Skye*. Memoir, Geological Survey Great Britain, HMSO, London.

Anon. (1971) *The Surveillance and Prediction of Volcanic Activity*. UNESCO, Paris.

Anon. (1976) *Disaster Prevention and Mitigation. 1, Volcanological Aspects*. United Nations, New York.

Baker, P.E. (1979) Geological aspects of volcano prediction. *Journal Geological Society*, 136, 341–346.

Beverage, J.P. and Cuthbertson, J.K. (1964) Hyperconcentrations of suspended sediment. *Proceedings American Society Civil Engineers, Journal Hydraulics Division*, 90, HY6, 117–128.

Bondal, C. and Robin, C. (1989) Volcan Popocatepetl: recent eruptive history and potential hazards and risks in future eruptions. In *Volcanic Hazards: Assessment and Monitoring*, Latter, J.H. (ed.), Springer-Verlag, Berlin, 110–128.

Booth, B. (1973) The Granadilla pumice deposit, southern Tenerife, Canary Islands. *Proceedings Geological Association*, 84, 353–370.

Booth, B. (1979) Assessing volcanic risk. *Journal Geological Society*, 136, 331–340.

Carey, S. and Sigurdsson, H. (1989) The intensity of Plinian eruptions. *Bulletin Volcanology*, 51, 28–40.

Cook, E.P. (1966) *Tufflavas and Ignimbrites*, Elsevier, New York.

Crandell, D.R., Booth, B., Kusumadinata, K., Shimozuru, D., Walker, G.P.L. and Westercamp, D. (1984) *Source Book for Volcanic Hazards Zonation*, UNESCO, Paris.

Decher, R.W. and Kinoshita, W.T. (1971) Geodetic measurements. *In The Surveillance and Prediction of Volcanic Activity*, UNESCO, Paris, 47–74.

Fedetov, S.A. (1985) Estimates of heat and pyroclast discharge of volcanic eruptions based upon the eruption cloud and steady plume observations. *Journal of Geodynamics*, 3, 275–302.

Fenner, C.N. (1923) The origin and mode of emplacement of the great tuff deposit of the Valley of Ten Thousand Smokes. *National Geographical Society America*, Technical Paper Katmai Series, No. 1, 74pp.

Forgione, G., Luongo, G. and Romano, R. (1989) Mt Etna (Sicily): Volcanic hazard assessment. In *Volcanic Hazards: Assessment and Monitoring*, Latter, J.H. (ed.), Springer-Verlag, Berlin, 137–150.

Fournier d'Albe, E.M. (1979) Objectives of volcanic monitoring and prediction. *Journal Geological Society*, 136, 321–326.

Francis, P.W. (1979) Infrared techniques for volcano monitoring and prediction – a review. *Journal Geological Society*, 136, 355–360.

Gorshkov, G.S. (1971) General introduction. In *The Surveillance and Prediction of Volcanic Activity*, UNESCO, Paris, 9–13.

Janda, R.J., Scott, K.M., Nolan, K.M. and Martinson, H.A. (1981) Lahar movement, effects and deposits. In *The 1980 Eruptions of Mount St Helens, Washington*, Lipman, P.W. and Mullineaux, D.R. (eds), US Geological Survey Professional Paper 1250, 461–478.

Kling, G.W., Clark, M.A., Compton, H.R., Devine, W.C., Evans, A.M., Humphrey, E.J., Koenigsberg, E.G., Lockwood, D.J.P., Tuttle, M.L. and Wagner, G.N. (1987) The Lake Nyos gas disaster in Cameroon, West Africa, *Science*, 236, 169–174.

Le Guern, F., Tazieff, H. and Faivre Pierret, R. (1982) An example of health hazard: people killed by gas during a phreatic eruption. *Bulletin Volcanologique*, 45, 153–156.

MacDonald, G.A. (1967) Forms and structures of extrusive basaltic rocks. In *Basalts*, Hess, H.H. and Poldervaart, A. (eds), Interscience, New York.

MacDonald, G.A. (1972) *Volcanoes*. Prentice-Hall, Englewood Cliffs, NJ.

Mori, J., McKee, C.O., Talai, B. and Itikarai, I. (1989) A summary of precursors to volcanic eruptions in Papua New Guinea. In *Volcanic Hazards: Assessment and Monitoring*, Latter, J.H. (ed.), Springer-Verlag, Berlin, 260–291.

Newhall, C.G. and Self, S. (1982) The volcanic explosivity index (VRI): an estimate of explosive magnitude for historical volcanism. *Journal Geophysical Research*, 87, 1231–1238.

Noguchi, K. and Kamiya, H. (1963) Prediction of volcanic eruption by measuring composition and amounts of gases. *Bulletin Volcanologique*, 26, 367–378.

Rittmann, A. (1962) *Volcanoes and their Activity*, translated by E.A. Vincent. Interscience, London.

Rothery, D.A. (1992) Monitoring and warning of volcanic eruptions by remote sensing. In *Geohazards: Natural and Man-Made*, McCall, G.J.H., Laming, D.J.C. and Scott, S.C. (eds), Chapman & Hall, London, 25–32.

Shimozuro, D. (1971) A seismological approach to the prediction of volcanic eruptions. In *The Surveillance and Prediction of Volcanic Activity*, UNESCO, Paris, 19–45.

Sparks, R.S.J. (1976) Grain size variations in ignimbrites and implications for the transport of pyroclastic flows. *Sedimentology*, 23, 147–188.

Suryo, I. and Clarke, M.G.C. (1985) The occurrence and mitigation of volcanic hazards in Indonesia as exemplified in the Mount Merapi, Mount Kelat and Mount Galungging volcanoes. *Quarterly Journal Engineering Geology*, 18, 79–98.

Tazieff, H. (1977) An exceptional eruption: Mt Niragonga, Jan 10, 1977. *Bulletin Volcanologique*, 40, 189–200.

Tazieff, H. 1979. What is to be forecast: outbreak of eruption or possible paroxysm? The example of the Guadeloupe Soufrière, *Journal Geological Society*, 136, 327–330.

Tazieff, H. (1988) Forecasting volcanic eruption disasters. In *Natural and Man-Made Hazards,* El-Sabh, M.I. and Murty, T.S. (eds), Riedel, Dordrecht, 751–772.

Tazieff, H. and Sabroux, J.C. (1983) *Forecasting Volcanic Events*, Elsevier, Amsterdam.

Thorarinsson, S. (1970) The Lakigigar eruptions of 1783. *Bulletin Volcanologique*, 33, 910–927.

Tilling, R.I. (1989) Volcanic hazards and their mitigation: progress and problems. *Reviews in Geophysics*, 27, 237–269.

Tsuya, H. (1955) Geological and petrological studies of Volcano Fiji. *Bulletin Earthquake Research Institute*, University of Tokyo, 33, 314–384.

Tyrrell, G.W. (1937) Flood basalts and fissure eruptions, *Bulletin Volcanologique*, 1, 89–111.

Verstappen, H.Th. (1992) Volcanic hazards in Columbia and Indonesia: lahars and related phenomena. In *Geohazards: Natural and Man-Made*, McCall, G.J.H., Laming, D.J.C. and Scott, S.C. (eds), Chapman & Hall, London, 33–42.

Voight, B., Glicken, H., Janda, R.J. and Douglass, P.M. (1981) Catastrophic rockslide avalanche of May 18th. In *The 1980 Eruption of Mount St Helens, Washington*, Lipman, P.W. and Mullineaux, D.R. (eds), US Geological Survey Professional Paper 1250, 347–377.

Chapter 3

Earthquake activity

3.1 Introduction

Although earthquakes have been reported from all parts of the world they are primarily associated with the edges of the plates that form the Earth's crust (Figure 3.1). The Earth's crust is being slowly displaced at the margins of the plates. Differential displacements give rise to elastic strains, which eventually exceed the strength of the rocks involved and faults then occur. The strained rocks rebound along the fault under the elastic stresses until the strain is partly or wholly dissipated.

Earthquake foci are confined within a limited zone of the Earth, the lower boundary of which is located at approximately 700 km, and they rarely occur at the Earth's surface. In fact, most earthquakes originate within the upper 25 km of the Earth. Because of its significance, the depth of focus has been used as the basis of a threefold classification of earthquakes: those occurring within the upper 70 km are referred to as shallow; those located between 70 and 300 km as intermediate; and those between 300 and 700 km as deep.

An earthquake propagates three types of shock wave. The first pulses that are recorded are termed primary or P waves. Sometimes they are referred to as push-and-pull waves since oscillation occurs to and fro in the direction of propagation of the wave. P waves are also called compression waves or longitudinal waves. The next pulses recorded are the S waves, sometimes referred to as secondary or shake waves. These waves oscillate at right angles to the direction of propagation and usually have a larger amplitude than P waves, but the latter travel more rapidly. The third type of vibration is known as the L wave. These waves travel from the focus of the earthquake to the epicentre, immediately above at the surface, and from there they radiate over the Earth's surface. Two types of L or surface wave can occur in a solid medium, namely, Rayleigh waves and Love waves. In the former, surface displacement occurs partly in the direction of propagation and partly in the vertical plane. They can only be generated in a uniform solid. Love waves occur in non-uniform solids and oscillate in a horizontal plane, normal to the direction of propagation. L waves are recorded after S waves.

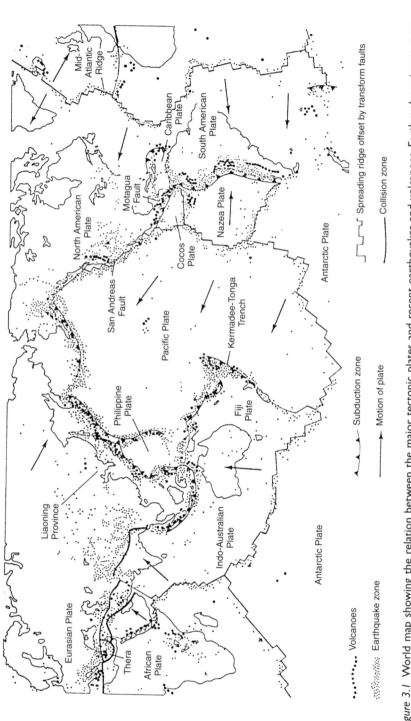

Figure 3.1 World map showing the relation between the major tectonic plates and recent earthquakes and volcanoes. Earthquake epicentres are denoted by the small dots, and the volcanoes by the large dots.

As P waves travel more quickly than S waves (in the crust $5.95–6.75$ km s^{-1} compared with $2.9–4.0$ km s^{-1}), the further they travel from the focus of an earthquake the greater is the time interval between them. Thus the distance of a recording station from the epicentre of an earthquake can be calculated from this time interval.

P waves are not as destructive as S or L waves, because they have a smaller amplitude, and the force that their customary vertical motion creates rarely exceeds the force of gravity. On the other hand, S waves may develop violent tangential vibrations strong enough to cause great destruction. The severity of an earthquake depends upon the amplitude and frequency of wave motion. S waves commonly have a higher frequency than L waves; nevertheless, the latter may be more powerful because of their larger amplitude.

Large earthquakes cause shock waves that travel throughout the world. Near the epicentre, the shocks are felt for only a matter of seconds, but with increasing distance from the epicentre the disturbance lasts for a progressively longer time. The vibrations of a shallow earthquake decrease rapidly in intensity away from the epicentre. Although the shocks of deep-seated earthquakes are usually weakly felt when they reach the surface, they extend over a much wider area.

It is assumed that the cause of most earthquakes is faulting. More than one rupture may occur along a fault, the second being triggered by the first. Furthermore, because it is energetically easier to make use of a pre-existing fault than initiate a new one, faults may be moved repeatedly by successive earthquakes. There seems reason to believe that great earthquakes may consist of a rapid succession of such breaks, giving the overall impression of continuous rupturing. Some areas are affected by earthquake swarms, that is, by a series of nearly equally large shocks accompanied by small ones. Surface faulting in a particular earthquake usually extends over only a part of an existing fault.

Initial movement may occur over a small area of a fault plane, subsequently being followed by slippage over a much larger surface. Such initial movements, which represent preliminary shattering of small obstructions along a fault zone, account for the foreshocks that precede an earthquake. When these have been overcome the main movement occurs, but complete stability is not restored immediately. The shift of strata involved in faulting relieves the main stress but develops new stresses in adjacent areas. Because stress is not relieved evenly everywhere, minor adjustments may arise along the fault plane and thereby generate aftershocks. The decrease in the strength of the aftershocks is irregular, and occasionally they may continue for a year or more after the principal shocks.

Although the spatial association of larger earthquakes with faults suggests that strain accumulates along the faults, the spatial association of smaller earthquakes does not necessarily lead to the same conclusion. For example, smaller earthquakes occur almost randomly throughout southern California. Moreover, many shallow-focus earthquakes in Japan, Italy and Yugoslavia have not been accompanied by observable faulting.

Displacement along fault zones may occur by slow, differential slippage, termed fault creep, as well as by sudden rupture. Fault creep may amount to several millimetres per year (Steinbrugge and Zacher, 1960). It is uncertain whether creep indicates that strain is being relieved along a major segment of a fault and as a result reducing the probability of a strong earthquake occurring, or that it represents an accumulation of excess strain in spite of slight relief. In fact, both situations might apply. Be that as it may, locations where creep is taking place should be identified and monitored for abnormal movements. Obviously, creep indicates that a fault is active and as such is a factor to be considered in land-use planning.

Seismic waves are recorded by a seismometer, which traces the oscillation caused by ground shaking on a drum, paper or magnetic tape to give a seismograph. There are various types of seismometer, and a seismic station may require an array of instruments, each of which is sensitive to a particular frequency range. Unfortunately, however, few useful data are obtained by seismic stations that are very close to the epicentre of major strong motions, as the instruments tend to be overwhelmed by the magnitude of the tremors. On the other hand, accelerometers are used to measure strong motions; indeed, they are only set in motion by strong tremors (see Section 3.5). Consequently, although accelerographs do not provide data on continuous small-scale seismic activity, they do provide useful records of major tremors. Earthquake activity is now monitored in a coordinated way by the World-Wide Network of Seismic Stations. There are over 120 stations at which recording is made continuously.

3.2 Intensity and magnitude of earthquakes

In earthquake regions it is necessary to establish the nature of the risk to structures. Thus the probability of the occurrence of earthquakes, their intensity and magnitude, and the likelihood of earthquake damage to a structure have to be evaluated. As soon as the intensity detrimental to a planned structure is established, the probability of an earthquake of this intensity in the given region should be estimated.

The severest earthquakes wreak destruction over areas of 2500 km^2 or more; however, most affect only tens of square kilometres. Earthquake intensity scales depend on human perceptibility and the destructivity of earthquakes. Whereas the degree of damage may be estimated correctly and objectively, the perceptibility of an earthquake depends on the location of the observer and his sensitivity. Several earthquake intensity scales have been proposed. The one given in Table 3.1 is the Mercalli scale, slightly modified by Richter (1956). The Medvedev–Sponheuer–Karnik (MSK) intensity scale (1964) is given in Table 3.2 for comparison. However, quantification of earthquake size by the amount of damage caused to structures at the epicentre is not really satisfactory. Not only are such surface effects dependent on local conditions, but a particularly

Table 3.1 Modified Mercalli scale, 1956 version, with Cancani's equivalent acceleration (these are not peak accelerations as instrumentally recorded). (After Wood and Neumann, 1931.)

Degrees	Description	Acceleration $mm\ s^{-2}$
I	Not felt. Only detected by seismographs.	< 2.5
II	Feeble. Felt by persons at rest, on upper floors, or favourably placed.	2.5–5.0
III	Slightly felt indoors. Hanging objects swing. Vibration like passing of light trucks. Duration estimated. May not be recognized as earthquake.	5.0–10
IV	Moderate. Hanging objects swing. Vibration like passing of heavy trucks, or sensation of a jolt like a heavy ball striking the walls. Standing motor cars rock. Windows, dishes, doors rattle. Glasses clink. Crockery clashes. In the upper range of IV wooden walls and frames creak.	10–25
V	Rather strong. Felt outdoors, direction estimated. Sleepers wakened. Liquids disturbed, some spilled. Small unstable objects displaced or upset. Doors swing, close, open. Shutters and pictures move. Pendulum clocks stop, start, change rate.	25–50
VI	Strong. Felt by all. Many frightened and run outdoors. Persons walk unsteadily. Windows, dishes, glassware broken. Ornaments, books, etc., fall off shelves. Pictures fall off walls. Furniture moved or overturned. Weak plaster and masonry cracked. Small bells ring (church, school). Trees, bushes shaken visibly or heard to rustle.	50–100
VII	Very strong. Difficult to stand. Noticed by drivers of motor cars. Hanging objects quiver. Furniture broken. Damage to masonry D, including cracks. Weak chimneys broken at roof line. Fall of plaster, loose bricks, stones, tiles, cornices, also unbraced parapets and architectural ornaments. Some cracks in masonry C. Waves on ponds, water turbid with mud. Small slides and caving-in along sand or gravel banks. Large bell rings. Concrete irrigation ditches damaged.	100–250
VIII	Destructive. Steering of motor cars affected. Damage to masonry C, partial collapse. Some damage to masonry B, none to masonry A. Fall of stucco and some masonry walls. Twisting, fall of chimneys factory stacks, monuments, towers, elevated tanks. Frame houses moved on foundations if not bolted down, loose panel walls thrown out, Decayed piling broken off. Branches broken from trees. Changes in flow or temperature of springs and wells. Cracks in wet ground and on steep slopes.	250–500
IX	Ruinous. General panic. Masonry D destroyed, masonry C heavily damaged, sometimes with complete collapse, masonry B seriously damaged. General damage to foundations. Frame structures, if not bolted, shifted off foundations. Frames cracked, serious damage to reservoirs. Underground pipes broken. Conspicuous cracks in ground. In alluviated areas sand and mud ejected, earthquake fountains, sand craters.	500–1000
X	Disastrous. Most masonry and frame structures destroyed with their foundations. Some well-built wooden structures and bridges destroyed. serious damage to dams, dykes, embankments. Large landslides. Water thrown on banks of canals, rivers, lakes, etc. Sand and mud shifted horizontally on beaches and flat land. Rail tracks bent slightly.	1000–2500
XI	Very disastrous. Rail tracks bent greatly. Underground pipelines completely out of service.	2500–5000
XII	Catastrophic. Damage nearly total. Large rock masses displaced. Lines of sight and level distorted. Objects thrown into the air.	> 5000

Table 3.2 Medvedev–Sponheuer–Karnik seismic intensity scale.

Classification of the scale

I Types of structure (buildings not anti-seismic)
  A: Buildings of fieldstone, rural structures, adobe houses, clay houses;
  B: Ordinary brick buildings, large block construction, half-timbered structures, structures of hewn blocks of stone;
  C: Pre-cast concrete skeleton construction, pre-cast large panel construction, well-built wooden structures.

2 Definition of quality
  Single, few: = 10%
  Many: = 20–50%
  Most: = 60%

3 Classification of damage to buildings

Grade 1: Slight damage:	Fine cracks in plaster, fall of small pieces of plaster.
Grade 2: Moderate damage:	Small cracks in walls, fall of fairly large pieces of plaster; pantiles slip off; cracks in chimney; parts of chimney fall down.
Grade 3: Heavy damage:	Large and deep cracks in walls; fall of chimneys.
Grade 4: Destruction:	Gaps in walls; parts or buildings may collapse; separate parts of the buildings lose their cohesion; inner walls and filled in walls of the frame collapse.
Grade 5: Total damage:	Total collapse of buildings.

4 Arrangement of the scale
  (a) Persons and surroundings
  (b) Structures
  (c) Nature

Intensity degrees

I   *Not noticeable*
  (a) The intensity of the vibration is below the limit of sensitivity; the tremor is detected and recorded by seismographs only
  (b) –
  (c) –

II *Scarcely noticeable (very slight)*
  (a) The vibration is felt only by individual people at rest in houses, especially on upper floors of buildings.
  (b) –
  (c) –

III *Weak*
  (a) The earthquake is felt indoors by a few people, outdoors only in favourable circumstances. The vibration is weak. Attentive observers notice a slight swinging of hanging objects, somewhat more heavily on upper floors.
  (b) –
  (c) –

IV *Largely observed*
  (a) The earthquake is felt indoors by many people, outdoors by a few. Here and there people awake, but are not frightened. The vibration is moderate. Windows, doors and dishes rattle; floors and walls creak; furniture begins to

shake; hanging objects swing slightly. Liquids in open vessels are slightly disturbed. In stationary motor cars, the shock is noticeable.

(b) –

(c) –

V   *Strong*

(a) The earthquake is felt indoors by most people, outdoors by many. Many sleeping people awake; a few run outdoors; animals become uneasy. Buildings tremble throughout; hanging objects swing considerably; pictures swing out of place; occasionally pendulum clocks stop; unstable objects may be overturned or shifted; open doors and windows are thrust open and slam back again; liquids spill in small amounts from well-filled open containers. The vibration is strong, sometimes resembling the fall of a heavy object in the building.

(b) Damage of Grade 1 in a few buildings of Type A is possible.

(c) Sometimes there is a change in the flow of springs.

VI   *Slight damage*

(a) The earthquake is felt by most indoors and outdoors; a few lose their balance; domestic animals run out of their stalls. In a few instances, dishes and glassware may break; books may fall down; heavy furniture may possibly move and small steeple bells may ring.

(b) Damage of grade 1 is sustained in single buildings of Type B and in many of type A; damage in a few buildings of Type A is of Grade 2.

(c) In a few cases, cracks up to a width of 1 cm are possible in wet ground; in mountains, occasional landslips may occur. Changes in the flow of springs and of levels of well water are observed.

VII   *Damage to buildings*

(a) Most people are frightened and run out of doors; many find it difficult to stand. The vibration is noticed by persons driving motor cars. Large bells ring.

(b) In many buildings of Type C, damage of grade 1 is caused; in many buildings of Type B, damage is of Grade 2; many buldings of Type A suffer damage of Grade 3, a few of Grade 4. In single instances, landslips of roadways occur on steep slopes; locally, cracks in roads and stone walls.

(c) Waves are formed on water, and water is made turbid by stirred-up mud. Water levels in wells change, and the flow of springs changes; in a few cases, dry springs have their flow restored, and existing springs stop flowing. In isolated instances, parts of sandy or gravelly banks slip off.

VIII   *Destruction of buildings*

(a) General fright; a few people show panic, also persons driving motor cars are disturbed. Here and there branches of trees break off; even heavy furniture moves and may overturn; hanging lamps are damaged in part.

(b) Many buildings of Type C suffer damage of Grade 2, and a few of Grade 3; many buildings of Type B suffer damage of Grade 3, and a few of Grade 4; many buldings of Type A suffer damage of Grade 4, and a few of Grade 5. Memorials and monuments move and twist; tombstones overturn; stone walls collapse.

(c) Small landslips occur in hollows and on banked roads on steep slopes; cracks form in the ground up to widths of several centimetres. New reservoirs come into existence; sometimes dry wells refill and existing wells become dry; in many cases, the flow and level of water in wells change.

IX   *General damage to buildings*

(a) General panic. Considerable damage to furniture; animals run to and fro in confusion and cry.

(b) Many buldings of Type C suffer damage of Grade 3, and a few of Grade 4; many

Table 3.2 Continued

buildings of Type B show damage of Grade 5. Monuments and columns fall. Reservoirs may show heavy damage. In individual cases, railway lines are bent, and roadways damaged.

(c) On flat land, overflow of water, sand and mud is often observed. Ground cracks of widths up to 10 cm form, in slopes and river banks more than 10 cm; furthermore, a large number of slight cracks occur in the ground. Falls of rock occur, and many landslides and earthflows. Large waves occur on water.

X   *General destruction of buildings*

(b) Many buildings of Type C suffer damage of Grade 4, and a few of Grade 5; many buildings of Type B show damage of Grade 5; Most of Type A collapse; dams, dykes and bridges may show severe to critical damage; railway lines are bent slightly; road pavement and asphalt show waves.

(c) In ground, cracks form up to widths of several decimetres, sometimes up to one metre; broad fissures occur parallel to watercourses; loose ground slides from steep slopes; considerable landslides are possible from river banks and steep coasts. In coastal areas, there is displacement of sand and mud; water from canals, lakes, rivers etc. is thrown on land; new lakes are formed.

XI *Catastrophe*

(b) Destruction of most, and collapse of many buildings of Type C; even well-built bridges and dams may be destroyed, and railway lines largely bent, thrusted or buckled; highways become unusable; underground pipes destroyed.

(c) Ground is fractured considerably by broad cracks and fissures, as well as by movement in horizontal and vertical directions; numerous landslides and falls of rock. The intensity of the earthquake requires special investigation.

XII *Landscape changes*

(b) Practically all structures above and below ground are heavily damaged or destroyed.

(c) The ground surface is radically changed. Considerable ground cracks with extensive vertical and horizontal movement are observed. Falls of rock and slumping of river banks occur over wide areas; lakes are dammed; waterfalls appear, and rivers are deflected. The intensity of the earthquake requires special investigation.

high level of damage (or intensity) can be achieved in either a small earthquake nearby or a large one further away.

The magnitude of an earthquake is an instrumentally measured quantity. In 1935, Richter devised a logarithmic scale for comparing the magnitudes of Californian earthquakes. He related the magnitude of a tectonic earthquake to the total amount of elastic energy released when overstrained rocks suddenly rebound and so generate shock waves.

The concept of magnitude has been widely extended and developed so that now a number of magnitudes are recognized, such as Richter local magnitude, M_R; local magnitude, M_L; body-wave magnitude, m_b; and surface-wave magnitude (long-period), M_s. Since each of these magnitudes relates to certain wave frequencies, the character of which changes with the energy released by an earthquake, none proves a satisfactory index of the largest seismic events. The moment magnitude, M, is used for these. This is based on the surface area of a fault displaced during an earthquake, its average length of movement and the rigidity of the rocks involved. The relationship between energy released, E, and magnitude, M, has been given by the following expression

$$\log E = a + bM \tag{3.1}$$

The values of the constants a and b have been modified several times as data have accumulated. In the case of surface and body waves the corresponding equations are respectively:

$$\log E = 12.24 + 1.44\, M_s \tag{3.2}$$

and

$$\log E = 4.78 + 2.57\, m_b \tag{3.3}$$

where E is given in ergs. The seismic moment, M_o, is related to the area of rupture, A, the rigidity modulus, μ, of the faulted rock, and the average dislocation, u, caused by an earthquake as follows:

$$M_o = Au\mu \tag{3.4}$$

In addition, seismic moment is related to the total energy of a fault rupture by the expression:

$$E = \sigma\, M_o/\mu \tag{3.5}$$

where σ is the average stress acting on a fault during an earthquake. The stress drop is the difference between the shear stress on the fault surface before and after rupture.

The largest earthquakes have had a magnitude of 8.9, and these release about 700 000 times as much energy as the earthquake at the threshold of damage. Earthquakes of magnitude 5.0 or greater generate sufficiently severe ground motions to be potentially damaging to structures. It has been estimated that in a typical year the Earth experiences two earthquakes over magnitude 7.8, seventeen with magnitudes between 7 and 7.8, and about a hundred between 6 and 7.

There is a rough relationship between the length of a fault break and the amount of displacement involved, and both are related to the magnitude of the resultant earthquake. Displacements range from a few millimetres to several metres. For instance, after a survey of fault displacement, King and Knopoff (1968) found that the maximum fault displacements were 6.55 m horizontally and 13.3 m vertically. They also showed that the lengths of ruptured faults varied from a few to about 600 km. As expected, the longer fault breaks have the greater displacements and generate the larger earthquakes. On the other hand, it has been shown by Ambraseys (1969) and Bonilla (1970) that the smaller the fault displacement, the greater the number of observed fault breaks. Bonilla *et al.* (1984) also showed that for the great majority of fault breaks the maximum displacement was less than 6 m and that the average displacement along the length of the fault was less than 50% of the maximum. The length of the fault break during a particular earthquake is generally only a fraction of the true length of the fault. Individual fault breaks during simple earthquakes have ranged in length from less than a kilometre to several hundred kilometres. What is more, fault breaks do not occur only in association with large and infrequent earthquakes but also occur in association with small shocks and continuous fault creep.

There is little information available on the frequency of breaking along active faults. All that can be said is that some master faults have suffered repeated movements – in some cases it has recurred in less than 100 years. By contrast, much longer intervals, totalling many thousands of years, have occurred between successive breaks. Therefore, because movement has not been recorded in association with a particular fault in an active area, it cannot be concluded that the fault is inactive.

Earthquakes resulting from displacement and energy release on one fault can sometimes trigger small displacements on other unrelated faults many kilometres distant. Breaks on subsidiary faults have occurred at distances as great as 25 km from the main fault, but with increasing distances from the main fault the amount of displacement decreases. For example, displacements on branch and subsidiary faults located more than 3 km from the main fault break are generally less than 20% of the main fault displacement.

Many seismologists believe that the duration of an earthquake is the most important factor as far as damage to or failure of structures, soils and slopes are concerned. Buildings may remain standing in earthquakes that last 30 s, whereas strong motions continuing for 100 s would bring about their collapse. For example, the duration of the main rupture along the fault responsible for the Kobe earthquake (January 1995; M = 7.2) was 11 s. However, the duration of strong motion was increased in some parts of the city located on soft soils to as long as 100 s, (Esper and Tachibana, 1998). What is important in hazard assessment is the prediction of the duration of seismic shaking above a critical ground acceleration threshold. The magnitude of an earthquake affects the duration much more than it affects the maximum acceleration, because the larger the magnitude, the greater the length of ruptured fault and hence the more extended the area from which the seismic waves are emitted. The idealized maximum ground accelerations in the vicinity of causative faults for earthquakes of various magnitudes are given in Table 3.3. Peak ground accelerations recorded 35 km from the causative fault responsible for the Kobe earthquake were 0.28g in the north–south direction, 2.7g in the east–west direction and 2.4g^2 in the vertical direction. In other words, the peak vertical acceleration

Table 3.3 Maximum ground accelerations and durations of strong phase shaking (after Housner, 1970).

Magnitude	Rupture length (km)	Maximum acceleration (%g)	Duration (s)
5.0		9	2
5.5	5 to 10	15	6
6.0	10 to 15	22	12
6.5	15 to 30	29	18
7.0	30 to 60	37	24
7.5	60 to 100	45	30
8.0	100 to 200	50	34
8.5	200 to 400	50	37

was approximately 0.9 times the peak horizontal acceleration, which exceeds typical values accorded in past earthquakes.

With increasing distance from a fault, the duration of shaking is longer and the intensity of shaking is less, the higher frequencies being attenuated more than the lower ones. The attenuation or reduction of acceleration amplitude that occurs as waves travel from a causative fault is the result of decreasing seismic energy attributable to dispersion that takes place as the outward-propagating waves occupy an increasing space. Seismic energy is also reduced due to absorption associated with internal damping in the rock material. Dispersion effects tend to increase the duration of strong shaking as the distance from the causative fault increases.

The physical properties of the soils and rocks through which seismic waves travel, as well as their geological structure, also influence surface ground motion. For example, if a plane wave traverses vertically through granite overlain by a thick, uniform deposit of alluvium, then theoretically the amplitude of the wave at the surface should be double that at the alluvium–granite contact. According to Ambraseys (1974), maximum acceleration within an earthquake source area may exceed $2g$ for competent bedrock. On the other hand, normally consolidated clays with low plasticity are incapable of transmitting accelerations greater than 0.1 to $0.15g$ to the surface, while clays with high plasticity allow accelerations of 0.25 to $0.35g$ to pass through. Saturated sandy clays and medium-dense sands may transmit 0.5 to $0.6g$, and in clean gravel and dry, dense sand accelerations may reach much higher values. However, Ambraseys maintained that the amplitude of the maximum ground velocities, and to some extent the duration of shaking or the rate at which energy flux is supplied to a structure, are more significant than ground accelerations as far as seismic problems are concerned. What is more, he pointed out that for all practical purposes there is no significant correlation between acceleration, magnitude and distance from an earthquake source in the epicentral area. Geological structure may enhance local shaking because of energy-focusing effects.

3.3 Effects of earthquakes

The most serious direct effect of an earthquake in terms of buildings and structures is ground shaking. As pointed out in Section 3.4, the type of ground conditions are important in this regard, buildings on firm bedrock suffering less than those on saturated alluvium. Nonetheless, buildings standing on firm rock can still be affected, so that susceptible buildings should not be located near to a fault trace. The type of construction also influences the amount of damage that occurs. Poorly constructed buildings and those that are not reinforced undergo the worst damage. Indeed, strong ground shaking can reduce cities to rubble if buildings are weakly constructed. For instance, on 29 February 1960, hundreds of old unreinforced masonry buildings and many younger but poorly constructed reinforced concrete structures were destroyed in Agadir, Morocco, by an earthquake of magnitude 5.9, intensity VII. Approximately 14 000 persons met their death out of a population of 33 000. A number of factors can influence the death toll in a major earthquake. The time of day determines whether large numbers of people will be in offices, factories or schools. Property losses can be enormous in metropolitan areas and this tends to rise dramatically with urban development. Fortunately, the ratio of loss of life to property damage tends to decline as more earthquake-resistant structures are built. In the Tokyo earthquake of 1 September 1923, 128 000 houses were destroyed by the earthquake shock. However, a further 447 000 were burnt out.

Ground displacement during an earthquake may be horizontal, vertical or oblique. Displacement is associated with rupture along faults due to the accumulation of strain. Once the elastic limit of the rocks involved is exceeded, then sudden movements occur. The displacement may be large; for example, horizontal ground displacement due to the earthquake of 18 April 1906 that affected San Francisco, amounted to 6.5 m along that particular segment of the San Andreas fault. The maximum vertical displacement was less than 1 m. However, ground rupture as a cause of damage to buildings can be avoided by not building on or near major active faults. On the other hand, it may prove impossible for roads, railways, canals or pipelines to avoid such faults. Regional crustal deformation manifested in uplift or subsidence may be associated with earthquakes. The movements that occurred at Niigata, Japan, prior to the earthquake of 16 June 1960 are referred to in Section 3.5. In the Alaskan earthquake of 27 March 1964, crustal deformation was due to faulting. Shorelines were uplifted by as much as 10 m in some places and sank by 2 m elsewhere. Some of the localities that underwent subsidence were subjected to flooding. Approximately 260 000 km^2 of coastal plain and adjoining sea floor were deformed. Furthermore, earthquakes may fissure the ground, this being most prevalent next to the fault trace. Fissures may gape up to 1 m wide at times.

Fault creep has been referred to in Section 3.1. Although it does not lead to personal injury, it can weaken and ultimately cause structures to fail. For instance, creep movement on the Hayward fault in California is responsible for the continuous fracturing of a water tunnel that supplies Berkeley. It also gives rise to the warping of railway tracks. Furthermore, it has been suggested that fault creep beneath the Baldwin Hill Reservoir, Los Angeles, was a contributory cause of its failure.

Landslides and other types of mass movement frequently accompany earthquakes, most occurring within 40 km of the earthquake concerned. Topography and preceding weather influence the likelihood of associated landslide occurrence. They can cause great loss of life and be extremely destructive. For instance, the earthquake of 31 May 1970 that affected Ancash, Peru, triggered a slide of mud, rock and ice on the north peak of Huascarán. The slide moved 3550 m downwards over a horizontal distance of 11 km at an average speed of 320 km h^{-1}. It buried the towns of Ranrahirca and Yunga, accounting for over one-third of the 67 000 deaths associated with the earthquake.

Loosely packed fine-grained sands and silt that are saturated may be subjected to liquefaction by earthquake shock. The resulting quick condition means that buildings can sink into the ground (see Section 3.4). In addition, lateral movement of such soils may be delineated by ruptured ground to form graben-like features, as happened at Turnagain Heights in Anchorage during the Alaskan earthquake of 27 March 1964 (Figure 3.2). Quick clays can also be liquefied by dynamic loading due to earthquakes.

Tsunamis are large seismic sea waves that are created by earthquakes. They

Figure 3.2 Subsidence at Government Hill School, Anchorage, Alaska, 27 March 1964. Soils failed and moved downslope, leaving part of the school unmoved, while the rest dropped into a graben.

can be extremely destructive along low-lying highly developed coasts (see Section 7.10). Seiches are also produced by earthquakes if the ground is tilted or if the period of the long vertical surface waves in the ground is in resonance with the period of free oscillation of the body of water. They are vertically oscillating standing waves.

3.4 Ground conditions and seismicity

The response of structures on different foundation materials has proved to be surprisingly varied. In general, structures not specifically designed for earthquake loadings have fared far worse on soft, saturated alluvium than on hard rock (Figure 3.3). In other words, amplitude and acceleration are much greater on deep alluvium than on rock, although, by contrast, rigid buildings may suffer less on alluvium than on rock. This is because the alluvium seems to have a cushioning effect and the motion may be changed to a gentle rocking, which is easier on such a building than the direct effect of earthquake motions experienced on harder ground. Conversely, with any kind of poor construction alluvial ground beneath the structure facilitates destruction, as happened in Kobe, where wooden houses built on alluvium sustained most damage. Intensity attenuation on rock is very rapid, whereas it is extremely slow on soft formations and speeds up only in the fringe area of the shock. Hence the character of intensity attenuation in any shock will depend largely on the surface geology of the shaken area. It is therefore important to try to relate the dynamic characteristics of a building to those of the subsoil on which it is founded. The vulnerability of a structure to damage is considerably enhanced if the natural frequency of vibration of the structure and the subsoil are the same.

Ground vibrations caused by earthquakes often lead to compaction of cohesionless soil and associated settlement of the ground surface (Ishihara and Okada, 1978). Loosely packed, saturated sands and silts tend to lose all strength and behave like fluids during strong earthquakes. When such materials are subjected to shock, densification occurs. During the relatively short time of an earthquake, drainage cannot be achieved and this densification therefore leads to the development of excessive pore water pressures, which cause the soil mass to act as a heavy fluid with practically no shear strength (Peck, 1979). In other words, a quick condition develops (Figure 3.4). Water moves upwards from the voids to the ground surface, where it emerges to form sand boils (see Figure 5.1). An approximately linear relationship exists between the relative density of sands and the stress required to cause initial liquefaction in a given number of stress cycles. Ground liquefaction has been noted in a number of earthquakes. For example, large-scale liquefaction of fills used in reclaimed areas of Osaka Bay occurred during the recent Kobe earthquake. Settling of up to 0.75 m was recorded and much damage was suffered by the port facilities and the adjoining industrial area.

If liquefaction occurs in a sloping soil mass, the entire mass will begin to

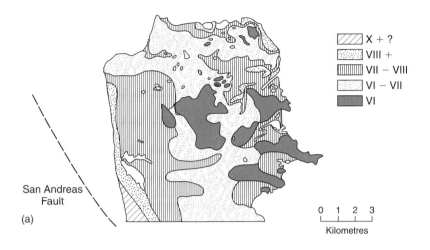

X + ?
VIII +
VII − VIII
VI − VII
VI

San Andreas
Fault

(a)

0 1 2 3
Kilometres

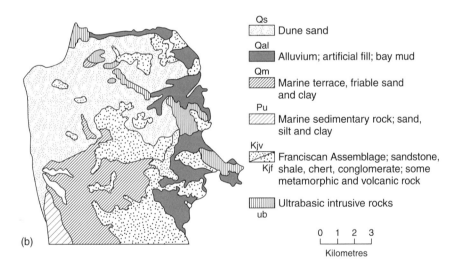

Qs
Dune sand

Qal
Alluvium; artificial fill; bay mud

Qm
Marine terrace, friable sand
and clay

Pu
Marine sedimentary rock; sand,
silt and clay

Kjv
Franciscan Assemblage; sandstone,
Kjf   shale, chert, conglomerate; some
metamorphic and volcanic rock

Ultrabasic intrusive rocks
ub

0 1 2 3
Kilometres

(b)

Figure 3.3 (a) Isoseismal lines on the San Francisco peninsula (based on the modified
Mercalli scale) drawn by H. O. Wood after the 1906 earthquake. (b) A generalized geological map of the San Francisco peninsula. Note the correlation
between geology and intensity (from Bolt, 1978).

move as a flow slide (Seed, 1970). Such slides develop in loose, saturated, cohesionless materials during earthquakes. In addition, loose saturated silts and
sands often occur as thin layers underlying firmer materials. In such instances,
liquefaction of the silt or sand during an earthquake may cause the overlying
material to slide over the liquefied layer. Structures on the main slide are frequently moved without suffering significant damage. However, a graben-like

Figure 3.4 Collapsed building in the Marina District of San Francisco as a result of the Loma Prieta earthquake of October 1989. The first storey failed and the second collapsed when the ground was liquefied, leaving only the third storey.

feature (see Figure 3.2) often forms at the head of the slide, and buildings located in this area are subjected to large differential settling and are often destroyed (Sokolenko, 1977). Buildings near the toe of the slide are commonly heaved upwards or even pushed over by the lateral thrust.

Clay soils, with the exception of quick clays, do not undergo liquefaction when subjected to earthquake activity, but under repeated cycles of loading large deformations can develop, although the peak strength remains about the same. Nonetheless, these deformations can reach the point where, for all practical purposes, the soil has failed. The damage caused by the 1985 earthquake that affected Mexico City was restricted almost exclusively to saturated deposits of clay (the Tacubaya Clay) that are part of an old lake bed. These deposits are characterized by low natural vibration frequencies and were affected by the low-frequency ground motions experienced at Mexico City (the epicentre was 370 km away in the Central American Trench, and high-frequency shocks tend to be attenuated rapidly with increasing distance from the epicentre). The clays amplified the shock waves between 8 and 50 times compared with the motions on solid rock in adjacent areas (Singh *et al.*, 1988). Those buildings that sustained the most severe damage were located mainly on that part of the old lake bed where the clay was more than 37 m thick. In fact, the severity of ground motion increased in relation to the thickness of the clay.

Fills may fail completely or slump severely, with associated longitudinal cracking, when subjected to vibration due to earthquakes. Canal banks have a long history of slope failure during earthquakes. The tendency for earthfills to slide downhill during earthquakes results in increased pressure on retaining walls, which therefore may be displaced.

It can be concluded from the foregoing paragraphs that microseismic observations can be used to establish seismic intensity increments for the basic categories of ground. Such an idea was first advanced by Reid (1908), who, after the San Francisco earthquake of 1906, introduced the concept of foundation coefficients for several major soil and rock types. The coefficient for the type of foundation that produced the least vibrational force as revealed by observed earthquake damage was designated unity. Estimates of the probable accelerations associated with these coefficients were also noted. Although it is now recognized, from the complex nature of strong-motion accelerograph records, that acceleration itself has little meaning unless the frequency is also given, there is reason to believe from later investigations that these coefficients provide a true picture of the relative earthquake intensities experienced on the types of foundation cited.

Such seismic intensity increments can be be obtained for different ground conditions by comparing the specific intensity changes for these different types of ground. The individual intensity increments for each type of ground have been related to a single standard (i.e. granite) (Table 3.4). A more recent series of intensity increments (Table 3.5), in which the standard ground conditions

Table 3.4 Seismic intensity increments.

Ground conditions	Reid (1908)	Medvedev (1965)
Granite	0	0
Limestone, sandstone	0–1.2	1–1.5
Gravel	1.2–2.1	1–1.6
Sand	1.2–2.1	1.2–1.8
Clay	1.5–2.0	1.2–2.1
Fill	2.1–3.4	2.3–3.0
Wet gravel sand	2.3	1.7–2.8
Clay		
Wet fill	3.5	3.3–3.9

Table 3.5 Average changes in intensity associated with different types of surface geology (from Degg, 1992).

Subsoil	Average change in intensity
Rock (e.g. granite, gneiss, basalt)	−1
Firm sediments	0
Loose sediments (e.g. sand, alluvial deposits)	+1
Wet sediments, artificially filled ground	+1.5

are firm unconsolidated sediments, has been proposed by the Munich Reinsurance Company and were quoted by Degg (1992).

3.5 Methods of seismic investigation

Unfortunately, there is no method at present of forecasting the exact location, size or time of an earthquake. However, it can be assumed that a probable prediction is reasonable and that past patterns of seismic activity will continue. Hence earthquake risk reports should take into account an appraisal of known faults, the distance from major faults, the number of recorded earthquakes, the history of damage, and an estimate of the magnitude and intensity of the strongest shock expected. The latter must take into consideration the ground conditions at the site.

3.5.1 Earthquake prediction

The purpose of earthquake forecasting is reduction of loss of life and damage to property (Mogi, 1985). Prediction should attempt to locate the earthquake event and area affected; the time of the event; and its magnitude and the probable distribution of damage. Earthquake prediction involves using studies of historical seismicity and the results of intensive monitoring of seismic and geological phenomena to establish the probability that a given magnitude of earthquake or intensity of damage will recur during a given period of time. Such predictions will help the development of safety measures to be taken in relation to the degree of risk (Table 3.6). In this way, earthquake alerts can be formed in terms of a staged gradation of prediction, from the long term to the immediate (Table 3.7).

Earthquake precursors are phenomena that precede major ground tremors; more specifically, they are anomalous values of such phenomena that may indicate the onset of an earthquake (Rikitake, 1979). Most important is the variation in seismic wave velocity. The P waves slow down (e.g. by up to 10% in the focal region) due to the minute fracturing in rocks during dilation but then accelerate as water occupies micro-fractures prior to an earthquake. As water is incompressible it is an effective transmitter of compressional waves, the elasticity of the rock mass being regained. Hence, a continuous record of such changes in shock wave velocity may be of value in earthquake prediction, as a quake may occur a day or so after the wave velocities return to normal. The period of time covered by these changes indicates the size of the earthquake that can be expected. For example, an earthquake of magnitude 5.4 may be preceded by a period of 4 months of lowered velocities, of magnitude 4 by only 2 months. Thus if the period of changes lasts for 14 years, it suggests a potentially violent earthquake of magnitude 7. Furthermore, the ratio of the P and S wave velocities decreases by up to 20% a few hours or days before a major quake (Scholz et al., 1973). This is primarily due to the reduction in the

Table 3.6 Safety measures based on earthquake predictions (after Savarenskij and Neresov, 1978).

Period of prediction	Buildings	Material assets	Safeguards for human life	Special measures
Operative (a few hours to one or two days)	Evacuate dangerous buildings; cease activities in places of public assembly	Evacuate the most important material assets	Allocate emergency equipment in the danger area; prepare medical establishments	Cut off electricity and gas mains; shut down nuclear reactors and dangerous chemical plants
In the short term (from 2 to 4 months)	Estimate probable damage; prepare public evacuation plans	Preserve major assets	Prepare emergency measures and medical establishments	Remove or safeguard hazardous substances; lower reservoir levels, etc.
In the long term (12 months)	Strengthen buildings of particular vulnerability to earthquakes		Plan emergency food stores; plan the use to be made of medical establishments	Transfer of hazardous substances to other places of storage

Table 3.7 Gradation of earthquake predictions (after Wyss, 1981).

Earthquake alert	Conditions and observations necessary
Stage 1	An approximately defined area is estimated to be more likely than surrounding seismic areas to experience a future earthquake (e.g. seismic gap or occurrence of at least one geophysical, geological or geodetic anomalous observation).
Stage 2	One of several crustal parameters show the beginnings of a long- to medium-term pattern of change known to have occurred before some other earthquakes. At least on of the prediction elements (location, size or time) is still poorly defined (e.g. occurrence time uncertainty is approximately equal to 50 percent of precursor time).
Stage 3	Changes in crustal parameters are observed which can be interpreted as indicating that the end of the long-term preparatory process is near (e.g. the anomalies return to normal). The three prediction elements are fairly well defined (e.g. occurrence time uncertainty is less than about 20 percent of precursor time).
Stage 4	In addition to the conditions of stage 3, an anomaly is measured that can be interpreted as a short-term precursor. Occurrence time uncertainty may range from hours to weeks.

compressional wave velocity. The duration, but not the amount of the decrease in velocity ratio, appears to be proportional to the earthquake magnitude. The velocity ratio returns to normal at the beginning of the 'critical' period immediately before an earthquake.

The dilatancy of rocks due to crustal deformation prior to an earthquake leads to an increase in their volume and results in ground shortening or lengthening, minor tilting, uplift or subsidence of the ground surface near an active fault. This ground movement can be measured by very accurate surveying or by high-precision tiltmeters and extensometers. Unfortunately, however, the point at which movements become critical is not always easy to detect. For example, Bolt (1993) referred to the uplift and subsidence that, for about 60 years, occurred along the coast of Japan, near Niigata, prior to the earthquake of 16 June 1964 (Figure 3.5). The rate of movement slowed in the late 1950s, and at the time of the earthquake a sudden subsidence was detected to the north alongside the epicentre. The amount of precursory uplift decreased gradually away from the epicentre and was not observed at distances exceeding 100 km. However, Bolt also mentioned the uplift at Palmdale in southern California, which since 1960 has amounted to over 350 mm; but no significant earthquakes have occurred in this period, although earthquakes have occurred prior to this period. Hence, he concluded that the best response was to intensify the various types of measurement in the area.

Other features brought about by rock dilation include a decrease in electrical resistivity and a change in magnetic susceptibility of the rock concerned,

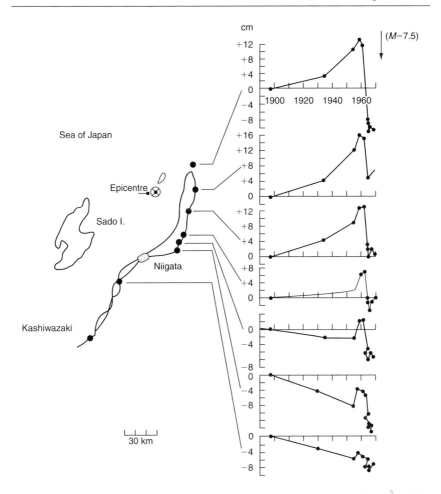

Figure 3.5 Surface deformations before the Niigata earthquake, Japan, determined by levelling measurements. Black circles show observation stations.

that is, the local magnetic field becomes stronger shortly before an earthquake. The decrease in resistivity is assumed to be due to the influx of water into cracks in the dilatant zone, flowing from its surroundings (Figure 3.6). Subsequently, the stress reduction attributable to the earthquake will allow cracks to close, thereby forcing water from the source region so that resistivity will rise. Changes in density are recorded by sensitive gravity meters, which are able to detect small changes. Changes in both magnetic and gravity measurements may help to locate hidden faults by indicating differences in the properties across sharp contacts.

The release of small quantities of the inert gases, notably radon, but also argon, helium, neon and xenon, takes place prior to an earthquake. The increase

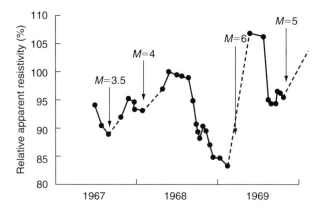

Figure 3.6 Electrical resistivity anomalies observed before earthquakes at Garm, USSR. Note the marked drop in resistivity in the half-year period before the earthquake of $M = 6$ and the subsequent increase during and after the earthquake.

in radon has been noted in well waters and has been attributed to increasing water flow, which carries radon with it into wells after dilation and cracking in neighbouring rocks (Figure 3.7). Hence, the movement and chemistry of groundwater may change before earthquakes. Stresses in saturated strata may cause spring discharge to increase, or the water in wells to alter its level or become turbid. During micro-fracturing due to an earthquake, migrating groundwater comes into contact with an increased surface area of rocks, dissolving radon in trace amounts. Stress in the rocks may speed the movement

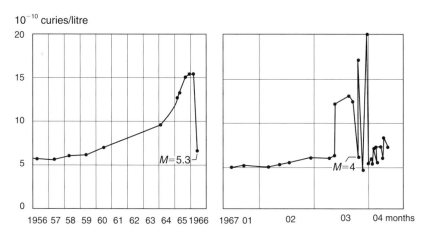

Figure 3.7 Radon changes in a mineral spring from a deep borehole in Tashkent before the Tashkent earthquake (left-hand graph) and before one of the aftershocks (right-hand graph).

of water in the phreatic zone and hence the migration of radon and halogens to the surface. Moreover, many active faults are zones of geothermal energy release, where it may be possible to relate the temperature of hot springs to incipient seismic activity. Obviously, it is important to implement a regular monitoring programme of stream and spring discharge, water quality and water temperature.

According to Raleigh *et al*. (1982), severe plate boundary earthquakes recur at intervals that may vary by as much as 50% from an average value. However, it would appear that there is some evidence that this interval may be proportional to the displacement caused by faulting when the previous earthquake occurred at the same location. If this is correct, then this will allow the recurrence interval to the next earthquake to be estimated more precisely, provided the long-term displacement is constant and well known.

Because damaging earthquakes in particular seismic areas cannot be predicted exactly in time, Bolt (1993) pointed out that the best strategy is to attempt to determine the probability that such an event will occur. He went on to note that a method of determining probability can be based on the elastic rebound theory of the cause of earthquakes. In other words, as strain increases the more likely it is that sudden slip will take place along faults and an earthquake will be generated. Recording geological and geodetic measurements should facilitate the determination of which segments of a fault are likely to slip in future. Hence, fault segments should be mapped accurately. It can be assumed that the earthquake that generates the largest magnitude would be that where movement occurs over the complete segment. Accordingly, it is necessary to determine which fault segments along an active fault have slipped in the past and to calculate the rate at which strain is accumulating in the area. Each time slippage has taken place can be related to the magnitude of the resultant earthquake, so the recurrence intervals between earthquakes of a given magnitude can be determined. From this, the range of probable occurrence of an earthquake of a particular size in a specific time period can be made.

3.5.2 Assessment of movement along faults

Most major fault zones do not consist of a single break but are made up of a great number of parallel and interfingering breaks, which may range over a kilometre apart. This indicates that displacements associated with intermittent great earthquakes have tended to migrate back and forth throughout the fault zone concerned. Likewise, fault displacements in future will not all take place along a single fracture. For example, in many areas of southern California geologically very recent displacements have been concentrated in narrow bands within much wider fault zones, and it seems likely that in such areas the next displacements will follow approximately the same traces as their immediate predecessors. Furthermore, in many areas the more abundant recurrent displacements have apparently taken place along the principal fault breaks, which suggests that these breaks extend to master faults located at depth and thus would have a higher probability of future movements.

Rock bending at faults means that strain is accumulating in rocks, possibly foreshadowing an earthquake and sudden fault displacement. Such tectonic movement can be measured with EDM equipment, by triangulation, by fault-movement quadrilaterals and by tiltmeters. It would appear that sudden strain release is more likely along a segment of a fault where the movement rate differs from that of adjacent areas or from its past rate, so this may aid earthquake prediction. Triangulation in critical fault zone areas provides precise survey control and a basis for determination of ground movements when observed in the future. Geodimeter measurement does not distinguish between movement due to fault slippage and that due to strain, primarily because the lines measured are long, from 13 to 32 km. Measurements over much shorter distances, 250 to 500 m, provide indication of slippage. These lines are arranged in quadrilaterals across a fault. Gradual vertical movements result in some tilting at the surface. Tiltmeters can detect tilting of 0.0001 mm along a 30.5 m horizontal line. Arrays of sesimographs, spread over large areas, are buried at various depths and arranged in two lines at right angles to each other. The arrays also can include electrical resistivity meters, magnetometers, gravity meters and strain gauges.

3.5.3 Aseismic investigation

Aseismic investigation prior to design of a structure provides the designer with some idea of the frequency and size of expected earthquakes, as well as the spectrum of anticipated earthquake motion. This spectrum is an envelope of maximum motions at each specific period of frequency at a particular location.

Aseismic investigations may vary according to the site and proposed structure. A geological map of the site should be prepared and the geological structure should be illustrated by sections. Note must be taken of the regional tectonics and particular attention must be paid to the investigation of recent faulting. The smallest earthquakes cause no visible damage, but the record of their epicentres may indicate the existence of an active fault. Precautions are obviously necessary in any region where conspicuously active faults have been mapped. If a region is a very seismic one, most places that are going to have earthquakes will already have had one.

It may be possible to assess the relative activity of a fault using geological, seismological and historical data and so classify it as active, potentially active, of uncertain activity, or inactive. Seismological data used to recognize active faults include historical or recent surface faulting with associated strong earthquakes; and tectonic fault creep or geodetic indications of fault movement. Ambraseys (1992) showed that in regions such as the Middle East, where low to moderate rates of seismic activity are experienced, historical (i.e. pre-twen-tieth century) data need to be included in hazard evaluations in order to enhance their meaning. A lack of known earthquakes, however, does not necessarily mean that a fault is not active. An active fault is indicated by geological features such as young deposits displaced or cut by faulting, fault scarps, fault rifts, pressure ridges, offset streams, enclosed depressions, fault valleys, and ground features such as open fissures, rejuvenated streams, folding or warping of young deposits, groundwater barriers in recent alluvium, *en echelon* faults, and fault paths on recent surfaces. Erosion features are not necessarily indicative of active faults but may be associated with some active zones. Historical sources of displacement of man-made structures such as roads, railways, etc. may provide further evidence of active faults.

Faults can be regarded as potentially active faults if there is no reliable report of historic surface faulting; faults are not known to cut or displace the most recent alluvial deposits but may be found in older alluvial deposits; geological features characteristic of active fault zones are subdued, eroded and discontinuous; water barriers may be present in older materials; and the geological setting in which the geometric relationship to active or potentially active faults suggests similar levels of activity. As far as seismic evidence is concerned, there may be alignment of some earthquake foci along the fault trace, but locations are assigned with a low degree of confidence.

Faults are of uncertain activity if data are insufficient to provide criteria definitive enough to establish fault activity. Additional studies are necessary if the fault is considered critical to a project.

In the case of inactive faults, there is no historical activity based on a thorough study of local sources. Geologically, features characteristic of active fault zones are not present in areas where faults are inactive, and geological evidence is available to indicate that the fault has not moved in the recent past. Although not a sufficient condition for inactivity, seismologically the fault concerned should not have been recognized as a source of earthquakes.

Areas that might be exposed to the hazards of surface faulting, fissuring, liquefaction, landslides, or other mass movement should be recognized. Drilling and sampling will aid the interpretation of the geological setting, and the samples will allow determination of the physical properties of the materials concerned. The values of the static and dynamic properties of the soils may be obtained by *in situ* testing. Geophysical surveys and aerial photographs may prove useful.

The ground can be force-vibrated by using a shaking machine and the response determined. By comparing this with the response of other known sites, a measure of the comparative sensitivity can be estimated. A refraction seismograph can be used to determine interval velocities, as well as cumulative velocities from the surface to any point in a drillhole. These uphole seismic surveys can be made with one explosion, using pick-ups distributed along a cable placed in the drillhole.

3.5.4 Accelerographs

Data relating to ground motion is essential for an understanding of the behaviour of structures during earthquakes. From the engineering point of view, the strong-motion earthquakes are the most important, since they damage or even destroy man-made structures. Records of shocks produced by such earthquakes are obtained by using ruggedly constructed seismometers called accelerometers. These are designed to operate only where earthquake vibrations are strong enough to actuate them. The accelerograph simultaneously records the two orthogonal horizontal and vertical components of ground acceleration as a function of time. The period and damping of the pick-ups are selected so that the recorded motions are proportional to the ground acceleration over the frequency range of about 0.06 to 25 cps, which encompasses the range of periods exhibited by typical engineering structures. The instrument has a resolution of the order of 0.001 g and is operative to about 1.0 g.

Figure 3.8 illustrates accelerograms obtained from an area subjected to strong shaking due to a destructive earthquake. There is a rapid initial rise to a strong, relatively uniform central phase of shaking for a certain duration, followed by a gradually decaying tail during which some very strong pulses of acceleration may still occur (Bender, 1984). The records exhibit about equal intensities of motion in the two horizontal directions, which is to be expected because of the almost random nature of the ground movement. The vertical component is normally somewhat less intense than the horizontal and is characterized by an

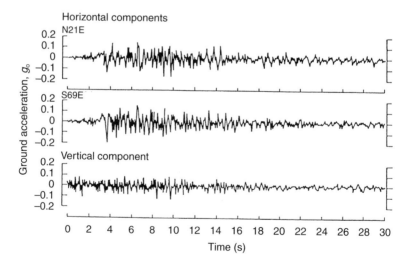

Figure 3.8 Ground acceleration recorded at Taft, California, approximately 40 km from the causative fault of the Arvin–Tehachapi earthquake (magnitude 7.7) of 21 July 1952. Recorded on approximately 6 m of alluvium overlying sedimentary rock (from Cherry, 1974, by permission of Kluwer Academic Publishers).

accentuation of the higher-frequency components when compared with the horizontal motions. The accelerometer was approximately 40 km from the causative fault. Accelerometers are placed on the surface and below ground level. They have also been installed in dams and at different levels in multi-storey buildings to compare structural motion at different heights with ground motion. Strong-motion records have provided information concerning the acceleration, displacement periods and duration of earthquakes.

3.6 Seismic hazard and risk

In earthquake-prone areas, any decision making relating to urban and regional planning, or for earthquake-resistant design, must be based on information concerning the characteristics of probable future earthquakes. However, the lack of reliable data to be used for design purposes has been a cause for concern in engineering seismology. Hence the engineer may on occasions accept an element of risk above what would otherwise be considered normal. This risk may, under certain conditions, prove economically acceptable for structures of relatively short life. However, for certain structures whose failure or damage during an earthquake may lead to disaster, this is not acceptable (Lomnitz and Rosenblueth, 1976). A considerable amount of informed judgement and technical evaluation is needed to determine what is an acceptable risk for a given project, and caution is required in the selection of earthquake design parameters. There are a number of methods for selecting appropriate earthquake design parameters, which may involve estimates of strong ground motions, but all of them depend on the quality of the input data. The estimation of the maximum probable earthquake ground motion, and when and where it will occur, is extremely difficult to assess.

Earthquake hazard is the probability of occurrence in a given area, during a given period of time, of ground motion capable of causing significant loss of property or life (Campbell, 1984). It may be expressed by indicating the probability of occurrence of certain ground accelerations, velocities and displacements, of ground movements of various durations, or any other physical parameter that adversely affects a structure. Earthquake risk may be defined in terms of the probability of the loss of life or property, or loss of function of structures or utilities. The assessment of that risk has to take into consideration the probability of occurrence of ground motions due to an earthquake, the value of the property and lives exposed to the hazard, and their vulnerability to damage or destruction, injury or death by ground motions associated with the hazard (Fournier d'Albe, 1982). In other words, the assessment of risk involves the assessment of value, vulnerability and hazard.

A hazard level specific to a site, using data acquired on the occurrence of earthquakes, is justified for projects in areas of significant seismicity. The preliminary source of information is the national earthquake code, if there is one. Most of these codes contain a map of the seismicity of different regions of the country. There is a diversity of earthquake maps of varied status and type (Table 3.8). If the preliminary survey shows that a significant seismic hazard exists, then the hazard evaluation procedures should be commensurate with both the type of project and the quantity and quality of information available.

National agencies such as the International Seismological Centre in Britain and the National Earthquake Information Service in the United States retain catalogues of earthquake events. Critical examination of the earthquake information is necessary. Well-reported events with instrumental data may be well located. However, prior to the inception of the World-Wide Seismograph Network in the mid-1960s, the location, magnitude and especially focal depth of earthquakes should be viewed with caution.

Ideally, complete data sets should be used to characterize the distribution of seismicity, and only well-located epicentres should be used to make even broad association with fault zones. Rarely is an epicentre located by historical studies to within a 5 km-radius circle, and unless an earthquake has its epicentre within a well-conditioned array of seismographs with intervening distances of less than around 50 km, instrumental location need be no better.

Once a catalogue of earthquakes that may have affected the area in question has been drawn up, and threshold levels and uncertainties of location have been established, it is possible to apply statistical techniques. These techniques can determine whether or not spatial and temporal distributions are random or whether there have been significant periods of quiescence and activity. They

Table 3.8 Classification of earthquake-related maps (after Skipp and Ambraseys, 1987).

Tectonic	Show tectonic units with indication of orogenic association, main structural units, major faulting and folding.
Seismotectonic	Show tectonic features relevant to earthquake generation: faults with indication of style and mobility, focal mechanisms, current crustal stress, neotectonic (Quaternary) features, vulcanicity.
Hazard expectancy	
Magnitude-based	Probabilistically based; show values of largest earthquake expected over given period within unit area.
Intensity-based	Probabilistically based (with or without tectonic control); show values of largest intensity expected over given period, usually expressed as contoured zoning.
Effective acceleration-based	Probabilistically based (with or without tectonic control); show effective peak acceleration within specified period at given confidence level.
Intensity zoning	Show maximum historical intensity, usually contoured.
Intensity potential	Show maximum potential intensity, contoured.
Seismicity index	Show zones from which coefficients can be selected in formulation of design earthquake input.

can also establish whether or not there is a valid zonation within a given distance from the site where departures from a background seismicity are significant.

Seismo-tectonic maps may be used to identify potentially active faults. However, such maps are rare, and even where they exist they may be so interpretative as to be highly contentious. Nonetheless, some tectonic appraisal of a site, making use of existing maps and published work, is necessary. Remote-sensing surveys should be examined. Geophysical maps (e.g. seismic, gravity and magnetic) may help.

3.7 Seismic zoning

Seismic zoning and micro-zoning provide a means of regional and local planning in relation to the reduction of seismic hazard, as well as being used in terms of earthquake-resistant design. While seismic zoning takes into account the distribution of earthquake hazard within a region, seismic micro-zoning defines the distribution of earthquake risk in each seismic zone. A seismic zoning map shows the zones of different seismic hazard in a particular area. Some seismic zoning maps summarize observations of past earthquake effects, the assumption being that the same pattern of seismic activity will be valid in the future, whereas others extrapolate from areas of past earthquakes to potential earthquake source areas (Karnik and Algermissen, 1978). Unfortunately, the first type of map cannot take account of any earthquake sources in areas that have been quiet during the period of observation, and the second category is more difficult to compile.

Seismic evidence obtained instrumentally and from the historical record can be used to produce maps of seismic zoning (see Table 3.8). Maximum hazard levels can be based on the assumption that future earthquakes will occur with the same maximum magnitudes and intensities recorded at a given location as in the past (Sokolenko, 1979). Hence, seismic zoning provides a broad picture of the earthquake hazard that can be involved in seismic regions and so has led to a reduction of earthquake risk. Detailed seismic zoning maps should take account of local engineering and geological characteristics, as well as the differences in the spectrum of seismic vibrations and, most important of all, the probability of the occurrence of earthquakes of various intensities. Expected intensity maps define the source areas of earthquakes and either the maximum intensity that can be expected at each location or the maximum intensity to be expected in a given time interval (Donovan *et al.*, 1978).

One of the problems involved in producing seismic zoning maps is that of choosing which parameters should be mapped. Usually, seismic zoning maps are linked to building codes and are commonly zoned in terms of macro-seismic intensity increments or related to seismic coefficients incorporated in a particular code. The building code may specify the variation of the coefficient in terms of ground conditions and type of structure. Other maps may distinguish zones of destruction (Figure 3.9). However, Karnik and Algermissen (1978) argued that there is a growing need for quantities related to earthquake-resistant design to be taken into account, such as maximum acceleration or peak particle velocity, predominant period of shaking or probability of occurrence. Algermissen *et al.* (1975) developed a technique for the probabilistic estimation of ground motion, that is, for deriving the maximum ground shaking at a particular point in a given number of years at a given level of probability. In this way, they were able to produce maps of horizontal acceleration for a particular time period.

There is no standard way of using the information for zoning in all seismic

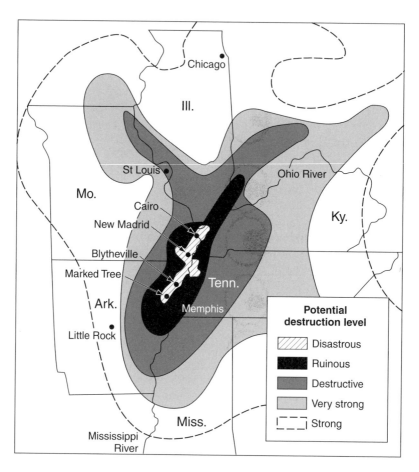

Figure 3.9 Zones forecasting likely destruction levels (to non-resistant structures) in the New Madrid earthquake region (courtesy of Federal Emergency Management Agency).

regions. What is more, in the compilation of these maps all engineering data concerning the surface manifestations of earthquakes must refer to identical ground conditions as they can have a strong influence on intensity. Geological investigations may be able to throw light upon the rate of strain within the crust and can delimit areas where faulting is widespread. However, geological data can only be used to give a qualitative assessment of seismic risk within an area.

A seismic zoning map showing relative risk, which distinguished four regions at different levels of hazard within the United States, was prepared from historical and geological evidence (Figure 3.10). This map was incorporated into the Uniform Building Code, of which it became legally a part, but it has now been superseded. In 1959, Richter published a zoning map of the United States based on seismo-statistical and geological data. Zoning was according to the modified Mercalli scale. The map distinguished regions of intensities 6, 7, 8 and 9, and it is assumed that earthquakes with intensities exceeding 9 are possible in the western zone (Figure 3.11). Much of the United States was considered highly seismic, about half occupying zones of intensity 8, and one-quarter zones of intensities 9 and 10. Consequently, Richter's map may be regarded as an overestimation. This is especially true of the eastern and central parts of the country. More recently, some notable changes have been made in the seismic zonation of the United States. Now probabilistic maps have been developed as the basis of seismic design provisions for building practice. As acceleration is a cause of damage to buildings and structures, the zonation is based on the spatial distribution of maximum seismic acceleration generated by past earthquakes (Figure 3.12). Such a map can be used for general site evaluation and initial design purposes.

Most studies of the distribution of damage attributable to earthquakes indicate that areas of severe damage are highly localized and that the degree of damage can change abruptly over short distances. These differences are frequently due to changes in soil conditions or local geology. Such behaviour has an important bearing on seismic micro-zoning, hence seismic micro-zoning maps can be used for detailed land-use planning and for insurance risk evaluation. These maps are obviously most detailed and more accurate where earthquakes have occurred quite frequently and the local variations in intensity have been recorded. A map of micro-seismic zoning for Los Angeles County is shown in Figure 3.13. It distinguishes three zones of ground response, active faults likely to give rise to ground rupture, and areas of potential landslide and liquefaction risk.

The spatial probability of a given magnitude or intensity at a certain recurrence interval can be portrayed by a seismic risk map. The risk can be expressed as the return period of a damaging earthquake or as the probability of widespread structural failure. An example of such a map for Israel is given in Figure 3.14. The hazard zonation uses a 50-year return period as this relates to the average life expectancy of a modern building, thereby ensuring that the

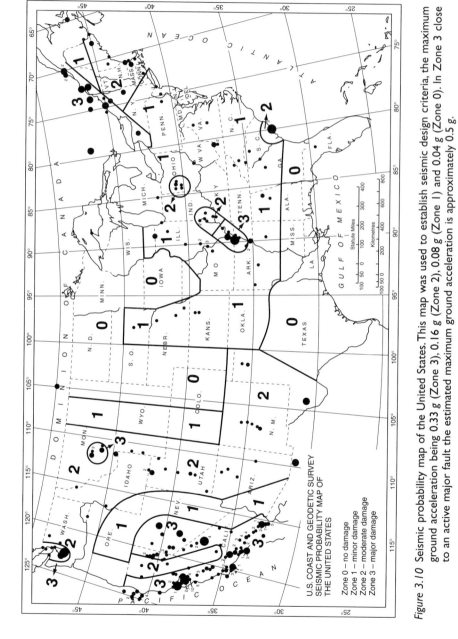

Figure 3.10 Seismic probability map of the United States. This map was used to establish seismic design criteria, the maximum ground acceleration being 0.33 g (Zone 3), 0.16 g (Zone 2), 0.08 g (Zone 1) and 0.04 g (Zone 0). In Zone 3 close to an active major fault the estimated maximum ground acceleration is approximately 0.5 g.

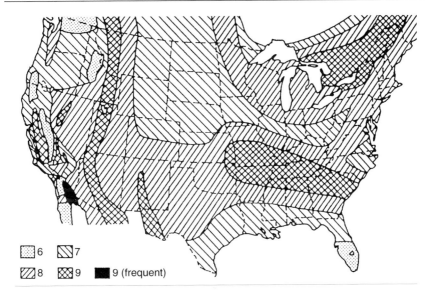

Figure 3.11 Seismic zoning of the United States (intensity ratings 6–9).

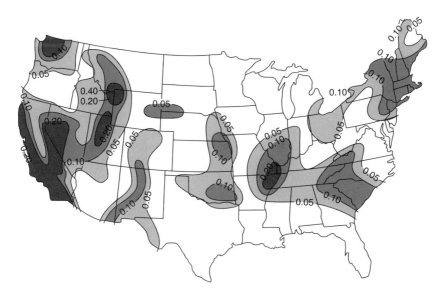

Figure 3.12 Seismic risk map for the United States. The contours indicate effective maximum acceleration levels (values are in decimal fractions of gravity) that might be expected (at a probability of 0.1) to be exceeded during a 50-year period.

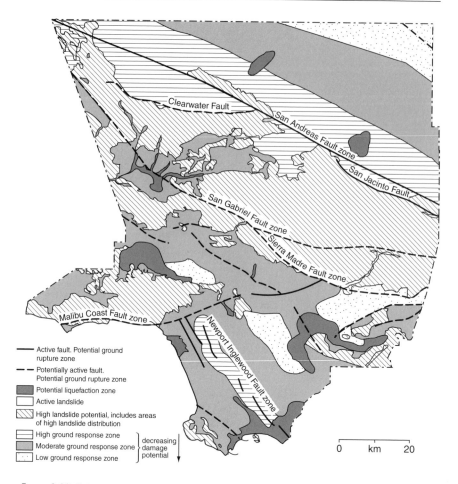

The map contains the following labels:

Clearwater Fault
San Andreas Fault zone
San Jacinto Fault
San Gabriel Fault zone
Sierra Madre Fault zone
Malibu Coast Fault zone
Newport Inglewood Fault zone

Legend:

— Active fault. Potential ground rupture zone
- - Potentially active fault. Potential ground rupture zone
Potential liquefaction zone
Active landslide
High landslide potential, includes areas of high landslide distribution
High ground response zone
Moderate ground response zone } decreasing damage potential
Low ground response zone

0   km   20

Figure 3.13 Seismic zoning map of Los Angeles County (Los Angeles County Department of Regional Planning).

hazard is assessed on a time scale that is meaningful (Degg, 1992). However, the zones show expected intensities in terms of firm consolidated sediments. In view of the influence of ground conditions on earthquake shocks, this represents a limitation. However, seismic intensity increments can be used to make adjustments (see Table 3.5). A qualitative assessment of the distribution of earthquake risk can be obtained by comparing the hazard risk map with one of population density. This indicates that earthquake risk in Israel is greatest along the coastal region, where dense concentrations of population combine with relatively high hazard. By contrast, although the Jordan rift and Jezreel valleys are areas of as high or higher hazard, the risk potential in these areas is lower because they have much lower population density.

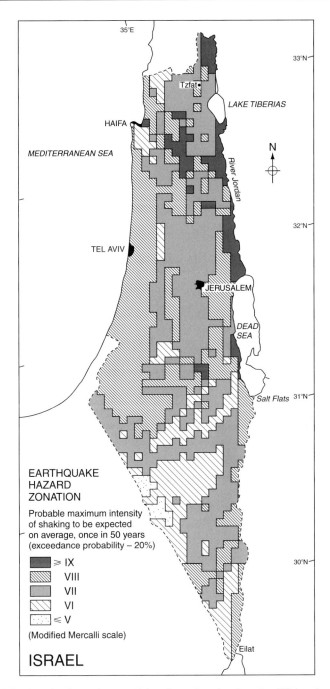

Figure 3.14 Earthquake hazard map of Israel produced using the ROA scheme (from Degg, 1992).

3.8 Earthquake analysis and design

3.8.1 Pseudostatic method

The pseudostatic method of earthquake analysis involves the seismic factor concept in building design. In other words, the stability of a structure is determined as for static loading conditions and the effects of an earthquake are taken into account by including in the computation an equivalent horizontal force acting on the structure. This means that the building is made strong enough to resist a horizontal force equal to a certain proportion of the weight of the building. Since weight is expressed in terms of the acceleration due to gravity, the horizontal force is expressed in the same way, and it is possible to compare this force with measured or computed accelerations of earthquakes. The horizontal force representing earthquake effects is expressed as the product of the weight of the structure and a seismic coefficient, k. In the United States, the value of the seismic coefficient is normally selected on the basis of the seismicity of the region; for example, in California the values range from 0.05 to 0.15. However, such values do not necessarily correspond with accelerations measured by strong-motion accelerometers.

One of the advantages of using a seismic factor is that it reduces a very complex problem in dynamics to one of simple statics that engineers can handle without difficulty. Although experience has shown this to be a practical approach it can underestimate seismic loading. Hence for certain types of structure it does not provide a satisfactory solution, and for many structures, especially high ones, the dynamic approach results in greater economy in construction costs without loss of building strength.

3.8.2 Dynamic behaviour of structures

The dynamic behaviour of a structure depends on the intensity of the loads acting upon it. For small-amplitude vibrations, structures behave linearly, but as the amplitude increases non-linear behaviour becomes increasingly important. In some types of structure, under increasing dynamic loads, cracks or local ruptures can occur that completely change the behaviour of the structure. Accordingly, knowledge of the true dynamic properties of actual structures is essential for earthquake design, so measurements should be taken when structures undergo strong vibration due to earthquakes. As a consequence, the Los Angeles Building Code, for example, requires that each new large building be instrumented with three strong-motion accelerometers, one in the basement, one at mid-height and one on the top floor. Figure 3.15 shows the motion recorded in a ten-storey building during an earthquake of magnitude 7.7. The building was constructed of reinforced concrete and founded on moderately firm alluvium. The motion recorded on the roof of the building indicates the natural period of vibration of the fundamental mode of the structure, whereas

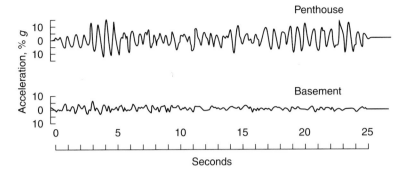

Figure 3.15 Earthquake-indicated vibrations of the ten-storey Hollywood Storage Building during the 21 July 1952 Tehachapi earthquake (epicentral distance 105 km). The exterior walls of the building had relatively few window openings and hence were relatively stiff (from Housner, 1970b).

the motion recorded in the basement shows no influence of building vibration. The resonance developed in multi-storey buildings shows that they are more susceptible to earthquake damage than single- or two-storey buildings. Buildings, towers and bridges must be designed so that their vibration frequencies are different from those of earthquakes.

3.8.3 Earthquake spectra and the design earthquake

A spectrum of a given earthquake can be obtained by plotting accelerations against the period values. The response spectrum is of special interest to the engineer and has been defined by Cherry (1974) as 'the maximum response of a linear, single degree of freedom, structure to a specific ground acceleration, plotted as a function of the natural period or frequency and damping of the structure.' It is possible to compute the forces developed by an earthquake once the spectrum is known. The maximum energy liberated by an earthquake provides a measure of maximum displacement, and hence of maximum base shear and maximum stress. When the maximum value of energy is plotted against period or frequency, the resultant curve is referred to as the energy response spectrum. If the maximum velocity or maximum absolute acceleration is plotted as a function of period or frequency, then the curves are termed the velocity response spectrum and the absolute acceleration spectrum, respectively.

The relationship that exists between maximum acceleration, velocity and displacement means that their spectra can be plotted on a single tripartite graph (Figure 3.16). The major peaks in the lightly damped curves indicate that those structures with natural periods which coincide with these peaks are strongly excited by the earthquake concerned. The spectrum intensity refers to the area under the velocity response spectrum over a specified period for a standard damping value. It provides a measure of the intensity of ground shaking in

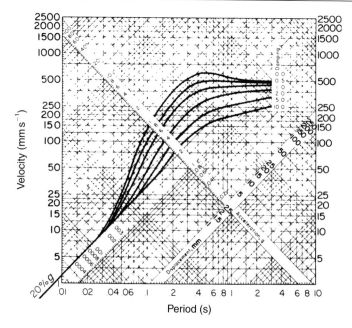

Figure 3.16 Combined plot of design spectrum giving acceleration (S_a), velocity (S_v) and displacement (S_d) as a function of period and damping. Scaled to 0.2 g acceleration at zero period (from Housner, 1970b).

that it expresses the average response of elastic structures to ground shaking. Consequently, the spectrum intensity indicates how much energy a structure has to absorb or dissipate in order to avoid earthquake damage.

The surface amplification spectrum refers to the theoretical behaviour of the ground calculated from measured properties and site geometry. The purpose of the development of this spectrum is to determine the effect of the surficial layers on an earthquake wave propagated in basement rock. Theoretically, a single wave progressing through stratified rock is reflected and refracted at each interface. The principal properties that affect the behaviour of this wave are density, shear-wave velocity and the thickness of the individual layers. These properties can be treated simply in a single-layer system at a fixed frequency.

The ground motion spectrum is a plot of the maximum amplitude of simple harmonic motion of ground motion. It is developed by multiplying basement motion by the amplification spectra, and is indicative of the behaviour of the surface of the ground in an earthquake. Basement rock motion is estimated by dividing the observed ground motion spectrum by the computed (theoretical) surface amplification spectrum where strong-motion records have been obtained.

The ultimate purpose of the above-mentioned techniques is to develop a design earthquake, that is, an estimate of the severest earthquake that is likely to occur within a 50- or 100-year period. More simply, it is an indication of the amount of motion, with its spectrum or frequency distribution. This is the critical factor required by the designer for dynamic analysis of the structure concerned and for determining its response to earthquake motion. The initial step in formulating a design earthquake at a given site is the estimation of the magnitudes of earthquakes that can be expected to affect the site during its lifetime. The extrapolated frequency–magnitude relations for southern California indicate that the largest expected earthquake per year has a magnitude of 6.2 and that the largest expected earthquake per 100 years has a magnitude of 8.2.

According to Housner (1970a), earthquake design criteria must be based on the probability of occurrence of strong ground shaking; the characteristics of the ground motion; the nature of the structural deformations; the behaviour of building materials when subjected to transient oscillatory strains; the nature of the building damage that might be sustained; and the cost of repairing the damage compared with the cost of providing additional earthquake resistance. The performance of buildings during earthquakes should guide the formulation of design criteria. Earthquake design criteria should specify the desired strength of structures in order to achieve a reasonably uniform factor of safety. This is usually done by means of a design spectrum. A design spectrum is not a specification of a particular earthquake ground motion; rather, it specifies the relative strengths of structures of different periods.

When establishing the design spectrum it is necessary to have some idea of the ground shaking that might be experienced at the location under consideration, so the earthquake history of the region should be examined, together with evidence of recent faulting or other indications of tectonic activity. Design spectrum curves must be consistent with previously recorded ground motions. Once the damping and allowable stresses are prescribed, the earthquake forces for which the structure is to be designed can be determined from the spectrum. In fact, the actual earthquake forces used in the design of a structure largely depend on the damping that the structure is assumed to have. Unfortunately, the damping that structures possess when vibrating strongly is not well known and therefore must be estimated.

The behaviour of an actual structure during an earthquake depends on its natural frequency and damping forces. In order to establish the earthquake design forces, models are vibrated at different vibration periods and the corresponding accelerations determined. Thus an idealized spectrum can be obtained applicable to any structure and any earthquake. Buildings have also been forced into vibration by means of shaking machines. The systematic analysis of the earthquake response of buildings can also be investigated by computer simulation.

3.9 Structural damage and its prevention

Earthquake damage to buildings depends upon how well they have been constructed and maintained; the seismic energy generated by the earthquake, as represented by the magnitude, duration and acceleration of strong motions in the ground; the ground response, which depends upon the type and character of the rocks and soils involved; and the distance from the epicentre of the earthquake (Ambraseys, 1988). Furthermore, an earthquake does not impart simple motion to a building but generally there are elastic impacts followed by forced and free vibrations of the structure. The damage potential of ground shaking is governed by the amplitude, duration and frequency of the waves, and these are unique to each earthquake. Few buildings are capable of resisting the displacement developed by acceleration exceeding 0.6 to 0.7 *g*. A general idea of the damage associated with different types of construction for different earthquake intensities is given in Table 3.9. If properly constructed, wood-frame buildings can withstand strong shaking. Buildings consisting of brick, stone or concrete bricks are not as resistant but may prove satisfactory if the mortar is good and they are reinforced with steel, especially at weak points. The primary reasons for the collapse of masonry buildings during earthquakes are their low tensile and shear strength, their high rigidity, low ductility, and low capacity for bearing reversed loads and the redistribution of stresses (Sachanski, 1978). Buildings with large windows can represent a hazard because the glass shatters as the building moves. Windows have been mounted in rubber in an attempt to reduce the likelihood of breaking during an earthquake.

Framed structures are flexible and capable of resisting large deformations and

Table 3.9 Earthquake loss-susceptibility data for different construction types (MM = modified Mercalli scale)(from Degg, 1992).

Construction type	Average damage (%) at intensity (MM):				
	VI	VII	VIII	IX	X
1 Adobe	8	22	50	100	100
2 Unreinforced masonry, non-seismic design	3.5	14	40	80	100
3 Reinforced concrete frames, non-seismic design	2.5	11	33	70	100
4 Steel frames, non-seismic design	1.8	6	18	40	60
5 Reinforced masonry, medium quality, non-seismic design	1.5	5.5	16	38	66
6 Reinforced concrete frames, seismic design	0.9	4	13	33	58
7 Shear wall structures, seismic design	0.6	2.3	7	17	30
8 Wooden structures, seismic design	0.5	2.8	8	15	23
9 Steel frames, seismic design	0.4	2	7	20	40
10 Reinforced masonry, high-quality seismic design	0.3	1.5	5	13	25

the redistribution of stresses (*ibid.*). The joints between columns and girders represent their weakest points. Joints in prefabricated-frame structural systems are also weak points. The concentrations of stress are therefore of major importance in determining behaviour under dynamic loading during an earthquake. Damage can occur as a result of buckling of columns and, in high-rise buildings, by lateral deformation causing an eccentricity of the vertical load and thereby generating further bending moments.

The seismic performance of a building is influenced by its mass and stiffness; its ability to absorb energy (i.e. its damping capacity); its stability; and its structural geometry and continuity. As such, the earthquake resistance of a building depends on an optimum combination of strength and flexibility, since it has to absorb and resist the impact of earthquake waves. Buildings that have been constructed of several types of material may be vulnerable to earthquake shocks.

All buildings have a certain natural period at which they tend to vibrate. If the earthquake waves have vibrations with the same period, the amplitude is increased and they will shake a building severely. If the period is very different, the effect will be much less severe. Adjoining buildings may have different fundamental periods and may vibrate out of phase with each other. If the buildings are less than 1 m apart, the resulting interference can cause mutual destruction. Progressive collapse may occur in this way if buildings are closely spaced down a hill. Taller buildings tend to have longer fundamental periods and to be more seriously damaged when located on soft ground that also has a long fundamental period. Low buildings with shorter fundamental periods are most vulnerable on rock since it has higher frequencies. It is possible to calculate the period a building will have when it is completed. Indeed, as noted previously, measurement of the period can be made by shaking a building. Alternatively, the fundamental periods of vibration of buildings can be determined by recording the small vibrations induced by wind and microseisms. The period itself depends upon the manner in which the weight is distributed in a building and upon the stiffness of the materials involved. The addition or removal of interior partitions or the storage of heavy goods may change the period considerably. The periods of earthquake waves can be determined by using strong-motion accelerometers. Basically, the analysis of the behaviour of a structure subjected to earthquake motion consists of modelling the structure, modelling the earthquake, investigating the model's response and comparing the results with the real structure (*ibid*).

Shallow foundations are sensitive to the vertical displacement component, especially if the structure is light and the water table high. Sometimes houses are thrown from their foundations. Deep foundations, however, are generally not hazardous in an earthquake. In some old buildings poorly attached front walls have fallen away, leaving the rest of the building little damaged (Figure 3.17). Overhanging cornices have fallen; indeed, this form of decoration should be abandoned in regions subject to earthquakes. Cracking of walls and plaster

Figure 3.17 Collapsed unit of the Pallante factory at Campania, Italy, due to an earthquake with a magnitude of 6.9. The building was poorly constructed.

is common. Chimneys frequently break at the roof line. A small earthquake may break pipes and the escaping water can cause appreciable damage. Generally, prevention of this type of damage is not difficult and, largely on the basis of statistics and experience, building codes have been and are still being developed that should eventually solve this particular problem without undue economic strain. In fact, most codes are directed towards ensuring safe and economical construction of ordinary buildings. However, they rarely give explicit guidance on the details of ground shaking (peak or effective peak ground acceleration or velocity, frequency, duration, etc.).

A well-constructed small building can be severely damaged by the fall of a larger and weaker adjacent building. Even in minor earthquakes, adjacent buildings may move with respect to each other. Generally, however, well-built, simple structures can often come through an earthquake unscathed. A reinforced concrete or steel-framed building possesses both strength and flexibility and, although there may be superficial damage to partitions and curtain walls, it usually remains substantially sound even after a large earthquake. Taking the Kobe earthquake of January 1995 as an example, those buildings that had base isolators were unaffected, and buildings that were designed according to the 1981 Building Standard Law generally performed well. This code mandated ductile structural steel reinforcement detailing. The most severe damage to reinforced structural steel buildings was to those that were constructed of non-ductile steel and concrete. Over 100 medium-rise buildings (around eight storeys in height) constructed during the 1960s and 1970s failed, most of the failures being shear failures of columns with very light transverse reinforcement. Mid-height, single-storey pancake collapse also occurred. However, most damage was recorded in wooden houses of one- or two-storey height, which were either unbraced or lightly braced. In fact, 55 000 houses collapsed, 32 000 were partially destroyed and an additional 7500 were damaged by fire (Esper and Tachibana, 1998).

In tall buildings inertial movement occurs, concentrating the force of shaking at ground level and generating destructive basal shear. This may cause the collapse of lower floors. Depending on the distribution of weight and strength in the building, other forms of inertia may cause collapse of the centre floors or the greater resonance experienced by the top floors may bring about their collapse (Figure 3.18). High-rise buildings are characterized by lower natural frequencies of vibration than low-rise buildings and consequently tend to be more sensitive to low-frequency ground motions, which predominate at distance from an earthquake epicentre. Hence, medium- to high-rise buildings, between six and twenty storeys, were the worst affected in Mexico City during the 1985 earthquake. The low-frequency ground motions caused many of the buildings to resonate, thereby prolonging and reinforcing the vibration within them. A point of concern is the increase in the number of high-rise buildings in cities, many of which are located in or adjacent to highly seismic zones, which means that they may begin to experience earthquake damage in the future.

Figure 3.18 Mid-floor failure of the Hotel de Carlo during the Mexico City earthquake of 19 September 1985.

The object lesson to be learned from destroyed and damaged buildings is, wherever possible, to avoid similar construction in the future. In addition, it is necessary to establish whether structures that have been subjected to earthquake and appear to have survived actually can withstand future earthquakes, that is, that they have not sustained hidden damage that would increase their likelihood of failure if and when they experienced subsequent strong ground motions.

In any large earthquake, community services such as water, gas, electricity, telecommunications and sewage disposal may be put out of action, and railways, highways and bridges may be damaged (Kubo and Katayama, 1978). For example, the Kobe earthquake badly damaged gas and water supply to the city, and these were not restored fully until mid-March of that year. By contrast, electricity and telecommunications performed relatively well. Both the elevated expressway and high-speed railway were badly damaged. In the former case, large hammerhead reinforced concrete piers sheared at their bases, causing a 500 m long section to collapse. The piers were lacking sufficient transverse reinforcement, having been constructed in 1968–69 under older seismic regulations. In addition, the original strength of the piers had been reduced by over 50% as a result of alkali aggregate reaction in the concrete, even though resin had been injected into cracks that had formed as a result of the reaction. Sections of the railway embankment failed and some overpasses collapsed. An elevated viaduct that carries the railway was severely damaged when some of the longer spans collapsed. Again, the viaduct was built in the 1960s. By contrast, many modern bridges in the area sustained little or no damage as these were constructed according to the latest codes.

Because nuclear reactors represent a potential hazard to public health and safety, their location and design are subject to strict governmental control. A site assessment for a nuclear power station must determine the long-term seismic history, and the properties and behaviour of the local soil and rock conditions. In the United States, it is normally necessary to design nuclear plants in relation to two types of earthquake. The first type of earthquake is the maximum credible or safe shutdown earthquake. This is defined as an earthquake that generates the maximum ground shaking for the structure so that the earthquake has only a small probability of occurrence during the life of the particular nuclear power station. Such stringent requirements are necessary to ensure the integrity of the containment walls of the reactor and the ability to close down the reactor safely if this becomes necessary. The second type of earthquake is the operating basis earthquake, which has a higher probability of occurrence than the maximum credible earthquake, having a possible occurrence interval of 100 years or more. Accordingly, the design is based on lower allowable stresses and for somewhat different combinations of conditions. In other words, the design must ensure that the dynamic loading attributable to such an earthquake does not prevent the continued operation of the station without undue risk to the health and safety of the employees and public.

3.10 Earthquakes and dams

Dams have to be designed to be safe under the normal and the design earthquake loading. Okamota (1978) noted that it is usually assumed that the design earthquake load is reduced by half when reservoirs are surcharged or empty and that a strong earthquake and an extraordinary flood are unlikely to occur at the same time. Two methods of analysis are used to ascertain the safety of dams in relation to earthquake shocks, namely, the pseudostatic method and the dynamic method. In the former method, the inertial force and the seismic water pressure on the dam are regarded as a static force. The magnitude of the inertial force is determined by the mass of the dam multiplied by a seismic coefficient. Seismic coefficients vary between 0.05 and 0.25 in the horizontal direction, depending on the type of dam and the seismicity of the dam site. The seismic coefficient in the vertical direction is taken as between 0 and 0.5 the horizontal coefficient.

Pseudostatic analysis of the Lower San Fernando dam in 1966 indicated that it would not fail in the event of an earthquake. However, on 9 February 1971, it was subjected to an earthquake that caused about 12 s of strong shaking, with a peak acceleration of about 0.5 g in the rock beneath the dam site. The upstream section, including 9.2 m of the crest, slid over 21 m into the reservoir (Figure 3.19). Fortunately the dam held – but only just! Consequently,

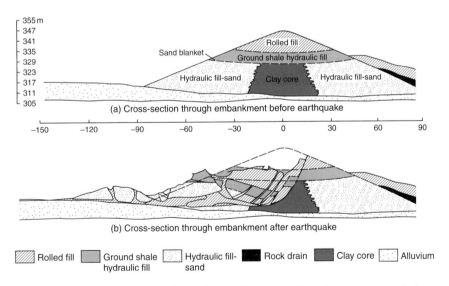

(a) Cross-section through embankment before earthquake

(b) Cross-section through embankment after earthquake

Rolled fill   Ground shale hydraulic fill   Hydraulic fill-sand   Rock drain   Clay core   Alluvium

Figure 3.19 Cross-section through the Lower San Fernando embankment dam before and after sliding took place due to the earthquake of 9 February 1971. There was no evidence of failure in the foundation soils, the slide occurring as a result of failure in a zone of soil about 6 m thick in the hydraulic fill near the base of the dam. The failure was accompanied by some liquefaction of the hydraulic sand fill (after Seed *et al.*, 1975).

Seed *et al.* (1975) suggested that the pseudostatic method appeared unsatisfactory for evaluating the seismic stability of earth dams.

Accordingly, dynamic analysis provides a more satisfactory basis for assessing the stability and deformation of embankment dams during earthquakes. In the dynamic method, the ground motion during earthquakes is specified initially, and then the vibration of the dam and of the reservoir water in response to ground motion are calculated. The inertial force and the dynamic water pressure applied to the dam are then determined.

Dams in regions subject to severe earthquakes should be flexible and self-healing, such as earth and earth–rock embankments. A concrete dam should not be considered where active faults cross the foundation, for fault movement can break the contact between foundation and dam. This allows full uplift pressure to act from beneath, thereby reducing shearing resistance along the base of the dam and causing failure by sliding on the foundation. Arch dams, although very resistant to earthquake shocks, are too brittle and slender to survive large displacements in the foundation or abutments. For example, an abrupt displacement of two opposing sectors of an arch dam, of the order of 0.25 to 0.5 m in almost any direction, may cause sudden complete failure. In such an instance, the concrete may be crushed or one end of the dam may be lifted off its abutment.

Earthquake-resistant earth dams are designed with a high freeboard and wider crest than those in stable regions. A higher camber and flatter slopes provide against slumping and sliding. The core of an earth dam must be larger, and in less high dams more plastic material makes self-healing more likely. If the maximum effective stress near the centre of the core exceeds 400 kPa in dams over 40 m in height, then lean saturated clays will tend to flow on dynamic loading, thereby closing cracks and fissures. Large filter zones of sand and gravel, with suitable transition characteristics, are arranged on the upstream and downstream sides of the core. Filters are also provided wherever different materials with different percolation characteristics are brought into contact with each other. This ensures control of leakage through transverse cracks in the core. Large outlets in such embankments are desirable to provide a means of lowering the reservoir quickly in an emergency.

For example, the Cedar Springs dam was completed in 1971 in an area of southern California where strong earthquakes are expected (Sherard *et al.*, 1974). Indeed, it is only 8 km from the San Andreas fault and recently active faults traverse its foundation. Accordingly, the dam consists of a zoned embankment with thick exterior shells of quarried rock and thick transitions of well-graded, coarse, sand–gravel mixture, with a silty sand zone separating the rolled clay core. The crest of the dam is 19 m wide, which is twice the usual width. The sand–gravel zone is too clean and cohesionless to allow an open crack to exist and its permeability, irrespective of the clay core, means that the quantity of water that could pass through would be safely controlled by the zones of rockfill. It represents an internal cohesionless transition zone for safe leakage control. Since the permeability of the transition zone will always be at least an order of magnitude less than that of the rockfill zone, the amount of leakage water will always fall well below the quantity that would tax the hydraulic capacity of the rockfill zone. The gradation of the transition zone is such that material cannot be washed from it into the voids of the rockfill zone. In such instances, the maximum leakage will emerge safely through the toe of the downstream rockfill zone. The clay of the core is highly non-dispersive and resistant to erosion.

The seriousness of earthquake hazard for many large dams, together with the importance of the project, dictate that seismic instrumentation programmes be set up. Bolt and Hudson (1975) discussed the manner in which dams could be instrumented so that the severity of earthquake ground motions and the dam response could be evaluated. They recommended that not less than four strong-motion accelerographs be installed at each site, two of which should be located so as to record earthquake motions in the foundation, and two used to measure the dam response. The latter two are usually situated near the top or on the crest of a dam.

3.11 Induced seismicity

Induced seismicity occurs where changes in the local stress conditions give rise to changes in strain in a rock mass. The sudden release of strain energy due to deformation and failure within a rock mass results in detectable earth movements. It is the action of man that causes the activity by bringing about these changes in the local stress conditions (McCann, 1988).

A number of events have occurred over the last 30 years or so that seem to support the idea that earthquakes might be triggered by artificially induced changes in the pore water pressures in rock masses. For example, Evans (1966) suggested that the earthquakes that affected Denver between April 1962 and November 1965, over 700 in number, but not exceeding M 4.3, were consequent upon the injection of waste fluids into a 3660 m deep disposal well. The Pre-Cambrian rocks that received this waste from the Rocky Mountain Arsenal consisted of highly fractured gneisses. It appears that the movements took place as the pore water pressures were raised by the injection of the waste. More specifically, rising pore water pressure reduces resistance to movement along fracture planes and thus elastic wave energy is released, generating an earthquake.

Similarly, the use of water injection to maximize the yield of crude oil can give rise to induced seismicity, presumably due to the injection of water leading to a reduction in the resistance to faulting consequent on the increase in pore water pressure. However, this would suggest that the rock masses involved are stressed to near failure stress prior to the injection of water. Gibbs *et al.* (1973) referred to seismic activity in the Rangely area, Colorado, associated with water injection in a nearby oilfield. They recorded 976 seismic events between 1962 and 1972, 320 of which exceeded magnitude 1 on the Richter scale. They described an apparent correlation between the number of earth tremors recorded and the quantity of water injected. Ten years later, Rothe and Lui (1983) recorded 31 earth tremors between March 1979 and March 1980 at the Sleepy Hollow oilfield in Nebraska. Again water had been injected to maximize the yield of oil. The range of magnitude varied between 0.6 and 2.9 on the Richter scale, and the source of the tremors coincided remarkably with the area of the oilfield.

More notably, a number of earthquakes have been recorded with their epicentres below or near large reservoirs. The evidence, however, is not conclusive. In the case of reservoirs this is not surprising, since so very few are instrumented to record local seismic events. But as dams are built higher and reservoirs impound larger volumes of water this cause-and-effect relationship needs to be resolved since induced earthquakes may cause serious damage, which needs to be avoided. Table 3.10 shows some of the world's largest reservoirs and dams, at some of which seismicity has been recorded.

The longest range of data has been provided by Lake Mead, Colorado, where over 10 000 small earthquakes have been recorded. A weak correlation between seismic activity and level of impounding has been reported at Lake Mead, and the epicentres are thought to have been located along faults. The most notable earthquake occurred at the Koyna reservoir, which is located not far from Bombay. Impounding began in 1963 and small shocks were recorded a few months later. These continued, gaining in intensity, until in December 1967 a shock with a magnitude of about 6.5 occurred. This caused significant damage and loss of life in the nearby village of Coynanagar. The dam was fissured in several places since the epicentre of the earthquake was near the dam. The focus was located at a depth somewhere between 10 and 20 km. Like Lake Mead, there is a correlation between fluctuating water level and seismicity. Figure 3.20 shows the number of earthquakes greater than magnitude 2 that were recorded in the first 10 years of life of the Kariba reservoir between Zimbabwe and Zambia. Most recently, tremors have been associated with the filling of the Katse reservoir, Lesotho (Bell and Haskins, 1997).

Seismic activity at reservoir sites may be attributable to water permeating the underlying strata, thereby increasing pore water pressure and decreasing the effective normal stresses so that shear strength along local faults is reduced. In addition, increased saturation may reduce the strength of rock masses sufficiently to facilitate the release of crustal strains. Such activity would not appear to be related to the weight of water stored, for some large impounded reservoirs have not been associated with noticeable seismic activity. Seismic activity may be initiated almost as soon as impounding begins, but the time of maximum activity sometimes shows an appreciable delay compared with the time when the reservoir reaches its maximum level.

Earthquakes associated with reservoir impounding tend to occur in swarms and have shallow foci and modest magnitudes, although they may sometimes

Table 3.10 Some large reservoirs in relation to seismic activity (after Lane, 1972).

Dam	Country	Type of dam	Height of dam (m)	Volume of dam ($m^3 \times 10^3$)	Capacity of reservoir ($m^3 \times 10^6$)	Seismicity after construction
Bhakra	India	Gravity	225	4130	9868	No record
Contra	Switzerland	Arch	230	660	86	Slight
Daniel Johnson (Manicopagan 5)	Canada	Multiple arch	214	2255	141 975	No record
Glen Canyon	USA	Arch	216	3747	33 304	Nil
Grancarevo	Yugoslavia	Arch	123	376	1277	Noticeable (M = 4)
Grande Dixence	Switzerland	Gravity	284	5957	400	Nil
Grandval	France	Multiple arch	88	180	292	Noticeable
Hoover	USA	Arch	221	3364	38 296	Noticeable (M = 5)
Kariba	Southern Rhodesia*	Arch	128	1032	160 368	Strong (M = 6)
Koyna	India	Gravity	104	1300	2780	Strong (M = 6.5)
Kremasta (Roi Paul)	Greece	Earth	160	7800	4750	Strong (M = 6.5)
Kurobegawa No. 4	Japan	Arch	186	1360	199	No record
Mangla	Pakistan	Earth	115	65 651	6358	Slight (M = 3.6)
Mauvoisin	Switzerland	Arch	237	2030	180	Slight
Monteynard	France	Arch	155	455	240	Noticeable (M = 4.9)
Oroville	USA	Earth	236	59 639	4298	Slight (M = 1.5)
Warragamba	Australia	Gravity	137	1233	2052	Noticeable

*Now Zimbabwe

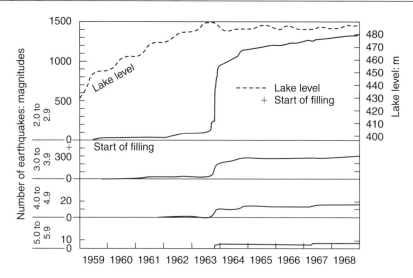

Figure 3.20   Kariba reservoir (Zimbabwe/Zambia): distribution of earthquake in time and in relation to lake level (after Lane, 1971).

cause destruction. They are characterized by an increased number of foreshocks and a slow decay of aftershocks. Reservoir impounding does not increase the maximum values of magnitudes. In fact, reservoirs probably tranform and induce seismic phenomena for which the stage is already set by existing tectonic features.

As far as monitoring of induced earthquakes from reservoir loading is concerned, a network of seismometers should be in operation prior to impounding to record the occurrence of small local earthquakes. The seismicity of the area has to be established before filling in order to distinguish whether local earthquakes are either a consequence of reservoir loading, or part of the general seismic pattern. Such information is also required to evaluate the probable size and location of future shocks.

Induced seismicity has been associated with many mining operations where changes in local stress conditions have given rise to corresponding changes in strain and deformation in the rock mass concerned. These changes have been responsible for movements that took place along discontinuities, which may be on a macroscopic or microscopic scale. In the former case they generate earth movements that are detectable at the ground surface. Such movements can cause damage to buildings, but generally this is minor, although very occasionally high local intensities are generated. Obviously, the extent of damage is related to the magnitude of the earth tremor and the distance from the source, as well as the nature of the surface rock and the strength of the structure. Nonetheless, such seismic events are often a cause of concern to the general public.

When mining, particularly hard rock at depth, violent rock failures, known as rockbursts, can occur in the workings. However, although all rockbursts generate seismic events, all seismic events associated with mining activity are not rockbursts. Cook (1976) suggested that the changes in stress consequent on mining were responsible for a sudden loss of stored strain energy, which results in brittle fracture in the rock mass in the excavation. These changes in strain energy may be related to a decrease in potential energy of the rock mass as the rock is mined. On the other hand, it may be that mining triggers latent seismic events in rock masses that are in a near-unstable condition.

Some damage has occurred to surface structures. For example, in 1976 an earth tremor of magnitude 5.1 damaged several buildings in Welkom in the gold mining area of the Orange Free State, South Africa. Neff *et al.* (1992) referred to severe cracking of some houses in a gold-mining township in western Transvaal; in fact, some houses had to be evacuated. The seismic tremors had magnitudes ranging up to 3.4. A seismic investigation was carried out which revealed that the response of the houses was dependent on the type of their foundation. In other words, the maximum peak particle velocity for a house on conventional spread footings was 40 mm s^{-1}, whereas a house on a light raft foundation with a compacted gravel mattress was 10 mm s^{-1}. These values compared with 20 mm s^{-1} for the maximum peak particle velocity of the ground surface. Hence, the rafts with gravel mattresses experienced an attenuation of ground vibration and were not damaged.

Seismic events have also been associated with coal mining in the United Kingdom. In 1975–77 earth tremors were recorded in Stoke-on-Trent, in the North Staffordshire coalfield, having magnitudes between 3 and 5. Further earth tremors occurred in 1980–81 (Kusznir *et al.*, 1982). Subsequently, Redmayne (1988) catalogued a number of earthquake events associated with British coalfields. He suggested that it seems likely that all deep coal mining is accompanied by such activity of varying magnitude, depending on geological conditions, local faulting, the local stress field, the rate and type of extraction, seam thickness, past mining in the area, and the nature of the overburden. Earth tremors were associated particularly with the South Wales, Staffordshire, Midlothian and Nottinghamshire coalfields. Redmayne was able to show a correlation between earth tremors and mining activity near Rosewell in the Midlothian coalfield, the tremors showing a marked decrease during holidays and weekends. The damage, if any, resulting from such tremors is minor.

Earth tremors have affected the Johannesburg area since 1894, eight years after the commencement of gold mining. The association of earth tremors with mining was established prior to the First World War. Early studies of seismic records of tremors in the Johannesburg area clearly showed that a peak occurred in the daily distribution of seismic events on weekdays around the time of blasting (Gane, 1939). The focal depths from which the tremors emanated were shown to be in close proximity to the position where mining was taking place. The magnitude of these events ranged from 2 to 4 on the Richter scale. It would appear that the seismic events increased in number and intensity up to 1953 and began to decline after 1965. The decline probably corresponded to the reduction in tonnage worked, although the decrease in seismic shocks was not proportional to the decrease in tonnage. These tremors are of short duration, seldom lasting much longer than one second. Ground vibration caused by such tremors is of higher frequency, smaller amplitude and shorter duration than those of tectonic earthquakes. Considerable development of microseismic monitoring instrumentation has taken place over recent years. Consequently, many mines now have sophisticated computer-controlled installations to monitor the occurrence of earth tremors.

References

Algermissen, S.T., Perkins, D., Isherwood, W., Gordon, D., Rengor, G. and Howard, C. (1975) Seismic risk evaluation of the Balkan region. *Survey of the Seismicity of the Balkan Region*, UNDP/UNESCO Report, US Geological Survey, Denver.

Ambraseys, N.N. (1969) Maximum intensity of ground movements caused by faulting. *Proceedings Fourth World Conference Earthquake Engineering*, Chile, 1, 154–162.

Ambraseys, N.N. (1974) Notes on engineering seismology. In *Engineering Seismology and Earthquake Engineering*, Solnes, J. (ed.), NATO Advanced Study Institutes Series, Applied Sciences, No. 3, 33–54.

Ambraseys, N.N. (1988) Engineering seismology. *International Journal Earthquake Engineering and Structural Dynamics*, 17, 1–106.

Ambraseys, N.N. (1992) Long-term seismic hazard in the eastern Mediterranean region. In *Geohazard: Natural and Man-Made*, McCall, G.J.H., Laming, D.J.C. and Scott, S.C. (eds), Chapman & Hall, London, 88–92.

Bender, B. (1984) Incorporating acceleration variability into seismic hazard analysis. *Bulletin Seismological Society America*, 74, 1451–1462.

Bonilla, M.G. (1970) Surface faulting and related effects. In *Earthquake Engineering*, Weigel, R.L. (ed.), Prentice-Hall, Englewood Cliffs, NJ.

Bonilla, M.G., Nash, R.K. and Lienkuemter, J.J. (1984) Statistical relationships covering earthquake magnitude, surface length and surface fault displacement. *Bulletin Seismological Society America*, 74, 2379–2411.

Bell, F.G. and Haskins, D.R. (1997) A geotechnical overview of the Katse Dam and Transfer Tunnel, Lesotho, with a note on basalt durability. *Engineering Geology*, 41, 175–198.

Bolt, B.A. (1978) *Earthquakes: A Primer*, W.H. Freeman, San Francisco.

Bolt, B.A. (1993) *Earthquakes*. W.H. Freeman, New York.

Bolt, B.A. and Hudson, D.E. (1975) Seismic instrumentation of dams. *Proceedings American Society Civil Engineers, Journal Geotechnical Engineering Division*, 101, 1095–1104.

Campbell, K.W. (1984) Probabilistic evaluation of seismic hazard for sites located near active faults. *Proceedings 8th World Conference Earthquake Engineering*, 1, 231–238.

Cherry, S. (1974) Design input for seismic analysis. In *Engineering Seismology and Earthquake Engineering*, Solnes, J. (ed.), NATO Advanced Study Institutes Series, Applied Sciences, No. 3, 151–162.

Cook, N.G.W. 1976. Seismicity associated with mining. *Engineering Geology*, 10, 99–122.

Degg, M.R. (1992) The ROA Earthquake Hazard Atlas project: recent work from the Middle East. In *Geohazards: Natural and Man-Made*, McCall, G.J.H., Laming, D.J.C. and Scott, S.C. (eds), Chapman & Hall, London, 93–104.

Donovan, N.C., Bolt, B.A. and Whitman, R.V. (1978) Development of expectancy maps and risk analysis. *Proceedings American Society Civil Engineers, Journal Structural Division*, 1, 1170–1192.

Esper, P. and Tachibanak, E. (1998) The lesson of Kobe earthquake. In *Geohazards and Engineering Geology, Engineering Geology Special Publication No. 14*, Geological Society, London.

Evans, D.M. (1966) Man-made earthquakes in Denver. *Geotimes*, 10, 11–18.

Fournier d'Albe, E. (1982) An approach to earthquake risk management. *Engineering Structures*, 4, 147–152.

Gane, P.G. (1939) A statistical study of the Witwatersrand earth tremors. *Journal Chemical, Metallurgy and Mining Society South Africa*, 40, 155–171.

Gibbs, J.F., Healy, J.H., Raleigh, G.B. and Cookley, J. (1973) Seismicity in Rangely, Colorado, area : 1962–1970. *Bulletin Seismological Society America*, 63, 1557–1570.

Hazzard, A.O. (1978) Earthquake and engineering: structural design, *The Consulting Engineer*, (May),14–23.

Housner, G.W. (1970a) Design spectrum. In *Earthquake Engineering*, Weigel, R.L. (ed.), Prentice-Hall, Englewood Cliffs, NJ, 93–106.

Housner, G.W. (1970b) Strong ground motion. In *Earthquake Engineering*, Weigel, R.L. (ed.), Prentice-Hall, Englewood Cliffs, NJ, 75–92.

Ishihara, R. and Okada, S. (1978) The effects of stress history on cyclic behaviour of sand. *Soils and Foundation*, 14, 31–45.

Karnik, V. and Algermissen, S.T. (1978) Seismic zoning. In *The Assessment and Mitigation of Earthquake Risk*. UNESCO Paris, 11–47.

King, C.Y. and Knopoff, L. 1968. Stress drop in earthquakes. *Bulletin Seismological Society America*, 58, 249–261.

Kubo, K. and Katayama, T. (1978) Earthquake resistant properties and design of public utilities. In *The Assessment and Mitigation of Earthquake Risk*. UNESCO, Paris, 171–184.

Kusznir, N.J., Al-Saigh, N.H. and Farmer, I.W. (1982) Induced seismicity resulting from roof caving and pillar failure in longwall mining. In *Strata Mechanics*, Farmer, I.W. (ed.), Elsevier, Amsterdam, 7–12.

Lane, R.G.T. (1972) Seismic activity at man-made reservoirs. *Proceedings Institution Civil Engineers*, Paper No. 7416, 15–24.

Lomnitz, C. and Rosenblueth, E. (1976) *Seismic Risk and Engineering Decisions*, Elsevier, Amsterdam.

McCann, D.M. 1988. Induced seismicity in engineering. In *Engineering Geology of Underground Movements, Engineering Geology Special Publication No. 5*, Bell, F.G., Culshaw, M.G., Cripps, J.C. and Lovell, M.A. (eds), Geological Society, London, 405–413.

Medvedev, S.V. (1965) *Engineering Seismology*. Israel Program for Scientific Translations, Jerusalem.

Medvedev, S., Sponheuer, W. and Karnik, V. (1965) *The MSK Intensity Scale*, Veroff Institute für Geodynamic, Jena, 48, 1–10.

Mogi, K. (1985) *Earthquake Prediction*. Academic Press. Orlando, Florida.

Neff, P.A., Wagener, R. von M. and Green, R.W.E. (1992) Houses damaged by mine tremors–distress repair and design. *Proceedings Symposium: Construction over Mined Areas*, Pretoria, South African Institution Civil Engineers, Yeoville, 133–137.

Okamota, S. (1978) Present trend of earthquake resistant design of large dams. In *The Assessment and Mitigation of Earthquake Risk*, UNESCO, Paris, 185–197.

Peck, R.B. (1979) Liquefaction potential: science versus practice. *Proceedings American Society Civil Engineers, Journal Geotechnical Engineering Division*, 105, 393–398.

Raleigh, C. B., Sieh, K. E., Sykes, L. R. and Anderson, D. L. (1982) Forecasting southern Californian earthquakes. *Science*, 217, 1097–1104.

Redmayne, D.W. (1988) Mining induced seismicity in U.K. coalfields identified on the BGS National Seismograph Network. In *Engineering Geology of Underground Movements, Engineering Geology Special Publication No. 5*, Bell, F.G., Culshaw, M.G., Cripps, J.C. and Lovell, M.A. (eds), Geological Society, London, 405–413.

Reid, H.F. (1908) *The Californian Earthquake of April 18, 1906*. Report of the State Earthquake Investigation Commission, Vol. 2, The Mechanics of the Earthquake, Carnegie Institute, Washington.

Richter, C.F. (1935) An instrumental earthquake scale. *Bulletin Seismological Society America*, 25, 1–32.

Richter, C.F. (1956) *Elementary Seismology*, W.H. Freeman, San Francisco.

Rikitake, T. (1979) Classification of earthquake precursors. *Tectonophysics*, 54, 293–309.

Rothe, G.H. and Lui, C. (1983) Possibility of induced seismicity in the vicinity of Sleepy Hollow oilfield, south western Nesbraska. *Bulletin Seismological Society America*, 73, 1357–1367.

Sachanski, S. (1978) Buildings: codes, materials, design. In *The Assessment and Mitigation of Earthquake Risk*, UNESCO, Paris, 157–170.

Savarenskij, E.F. and Neresov, I.L. (1978) Earthquake prediction. In *The Assessment and Mitigation of Earthquake Risk*, UNESCO, Paris, 66–90.

Scholtz, C.H., Sykes, L.R. and Aggarval, Y.P. (1973) Earthquake prediction: a physical basis. *Science*, 181, No. 4102, 803–810.

Seed, H.B. (1970) Soil problems and soil behaviour. In *Earthquake Engineering*, Wiegel, R.L. (ed.), Prentice-Hall, Englewood Cliffs, NJ, 227–252.

Seed, H.B., Lee, K.L., Idriss, M.L. and Madisi, F.I. (1975) The slides in the San Fernando Dams during the earthquake of February 9th, 1971. *Proceedings American Society Civil Engineers, Journal Geotechnical Engineering*, 111, 651–688.

Sherard, J.L., Cliff, L.S. and Allen, L.R. (1974) Potentially active faults in dam foundations. *Geotechnique*, 24, 367–429.

Singh, S.K., Mena, E. and Castro, R. (1988) Some aspects of source characteristics of the 19 September 1985 Michoacan earthquake and ground motion amplification in and near Mexico City from strong motion data. *Bulletin Seismological Society America*, 78, 451–477.

Skipp, B.O. and Ambraseys, N.N. (1987) Engineering seismology. In *Ground Engineer's Reference Book*, Bell, F.G. (ed.), Butterworths, London, 18/1–18/26.

Sokolenko, V.P. (1977) Landslides and collapses in seismic zones and their prediction. *Bulletin Association Engineering Geologists*, 15, 4–8.

Sokolenko, V.P. (1979) Mapping the after effects of disastrous earthquakes and estimation of hazard for engineering construction. *Bulletin Association Engineering Geologists*, 19, 138–142.

Steinbrugge, K.V. and Zacher, E.G. (1960) Creep on the San Andreas fault: fault creep and property damage. *Bulletin Seismological Society America*, 50, 389–398.

Wood, H. O. and Neumann, F. (1931) The modified Mercalli scale of 1931. *Bulletin American Seismological Society*, 21, 277–283.

Wyss, M. (1981) Recent earthquake prediction research in the United States. In *Current Research in Earthquake Prediction*, Rikitake, T. (ed.), Reidel, Dordrecht, 81–127.

Chapter 4

Mass movements

4.1 Soil creep and valley bulging

Mass movements on slopes can range in magnitude from soil creep on the one hand to instantaneous and colossal landslides on the other. Sharpe (1938) defined creep as the slow downslope movement of superficial rock or soil debris, which is usually imperceptible except by observations of long duration. Walker *et al.* (1987) suggested that creep could be regarded as mass movement that occurs at less than 0.06 m per year. Creep is a more or less continuous process and a distinctly surface phenomenon. It occurs on slopes with gradients somewhat in excess of the angle of repose of the material involved. Like landslip, its principal cause is gravity, although it may be influenced by seasonal changes in temperature, and by swelling and shrinkage in surface rocks. Other factors that contribute towards creep include interstitial rain washing, ice crystals heaving stones and particles during frost, and the wedging action of rootlets. The liberation of stored strain energy in the weathered zone, particularly of overconsolidated clays with strong diagenetic bonds, is another contributory cause of creep (Bjerrum, 1967). Although creep movement is exceedingly slow, there are occasions on record when it has carried structures with it.

Evidence of soil creep may be found on almost every soil-covered slope. For example, it occurs in the form of small terracettes, downslope tilting of poles, the curving downslope of trees and soil accumulation on the uphill sides of walls. Indeed, walls may be displaced or broken, and sometimes roads may be moved out of alignment. The rate of movement depends not only on climatic conditions and the angle of slope but also on the soil type and parent material.

Talus (scree) creep occurs wherever a steep talus exists. Its movement is quickest and slowest in cold and arid regions, respectively.

Solifluction is a form of creep that occurs in cold climates or high altitudes where masses of saturated rock waste move downslope. Generally, the bulk of the moving mass consists of fine debris, but blocks of appreciable size may also be moved. Saturation may be due to water from either rain or melting snow. Moreover, in periglacial regions associated with permafrost water cannot drain into the ground, since it is frozen permanently. Solifluction differs from mudflow in that it moves much more slowly, the movement is continuous and it occurs over the whole slope. Solifluction processes vary at different altitudes owing to the progressive comminution of materials in their downward migration and to differences in the growth of vegetation. At higher elevations surfaces are irregular and terraces are common, but at lower elevations the most common solifluction phenomena are continuous aprons of detritus that skirt the bases of all the more prominent relief features.

Valley bulges consist of folds formed by mass movement of argillaceous material in valley bottoms, the argillaceous material being overlain by thick competent strata. The amplitude of the fold can reach 30 m in those instances where a single anticline occurs along the line of the valley. Alternatively, the valley floor may be bordered by a pair of reverse faults or a belt of small-scale folding.

Valley bulging was noted in the Shipton valley in Northamptonshire by Hollingworth *et al.* (1944). There the Lias Clay exposed in the valley bottom bulges up to form anticlines with dips towards the valley sides. The clays are overlain by ferruginous sandstones and limestones. The authors explained these

Figure 4.1 Folded, interbedded sandstones and shales of Namurian age produced by valley bulging and revealed during excavations for the Howden Dam, near Sheffield, England, in 1934 (courtesy of Severn–Trent Water Authority).

features as stress-relief phenomena, that is, as stream erosion proceeded in the valley the excess loading on the sides caused the clay to squeeze out towards the area of minimum loading. This caused the rocks in the valley to bulge upwards.

However, other factors may also be involved in the development of valley bulging, such as high piezometric pressures, swelling clays or shales, and rebound adjustments of the stress field due to valley loading and excavation by ice. Notable examples of valley bulges occur in some of the valleys carved in Millstone Grit country in the Pennine area (Figure 4.1).

The valleyward movement of argillaceous material results in cambering of the overlying competent strata, blocks of which become detached and move down the hillside. Fracturing of cambered strata produces deep debris-filled cracks or 'gulls', which run parallel to the trend of the valley. Some gulls may be several metres wide. Small gulls are sometimes found in relatively flat areas away from the slopes with which they are associated.

4.2 Causes of landslides

Varnes (1978) defined landslides as the downward and outward movement of slope-forming materials composed of rocks, soils or artificial fills. The displaced material has well-defined boundaries. Creep, solifluction and avalanching were excluded from his definition. Movement may take place by falling, sliding or flowing, or some combination of these factors. This movement generally involves the development of a slip surface between the separating and remaining masses. However, rockfalls, topples and debris flows involve little or no true sliding on a slide surface. The majority of stresses found in most slopes are the sum of the gravitational stress from the self-weight of the material plus the residual stress.

In most landslides a number of causes contribute towards movement, and any attempt to decide which one finally produced the failure is not only difficult but pointless. Often the final factor is nothing more than a trigger mechanism that set in motion a mass which was already on the verge of failure. Hence elements that influence slope stability are numerous and varied, and they interact in complex and often subtle ways. Basically, however, landslides occur because the forces creating movement, the disturbing forces (M_D) exceed those resisting it, the resisting forces (M_R) that is, the shear strength of the material concerned. In general terms, therefore, the stability of a slope may be defined by a factor of safety (F) where:

$$F = M_R/M_D \qquad (4.1)$$

If the factor of safety exceeds 1, then the slope is stable, whereas if it is less than 1, the slope is unstable.

The lithology of soils and rocks influences their physical and chemical behaviour, which in turn affects the shear strength, permeability and durability, all three of which influence slope stability.

The causes of landslides were grouped into two categories by Terzaghi (1950), namely, internal causes and external causes (Table 4.1). The former include those mechanisms within the mass that bring about a reduction of its shear strength to a point below the external forces imposed on the mass by its environment, thus inducing failure. External mechanisms are those outside the mass involved that are responsible for overcoming its internal shear strength, thereby causing it to fail.

The common force tending to generate movements on slopes is, of course, gravity. Generally, the steeper the slope, the greater is the likelihood that landslides will occur. Obviously, there is no universal threshold value at which slides take place, since this must be related to the ground conditions. Many steep slopes on competent rock are more stable than comparatively gentle slopes on weak material. Nonetheless, numerous authors have reported critical angles below which sliding does not occur in particular soil or weak rock types (Walker *et al.*, 1987). Carrara *et al.* (1977) also related slope stability to slope aspect (i.e. the direction in which a slope faces), and to its curvature both downslope and across-slope.

Climatic conditions act in a number or ways to promote the occurrence of landslides. Rainfall is the most important climatic factor. Landslides can be triggered by rainfall if some threshold intensity is exceeded so that pore water pressures are increased by a required amount (Olivier *et al.*, 1994). The process is also affected by the duration of the rainfall, by the slope angle, the distribution of shear strength and permeability, particularly within the regolith, the effects of soil pipes, variations in regolith thickness, and antecedent weather and pore water pressure conditions. Jahns (1978) suggested that larger, deeper slides were more likely to occur during long continuous wet periods, while shallow slides were more characteristic of short-duration, high-intensity rainfall. A characteristic of the landslides in the London Clay at Southend, Essex, is their highly episodic nature in that none will occur for several years and then numerous slides will take place at more or less the same time. The slides tend to develop when the soil moisture deficit falls to zero (Hutchinson, 1992). Types and severity of landsliding vary notably from one climatic region to another. For instance, frost action in arctic regions may lead to rockfalls, and thawing of permafrost may induce debris slides and flows. Damaging debris slides may be triggered by the heavy rainfall of thunderstorms in semi-arid regions.

The influence of vegetation on slope stability has been examined by several authors and is summarized in Table 4.2. Greenway (1987) provided details of how to assess the relative importance of the various factors outlined in Table 4.2 in relation to slope stability. According to Varnes (1984), distressed vegetation may help to demonstrate slope movement, either in the field or on aerial photographs. Removal of vegetative cover alters the hydrological and

Table 4.1 Processes leading to landslides (after Terzaghi, 1950).

Name of agent	Event or process that brings agent into action	Mode of action of agent	Slope materials most sensitive to action	Physical nature of significant actions of agent	Effects on equilibrium conditions of slope
Transporting agent	Construction operations or erosion	(1) Increase of height or rise of slope	Every material	Changes state of stress in slope-forming material	Increase of shearing stresses
			Stiff fissured clay, shale	Changes state of stress and causes opening of joints	Increases shearing stresses and initiates process 8
Tectonic stresses	Tectonic movements	(2) Large-scale deformations of Earth's crust	Every material	Increases slope angle	Increases of shearing stresses
Tectonic stresses or explosives	Earthquakes or blasting	(3) High-frequency vibrations	Every material	Produces transitory change of stress	Decrease of cohesion and increase of shearing stresses
			Loess, slightly cemented sands, and gravel	Damages intergranular bonds	
			Medium or fine loose sand in saturated state	Initiates rearrangement of grains	Spontaneous liquefaction
Weight of slope-forming material	Process that created the slope	(4) Creep on slope	Stiff fissured clay, shale remnants of old slides	Opens up closed joints, produces new ones	Reduces cohesion, accelerates process 8
		(5) Creep in weak stratum below foot of slope	Rigid materials resting on plastic ones		
Water	Rain or melting snow	(6) Displacement of air in voids	Moist sand	Increases pore water pressure	Decrease of frictional resistance
		(7) Displacement of air in open joints	Jointed rock, shale		
		(8) Reduction of capillary pressure associated with swelling	Stiff fissured clay and some shales	Causes swelling	Decrease of cohesion

Table 4.1 Continued

Name of agent	Event or process that brings agent into action	Mode of action of agent	Slope materials most sensitive to action	Physical nature of significant actions of agent	Effects on equilibrium conditions of slope
	Rains or melting snow	(9) Chemical weathering	Rock of any kind	Weakens intergranular bonds (chemical weathering)	} Decrease of cohesion
	Frost	(10) Expansion of water due to freezing	Jointed rock	Widens existing joints, produces new ones	Decrease of frictional resistance
		(11) Formation and subsequent melting of ice layers	Silt and silty sand	Increases water content of soil in frozen top-layer	Decrease of frictional resistance
	Dry spell	(12) Shrinkage	Clay	Produces shrinkage cracks	Decrease of cohesion
	Rapid drawdown	(13) Produces seepage towards foot of slope	Fine sand, silt, previously drained	Produces excess pore water pressure	Decrease of frictional resistance
	Rapid change of elevation of water table	(14) Initiates rearrangement of grains	Medium or fine loose sand in saturated state	Spontaneous increase of pore water pressure	Spontaneous liquefaction
	Rise of water table in distant aquifer	(15) Causes a rise of piezometric surface in slope-forming material	Silt or sand layers between or below clay layers	Increases pore water pressure	Decrease of frictional resistance
	Seepage from artificial source of water (reservoir or canal)	(16) Seepage towards slope	Saturated silt	Increases pore water pressure	Decrease of frictional resistance
		(17) Displaces air in the voids	Moist, fine sand	Eliminates surface tension	} Decrease of cohesion
		(18) Removes soluble binder	Loess	Destroys intergranular bond	
		(19) Subsurface erosion	Fine sand or silt	Undermines the slope	Increase of shearing stress

Table 4.2   Effects of vegetation on slope stability (after Greenway, 1987).

Hydrological mechanisms	Influence
1 Foliage intercepts rainfall, causing absorptive and evaporative losses that reduce rainfall available for infiltration.	B
2 Roots and stems increase the roughness of the ground surface and the permeability of the soil, leading to increased infiltration capacity.	A
3 Roots extract moisture from the soil, which is lost to the atmosphere via transpiration, leading to lower pore water pressures.	B
4 Depletion of soil moisture may accentuate desiccation cracking in the soil, resulting in higher infiltration capacity.	A
Mechanical mechanisms	
5 Roots reinforce the soil, increasing soil shear strength.	B
6 Tree roots may anchor into firm strata, providing support to the upslope soil mantle through buttressing and arching.	B
7 Weight of trees surcharges the slope, increasing normal and downhill force components.	A/B
8 Vegetation exposed to the wind transmits dynamic forces into the slope.	A
9 Roots bind soil particles at the ground surface, reducing their susceptibility to erosion.	B

Note: A – Adverse to stability; B – Beneficial to stability.

hydrogeological conditions of a slope, and frequently leads to accelerated run-off, and consequent increased erosion and increased probability of slides and debris flow occurring.

An increase in the weight of slope material means that shearing stresses are increased, leading to a decrease in the stability of a slope, which may ultimately give rise to a slide. This can be brought about by natural or artificial (man-made) activity. For instance, removal of support from the toe of a slope, either by erosion or excavation, is a frequent cause of slides, as is overloading the top of a slope. Such slides are external slides in that an external factor causes failure.

Submerged and most partly submerged slopes are comparatively stable as the water pressure acting on the surface of the slope reduces the shearing stresses. If, however, the water level falls (as it does due to rapid drainage of a reservoir or due to tidal effects) the stabilizing influence of the water disappears. If the slopes consist of cohesive soils, the water table will not be lowered at the same rate as the body of water. This means that the slope is temporarily overloaded with excess pore water, which may lead to failure. Thus rapid drawdown may be critical for slope stability (Morgenstern, 1963).

In many parts of the world, marine erosion on near-present coastlines was halted by the glacio-eustatic lowering of sea level during Pleistocene times and recommenced on subsequent recovery. For example, landslides around the English coast were generally reactivated by rising sea levels in Flandrian times, that is, 8000 to 4000 years BP. Hutchinson (1992) stated that once the sea level became reasonably constant, erosion continued at a steady pace, giving rise to coastal landslides. A cyclic situation then develops in which landslide material is removed by the sea and the cliffs are steepened, which leads to further landsliding. Hence extended periods of slow movement are succeeded by sudden first-time failures. Previously, Hutchinson (1973) had noted that the cliffs developed in London Clay at Warren Point, Isle of Sheppey, have a landslip cycle of approximately 40 years. By contrast, the coastal landslide cycle for the harder Cretaceous rocks forming the Undercliff, Isle of Wight, is about 6000 years (Hutchinson *et al.*, 1991). Coastal landslides in softer rocks have frequently occurred in association with major coastal surges.

Other external mechanisms include earthquakes (see Chapter 3) and other shocks and vibrations. Earthquake shocks and vibrations in granular soils not only increase the external stresses on slope material but they can also cause a reduction in the pore space, which effectively increases pore water pressures. The area affected by landslides caused by earthquakes is influenced by the magnitude of the earthquake. For example, Keefer (1984) suggested that an earthquake with a Richter magnitude of 4 probably would not generate landslides, whereas a magnitude of 9.2 would cause landslides to take place over an area as large as 500 000 km^2. He further suggested that rockfalls, rock slides, soil falls and soil slides are triggered by the weaker seismic tremors, whereas deepseated slides and earthflows are generally the result of stronger earthquakes. Materials that are particularly susceptible to earthquake motions include loess, volcanic ash on steep slopes, saturated sands of low density, quick clays and loose boulders on slopes. The most severe losses of life have generally been caused by earthquake-induced landslides; for example, one that occurred in Kansu Province, China, in 1920 killed around 200 000 people.

Internal slides are generally caused by an increase in pore water pressures within the slope material, which causes a reduction in the effective shear strength. Indeed, it is generally agreed that in most landslides groundwater constitutes the most important single contributory cause. Therefore identification of the source and amount of water, water movement and development of excess pore water pressure is important. An increase in water content also means an increase in the weight of the slope material or its bulk density, which can induce slope failure. Significant volume changes may occur in some materials, notably clays, on wetting and drying out. Not only does this weaken the clay by developing desiccation cracks within it, but the enclosing strata may also be affected adversely. Rises in the levels of water tables because of shortduration, intense rainfall or prolonged rainfall of lower intensity are a major cause of landslides (Bell, 1994). At times the water table may perch on the failure surface of a landslide. Seepage forces within a granular soil can produce a reduction in strength by reducing the number of contacts between grains. Water can also weaken slope material by causing minerals to alter or by bringing about their solution. However, according to Terzaghi (1950), in humid climates at least, there is always enough moisture in the ground to act as a lubricant for movement. He therefore dismissed the idea that slides that developed after rainfall were principally due to the lubricating effect of excess water. In such instances slides occur due to the consequent increase in pore water pressures.

Precipitation is a time-dependent factor that influences pore water pressure. Shallow slides can develop after high rainfall. For example, proximate rainfall, according to Brand et al. (1984), tends to trigger shallow slides in Hong Kong, whereas antecedent rainfall is an important factor for deeper slides. Another factor that influences saturated low-permeability soils is the long duration for equalization of pore water pressures after unloading. In terms, for example, of the London Clay, it has been suggested by Vaughan and Walbanke (1973) that equilibration of pore water pressure could take thousands of years. Hence this can exercise significant control on the delayed failure of certain slopes such as those developed in heavily overconsolidated clay.

Weathering can effect a reduction in strength of slope material, leading to sliding. The necessary breakdown of equilibrium to initiate sliding may take decades. For example, Chandler (1974) quoted a case of slope failure in Lias Clay in Northamptonshire, which was primarily due to swelling in the clay. It took 43 years to reduce the strength of the clay below the critical level at which sliding occurred. Indeed, in relatively impermeable cohesive soils the swelling process is probably the most important factor leading to a loss of strength and therefore to delayed failure. On the other hand, progressive softening of fissured clays involved in a constructed slope can occur while construction is taking place.

4.3 Classification of landslides

There are many classifications of landslides, which to some extent is due to the complexity of slope movements (Hansen, 1984). However, by far the most widely used classification is that of Varnes (1978). Varnes classified landslides according to the type of movement undergone on the one hand and the type of materials involved on the other (Figure 4.2). Types of movement were grouped into falls, slides and flows. The materials concerned were simply grouped as rocks and soils. Obviously, one type of slope failure may grade into another; for example, slides often turn into flows. Complex slope movements are those in which there is a combination of two or more principal types of movement. Multiple movements are those in which repeated failures of the same type occur in succession, and compound movements are those in which the failure surface is formed of a combination of curved and planar sections. Skempton and Hutchinson (1969) proposed a geotechnical classification of land-slides (Table 4.3) based on two criteria that should be considered in relation to stability analysis, namely, the history of slope movement and the time to failure. A summary of the features that can aid the recognition of slope movement is given in Table 4.4.

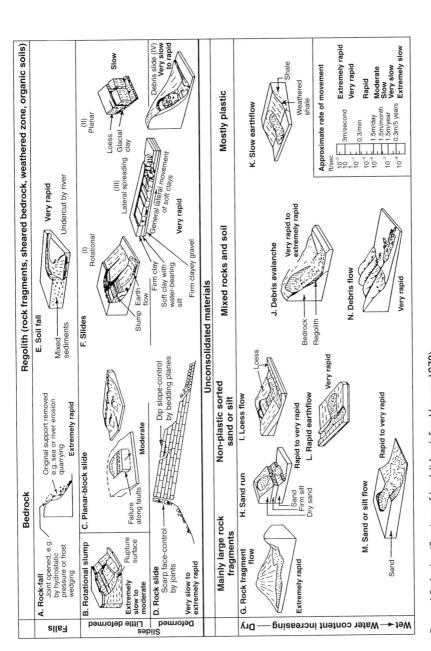

Figure 4.2 A classification of landslides (after Varnes, 1978).

Table 4.3　Geotechnical classification of landslides (after Skempton and Hutchinson, 1969).

Soil fabric conditions (affecting cohesion and internal friction)	Pore fluid pressure conditions on slope surfaces (affecting pore water pressures)
1 First-time slides in previously unsheared ground: soil fabric tends to be random (or oriented as a result of depositional history) and shear strength parameters are at peak or between peak and residual values.	A Short-term (undrained) – no equalization or excess water pressure set by changes in total stress.
2 Slides on pre-existing shears associated with: a) previous landslides b) colluvium c) periglacial solifluction d) other freeze–thaw processes e) tectonics f) lateral expansion	B Intermediate – partial equalization of excess pore water pressures. C Long-term (drained) – complete equalization of excess pore water pressures to steady seepage values.
In these cases, the soil fabric surface is highly oriented in the slip direction and shear strength parameters are at or about residual values.	Note that combinations of A, B and C can occur at different times in the same landslide: for example, a particularly dangerous type of slide is that in which long-term, steady seepage conditions (C) exist up to failure but during failure, undrained conditions (A) apply; that is, a drained/undrained failure.

4.3.1 Falls

Falls are very common. The moving mass travels mostly through the air by free fall, saltation or rolling, with little or no interaction between the moving fragments. Movements are very rapid and may not be preceded by minor movements. A rockfall event involves a single block or group of blocks that become detached from a rock face; each block may be a falling block behaving more or less independently of other blocks. Blocks may be broken during the fall. There is temporary loss of ground contact and high acceleration during the descent, with blocks attaining significant kinetic energy. Blocks accumulate at the bottom of a slope as scree deposit. If a rockfall is active or very recent, then the slope from which it was derived is scarped. Frost–thaw action is one of the major causes of rockfall.

Toppling failure is a special type of rockfall, which can involve considerable volumes of rock. The danger of slope toppling increases with increasing discontinuity angle, and steep slopes in vertically jointed rocks frequently exhibit signs of toppling failure.

Skempton and Hutchinson (1969) noted that falls in clay soils generally represent short-term failures that originate from a tension crack in newly exposed slopes.

Table 4.4 Features that aid in recognition of common types of slope movement (after Rib and Liang, 1978).

Type of motion	Kind of material	Parts surrounding slide			Parts that have moved			
		Crown	Main scarp	Flanks	Head	Body	Foot	Toe
Falls, topples	Rock	Consists of loose rock; probably has cracks behind scarp; has irregular shape controlled by local joint system	Is usually mostly vertical, irregular, bare, and fresh; usually consists of joint or fault surfaceS	Are mostly bare edges of rock	Is usually not well defined, consists of fallen material that forms heap of rock next to scarp	Falls: Has irregular surface of jumbled rock that slopes away from scarp and that, if large and if trees or material of contrasting colours are included may show direction of movement radial from scarp; may contain depressions	Is commonly buried; if visible, generally shows evidence of reason for failure, such as prominent joint or bedding surface, underlying weak rock, or banks undercut by water	Is irregular pile of debris or talus if slide is small; may have rounded outline and consist of broad, curved transverse ridge if slide if large
	Soil	Has cracks behind scarp	Is nearly vertical, fresh, active, and spalling on surface	Are often nearly vertical	Is usually not well defined; consists of fallen material that forms heap of rock next to scarp	Topples: Consists of unit or units tilted away from crown: Is irregular	Is commonly buried; if visible generally shows evidence of reason for failure, such as prominent joint or bedding surface, underlying weak rock, or banks undercut by water	Is irregular
Slides, rotational slump	Soil	Has numerous cracks that are mostly curved concave toward slide	Is steep, bare, concave toward slide, and commonly high; may show striae and furrows on surface running from crown to head; may be vertical in upper part	Have striae with strong vertical component near head and strong horizontal component near foot; have scarp height that decreases toward foot; may be higher than original ground surface between foot and toe; have en echelon cracks that outline slide in early stages	Has remnants of land surface flatter than original slope or even tilted into hill, creating at base of main scarp depressions in which perimeter ponds form; has transverse cracks, minor scarps, grabens, fault blocks, bedding attitude different from surrounding area, and trees that lean uphill	Consists of original slump blocks generally broken into smaller masses; has longitudinal cracks, pressure ridges, and occasional overthrusting; commonly develops small pond just above foot	Commonly has transverse cracks developing over foot line and transverse pressure ridges developing below foot line; has zone of uplift; no large individual blocks, and trees that lean downhill	Is often a zone of earthflow of lobate form in which material is rolled over and buried, has trees that lie flat or at various angles and are mixed into the material
	Rock	Has cracks that tend to follow fracture pattern in original rock	Is steep, bare, concave toward slide, and commonly high; may show striae and furrows on surface running from crown to head; may be vertical in upper part	Have striae with strong vertical component near head and strong horizontal component near foot; have scarp height that decreases toward foot; may be higher than original ground surface between foot and toe; have en echelon cracks that outline slide in early stages	Has remnants of land surface flatter than original slope or even tilted into hill, creating at base of main scarp depressions in which perimeter ponds form; has transverse cracks, minor scarps, grabens, fault blocks, bedding attitude different from surrounding area, and trees that lean uphill	Consists of original slump blocks somewhat broken up; has little plastic deformation; has longitudinal cracks, pressure ridges, and occasional overthrusting; commonly develops small pond just above foot	Commonly has transverse cracks developing over foot line and transverse pressure ridges developing below foot line; has zone of uplift, no large individual blocks, and trees that lean downhill	Has little or no earthflow; is often nearly straight and close to foot; may have steep front

Table 4.4 Continued

	Parts surrounding slide				Parts that have moved			
Type of motion	Kind of material	Crown	Main scarp	Flanks	Head	Body	Foot	Toe
Translational block	Rock or soil	Has cracks, most of which are nearly vertical and tend to follow contour of slope	Is nearly vertical in upper part and nearly plane and gently to steeply inclined in lower part		Ils relatively undisturbed and has no rotation	Is usually composed of single or few units, is undisturbed except for common tension cracks that show little or no vertical displacement	Has none and no zone of uplift	Ploughs or overrides ground surfaces
Rock	Rock	Contains loose rock; has cracks between blocks	Is usually stepped according to spacing of joints or bedding planes; has irregular surface in upper part and is gently to steeply inclined in lower part; may be nearly planar or composed of rock chutes	Are irregular	Has many blocks of rock	Has rough surface of many blocks, some of which may be in approximately their original attitude but lower if movement was slow translation	Usually has none	Consists of accumulation of rock fragments
Dry flows	Rock	Consists of loose rock; probably has cracks behind scarp; has irregular shape controlled by local joint system	Is usually almost vertical irregular, bare, and fresh; usually consists of joint or fault surfaces	Are mostly bare edges of rock	Has none	Has irregular surface of jumbled rock fragments sloping down from source region and generally extending far out on valley floor; shows lobate transverse ridges and valleys	Has none	Composed of tongues; may override low ridges in valley
	Soil	Has no cracks	Is funnel-shaped at angle of repose	Have continuous curve into main scarp	Usually has none	Is conical heap of soil, equal in volume to head region	Has none	
Wet debris avalanche, debris flow	Soil	Has no cracks	Typically has serrated or V-shaped upper part; is long and narrow, bare, and commonly striated	Are steep and irregular in upper part; may have levees built up in lower parts	May have none	Consists of large blocks pushed along in a matrix of finer material, has flow lines, follows drainage ways and can make sharp turns, is very long compared with breadth	Is absent or buried in debris	Spreads laterally in lobes, if dry, may have a steep front about a metre high
Earthflow Sand flow	Soil	May have a few cracks	Is concave towards slide, in some types is nearly circular and slide issues through narrow orifice	Are curved; have steep sides	Commonly consists of a slump block	Is broken into many small pieces, shows flow structure	Has none	Is spreading and lobate, consists of material rolled over and buried; has trees that lie flat or at various angles and are mixed into toe material
Silt flow	Soil	Has few cracks	Is steep and concave towards slide; may have variety of shapes in outline nearly straight, gentle arc, circular or bottle-shaped	Commonly diverge in direction of movement	Is generally under water	Spreads out on underwater floor	Has none	Is spreading and lobate

4.3.2 Slides

In true slides, the movement results from shear failure along one or several surfaces, such surfaces offering the least resistance to movement. The mass involved may or may not experience considerable deformation. One of the most common types of slide occurs in clay soils, where the slip surface is approximately spoon-shaped. Such slides are referred to as rotational slides (Figure 4.3). They are commonly deep-seated (depth/length ratio = 0.15–0.33). Backward rotation of the failed mass is the dominant characteristic, and the failed material remains intact to the extent that only one or a few discrete blocks are likely to form.

Although the slip surface is concave upwards it seldom approximates to a circular arc of uniform curvature. For instance, if the shear strength of the soil is lower in the horizontal than vertical direction, the arc may flatten out; if the soil conditions are reversed, then the converse may apply. What is more, the shape of the slip surface is very much influenced by the existing discontinuity pattern.

Rotational slides usually develop from tension scars in the upper part of a slope, the movement being more or less rotational about an axis located above the slope. The tension cracks at the head of a rotational slide are generally concentric and parallel to the main scar. Undrained depressions and perimeter lakes, bounded upwards by the main scar, characterize the head regions of many rotational slides.

When the scar at the head of a rotational slide is almost vertical and unsupported, then further failure is usually just a matter of time. As a consequence, successive rotational slides occur until the slope is stabilized. These are retrogressive slides and they develop in a headward direction. All multiple

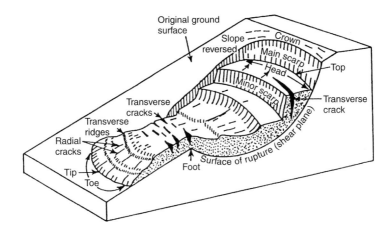

Figure 4.3 The main features of a rotational slide (after Varnes, 1978).

retrogressive slides have a common basal shear surface in which the individual planes of failure are combined.

Non-circular slips occur in overconsolidated clays in which weathering has led to the development of quasi-planar slide surfaces, or in unweathered structurally anisotropic clays. Both circular and non-circular shallow rotational slips tend to form on moderately inclined slopes in weathered or colluvial clays.

Translational slides occur in inclined stratified deposits, the movement occurring along a planar surface, frequently a bedding plane. The mass involved in the movement becomes dislodged because the force of gravity overcomes the frictional resistance along the potential slip surface, the mass having been detached from the parent rock by a prominent discontinuity such as a major joint. Slab slides, in which the slip surface is roughly parallel to the ground surface, are a common type of translational slide. Such a slide may progress almost indefinitely if the slip surface is sufficiently inclined and the resistance along it is less than the driving force. Slab slides can occur on gentler surfaces than rotational slides and may be more extensive.

According to Skempton and Hutchinson (1969), compound and translational slides develop in clay deposits when rotation is inhibited by an underlying planar feature, such as a bedding plane or the base of a weathered boundary layer. Translational slides tend to be more superficial than compound slides, being governed by more shallow inhomogeneities. Clay that is subjected to part rotational, part translational sliding is often distorted and broken. Block slides may develop in the more lithified, jointed deposits of clay, blocks of clay first separating and then sliding on well-defined bedding, joint or fault planes. Slab slides are characteristic of more weathered clay slopes of low inclination. Material moves en masse with little internal distortion.

Weathered mantle and colluvial materials are particularly prone to slab failure, which rarely occurs with depth/length ratios greater than 0.1. If a sufficient number of overlapping slips develop, they may form a shallow translational retrogressive slide.

Failures that involve lateral spreading may develop in clays, quick clays and varved clays. This type of failure is due to high pore water pressure in a more permeable zone at relatively shallow depth, dissipation of pore water pressure leading to the mobilization of the clay above. The movement is usually complex, being predominantly translational, although rotation and liquefaction, and consequent flow may also be involved. Such masses, however, generally move over a planar surface and may split into a number of semi-independent units. Like other landslides, these are generally sudden failures, although sometimes movement can take place slowly.

Rock slides and debris slides are usually the result of a gradual weakening of the bonds within a rock mass and are generally translational in character. Most rock slides are controlled by the discontinuity patterns within the parent rock. Water is seldom an important direct factor in causing rock slides, although it may weaken bonding along joints and bedding planes. Freeze–thaw action, however, is an important cause. Rock slides commonly occur on steep slopes and most are of single rather than multiple occurrence. They are composed of rock boulders. Individual fragments may be very large and may move great distances from their source. Debris slides are usually restricted to the weathered zone or to surficial talus. With increasing water content debris slides grade into mudflows. These slides are often limited by the contact between the loose material and underlying firm bedrock.

4.3.3 Flows

In a flow the movement resembles that of a viscous fluid (Bishop, 1973). In other words, as movement downslope continues, intergranular movements become more important than shear surface movements. Slip surfaces are usually not visible or are short-lived, and the boundary between the flow and the material over which it moves may be sharp or may be represented by a zone of plastic flow. Some content of water is necessary for most types of flow movement, but dry flows can and do occur. Consequently, the range of water content in flows must be regarded as ranging from dry at one extreme to saturated at the other. Dry flows, which consist predominantly of rock fragments, are simply referred to as rock fragment flows or rock avalanches and generally result from a rock slide or rockfall turning into a flow. They are generally very rapid and short-lived, and are frequently composed mainly of silt or sand. As would be expected, they are of frequent occurrence in rugged mountainous regions, where they usually involve the movement of many millions of tonnes of material. Wet flows occur when fine-grained soils, with or without coarse debris, become mobilized by an excess of water. They may be of great length.

Progressive failure is rapid in debris avalanches and the whole mass, either because it is quite wet or is on a steep slope, moves downwards, often along a stream channel, and it advances well beyond the foot of a slope. Lumb (1975) reported speeds of 30 m s^{-1} for debris avalanches in Hong Kong. The main characteristics of many slips that occur in the residual soils (mainly decomposed granite) of Hong Kong are the rapid fall of debris (once movement starts the whole mass separates from the main slope within minutes) and the shallow depth of the slide, usually less than 3 m. The ratio of thickness to length of the scar is usually less than 1.5. There is rarely any prior warning that a slip is imminent. The prime cause of failure is direct infiltration of rainwater into the surface zones of slopes, leading to soil saturation and its loss of effective cohesion. Debris avalanches are generally long and narrow, and frequently leave V-shaped scars tapering headwards. These gullies often become the sites of further movement.

Debris flows are distinguished from mudflows on the basis of particle size, the former containing a high percentage of coarse fragments, while the latter consist of at least 50% sand-size or less (Figure 4.4). Almost invariably, debris flows follow unusually heavy rainfall or the sudden thaw of frozen ground. These

Figure 4.4 Debris flow, Arthur's Pass, South Island, New Zealand.

flows are of high density, perhaps 60 to 70% solids by weight, and are capable of carrying large boulders. Like debris avalanches, they commonly cut V-shaped channels, at the sides of which coarser material may accumulate as the more fluid central area moves down-channel. Debris may move over many kilometres.

Mudflows may develop when a rapidly moving stream of storm water mixes with a sufficient quantity of debris to form a pasty mass (Figure 4.5). Because such mudflows frequently occur along the same courses, they should be kept under observation when significant damage is likely to result. Mudflows frequently move at rates ranging between 10 and 100 m min^{-1} and can travel over slopes inclined at 1° or less, although they usually develop on slopes with shallow inclinations, that is, between 5 and 15°. Skempton and Hutchinson (1969) observed that mudflows also develop along discretely sheared boundaries in fissured clays and varved or laminated fluvio-glacial deposits where the

Figure 4.5 House wrecked by mudflow, Westville, South Africa.

ingress of water has led to softening at the shear zone. Movement involves the development of forward thrusts due to undrained loading of the rear part of the mudflow, where the basal shear surface is inclined steeply downwards. A mudflow continues to move down shallow slopes due to this undrained loading which is implemented by frequent small falls or slips of material from a steep rear scarp on to the head of the moving mass. This not only aids instability by loading but it also raises the pore water pressures along the back part of the slip surface (Hutchinson and Bhandari, 1971; Bromhead, 1978).

An earthflow involves mostly cohesive or fine-grained material, which may move slowly or rapidly. The speed of movement is to some extent dependent on water content in that the higher the content, the faster the movement. Slowly moving earthflows may continue to move for several years. These flows generally develop as a result of a build-up of pore water pressure, so that part of the weight of the material is supported by interstitial water with consequent decrease in shearing resistance. If the material is saturated, a bulging frontal lobe is formed and this may split into a number of tongues, which advance with a steady rolling motion. Earthflows frequently form the spreading toes of rotational slides due to the material being softened by the ingress of water. Skempton and Hutchinson (1969) restricted the term 'earthflow' to slow movements of softened weathered debris, as forms at the toe of a slide. They maintained that movement was transitional between a slide and a flow, and that earthflows accommodated less breakdown than mudflows.

4.4 Landslides in soils

Displacement in soil, usually along a well-defined plane of failure, occurs when shear stress rises to the value of shear strength. The shear strength of the material along the slip surface is reduced to its residual value, so that subsequent movement can take place at a lower level of stress. The residual strength of a soil is of fundamental importance as far as the behaviour of landslides is concerned (Skempton, 1964; Skempton and Early, 1972), and in progressive failure (Bishop, 1971).

A slope in dry frictional soil should be stable provided its inclination is less than the angle of repose. Slope failure tends to be caused by the influence of water. For instance, seepage of groundwater through a deposit of sand in which slopes exist can cause them to fail. Failure on a slope composed of granular soil involves the translational movement of a shallow surface layer. The slip is often appreciably longer than it is deep. This is because the strength of granular soils increases rapidly with depth. If, as is generally the case, there is a reduction in the density of the granular soil along the slip surface, then the peak strength is reduced ultimately to the residual strength. The soil will continue shearing without further change in volume once it has reached its residual strength. Although shallow slips are common, deep-seated shear slides can occur in granular soils. They are usually due either to rapid drawdown or to the placement of heavy loads at the top of the slope.

In cohesive soils, slope and height are interdependent and can be determined when the shear characteristics of the material are known. Because of their water-retaining capacity (due to their low permeability), pore water pressures are developed in cohesive soils. These pore water pressures reduce the strength of an element of the failure surface within a slope in cohesive soil, and the pore water pressure at that point needs to be determined to obtain the total and effective stress. This effective stress is then used as the normal stress in a shear box or triaxial test to assess the shear strength of the clay concerned. Skempton (1964) showed that on a stable slope in clay the resistance offered along a slip surface, that is, its shear strength (s) is given by

$$s = c' + (\sigma - u)\tan \phi' \tag{4.2}$$

where c' = cohesion intercept, ϕ' = angle of shearing resistance (these are average values around the slip surface and are expressed in terms of effective stress), σ = total overburden pressure, u = pore water pressure. In a stable slope only part of the total available shear resistance along a potential slip surface will be mobilized to balance the total shear force (τ), hence

$$\Sigma\tau = \Sigma c'/F + \Sigma(\sigma - u)\tan \phi'/F \tag{4.3}$$

If the total shear force equals the total shear strength a slip can occur (that is, F = 1.0).

Cohesive soils, especially in the short-term conditions, may exhibit relatively uniform strength with increasing depth. As a result, slope failures, particularly short-term failures, may be comparatively deep-seated, with roughly circular slip surfaces. This type of failure is typical of relatively small slopes. Landslides on larger slopes are often non-circular failure surfaces following bedding planes or other weak horizons.

When shear failure occurs for the first time in an unfissured clay the undisturbed material lies very nearly at its peak strength, and it is the effective peak friction angle (ϕ'_p) and the effective peak cohesion (c'_p) that are used in analysis. In slopes excavated in fissured overconsolidated clay, although stable initially, there is a steady decrease in the strength of the clay towards a long-term residual condition. During the intermediate stages swelling and softening take place, due to the dissipation of residual stress and the ingress of water along fissures that open on exposure. Large strains can occur locally due to the presence of the fissures, and considerable non-uniformity of shear stress along the potential failure surface and local over-stressing leads to progressive slope failure. Indeed, Skempton and DeLory (1957) maintained that reorientation of platey minerals can occur along shear surfaces. In 1964, Skempton showed that if a clay is fissured, then initial sliding occurs at a value below peak strength. Residual strength, however, is reached only after considerable slip movement has taken place, so the strength relevant to first-time slips lies between the peak and residual values. Skempton accordingly introduced the term 'residual factor' (R), which he defined as:

$$R = \frac{\text{peak shear strength} - \text{mean shear stress at failure}}{\text{peak shear strength} - \text{residual shear strength}} \qquad (4.4)$$

The residual factor represents the proportion of the slip surface over which the strength has deteriorated to the residual value. The residual factor is therefore used in analysis where the residual strength is not developed over the entire slip surface. It allows the mean shear stress at failure to be used in the calculation of the factor of safety. Skempton gave a value for the residual factor of 0.08 for an unweathered till, while that for weathered London Clay ranged from 0.56 to 0.8. The value for the till indicates that slip would take place only when the applied shear stress approaches the peak value for the clay. The values for the London Clay show that from 56 to 80% of the length of the slip surface would have reached its residual strength by the time the slip occurred.

Chandler (1977) showed that where drainage is impeded in cohesive soils, the threshold slope for landsliding approximates to half the angle of residual shearing resistance (ϕ'_r). However, he went on to point out that as the normal stresses increase, the value of ϕ'_r frequently falls. This means that where there is a potential for deep-seated landsliding on clay slopes, the range of normal stress will be large and there will not be a unique value of ϕ'_r. As a consequence, larger landslides will move on flatter slopes than small slides, other factors being equal.

Bjerrum (1967) maintained that failures in overconsolidated plastic clays and clay shales are preceded by the development of a surface on which sliding is continuous. This mechanism of progressive failure gradually reduces the strength of the soil from the peak to the residual value. He went on to state that for this sliding surface to develop, then it is necessary that

- the internal lateral stresses are high enough to cause stress concentrations in front of an advancing sliding surface, where the shear stresses exceed peak shear strength;
- the clay contains enough recoverable strain energy to produce the required amount of expansion in the direction of sliding to strain the clay in the zone of failure; and
- the residual shear strength is relatively low compared with the peak shear strength.

Bjerrum (*ibid.*) also noted that the magnitude of lateral stresses in overconsolidated clays is largely dependent on strain energy. In this context, he recognized two types of clay, namely, those in which the strain energy is recovered simultaneously with a change of stress and those in which it is not. In the latter type of clay, strain energy is not immediately available, because diagenetic bonds were formed when the deposit was carrying maximum overburden. These bonds 'welded' the contacts between particles and prevented bent particles form straightening when the load was reduced by subsequent erosion.

Weathering, however, does lead to the breakdown of these bonds, when the clay is exposed at the surface. Hence the strain energy is liberated gradually. Consequently, the behaviour of overconsolidated clays depends on whether the diagenetic bonds are weak or strong. In the former case most of the strain energy is recovered on unloading. Swelling in the clay is more or less unrestricted and the ratio of horizontal to vertical effective stress increases during rebound. Weathering has little effect on the upper layer of such clay. By contrast, swelling is restricted in overconsolidated clays with strong diagenetic bonds because recoverable strain energy is not released on unloading. The horizontal effective stress is relatively small. Weathering, however, does give rise to substantial swelling and to an increase in effective stress parallel to the surface, with expansion occurring in the same direction.

Hence, the development of progressive failure in overconsolidated clays depends very much upon the time at which the stored recoverable strain energy is liberated. Those deposits with strong diagenetic bonds that are subjected to weathering are the most dangerous, since this leads to a large amount of stored strain energy being released. This in turn gives rise to high lateral stresses and, as remarked, to a notable expansion parallel to the surface. The behaviour of overconsolidated clay with weak bonds is more or less the same in unweathered and weathered condition and can prove suspect. If overconsolidated clay is unweathered, it is least dangerous.

Bjerrum (*ibid.*) illustrated this order of susceptibility by an analysis of 60 slides in overconsolidated clays. He found that 55% had occurred in the weathered material of those clays with strong diagenetic bonds, 35% in clays with weak bonds and the remainder in unweathered clays with strong diagenetic bonds. He also suggested that progressive failure becomes more likely as the plasticity index increases.

4.5 Landslides in rock masses

Landsliding of steep slopes in hard, unweathered rock (defined as rock with an unconfined compressive strength of 35 MPa and over) is largely dependent on the incidence, orientation and nature of the discontinuities present. It is only in very high slopes and/or weak rocks that failure in intact material becomes significant. Data relating to the spatial relationships between discontinuities afford some indication of the modes of failure that may occur, and information relating to the shear strength along discontinuities is required for use in stability analysis. As in soil, the shearing resistance of rock with a random pattern of jointing can be obtained from the Coulomb equation. The value of the angle of shearing resistance (ϕ) depends on the type and degree of interlock between the blocks on each side of the surface of sliding, but in such rock masses interlocking is independent of the orientation of the surface of sliding.

In a bedded and jointed rock mass, if the bedding planes are inclined, the critical slope angle depends upon their orientation in relation to the slope and the orientation of the joints. The relation between the angle of shearing resistance (ϕ) along a discontinuity, at which sliding would occur under gravity, and the inclination of the discontinuity (a) is important. If $a < \phi$ the slope is stable at any angle, while if $\phi < a$, then gravity would induce movement along the discontinuity surface and the slope would not exceed the critical angle, which would have a maximum value equal to the inclination of the discontinuities. It must be borne in mind, however, that rock masses are generally interrupted by more than one set of discontinuities.

Hard rock masses are liable to sudden and violent failure if their peak strength is exceeded in an excessively steep or high slope. On the other hand, soft materials, which exhibit small differences between peak and residual strengths, tend to fail by gradual sliding. The relative sensitivity of the factor of safety to the variation in importance of each parameter that influences the stability of slopes depends initially on the height of the slope. For example, Richards *et al.* (1978) graded each parameter concerned, in order of importance with respect to their effect on the factor of safety, in relation to slopes with heights of 10, 100 and 1000 m. The results are shown in Table 4.5 and Figure 4.6. The heights and angles of slopes in hard rocks can be estimated roughly from Figure 4.6. The joint inclination is always the most important

Table 4.5   Sensitivity of factor of safety to various parameters (after Richards et al., 1978).

Function affecting the factor of safety	Probable range of magnitude
Unit weight	0–300 kN m³
Cohesion	15–30 kPa
Water pressure	0–H metres
Friction angle	0–60°
Joint inclination	10–50°

Order of importance of functions

Rank	Slope height		
	10 m	100 m	1000 m
1	Joint inclination	Joint inclination	Joint inclination
2	Cohesion	Friction angle	Friction angle
3	Unit weight	Cohesion	Water pressure
4	Friction angle	Water pressure	Cohesion
5	Water pressure	Unit weight	Unit weight

parameter as far as slope stability is concerned. Friction is the next most important parameter for slopes of medium and large height, whereas unit weight is more important than friction for small slopes. Cohesion becomes less significant with increasing slope height, while the converse is true as far as the effects of water pressure are concerned.

The shear strength along a joint is mainly attributable to the basic frictional resistance that can be mobilized on opposing joint surfaces. Normally, the basic friction angle (ϕ_b) approximates to the residual strength of the discontinuity. An additional resistance is consequent upon the roughness of the joint surface. Shearing at low normal stresses occurs when the asperities are overridden, at higher confining conditions and stresses, then they are sheared through. Barton (1974) proposed that the shear strength (τ) of a joint surface could be represented by the following expression:

$$\tau = \sigma_n \tan (JRC \log_{10} (JCS/\sigma_n) + \phi_b) \tag{4.5}$$

where σ_n is the effective normal stress, JRC is the joint roughness coefficient and JCS is the joint compressive strength. The shear strength along a discontinuity is also influenced by the presence and type of any fill material and by the degree of weathering undergone along the discontinuity.

According to Hoek and Bray (1981), in most hard rock masses neither the angle of friction nor the cohesion is dependent upon moisture content to a significant degree. Consequently, any reduction in shear strength is almost solely attributable to a reduction in normal stress across the failure plane and it is

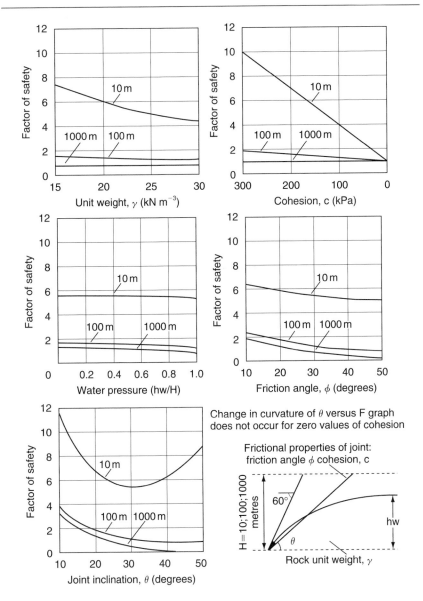

Figure 4.6 Sensitivity analysis for slope stability calculations: hw = height of water in slope; H = height of slope (after Richards et al., 1978).

water pressure, rather than moisture content, that influences the strength characteristics of a rock mass.

The principal types of failure that are generated in rock slopes are rotational, translational and toppling modes (Figure 4.7). Rotational failures normally

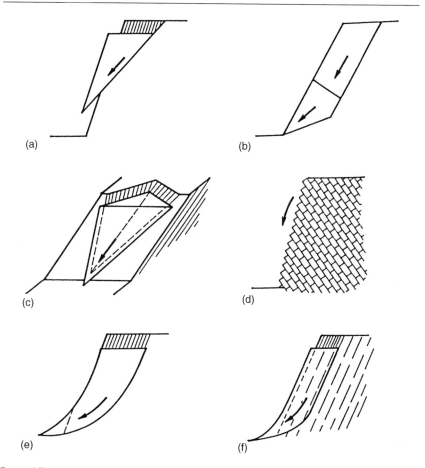

Figure 4.7 Idealized failure mechanisms in rock slopes: (a) plane; (b) active and passive blocks; (c) wedge; (d) toppling; (e) circular; (f) non-circular.

occur only in structureless overburden, highly weathered material or very high slopes in closely pointed rock. They may develop either circular or non-circular failure surfaces. Circular failures take place where rock masses are intensely fractured, or where the stresses involved override the influence of the discontinuities in the rock mass. Relict jointing, along which sliding may take place, may persist in highly weathered materials. These failure surfaces are often intermediate in geometry between planar and circular slides.

There are three kinds of translational failures: plane failure, active and passive block failure, and wedge failure. Plane failures are a common type of translational failure and occur by sliding along a single plane that daylights in the slope face (that is, the dip of the failure plane is lower than that of the slope). When considered in isolation, a single block may be stable. Forces imposed by unstable adjacent blocks may give rise to active and passive block failures (Figure 4.7). In wedge failure, two planar discontinuities intersect, the wedge so formed daylighting into the face. In other words, failure may occur if the line of intersection of both planes dips into the slope at an angle lower than that of the slope.

Toppling failure is generally associated with steep slopes in which the jointing is near vertical. It involves the overturning of individual blocks and is therefore governed by discontinuity spacing as well as orientation. The likelihood of toppling increases with increasing inclination of the discontinuities. Water pressure within discontinuities helps to promote the development of toppling.

Rockfalls have been referred to above. Failures involving relatively small volumes of rock often pose greater problems, particularly in areas where steep terrain is in close proximity to developed areas or transportation corridors. For instance, Martin (1988) reported that rockfalls, small rock slides and ravelling are the most chronic problems on transportation routes in mountainous areas of North America; millions of dollars are spent annually on maintenance and remedial measures to provide protection against such hazards.

4.6 A brief note on slope stability analysis

An analysis of stability should determine under what conditions a slope will remain stable, the stability being expressed in terms of the factor of safety. The stability improves as the value of the factor of safety increases above unity. A soil or rock mass under given loading should have an adequate factor of safety with respect to shear failure, and deformation under given loads should not exceed certain tolerable limits.

Analysis provides an evaluation of slope stability by means of quantitative assessment of slope stability behaviour and thereby offers an important input for the design and hazard assessment of slopes. Morgenstern (1992) distinguished between analyses used to determine pre-failure conditions and post-failure conditions. The purpose of pre-failure analysis is to evaluate the safety of a slope, whereas post-failure or back-analysis provides an explanation of a landslide event. The back-analysis provides the data for the design of any remedial measures.

Several methods are available for analysis of the stability of slopes. Most of these may be classed as limit equilibrium methods, in which the basic assumption is that the failure criterion is satisfied along the assumed path of failure. A free mass is taken from the slope and starting from known or assumed values of the forces acting upon the mass, calculation is made of the shear resistance required for equilibrium. This shear resistance is then compared with the estimated or available shear strength to give an indication of the factor of safety.

The two-dimensional methods of analysis have been summarized by Fredlund (1984) and Nash (1987). Computer programs are widely available that allow the use of the most rigorous methods. As Morgenstern (1992) pointed out, two-dimensional methods involve analytical solutions for simplified homogeneous profiles, and numerical methods, primarily based on the method of slices, are used for more complex sections. However, more recently three-dimensional methods of limit equilibrium analysis have been developed (Gens *et al.*, 1988). The use of three-dimensional methods reduces the degree of error in the results of analysis. The three-dimensional equivalent of the method of slices has been termed the method of columns and was developed by Hovland (1977). According to Morgenstern (1992), one of the most useful methods, which is an extension to three dimensions of Bishop's simplified method of analysis, has been proposed by Hungr (1987).

The drawbacks relating to the reliability of limit equilibrium analysis are attributable to difficulties inherent in site investigation, and material sampling and testing. The two most important parameters in limit equilibrium analysis are pore water pressure distribution and the shear strength of the material. A site investigation and associated laboratory testing programme cannot truly convey the operational conditions that govern stability. Hence the pore water pressure distributions and shear strength used in an analysis must be modified in some representative manner. Pore water pressures may be affected by rainfall, and their equilibration in materials of low permeability is of long duration. This, together with the long-term reduction in strength of many earth materials, poses problems in terms of the values to be used for stability analyses. In fact, the peak strength as obtained by laboratory testing is not operational over the complete slope when it fails, part of the slope having been stressed beyond the peak. As such, Morgenstern (1992) indicated that deformational analyses are required for this type of problem, coupled with limit equilibrium analyses to obtain an overall assessment of stability. However, in most instances where slope stability has to be determined, if the factor of safety is chosen correctly, then the deformation will be acceptable and the slope will behave as predicted. Progressive failure resulting from strain weakening can be handled by finite element methods (Chan and Morgenstern, 1987; Potts *et al.*, 1990).

Translational or plane failures frequently occur in rock masses. The forces acting on a block in a translational slide (Figure 4.8) are the gravitational weight (W), the disturbing force acting down the slide plane ($W \sin \beta$), the normal force acting across the sliding plane ($W \cos \beta$) and the shearing resistance of the surface between the block and the plane (R). According to Hoek (1970), the shearing resistance is given by:

$$R = cA + W \cos \beta \tan \phi \qquad (4.6)$$

where A is the base area of the block. The condition of limiting equilibrium occurs when:

$$W \sin \beta = cA + b \cos \beta \tan \phi \qquad (4.7)$$

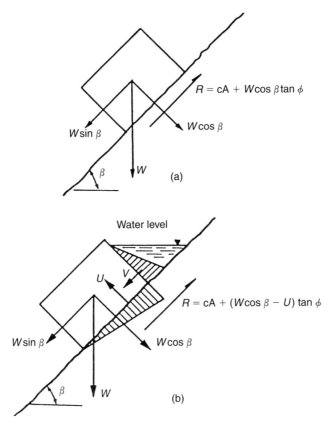

Figure 4.8 (a) Forces acting on a block resting on an inclined plane: W = gravitational weight of the block; $W \sin \beta$ = disturbing force acting down the plane; $W \cos \beta$ = normal force acting across the sliding plane; R = shear resistance of the surface between the block and the plane. (b) Forces acting on a block resting on an inclined plane with water trapped behind the block (after Hoek, 1970).

However, the above expression does not take into account the presence of water within a slope. If, for instance, water is trapped behind the upper face of the block, water pressure distributions are set up along the face and base of the block. The resultant forces U and V act in the directions shown in Figure 4.8, and the limiting equation becomes:

$$W \sin \beta + V = cA + (W \cos \beta - U) \tan \phi \qquad (4.8)$$

The limiting equilibrium occurs at slope angle a, which is less than the angle of friction, as both water forces (U, V) act in directions that tend to cause instability. The factor of safety (F) then becomes:

$$F = \frac{cA + (W \cos \beta - U) \tan \phi}{W \sin \beta + V}$$
(4.9)

There are an infinite number of possibilities for producing limiting equilibrium when sliding occurs on two planes with different frictional values. The factor of safety for wedges is often taken as the value that yields the same factor of safety on both planes. According to Richards *et al.* (1978), this is very arbitrary and probably unrealistic. They suggested that it was more useful to consider the sensitivity of the wedge to possible changes in shear strength characteristics and water pressure conditions.

Hoek (1971) maintained that the condition for toppling is defined by the position of the weight vector in relation to the base of the block. If the weight vector, which passes through the centre of gravity of the block, falls outside the base of the block, toppling will occur (Figure 4.9).

One of the most important aspects of rock slope analysis is the systematic collection and presentation of geological data so that they can be readily evaluated and incorporated into stability analyses. Spherical projections provide a convenient method for the presentation of geological data. The use of spheri-

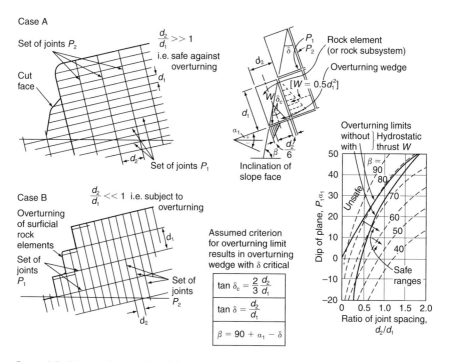

Figure 4.9 Criteria for toppling failure in two dimensions in terms of the mean spacing and orientation of the joint sets, the slope angle and the presence of hydrostatic forces (after Richards et al., 1978).

cal projections, commonly the Schmidt net, means that the traces of planes on the surface of the 'reference sphere' can be used to define the dips and dip directions of the planes. In other words, the inclination and orientation of a particular plane are represented by a great circle or a pole, normal to the plane, which are drawn on an overlay placed over the stereonet. Hoek and Bray (1981) illustrated how to plot great circles and poles using the stereonet technique and have helped to pioneer the use of this technique for analysis of the stability of rock slopes. Different types of slope failure are associated with different geological structures, and these give different general patterns when analysed by stereonet methods (Figure 4.10).

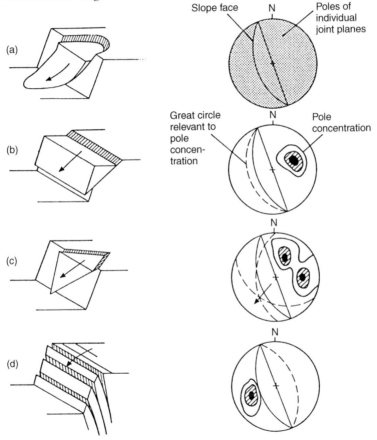

Figure 4.10 Representation of structural data concerning four possible slope failure modes plotted on equal-area stereonets as pole, which are contoured to show relative concentration, and great circles: (a) circular failure in heavily jointed rock with no identifiable structural plan; (b) plane failure in highly ordered structure such as slate; (c) wedge failure on two intersecting sets of joints; (d) toppling failure caused by steeply dipping joints (after Hoek and Bray, 1981).

4.7 Monitoring slopes

Dunnicliff (1992) emphasized that a monitoring programme for landslides should begin with defining the objective and end with planning how the results of the programme are to be implemented. He noted a number of steps that should be followed in the planning and execution of such a programme. One of the primary requirements is to assess the existing conditions, in particular to determine whether or not the landslide is active, and if it is, where it is moving and the rate at which it is moving. Obviously, the causes of movement must be determined if satisfactory remedial treatment is to be undertaken. If the assessment indicates that the landslide is active, or potentially unstable, then three choices are available. The first is to do nothing and accept the consequences of failure. Second, a monitoring programme can be put into effect to warn of instability so that remedial treatment can be undertaken prior to failure occurring. Third, the slope can be stabilized and a monitoring programme installed to verify that the slope is stable.

Small movements usually precede slope failure, particularly a catastrophic failure, and accelerating displacement frequently precedes collapse. If these initial small movements are detected in sufficient time, remedial action can be taken to prevent or control further movement. A slope monitoring system provides a means of early warning, and involves the use of sensitive instruments. Other adverse conditions that give rise to instability, notably excess pore water pressures, also require recording.

When there is a lack of adequate data uncertainties are likely to arise in design. Under such circumstances, if the stability of a slope is in doubt, then the expense of a monitoring programme may be justified, provided that remedial measures following the detection of incipient failure are feasible, and that the cost of monitoring and remedial action is less than the cost if a slope failure were to occur (Franklin and Denton, 1973). Even if a complete picture of the ground conditions is available, the analytical methods may not be able to deal with the complexity of a real situation. Consequently, data must be simplified into an idealized model, with resulting loss of accuracy. Such uncertainties are normally taken account of in the selection of the factor of safety. Monitoring can justify the use of a lower factor of safety than would otherwise be permissible, provided that it is accompanied by contingency plans for remedial action should the slope in question prove unstable. Accordingly, the cost of a monitoring system has to be measured against the cost of operating at an uneconomically high safety factor, necessitating either flatter slopes or expensive remedial work. The total value of the project concerned and the cost and effect on the project if slopes fail also have to be considered.

One of the first steps in the planning and design of a monitoring system is to assess the extent and depth of potentially unstable rock material, and to determine the factors of safety against sliding for various modes of failure. This indicates whether there is a problem of slope stability, and aids the choice of instrumentation and its location within the slope to be made.

4.7.1 Monitoring movement

Monitoring of movement provides a direct check on the stability of a slope. Instruments indicate the location, direction and maximum depth of movements and their results help to determine the extent and depth of treatment that is necessary. What is more, the same instruments can then be used to determine the effect of this treatment. Monitoring of surface movement can be done by conventional surveying techniques, the use of electronic distance measurement or laser equipment providing accurate results. Surveys should be designed to suit the topography and the anticipated directions of movements. Surveying should extend beyond the limits of possible movement into the surrounding stable area. In this way, any development of surface strain in advance of the appearance of tension cracks can be detected, as can any toe heave. Automated slope monitoring procedures, using total station surveying instruments, can be programmed to take various measurements across a slope (Tran-Duc *et al.*, 1992).

Surface contouring and coordinate-fixing photogrammetric methods generally give a lesser accuracy than conventional survey or electronic distance measurement but offer a complete picture rather than a set of pre-located targets (Planicka and Nosek, 1970). However, precise results can be obtained when close-up photographs are taken from ground stations and are measured in a stereo comparator. Movements may be revealed by an examination of a sequence of photographs taken at suitable time intervals. Photographs can be used to evaluate pre-existing ground topography, for back-analysis of previous landslips and as a basis for engineering geological mapping.

The appearance of tension cracks at the crest of a slope may provide the first indication of instability. Crack measurements, that is, their width and vertical offsets, should be taken since they may provide an indication of slope behaviour. Dunnicliffe (1988) described a number of ways in which cracks can be measured. These included using a survey tape, a tensional wire crack gauge or a surface extensometer. The latter consists of a sleeved rod that spans between anchor points on each side of the tension crack, and a mechanical or electrical transducer.

Single-point tiltmeters can also be used to monitor surface movements when the surface of the landslide has a rotational component. Portable tiltmeters can be used with a series of reference places, or tiltmeters can be left in place and connected to a data logger. Dunnicliffe (1992) mentioned multi-point liquid-level gauges that have been employed to detect slope movements.

Measurements of subsurface horizontal movements are more important than measurements of subsurface vertical movements. Subsurface movements can be recorded by using extensometers, inclinometers and deflectometers. Borehole extensometers are used to measure vertical displacement of the ground at different depths. Fixed borehole extensometers include the single- and multi-rod types. A single-rod extensometer is anchored in a borehole, and movement between the rod and the reference sleeve is monitored. However, the rod can bind in the sleeve if significant deformation occurs perpendicular to the length of the extensometer. Their use is therefore limited to where the borehole is almost parallel to the expected direction of movement. Hence it is rarely practical to install them from the ground surface. Multiple-rod installations monitor displacements at various depths using rods of varying lengths. Each rod is isolated by a close-fitting sleeve and the complete assembly is grouted into place, fixing the anchors to the ground while allowing free movement of each rod within its sleeve (Figure 4.11a). A precise borehole extensometer essentially consists of circular magnets embedded in the ground, which act as markers, and reed switch sensors move in a central access tube to locate the positions of the magnets (Figure 4.11b). A slope extensometer is a multi-point borehole extensometer that uses

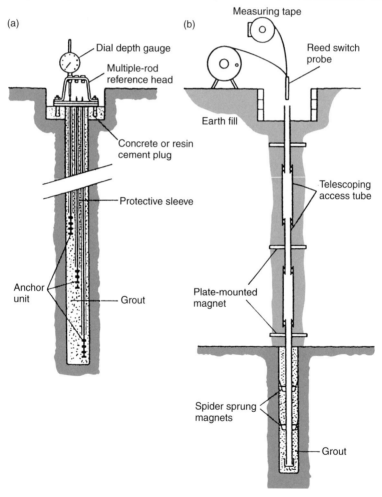

Figure 4.11 (a) Multiple-rod extensometer; (b) magnetic-probe extensometer (courtesy of Soil Instruments Ltd).

tensioned wires instead of rods to monitor deformation perpendicular to the axis of the borehole. Up to 10 anchors and wires can be installed in a borehole. At the head of the extensometer each wire passes over a pulley and is attached to a weight. When shear deformation occurs between two anchors, there is no vertical movements of the weights for those anchors above the shear zone. On the other hand, weights attached to anchors below the shear zone move downwards as the wires are displaced by shear. However, the precision with which the shear zone can be located is much less than for inclinometers.

An inclinometer is used to measure horizontal movements below ground (Mikkelsen, 1996). High-accuracy inclinometer measurements frequently represent the initial data relating to subsurface movement. Inclinometers can detect differential movements of 0.17 to 3.4 mm per 10 m run of hole. Inclinometers designed for permanent installation in a hole usually comprise a chain of pivoted rods. Angular movements between rods can be measured at the pivot points. Another type of permanently installed inclinometer uses a flexible metal strip on to which resistance strain gauges, which record any bending in the strip induced by ground movements, are bonded. Fixed-position inclinometers monitor differential lateral movement between the borehole collar and a deep datum, and are most frequently used in slope stability work. Large-diameter inclinometer casing should be used for monitoring landslides so as to provide a longer period over which readings can be taken as shear deformation occurs. Probe inclinometers are inserted into a special casing in a borehole each time a set of readings is required (Figure 4.12). They incorporate a pendulum, the deflection of which indicates movement. Automatic remote winching systems have been developed, which permit the inclinometer probe to be lowered and raised at predetermined intervals and the data transferred to truck-mounted recorders (Lollino, 1992). An in-place inclinometer consists of a series of gravity-sensing transducers (tiltmeters), positioned at intervals along the casing, joined by articulated rods. Borehole deformation data are determined from the distances between the transducers and the measured changes in inclination.

Multiple deflectometers operate on a similar principle to in-place inclinometers, but rotation is measured by angle transducers instead of tilt transducers. They are usually installed within inclinometer casing. These deflectometers can detect shear deformation across a borehole at any inclination.

4.7.2 Monitoring load

Anchors, rock bolts and retaining walls, although designed to a prescribed working load, only develop this load as the material they support starts to move against them. Monitoring of loads and pressures indicates whether the support system has been adequately designed and can also show whether the slope is progressing towards a more stable or an unstable condition. Loads on rock anchors and rock bolts can be monitored by load cells.

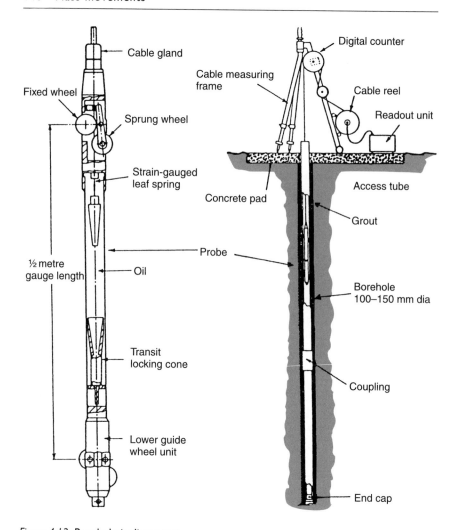

Figure 4.12 Borehole inclinometer.

Contact pressures on retaining walls can be recorded by pressure cells (Thomas and Ward, 1969). A flat jack directly records changes in pressure. Stress may be measured directly by a hydraulically operated diaphragm earth pressure cell such as the Glotzl cell.

4.7.3 Monitoring groundwater

Groundwater is one of the most influential factors governing the stability of slopes. Instability problems may be associated with either excessive discharge or excessive pore water pressure. Pore water pressures are recorded by a piezometer, the simplest type comprising a standpipe installed in a borehole (Figure 4.13). If the minimum head recorded is less than 8 m below ground level, 'closed system' piezometers connected to mercury manometers are normally used. Pressure transducers are necessary where greater heads have to be measured, especially in low-permeability ground. These instruments respond to changes of pressure acting on a flexible diaphragm by recording diaphragm deflections, which are converted to pressure values. Only very small quantities of water are required to produce full-scale deflection, and such piezometers are

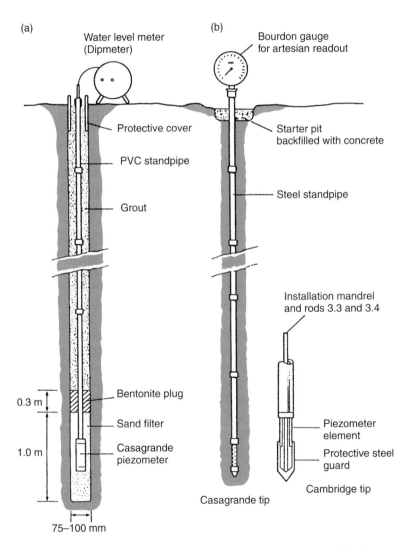

Figure 4.13 Standard piezometers: (a) borehole standpipe piezometer; (b) drive-in standpipe piezometer (courtesy of Soil Instruments Ltd).

therefore especially helpful when an almost instantaneous response is required.

There are important differences between the monitoring of water pressures in rock and in soil. Usually in rock the majority of flow takes place via discontinuities rather than through intergranular pore space. The predominance of fissure flow means that piezometer heads in rock slopes often vary considerably from point to point and therefore a sufficient number of piezometers must be installed to define the overall conditions. They should be located with reference to the geology, especially with regard to the intersection of major discontinuities in rock masses. This can be facilitated by examining the fracture index, by inspecting the drillhole with a television camera, by packer testing, or by logging the velocity of flow in the drillhole using micro-propeller or dilution methods. The piezometer test section in rock, that is, the permeable filter material between sections of grouted hole, may need to be as long as 4 m in order to incorporate a representative number of water-bearing fissures.

4.7.4 Monitoring acoustic emissions or noise

Movements in rock or soil masses are accompanied by the generation of acoustic emissions or noise. The detection of acoustic emissions is most effective when the amplitude of the signals is high. Hence detection is more likely in rock masses or cohesionless soils than cohesive soils. Obviously, when a slope collapses noise is audible, but sub-audible noises are produced at earlier stages in the development of instability. Normally, the rate of these micro-seismic occurrences increases rapidly with the development of instability. Such noises can be picked up by an array of geophones located in the vicinity of the slope or in shallow boreholes. Most movements generating noise originate near or along the plane of failure, so seismic detection helps to locate the depth and extent of the surface of sliding.

4.8 Landslide hazard, investigation and mapping

Landslide hazard refers to the probability of a landslide of a given size occurring within a specified period of time within a particular area. The associated risk is the associated loss of lives or damage to property. Consequently, landslide hazard has to be assessed before landslide risk can be estimated.

Varnes (1984), writing on landslide hazard, noted that slope failures are normally not as spectacular or as costly as major floods or earthquakes, yet they are more widespread. Furthermore, much of the damage and sometimes a significant proportion of the loss of life due, for example, to earthquakes is attributable to landslides. Indeed, Schuster and Fleming (1986) estimated that landslides in the United States cause between 25 and 50 deaths and $1 to $2 billion in economic losses annually. The costs of landslides are both direct and indirect, and range from the expense of clean-up operations and repair or replacement of structures to lost tax revenues, and reduced productivity and property values. Nonetheless, landslides are considered to be one of the most potentially predictable of geological hazards, and slope failures tend to affect discrete areas of land. Furthermore, landslides, of all the geological hazards, are perhaps the most amenable to avoidance, prevention and correction measures. Be that as it may, and continuing with the United States as an example, Wold and Jochim (1989) pointed out that in spite of the availability of successful techniques for landslide management and control, landslide loses are increasing. This is primarily because of the increasing pressure of development in areas of hazardous terrain, as well as the failure of state authorities and private developers to recognize landslide hazards and so apply appropriate measures for their mitigation. This is in spite of the fact that there is overwhelming evidence that landslide hazard mitigation programmes serve both public and private interests by saving the cost of implementation many times over.

Investigations of slope stability involves, according to Crozier (1984), selection of specific criteria upon which stability assessment is to be based; recognition and measurement of field evidence for instability; assessment and classification of the degree of stability; and mapping of stability conditions. If a large area is involved, then field investigation and mapping are divided into land unit areas. In terms of the assessment of stability, the frequency with which a slope fails, together with the potential for failure, are fundamental criteria, as are the magnitude, rate and type of movement. Crozier recognized six stability classes based on frequency or potential landslide criteria (Table 4.6). Different land-use activities can tolerate different degrees of landslide activity.

Table 4.6 Slope stability classification: frequency and potential criteria (after Crozier, 1984).

Class I	Slopes with active landslides. Material is continually moving, and landslide forms are fresh and well defined. Movement may be continuous or seasonal.
Class II	Slopes frequently subject to new or renewed landslide activity. Movement is not a regular, seasonal phenomenon. Triggering of landslides results from events with recurrence intervals of up to five years.
Class III	Slopes infrequently subject to new or renewed landslide activity. Triggering of landslides results from events with recurrence intervals greater than five years.
Class IV	Slopes with evidence of previous landslide activity but which have not undergone movement in the preceding 100 years.
	Subclass IVa: Erosional forms still evident,
	Subclass IVb: Erosional forms no longer present – previous activity indicated by landslide deposits
Class V	Slopes which show no evidence of previous landslide activity but which are considered likely to develop landslides in the future. Landslide potential indicated by stress analysis or analogy with other slopes.
Class VI	Slopes which show no evidence of previous landslide activity and which by stress analysis or analogy with other slopes are considered stable.

Hutchinson (1988) maintained that it is important to recognize landslides as first-time slides or those that occur along previously formed slip surfaces on the one hand and whether they develop under undrained, drained–undrained or drained conditions. He indicated that the degree of brittleness largely controls run-out. Brittleness and run-out are usually greatest in first-time slides, particularly if they also involve drained–undrained behaviour. On pre-existing shear zones brittleness is normally low or zero. Hence renewal of movement along such shear zones generally takes place slowly and is accompanied by little run-out.

Although particular slopes that lack a history of previous landslides can be designated unstable, most stability assessment is based on the recognition of features that indicate former or active mass movement. Crozier (1984) listed the features that indicate whether landsliding areas can be regarded as active or inactive (Table 4.7). However, different zones within a landslide may become stabilized at different times. As slopes that have moved previously are likely

Table 4.7   Features indicating active and inactive landslides (after Crozier, 1984).

Active	Inactive
Scarps, terraces and crevices with sharp edges	Scarps, terraces and crevices with rounded edges
Crevices and depressions without secondary infilling	Crevices and depressions infilled with secondary deposits
Secondary mass movement on scarp faces	No secondary mass movement on scarp faces
Surface of rupture and marginal shear planes show fresh slickensides and striations	Surface of rupture and marginal shear planes show old or no slickensides and striations
Fresh fractured surfaces on blocks	Weathering on fractured surfaces on blocks
Disarranged drainage system; many ponds and undrained depressions	Integrated drainage system
Pressure ridges in contact with slide margin	Marginal fissures and abandoned levees
No soil development on exposed surface of rupture	Soil development on exposed surface of rupture
Presence of fast-growing vegetation	Presence of slow-growing vegetation
Distinct vegetation differences 'on' and 'off' slide	No distinction between vegetation 'on' and 'off' slide
Tilted trees with no new vertical growth	Tilted trees with new vertical growth above inclined trunk
No new supportive, secondary tissue on trunks	New supportive, secondary tissue on trunks

to move again, the identification of landslide features plays an integral part in the prediction of potential instability. As pointed out above, the occurrence of a landslide can rarely be attributed to one factor alone, so it is important to establish the critical combination of factors.

Forecasting the time of occurrence of a failure is generally considerably more difficult than assessing the degree of stability of a slope, because of the precision with which the instability process needs to be known. Unfortunately, the history of landslide activity is rarely sufficiently well documented for direct assessment of landslide recurrence intervals. The recurrence interval is the average time between landslides of the same magnitude. If the variation in recurrence interval is plotted against the magnitude of the event, then a recurrence curve is produced. Methods of determining recurrence intervals include the use of radio-isotopes, infrared photography, dendrochronology, debris mapping and tree damage assessments. Recurrence intervals of landslides are often difficult to evaluate unless they can be correlated with some other factor, such as rainfall, for which recurrence data exist. Nonetheless, catastrophic phenomena with a recurrence interval in excess of 100 years are extremely difficult to evaluate. In the case of rainstorm-triggered events it may be possible, by examination of the record, to establish the minimum rainfall intensity required to generate landslides. In this way, the recurrence interval can be derived from the meteorological record. Even so, this method suffers drawbacks in terms of the duration of climatic records available and the assumption of climatic constancy. Furthermore, other factors influence slope stability.

The magnitude of movement can be derived from aerial photographs of a landslide or from field investigation. However, in many unstable areas, the form and limits to the erosional and depositional parts of a landslide are difficult to define. As a result, accurate measurements of volume cannot be determined.

The rate and depth of movement along with the amount of deformation are important criteria in determining the stability of an area for land use. Rates of movement that exceed 1.5 m annually probably cannot be tolerated by buildings or routeways.

Most landslide investigations are local and site-specific in character, being concerned with establishing the nature and degree of stability of a certain slope or slope failure or group of these. Such investigations involve desk studies; geomorphological mapping from aerial photographs or satellite imagery, as well as mapping in the field; subsurface investigation by trenches, shafts or boreholes; sampling and testing, especially of slip-surface material; monitoring of surface movements; monitoring pore water pressures with piezometers; and analysis of data obtained. There are a number of data sources from which background information can be obtained. These include reports, records, papers, topographical and geological maps, and aerial photographs and remote-sensing imagery. Field investigation, monitoring, sampling and laboratory testing may provide more accurate and therefore more valuable data.

Existing landslides can be mapped directly, primarily by use of aerial photographs or imagery, with field checks made when necessary. The simplest form of landslide mapping is a record of landslide locations. Landslides should be classified as active or inactive, historical data being of importance, as well as recent evidence of movement and freshness of form. Problems of recognition of ancient landslides may arise as they may have little surface expression or be hidden by a subsequently developed slope mantle.

The chief advantage of aerial photography over field investigation is that terrain characteristics can be seen in relation to each other over a large area at the same time. This permits the boundaries of large landslides to be determined, together with an appreciation of the environmental setting in which the landslide occurs. Photographic coverage made at intervals over a number of years facilitates the determination of changes in topography and allows the approximate establishment of times when landslide events occurred. The data derived from aerial photographs should be checked in the field.

Digitization of data from aerial photographs or remote-sensing imagery enables this to be handled in geographical information systems (GIS), through which the landslide-controlling factors can be evaluated (Figure 4.14; Carrara *et al.*, 1990). Expert systems and artificial intelligence are also being employed (Wislocki and Bentley, 1991).

Quantitative results appear to be obtainable in forested areas by using airborne vertical laser sensing. The airborne laser topography profiler (ALTP) system is operated from a helicopter with a laser pulse frequency of 2 kHz. This provides a reading about every 20 mm along the flight path, the aim being to penetrate the forest canopy with a proportion of readings.

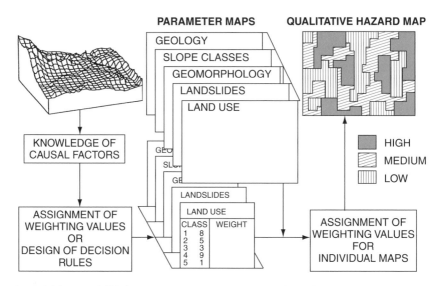

Figure 4.14 Use of GIS for qualitative map combination (after Soeters and van Westen, 1996).

Hutchinson (1992) suggested that the terrain evaluation approach seemed the best way of extrapolating available subsurface data and ensuring that physical insights are fully considered in arriving at a landslide hazard assessment. In Hong Kong, terrain evaluation constitutes an important part of the GASP (Geotechnical Areas Studies Programme), in which GLUM (Geotechnical Land Use Map) classes define areas of varying geotechnical limitations on development (Brand, 1988).

A useful starting point in landslide investigation, as well as an efficient method of data collection, is provided by the use of checklists, such as suggested by Cooke and Doornkamp (1990). A landslide inventory form was devised by Carrara and Merenda (1976) for use in mapping programmes (Figure 4.15). The principal advantage of using checklists or inventories is that each of the main categories of influencing and controlling parameters can be systematically examined. It also enables each separate slope unit to be classified according to a stability rating. The checklist can be used either during an investigation of aerial photographs or during a field survey and provides a systematic examination of the main factors influencing mass movement. The more boxes that are ticked on the right-hand side of the checklist, the more the slope concerned is approaching an unstable state.

Because most landslides occur in areas previously affected by instability and because few occur without prior warning, Cotecchia (1978) emphasized the importance of carrying out careful surveys of areas that appear potentially unstable and of making systematic records of the relevant phenomena. He provided a review of the techniques involved in mapping mass movements, as well as itemizing which data should be included on such maps. He maintained that the ultimate aim should be the production of maps of landslide hazard zoning. Landslide hazard maps delineate areas that probably will be affected by slope instability within a given period of time. Landslide risk maps attempt to quantify the vulnerability of an area either in relation to the probability of occurrence of a landslide or to the likelihood of damage to property or injury or death to person.

Generally, the purpose of mapping landslide hazards is to locate problem areas and to help understand why, when and where landslides are likely to occur. Accordingly to Varnes (1984), three basic principles have guided the production of landslide hazard zonation maps. The first assumption involves the concept that slope failures in the future will take place for reasons similar to those that gave rise to failures in the past. In fact, the evidence of past slope instability is frequently the best guide to future behaviour of a slope. However, the absence of landslides in the past or present does not mean that they will not occur in the future. Second, the basic causes of slope instability are fairly well known, so most can be recognized and mapped. It is often possible to estimate the relative contribution to slope instability of the conditions and processes that are responsible for slope instability once they have been identified. Third, in this way the degree of potential hazard can be assessed, depending on the

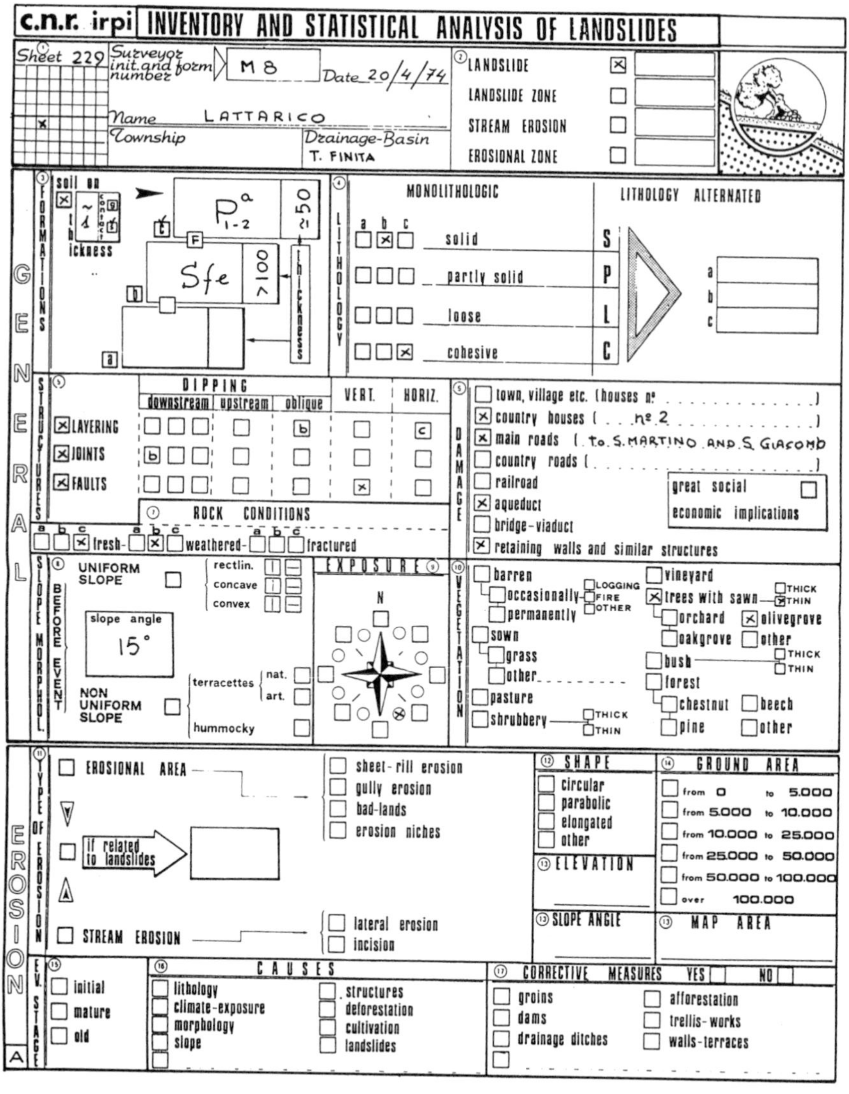

Figure 4.15 Form for data collection with an example of its use (after Carrara and Merenda, 1976).

correct recognition of the number of failure-inducing factors present, their severity and interaction. In addition, Hutchinson (1992) suggested that the various types of landslide can generally be identified and classified morphologically, geologically and geotechnically. Data processing can range from subjective evaluation to sophisticated processing by computers.

Walker *et al.* (1987) itemized a number of questions that require answers

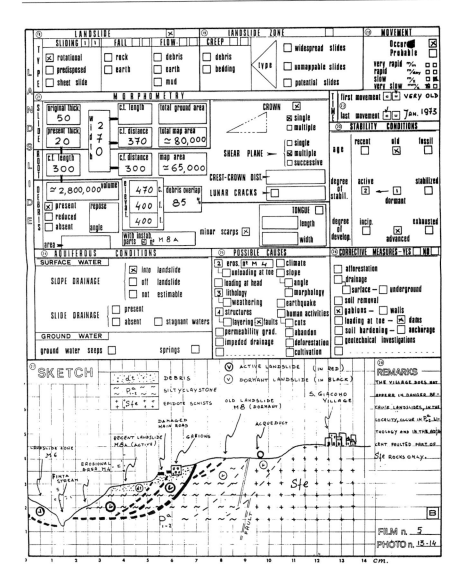

Figure 4.15 Continued

when designing a landslide hazard mapping scheme. These included, should the map be wholly concerned with actual or potential landslides; should the emphasis be placed on the nature of the slope failure, the extent of instability, the frequency of occurrence or the consequences of slope failure; and should the map be concerned with hazard or risk? The answers help to determine how extensive the investigation needs to be and the scale at which the mapping will be undertaken. Regional maps tend to be general-purpose, whereas community maps or site maps are often concerned with more specific problems. In the latter case, it is more cost-effective to represent only the necessary data for the particular purpose.

Many landslide maps show only the hazards known at a particular time, whereas others provide an indication of the possibility of landslide occurrence (Figure 4.16). The latter are landslide-susceptibility maps and involve some estimation of relative risk. An assessment of the level of risk attributable to landslide occurrence for a particular area involves classification of the data obtained into risk groups. Most risk classifications recognize low, medium and high levels of risk, but the categories are usually poorly defined. Walker *et al.* (*ibid.*) provided a summary of a number of risk classifications and stability ratings of slopes that have been used in Australia. However, these examples tend to be specific to the areas for which they were developed and so cannot be applied readily to other areas.

Huma and Radulescu (1978) described the use of computer techniques to produce maps of slope stability. The necessary data were obtained from aerial photographs, as well as field and laboratory investigations. The data included a survey of the lithology and mechanical characteristics of the rocks concerned, their structural and hydrogeological conditions, the slope angle, and the amount of vegetation cover and exposure. A computer was then used to assess the data in terms of slope stability and to plot a map of land stability zones.

Multivariate landslide susceptibility maps produced by the statistical evaluation of the physical factors that influence slope instability are generally based on grid cells and produced by computer. Cells are usually square with sides between 60 and several hundred metres in length. The end-result of the assessment of the various data is normally a numerical rating for each cell, which forms the basis of landslide hazard assessment (Figure 4.17). However, grid cells of fixed size have the disadvantage of often relating poorly to geomorphological slope units. Consequently, later work, facilitated by the use of GIS, has associated land characteristics with geomorphological units (Carrara *et al.*, 1990). Discriminant analysis has been used, employing combinations of measured parameters, to distinguish stable from unstable slopes (Payne, 1985).

Moser (1978) suggested that various geotechnical maps could be produced in order to make an assessment of the necessary treatment of mass movements in upland areas. These included maps showing the classification of mass movements, the type and thickness of soil, and the type of bedrock (in both latter cases the geotechnical properties of the materials are investigated).

Effective landslide hazard management has done much to reduce economic and social losses due to slope failure by avoiding the hazards or by reducing the damage potential (Schuster, 1992). This has been accomplished by restrictions placed on development in landslide-prone areas; application of excavation grading, landscaping and construction codes; use of remedial measures to prevent or control slope failures; and landslide warning systems.

Future policy should develop and promote techniques and initiate hazard recognition and reduction schemes as a preventive measure. Risk reduction can be achieved with reference to either the process system or the degree of vulnerability of the land use. The final choice of risk reduction measures will depend on the type of development, either in existence or proposed, and the type, magnitude and time scale of the hazardous process.

Kockelman (1986) dealt with some of the techniques used to reduce landslide hazards (Table 4.8). These may be used in a variety of combinations to help to solve both existing and potential landslide problems. The techniques are generally applicable to all types of surface ground failure, including flows, slides and falls. The effectiveness of each hazard reduction technique varies with time, place and persons involved in the planning and implementing of the

Table 4.8   Some techniques for reducing landslide hazards (after Kockelman, 1986).

Discouraging new developments in hazardous areas by:
   Disclosing the hazard to real-estate buyers
   Posting warnings of potential hazards
   Adopting utility and public facility service area policies
   Informing and educating the public
   Making a public record of hazards

Removing or converting existing development through:
   Acquiring or exchanging hazardous properties
   Discontinuing non-conforming uses
   Reconstructing damaged areas after landslides
   Removing unsafe structures
   Clearing and redeveloping blighted areas before landslides

Providing financial incentives or disincentives by:
   Conditioning federal and state financial assistance
   Clarifying the legal liability of property owners
   Adopting lending policies that reflect risk of loss
   Requiring insurance related to level of hazard
   Providing tax credits or lower assessments to property owners

Regulating new development in hazardous areas by:
   Enacting grading ordinances
   Adopting hillside development regulations
   Amending land-use zoning districts and regulations
   Enacting sanitary ordinances
   Creating special hazard reduction zones and regulations
   Enacting subdivision ordinances
   Placing moratoriums on rebuilding

Protecting existing development by:
   Controlling landslides and slumps
   Controlling mudflows and debris flows
   Controlling rockfalls
   Creating improvement districts that assess costs to beneficiaries
   Operating monitoring, warning, and evacuating systems

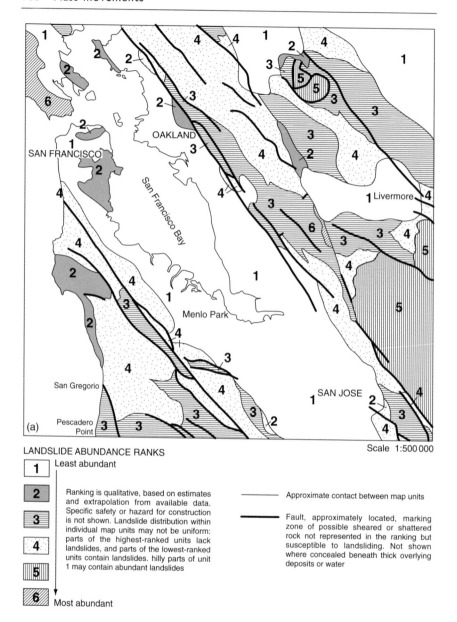

LANDSLIDE ABUNDANCE RANKS   Scale 1:500 000

| 1 | Least abundant |

| 2 | Ranking is qualitative, based on estimates and extrapolation from available data. Specific safety or hazard for construction is not shown. Landslide distribution within individual map units may not be uniform: parts of the highest-ranked units lack landslides, and parts of the lowest-ranked units contain landslides. hilly parts of unit 1 may contain abundant landslides |

—————— Approximate contact between map units

—————— Fault, approximately located, marking zone of possible sheared or shattered rock not represented in the ranking but susceptible to landsliding. Not shown where concealed beneath thick overlying deposits or water

| 6 | Most abundant |

Figure 4.16 (a) Map of relative abundance of landslides in the San Francisco Bay region (from Radbruch and Wentworth, 1971).

programme for reducing the hazard. The control of the landslide hazard system is easier for new developments as vulnerability can be restricted.

There are several methods of discouraging development in hazardous areas. These include disclosing hazards to real estate buyers, posting warning signs,

(b)

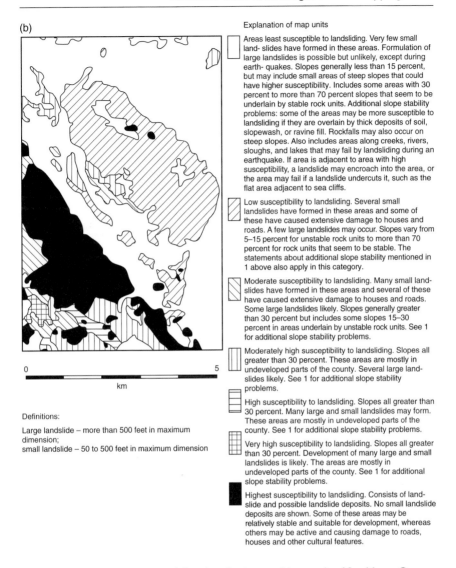

Explanation of map units

Areas least susceptible to landsliding. Very few small land- slides have formed in these areas. Formulation of large landslides is possible but unlikely, except during earth- quakes. Slopes generally less than 15 percent, but may include small areas of steep slopes that could have higher susceptibility. Includes some areas with 30 percent to more than 70 percent slopes that seem to be underlain by stable rock units. Additional slope stability problems: some of the areas may be more susceptible to landsliding if they are overlain by thick deposits of soil, slopewash, or ravine fill. Rockfalls may also occur on steep slopes. Also includes areas along creeks, rivers, sloughs, and lakes that may fail by landsliding during an earthquake. If area is adjacent to area with high susceptibility, a landslide may encroach into the area, or the area may fail if a landslide undercuts it, such as the flat area adjacent to sea cliffs.

Low susceptibility to landsliding. Several small landslides have formed in these areas and some of these have caused extensive damage to houses and roads. A few large landslides may occur. Slopes vary from 5–15 percent for unstable rock units to more than 70 percent for rock units that seem to be stable. The statements about additional slope stability mentioned in 1 above also apply in this category.

Moderate susceptibility to landsliding. Many small land-slides have formed in these areas and several of these have caused extensive damage to houses and roads. Some large landslides likely. Slopes generally greater than 30 percent but includes some slopes 15–30 percent in areas underlain by unstable rock units. See 1 for additional slope stability problems.

Moderately high susceptibility to landsliding. Slopes all greater than 30 percent. These areas are mostly in undeveloped parts of the county. Several large land-slides likely. See 1 for additional slope stability problems.

High susceptibility to landsliding. Slopes all greater than 30 percent. Many large and small landslides may form. These areas are mostly in undeveloped parts of the county. See 1 for additional slope stability problems.

Very high susceptibility to landsliding. Slopes all greater than 30 percent. Development of many large and small landslides is likely. The areas are mostly in undeveloped parts of the county. See 1 for additional slope stability problems.

Highest susceptibility to landsliding. Consists of land-slide and possible landslide deposits. No small landslide deposits are shown. Some of these areas may be relatively stable and suitable for development, whereas others may be active and causing damage to roads, houses and other cultural features.

0   5

km

Definitions:

Large landslide – more than 500 feet in maximum dimension;
small landslide – 50 to 500 feet in maximum dimension

Figure 4.16 (b) Landslide susceptibility classification used in a study of San Mateo County, California (from Brabb *et al.*, 1972).

adopting appropriate utility and public facility service area policies, informing and educating the public, and recording the hazard in public documents.

Recurrent damage from landslides can be avoided by permanently evacuating areas that continue to have slope failures. If there are obvious indications of instability, evacuations must be considered. This may be easier if the choice is on an individual scale, but social and political forces, including hazard perception, may restrict whole settlements from relocating, although it should be

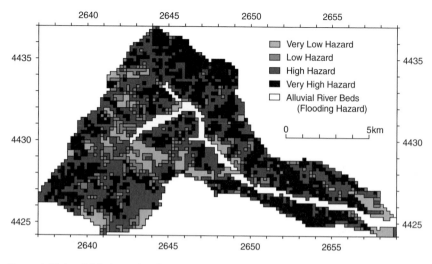

Figure 4.17 Landslide hazard and erosion zones, Ferro River basin, Calabria, southern Italy (from Carrara *et al.,* 1978).

forced if necessary. Structures may be removed or converted to a use that is less vulnerable to landslide damage. Techniques for removal or conversion include the acquiring or exchanging of hazardous areas and relocating their occupants, incorporating non-conforming use provisions into zoning ordinances, clearing and redeveloping damaged areas, removing unsafe structures and public nuisances, and urban redevelopment.

Financial techniques for encouraging or discouraging development in landslide areas include amending government assistance programmes, increasing awareness of legal liability, adopting appropriate lending policies, requiring insurance, and providing tax credits or lower assessments.

Various types of land-use and land-development regulations can be used to reduce landslide hazards. They are often the most economical and most effective means available to a local government. It is unrealistic to assume that development can be indirectly discouraged for a long period and other techniques, such as protecting existing development or purchasing hazardous areas, can be very costly. Development can be prohibited, restricted or regulated in landslide areas. They can be used as open spaces, or the density of development can be kept to a minimum to reduce its effects and the potential for damage. Zoning and subdivision regulations, as well as moratoriums on rebuilding, can be used to meet these objectives.

Property damage from landslides often leads to a demand for costly public works to provide protection for existing developments. This demand is usually limited to smaller landslide areas because of the costs involved, and the necessity for careful and accurate engineering design, construction and maintenance.

Potentially unstable land can be monitored so that residents can be warned and, if necessary, evacuated. Immediate relaying of information is vital in areas where landslides happen rapidly.

4.9 Methods of slope control and stabilization

As the same preventive or corrective work cannot always be applied to different types of slide, it is important to identify the type of slide that is likely to take place, or that has taken place. In this context, however, it is important to bear in mind that landslides may change in character and that they are usually complex, frequently changing their physical characteristics as time proceeds. When it comes to the correction of a landslide, as opposed to its prevention, since the limits and extent of the slide are generally well defined, the seriousness of the problem can be assessed. Nevertheless, in such instances consideration must be given to the stability of the area immediately adjoining the slide. Obviously, any corrective treatment must not adversely affect the stability of the area around the slide.

If landslides are to be prevented, then areas of potential landsliding must first be identified, as must their type and possible amount of movement. Then if the hazard is sufficiently real the engineer can devise a method of preventive treatment. Economic considerations, however, cannot be disregarded. In this respect it is seldom economical to design cut slopes sufficiently flat to preclude the possibility of landslides, and indeed many roads in upland terrain could not be constructed with the finance available without accepting some risk of landslides. All the same, this is no justification for lack of thorough investigation and adoption of all economic means of slide prevention.

Slide prevention may be brought about by reducing the activating forces, by increasing the forces resisting movement or by avoiding or eliminating the slide. In the first case, reduction of the activating forces can be accomplished by removing material from that part of the slide that provides the force that will give rise to movement and by drainage, which reduces pore water pressures and bulk density. Drainage also brings about an increase in shearing resistance.

The most frequently used methods of slope stabilization include retention systems, buttresses, slope modification and drainage. Properly designed retention systems can be used to stabilize most types of slope where large volumes of earth materials are not involved and where lack of space excludes slope modification. The control of subsurface water frequently represents a major component in slope stabilization works (Walker and Mohen, 1987).

4.9.1 Rockfall treatment

It is rarely economic to design a rock slope so that no subsequent rockfalls occur. Therefore, except where absolute security is essential, slopes should be designed to allow small falls of rock under controlled conditions. For an economic design, about 10% of the slope area may require some treatment at a later date. Subsequent slope treatment may take the form of a reduction in the overall slope angle so as to increase the factor of safety. Obviously, care must be taken to avoid damaging the slope when it is being trimmed by further blasting. Care should also be taken to maintain a constant slope line. One of the prerequisites for safe rock slopes involves sealing of loose blocks.

The design of systems to prevent rockfall requires data concerning trajectory (height of bounce), velocity, impact energy and total volume of accumulation. Computer programs are available that can model the rockfall behaviour of a slope (Pfeiffer *et al.*, 1990).

Ritchie (1963) described the use of a chain-link fence suspended from a cable mounted on a compressive spring to absorb the impact of rocks. Single-mesh fencing supported by rigid posts will contain small rockfalls. Larger, heavy-duty catch fences or nets are required for larger rockfalls. Rolling rocks up to 0.6 m in diameter can be restrained by a chain-link fence, but this can suffer severe damage when hit by rocks of this size and is not able to stop larger rocks. Fences have been developed so that when rocks collide with them, the nets engage energy-absorbing friction brakes, which extend the time of collision and in this way increase the capacity of the nets to restrain falling rocks. The braking device is incorporated into the tie-backs and cables in the catch fence. The brakes consist of a cable loop and a friction brake, which needs to be reset after each significant rockfall.

Wire meshing of rock slopes is one of the most effective methods of preventing rockfalls from steep slopes. The wire panels are laced together with binding wire. If a more robust method of linking panels is required, then horizontal and vertical steel cables can be shackled to the mesh and fixed to hooks and dowels. The latter provide a strong cable grid at 2 to 3 m centres, offering a greater resistance to large-scale block movement.

The use of cable lashing and cable nets to restrain loose rock blocks was referred to by Piteau and Peckover (1978). Areas of potential instability can be covered with mesh fixed with light cables. High-capacity horizontal cables are then strung across the block using anchoring and tensioning methods.

Rock traps in the form of a ditch and/or barrier can be installed at the foot of a slope. Ritchie (1967) provided a guide for the dimensions of such ditches (Figure 4.18). These dimensions can be reduced if the bottom of the ditch is filled with gravel, if a barrier is also used, if the face is netted or if it is excavated in soft rocks. Benches can also act as traps to retain rockfall, especially if a barrier is placed at their edge. Wire mesh suspended from the top of the face provides another method of controlling rockfall. Where a road or railway passes along the foot of a steep slope, then protection from rockfall is afforded by the construction of a rigid canopy from the face (Figure 4.19).

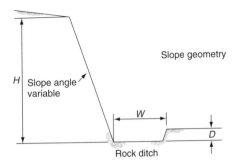

Variables in ditch design for rockfall areas

Rock slope (°) angle	Height (m) H	Ditch width (m) W	Ditch depth (m) D	Notes
90	5–10	3	1	
	10–20	5	1.5	1
	>20	7	1.5	1
75	5–10	3	1	
	10–20	5	1.5	1
	20–30	7	2	2
	>30	8	2	2
65	5–10	3	1.5	1
	10–20	5	2	2
	20–30	7	2	2
	>30	8	2.5	2
55	0–10	3	1	
	10–20	5	1.5	1
	>20			
45	0–10	3	1	
	10–20	3	2	1
	>20	5	2	2

Notes:
1. If dimension D is greater than 1, rock retaining fence should be used if ditch adjacent to highway.
2. Ditch dimension may be reduced to 1.5 if rock fence used.

Figure 4.18 Design of ditches for rockfall protection (after Ritchie, 1967).

4.9.2 Alteration of slope geometry

Altering the geometry of a slope often is the most efficient way of increasing the factor of safety, especially in deep-seated slides (Leventhal and Mostyn, 1987). However, such an approach may not be easy to adopt for long translational slides where there is no obvious toe or crest, where the geometry is determined by engineering constraints or where the unstable area is so complex that a change in topography that improves the stability of one area adversely affects the stability of another. Unstable material can be removed and, if necessary, replaced by stronger material. Alternatively, unstable material can be removed

Figure 4.19 Canopy constructed over road in area subject to rockfalls, British Columbia.

from near the crest of the slope and material added to the toe. In fact, it is usually more practical to load the toe. The material used for the construction of the toe fill should be free-draining, or adequate internal drainage should be incorporated.

Although partial removal is suitable for dealing with most types of mass movement, for some types it is inappropriate. For example, removal of material from the head has little influence on flows or slab slides. On the other hand, this treatment is eminently suitable for rotational slips (Bell and Maud, 1996). Slope flattening, however, is rarely applicable to rotational or slab slides. Slope reduction may be necessary in order to stabilize the toe of a slope and so prevent successive undermining with consequent spread of failure upslope. The size of a landslide obviously influences the excavation method, and only relatively small slides can be completely removed economically.

Benching brings about stability by dividing a slope into segments. Benches ideally should be over 5 m wide to allow access for inspection and therefore should be kept clear. If rock faces are to be scaled efficiently, benches should not be higher than 12 m. Drainage systems can be installed on benches.

4.9.3 Reinforcement of slopes

If some form of reinforcement is required to provide support for a slope, then it is advisable to install it as quickly as possible after excavation, in other words before the strength available in the first stages of joint dilation is lost during slope heave.

Dentition refers to masonry or concrete infill placed in fissures or cavities in a rock slope (Figure 4.20). The use of the same rock material for the masonry as forms the slope provides a more attractive finish than otherwise. It is often necessary to remove soft material from fissures and pack the void with permeable material prior to constructing the dentition. Drainage should be provided through the latter.

Thin to medium bedded rocks dipping parallel to the slope can be held in place by steel dowels, which are up to 2 m in length. Holes are drilled beneath the slip surface and are normal to the bedding. The dowels are grouted into place and they are not stressed. They are used where low loads are needed to increase stability and where the joint surfaces are at least moderately rough.

Figure 4.20 Rock bolts, shotcrete and dentition used to stabilize a cutting in limestone, North Wales.

Deformation in the rock mass stretches the untensioned dowel until sufficient stress is developed to prevent further strain.

Rock bolts can be used as reinforcement to enhance the stability of slopes in jointed rock masses (Figure 4.20). They provide additional strength on critical planes of weakness within the rock mass. Rock bolts inclined to the potential plane of failure provide greater resistance than those installed normal to the plane. Hence the design of rock bolt systems depends on prior knowledge of the potential failure mode of the slope. Design charts can be used to estimate the amount of support that has to be provided by the installation of rock bolts. Rock bolts may be up to 8 m in length, with a tensile working load of up to 100 kN. They are put in tension so that the compression induced in the rock mass improves shearing resistance on potential failure planes. Bearing plates, light steel sections or steel mesh may be used between bolts to support the rock face.

Tensioned rock bolts are placed over surfaces of potential rock failure and anchored in stable rock beneath. The tension applied to the bolts give rise to increased normal stress in the direction of the bolts and decreased shear stress on the failure surface. The anchorage is provided by mechanical or grouted anchors. In the single-bore multiple-anchor (SBMA) system each anchor has its own tendon and is located at different positions in the drillhole, thereby transferring its load to a discrete length of the anchor bore (Barley, 1991). This permits the simultaneous mobilization of almost the entire ground strength throughout the length of the drillhole.

When using rock bolts, consideration should be given to the influence of uplift and pore water pressures due to water in fractures, and also to the effect of this water freezing. In order to counteract these factors, the fractures should be grouted and the slope drained by inclined drillholes.

Rock anchors are used for major stabilization works, especially in conjunction with retaining structures, and may exceed 30 m in length. As far as rock anchors are concerned, Littlejohn (1990) noted that there has been a trend towards higher load capacities for individual and concentrated groups of anchors. For example, pre-stressing of the order of 200 t m^{-1} has been used at dam sites, which means that the capacity of an individual anchor is well in excess of 1000 t. In fact, working capacities of several thousand tonnes are being seriously considered. Because the stress levels are far greater than those involved in rock bolting, anchor loads are more dependent upon rock type and structure.

When a single anchor fails in a homogeneous rock mass an inverted cone of rock is pulled out of the rock mass. Normally the uplift capacity is equated to the weight of the rock cone involved and provides a method of calculating the ultimate resistance to withdrawal. However, the presence of discontinuities within a rock mass means that this simple cone approach has to be modified. Shear strength generally plays a major part in the ultimate resistance to withdrawal, as does the surface area of the cone (Littlejohn and Bruce, 1975).

Because of the risk of laminar failure or excessive anchor movement, the lengths of closely spaced anchors can be staggered in order to reduce the intensity of stress across discontinuities. Anchors can also be installed at different inclinations in order to dissipate the load within a rock mass.

Gunite or shotcrete is frequently used to preserve the integrity of a rock face by sealing the surface and inhibiting the action of weathering. The former is pneumatically applied mortar, the latter pneumatically applied concrete. The force of impact on the rock surface compacts the material. Compressive strengths of up to 40 MPa can be developed within 28 days, and flexural strengths varying between 550 and 700 kPa are normal within this period of time. The modulus of elasticity exceeds 35 000 MPa and shrinkage varies between 0.03 and 0.106%. A relatively dry mixture is normally used so that the material is self-supporting. Provided that the rock surface is properly prepared, the bond with rock is good. Gunite/shotcrete adapts to the surface configuration and can be coloured to match the colour of the surrounding rocks. Coatings may be reinforced with wire mesh and/or used in combination with rock bolts. Groundwater must be allowed to drain through the protective cover, otherwise it may be affected by frost action and groundwater pressures within the rock mass. It is generally considered that such surface treatment offers negligible support to the overall slope structure. Heavily fractured rocks may be grouted in order to stabilize them.

4.9.4 Restraining structures

Restraining structures control sliding by increasing the resistance to movement. They include retaining walls, cribs, gabions, buttresses and piling. The following minimum information is required to determine the type and size of a restraining structure:

- The boundaries and depth of the unstable area, its moisture content and its relative stability. For example, excessive pore water pressures are likely to give difficulties in designing retaining walls.
- The type of slide that is likely to develop or has occurred.
- The foundation conditions, since restraining structures require a satisfactory anchorage.

Retaining walls are often used where there is a lack of space for the full development of a slope, such as along many roads and railways. As retaining walls are subjected to unfavourable loading, a large wall width is necessary to increase slope stability, which means that they are expensive. Retaining structures should be designed for a predetermined load, which they are to transmit to the foundation of known bearing capacity. Retaining walls are located at the foot of a slope and should include adequate provision for drainage: for example, weep holes through the wall and pipe drainage in any backfill. This will not only prevent the build-up of pore water pressures but it will also reduce the effects of frost. Nonetheless, there are certain limitations that must be considered before retaining walls are used as a method of landslide control. These involve the ability of the structure to resist shearing action, overturning and sliding on or below the base of the structure.

The use of gravity walls to stabilize a slope is generally restricted in terms of height in that free-standing gravity walls have an upper limit of about 10 m and slides of only modest proportions can be prevented or stabilized using this type of structure (Morgenstern, 1987). Free-standing gravity walls usually require their bedrock foundation to be located at shallow depth. Hence stabilization of a slope where the failure surface is deep or where the forces are larger than can be carried by a gravity wall can be brought about by the installation of anchors founded below the failure surface. Pre-stressed anchor walls actively oppose the movement of the soil or rock mass. Walls can also be formed of contiguous piles.

Reinforced earth can be used for retaining earth slopes (Figure 4.21). Reinforced soil structures have advantages over traditional retaining walls in that they are flexible and so can tolerate large deformations, are resistant to seismic loadings and are often less costly to construct. Such structures provide a high degree of structural damping, which absorbs the dynamic energy associated with earthquakes. Thus reinforced earth can be used on poor ground where conventional alternatives would require expensive foundations. Reinforced earth structures are constructed by erecting facing panels at the face

Figure 4.21 Use of reinforced earth on a road embankment near Loch Lomond, Scotland.

of the wall at the same time as the earth is placed. Strips of galvanized steel are fixed to the panels. The system relies on the transfer of shear forces to mobilize the tensile capacity of closely spaced reinforcing strips. However, galvanized steel is subject to corrosion with time and so its inclusion is restricted to granular backfills, which are free-draining. Epoxy-coated steel reinforcements offer higher resistance to corrosion. Geosynthetic materials, notably geogrids, are also used in reinforced soil structures; however, they possess low stiffness relative to steel, so the amount of deformation required to develop maximum shear strength may exceed the allowable deformation of the soil structure. Grid reinforcement systems consist of polymer or metallic elements arranged in rectangular grids, metallic bar mats and wire mesh (Mitchell and Christopher, 1990). The passive resistance developed on the cross-members means that grids are more resistant to pull-out than strips. However, full passive resistance develops only after large deformations (50–100 mm). Sheet reinforcement commonly consists of geotextiles that are placed horizontally between layers of granular soil. Facing elements are formed by wrapping the geotextile around the face of the soil and covering the exposed geofabric with gunite, shotcrete or asphalt emulsion. In anchored earth, passive resistance is developed against the anchors at the ends of the reinforcing bars, the other end of the bars being attached to concrete facing panels.

Soil nailing has been used to retain slopes, the nails consisting of steel bars, metal rods or metal tubes, which are driven into *in situ* soil or soft rock or grouted into bored holes. The nails are passive elements that are not post-tensioned. Normally one nail is used for each 1 to 6 m^2 of ground surface. The ground surface between the nails is covered with a layer of shotcrete reinforced with wire mesh. Soil nailing is most effective in dense granular soils and low-plasticity stiff silty clays. It is generally not cost-effective, according to Mitchell and Christopher (1990), in loose granular soils, poorly graded soils, soft cohesive soils and highly plastic clays.

The root pile system utilizes micropiles to form a monolithic block of reinforced soils, which extends beneath the critical failure surface (Lizzi, 1977). The reinforcement provided by the micropiles is strongly influenced by their three-dimensional, root-like geometrical arrangement. Root piles are cast-in-place reinforced concrete piles that vary in diameter form 75 to 300 mm.

Cribs may be constructed of pre-cast reinforced concrete or steel units set up in cells that are filled with gravel or stone (Figure 4.22). Wooden cribs have been used, particularly in the United States. Crib walls offer rapid and easy construction even in difficult terrain. The system is reasonably flexible due to the segmental nature of the elements that comprise the walls and is therefore not particulary sensitive to differential settling. Plant growth can occur on the faces of crib walls, which masks their presence. They only serve to support shallow translation slides.

Gabions consist of strong wire mesh surrounding placed stones (see Figure 6.18). Like crib walls, they can also be constructed readily, especially in

Figure 4.22 Crib wall in Hamilton, North Island, New Zealand.

difficult terrain, and they are flexible. The gabion filling provides for good subsurface drainage conditions in the vicinity of the wall, and filtration protection between the gabions and the wall backfill or soil can be afforded by geotextiles.

Concrete buttresses have occasionally been used to support large blocks or rock, usually where they overhang. Piles have been used as a method of controlling landslides, particularly in the United States, but they have not always been successful. Indeed, they are perhaps the most controversial restraining method used. Piles can increase the shearing resistance of a soil mass, but on the other hand they may be ineffective if soil moves between them; in fact, they may be overturned or sheared. Obviously, piles must not be driven into soils that become quick upon vibration. Loading the base of a slope with rockfill will offer resistance against sliding and may act as a drainage filter.

Relocation and bridging are two means of avoiding landslip areas. In some cases relocation of a cutting for a road may be more economic than correcting large-scale mass movements. By contrast, bridging a landslide is rarely practical because of the cost involved. It is only resorted to on steep slopes, that is, steeper than 2 in 1, where relocation is not feasible. The dimensions of the slide are also important; for example, spans beyond 100 m are generally uneconomic. If piers are used in the slipped material, then they should be founded in stable rock. The piers should not hinder movement within the mass, so that excessive lateral thrust is not built up against them. The bridge abutments must be located in firm material. Tunnels can be excavated to allow traffic to avoid slopes prone to failure.

4.9.5 Drainage

Drainage is the most generally applicable method for improving the stability of slopes or for the corrective treatment of slides, regardless of type, since it reduces the effectiveness of one of the principal causes of instability, namely, excessive pore water pressure. In rock masses, groundwater also tends to reduce the shear strength along discontinuities. Moreover, drainage is the only economic way of dealing with slides involving the movement of several million cubic metres.

The surface of a landslide is generally uneven, hummocky and traversed by deep fissures. This is particularly the case when the slipped area consists of a number of slices. Water collects in depressions and fissures, and pools and boggy areas are formed. In such cases the first remedial measure to be carried out is surface drainage. Surface run-off should not be allowed to flow unrestrained over a slope. This is usually prevented by the installation of a drainage ditch at the top of an excavated slope to collect water drainage from above. The ditch, especially in soils, should be lined to prevent erosion, otherwise it will act as a tension crack. It may be filled with cobble aggregate. Herringbone ditch drainage is usually employed to convey water from the surfaces of slopes. These drainage ditches lead into an interceptor drain at the foot of the slope (Figure 4.23). Infiltration can be lowered by sealing the cracks in a slope by regrading or filling with cement, bitumen or clay. A surface covering has a similar purpose and function. For example, the slope may be covered with granular material resting upon filter fabric.

Water can be prevented from reaching a zone of potential instability by a cut-off. Cut-offs can take the form of a trench backfilled with asphalt or concrete, sheet piling, a grout curtain or a well curtain whereby water is pumped from a row of vertical wells. Exploration boreholes can subsequently be used as pumping wells. Such barriers may be considered where there is a likelihood of internal erosion of soft material taking place due to the increased flow of water attributable to drainage measures.

Trench drains are filled with free-draining materials and may be lined with geotextiles. They are used for shallow subsurface drainage. Lew and Graham

Figure 4.23 Drainage of a slope by gravel-filled drains.

(1988) described the use of sand drains, 0.85 m in diameter, extending from the base of a trench drain, to stabilize a slope.

Support and drainage may be afforded by counterfort-drains, where an excavation is made in sidelong ground likely to undergo shallow, parallel slides. Deep trenches are cut into the slope, lined with filter fabrics, and filled with granular filter material. The granular fill in each trench acts as a supporting buttress or counterfort, as well as providing drainage. However, counterfort-drains must extend beneath the potential failure zone, otherwise they merely add unwelcome weight to the slipping mass.

Deep wells are used to drain slopes where the depths involved are too deep for the construction of trench drains. For example, Bianco and Bruce (1991) described the use of large-diameter vertical wells (up to 2 m diameter), spaced at 5 to 20 m centres and extending to depths of 50 m. The wells were connected to each other near their bases by means of horizontal drillholes. The latter were lined with PVC pipe, 100 mm in diameter, and served as gravity collector drains. Two out of every three wells were filled with face-draining material, the third remaining open for monitoring flow rates. Usually the collected water flows from the base of the drainage wells under gravity, but occasionally pumps may be installed in the bottom of wells to remove the water.

Successful use of subsurface drainage depends on tapping the source of water, the presence of permeable material, which aids free drainage, the location of the drain on relatively unyielding material to ensure continuous operation (flexible PVC drains are now frequently used) and the installation of a filter to min-imize silting in the drainage channel. Drainage galleries are costly to construct, generally being placed by tunnelling techniques, and in slipped areas may expe-rience caving. Most galleries are designed to drain by gravity. In certain cases, collector galleries may be needed from which to pump water. According to Zaruba and Mencl (1982), they should be backfilled with stone to ensure their drainage capacity if partially deformed by subsequent movements. These two authors maintained that galleries are indispensable in the case of large slipped masses where drainage has to be carried out over lengths of 200 m or more. Drain holes may be made around the perimeter of a gallery to enhance drainage.

Subhorizontal drainage holes are much cheaper than galleries and are satis-factory over short lengths, but it is more difficult to intercept water-bearing layers with them. Subhorizontal drains can be inserted from the ground sur-face or by drilling from drainage galleries, large-diameter wells or caissons. The drain hole is typically 120 to 150 mm in diameter (Sembenelli, 1988). The drain holes are lined with slotted PVC pipe. The pipes are lined on the out-side with a geotextile covering, which acts as a filter. When individual benches are drained by horizontal holes, the latter should lead into a properly graded interceptor trench, which is lined with impermeable material.

Schuster (1992) mentioned electro-osmotic dewatering, vacuum dewatering and siphon drainage as techniques that have been employed for slope stabi-lization. However, their use is very infrequent.

Afforestation may help to stabilize shallow slides, but it cannot prevent fur-ther movements occurring in large landslip areas. It can, however, lower infil-tration. The most satisfactory trees are those that consume most water and have high transpiration rates, which means that deciduous trees are better than conifers. The root system helps to bind soil together, and foliage and plant residues reduce the impact of rainfall.

References

Barley, A.D. (1991) Slope stabilization by new ground anchorage systems in rocks and soils. *Proceedings International Conference on Slope Stability Engineering: Developments and Applications*, Isle of Wight, Thomas Telford Press, London, 335–340.

Barton, N. (1974) Estimating the shear strength of rock joints. *Proceedings Third International Congress Rock Mechanics (ISRM)*, Denver, 2, 219–220.

Bell, F.G. (1994) Floods and landslides in Natal and notably the greater Durban region, September 1987: a retrospective view. *Bulletin Association Engineering Geologists*, 31, 59–74.

Bell, F.G. and Maud, R.R. (1996) Landslides associated with the Pietermaritzburg Formation in the greater Durban area, South Africa. *Environmental and Engineering Geoscience*, 2, 557–573.

Bianco, B. and Bruce, D.A. (1991) Large landslide stabilization by deep drainage wells. *Proceedings International Conference on Slope Stability Engineering: Developments and Applications*, Isle of Wight, Thomas Telford Press, London, 319–326.

Bishop, A.W. (1971) The influence of progressive failure on the choice of the method of stability analysis. *Geotechnique*, 169–172.

Bishop, A.W. (1973) The stability of tips and spoil heaps. *Quarterly Journal Engineering Geology*, 6, 335–376.

Bjerrum, L. (1967) Progressive failure in slopes of overconsolidated plastic clay and clay shales. *Proceedings American Society Civil Engineers, Journal Soil Mechanics Foundations*, 93, 1–49.

Brabb, E.E., Pampeyan, E.H. and Bonilla, M.G. (1972) Landslide susceptibility in San Mateo County, California. US Geological Survey, Miscellaneous Field Studies Map MF-360.

Brand, E.W. (1988) Landslide risk assessment in Hong Kong. *Proceedings Fifth International Symposium on Landslides*, Lausanne, Bonnard, C. (ed.), Balkema, Rotterdam, 2, 1059–1074.

Brand, E.W., Premchitt, J. and Philipson, H.B. (1984) Relationship between rainfall and landslides in Hong Kong. *Proceedings Fourth International Symposium on Landslides*, Toronto, 1, 377–384.

Bromhead, E.N. (1978) Large landslides in London Clay at Herne Bay, Kent. *Quarterly Journal Engineering Geology*, 11, 291–304.

Carrara, A. and Merenda, L. (1976) Landslide inventory in northern Calabria, southern Italy. *Bulletin Geological Society America*, 87, 1153–1162.

Carrara, A., Catalano, E., Sorriso Valvo, M., Reali, C., Meremda, L. and Rizzo, V. (1977) Landslide morphology and topology in two zones, Calabria, Italy. *Bulletin International Association Engineering Geology*, 16, 8–13.

Carrara, A., Cardinali, M., Detti, R., Guzzetti, F., Pasqui, V. and Reichenbach, P. (1990) Geographical Information Systems and multivariate models in landslide hazard evaluation. *Proceedings Sixth International Conference and Field Workshop on Landslides: Alps '90*, Milan, Cancelli, A. (ed.), 17–28.

Chan, D.H. and Morgenstern, N.R. (1987) Analysis of progressive deformation of the Edmonton Convention Centre excavation. *Canadian Geotechnical Journal*, 24, 430–440.

Chandler, R.J. (1974) Lias Clay: the long term stability of cutting slopes. *Geotechnique*, 24, 21–38.

Chandler, R.J. (1977) The application of soil mechanics methods to the study of slopes. In *Applied Geomorphology*, Hails, J.R. (ed.), Elsevier, Amsterdam, 157–182.

Cooke, R.M. and Doornkamp, J.C. (1990) *Geomorphology in Environmental Management*, second edition. Clarendon Press, Oxford.

Cotecchia, V. (1978) Systematic reconnaissance mapping and registration of slope movements. *Bulletin International Association Engineering Geology*, 17, 5–37.

Crozier, M.J. (1984) Field assessment of slope instability. In *Slope Instability*, Brunsden, D. and Prior, D.B. (eds), Wiley–Interscience, Chichester, 103–142.

Dunnicliff, J. (1988) *Geotechnical Instrumentation for Monitoring Field Performance*. Wiley, New York.

Dunnicliff, J. (1992) Monitoring and instrumentation of landslides. *Proceedings Sixth International Symposium on Landslides*, Christchurch, Bell, D.H. (ed.), Balkema, Rotterdam, 3, 1881–1886.

Franklin, J.A. and Denton, P.E. (1973) The monitoring of rock slopes. *Quarterly Journal Engineering Geology*, 6, 259–286.

Fredlund, D.G. (1984) Analytical methods for slope analysis. *Proceedings Fourth International Symposium on Landslides*, Toronto, 1, 229–250.

Gens, A., Hutchinson, J.N. and Cavounidis, S. (1988) Three dimensional analysis of slides in cohesive soils, *Geotechnique*, 38, 1–23.

Greenway, D.R. (1987) Vegetation and slope stability. In *Slope Stability*, Anderson, M.G. and Richards, K.S. (eds), Wiley, New York, 187–230.

Hansen, A. (1984) Landslide hazard analysis. In *Slope Instability*, Brunsden, D. and Prior, D.B. (eds), Wiley–Interscience, Chichester, 523–602.

Hoek, E. (1970) Estimating the stability of excavated slopes in opencast mines. *Transactions Institution Mining and Metallurgy, Section A, Mining Industry*, 79, A109–A132.

Hoek, E. (1971) The influence of structure on the stability of rock slopes. *Proceedings First Symposium on Stability in Open Pit Mining*, Vancouver, American Institute Mining Engineers, 49–63.

Hoek, E. and Bray, J.W. (1981) *Rock Slope Engineering*. Institution Mining and Metallurgy, London.

Hollingworth, S.E., Taylor, J.H. and Kellaway, G.A. (1944) Large-scale superficial structures in the Northampton Ironstone field. *Quarterly Journal Geological Society*, 100, 1–44.

Hovland, J. (1977) Three dimensional slope stability analysis method. *Proceedings American Society Civil Engineers, Journal Geotechnical Engineering Division*, 103, 971–986.

Huma, I. and Radulescu, D. (1978) Automatic production of thematic maps of slope stability. *Bulletin International Association Engineering Geology*, 17, 95–99.

Hungr, O. (1987) An extension of Bishop's simplified method of slope stability analysis to three dimensions. *Geotechnique*, 37, 113–117.

Hutchinson, J.N. (1973) The response of the London Clay cliffs to differing rates of toe erosion. *Geologicia Applicata e Idrogeologia*, 8, 221–239.

Hutchinson, J.N. (1988) Morphological and geotechnical parameters of landslides in relation to geology and hydrology. *Proceedings Fifth International Symposium on Landslides*, Lausanne, Bonnard, C. (ed.), Balkema, Rotterdam, 1, 3–35.

Hutchinson, J.N. (1992) Landslide hazard assessment. *Proceedings Sixth International Symposium on Landslides*, Christchurch, Bell, D.H. (ed.), Balkema, Rotterdam, 3, 1805–1841.

Hutchinson, J.N. and Bhandari, R.K. (1971) Undrained loading, a fundamental mechanism of mudflows and other mass movements. *Geotechnique*, 21, 353–358.

Hutchinson, J.N., Bromhead, E.N. and Chandler, M.P. (1991) Investigation of coastal landslides at St Catherine's Point, Isle of Wight. *Proceedings International Conference on Slope Stability Engineering: Applications and Developments*, Isle of Wight, Thomas Telford Press, London, 151–161.

Jahns, R.H. (1978) Geophysical predictions. In *Landslides: Analysis and Control*, Schuster, R.L. and Krizek, R.J. (eds), Transportation Research Board, Special Report 176, National Academy of Sciences, Washington, DC, 58–65.

Keefer, D.K. (1984) Landslides caused by earthquakes. *Bulletin American Geological Society*, 95, 406–421.

Kockelman, W.J. (1986) Some techniques for reducing landslide hazards. *Bulletin Association Engineering Geologists*, 23, 29–52.

Leventhal, A.R. and Mostyn, G.R. (1987) Slope stabilization techniques and their application. *Proceedings Extension Course on Soil Slope Stability and Stabilization*, Sydney, Walker, B.F. and Fell, R. (eds), Balkema, Rotterdam, 121–181.

Lew, K.V. and Graham, J. (1988) Riverbank stabilization by drains in plastic clay. In *Proceedings Fifth International Symposium on Landslides*, Lausanne, Bonnard, C. (ed.), 2, 939–944.

Littlejohn, G.S. (1990) Ground anchorage practice. In *Design and Performance of Earth Retaining Structures*, Lambe, P.C. and Hansen, L.A. (eds), Geotechnical Special Publication 25, American Society Civil Engineers, New York, 692–733.

Littlejohn, G.S. and Bruce, D.A. (1975) Rock anchors: state-of-the-art. *Ground Engineering*, **8**, 3, 25–32; 4, 41–48; 6, 36–45.

Lizzi, F. (1977) Practical engineering in structurally complex formations. The in situ reinforced earth. In *Proceedings International Symposium on the Geotechnics of Structurally Complex Formations*, Capri, 327–333.

Lollino, G. (1992) Automated inclinometric system. *Proceedings Sixth International Symposium on Landslides*, Christchurch, Bell, D.H. (ed.), 2, 1147–1150.

Lumb, P. (1975) Slope failures in Hong Kong. *Quarterly Journal Engineering Geology*, **8**, 31–65.

Martin, D.C. (1988) Rockfall control: an update: technical note. *Bulletin Association Engineering Geologists*, **25**, 137–144.

Mikkelsen, P.E. (1996) Field instrumentation. In *Landslides: Investigation and Mitigation*, Turner, A.K. and Schuster, R.L. (eds), Transportation Research Board, Special Report 247, National Research Council, Washington, DC, 278–318.

Mitchell, J.K. and Christopher, B.R. (1990) North American practice in reinforced soil systems. In *Design and Performance of Earth Retaining Structures*, Lambe, P.C. and Hansen, L.A. (eds), Geotechnical Special Publication 25, American Society Civil Engineers, New York, 322–346.

Morgenstern, N.R. (1963) Stability charts for earth slopes during rapid downdraw. *Geotechnique*, **13**, 121–131.

Morgenstern, N.R. (1992) The role of analysis in the evaluation of slope stability. *Proceedings Sixth International Symposium on Landslides*, Christchurch, Bell, D.H. (ed.), Balkema, Rotterdam, 3, 1615–1629.

Moser, M. (1978) Proposals for geotechnical maps concerning slope stability potential in mountain watersheds. *Bulletin International Association Engineering Geology*, 17, 100–108.

Nash, D.F.T. (1987) A comparative review of limit equilibrium methods of stability analysis. In *Slope Stability*, Anderson, M.G. and Richards, K.S. (eds), Wiley, New York.

Olivier, M., Bell, F.G. and Jermy, C.A. (1994) The effect of rainfall on slope failure, with examples from the greater Durban area. *Proceedings Seventh Congress International Association Engineering Geology*, Lisbon, Balkema, Rotterdam, 3, 1629–1636.

Payne, H.R. (1985) Hazard assessment and rating methods. In *Landslides in South Wales Coalfield*, Morgan, C.S. (ed.), Polytechnic of Wales, Pontypridd, 59–71.

Pfeiffer, T.J., Higgins, J.D. and Turner, A.K. (1990) Computer aided rockfall hazard analysis. *Proceedings Sixth Congress International Association Engineering Geology*, Amsterdam, Balkema, Rotterdam, 93–103.

Piteau, D.R. and Peckover, F.L. (1978) Rock slope engineering. In *Landslides: Analysis and Control*, Schuster, R.L. and Krizek, R.J. (eds), Transportation Research Board, Special Report 176, National Academy of Sciences, Washington, DC, 192–228.

Planicka, A. and Nosek, L. (1970) Terrestrial photogrammetry in measurement of deformations in rockfill dams. *Proceedings Tenth International Congress on Large Dams*, Montreal, 3, 207–215.

Potts, D.M., Dounias, G.T. and Vaughan, P.R. (1990) Finite element analysis of Carsington embankment. *Geotechnique*, 40, 79–102.

Radbruch, D.H. and Wentworth, C.M. (1971) Estimated relative abundance of landslides in the San Francisco Bay Region, California. US Department of the Interior, Geological Survey, Menlo Park.

Rib, H.T. and Liang, T. (1978) Recognition and identification. In *Landslides: Analysis and Control*, Schuster, R.L. and Krizek, R.J. (eds), Transportation Research Board, Special Report 176, National Academy of Sciences, Washington, DC, 34–80.

Richards, L.R., Whittle, R.A. and Ley, G.M.M. (1978) Appraisal of stability conditions in rock slopes. In *Foundation Engineering in Difficult Ground*, Bell, F.G. (ed.), Butterworths, London, 449–512.

Ritchie, A.M. (1963) The evaluation of rockfall and its control. *Highway Research Board*, Record No. 17, Washington, DC, 13–28.

Schuster, R.L. (1992) Recent advances in slope stabilization. *Proceedings Sixth International Symposium on Landslides*, Christchurch, Bell, D.H. (ed.), Balkema, Rotterdam, 3, 1715–1745.

Schuster, R.L. and Fleming, R.W. (1986) Economic losses and fatalities due to landslides. *Bulletin Association Engineering Geologists*, 23, 11–28.

Sembenelli, P. (1988) Stabilization and drainage. *Proceedings Fifth International Symposium on Landslides*, Lausanne, Bonnard, C. (ed.), Balkema, Rotterdam, 2, 813–819.

Sharpe, C.F.S. (1938) *Landslides and Related Phenomena*. Columbia University Press, New York.

Skempton, A.W. (1964) Long-term stability of clay slopes. *Geotechnique*, 14, 77–101.

Skempton, A.W. and DeLory, F.A. (1957) Stability of natural slopes in London Clay. *Proceedings Fourth International Conference on Soil Mechanics and Foundation Engineering*, London, 2, 378–381.

Skempton, A.W. and Early, K.R. (1972) Investigations of the landslide at Walton's Wood, Staffordshire. *Quarterly Journal Engineering Geology*, 5, 19–42.

Skempton, A.W. and Hutchinson, J.N. (1969) Stability of natural slopes and embankment foundations. *Proceedings Seventh International Conference Soil Mechanics and Foundation Engineering*, Mexico City, State-of-the-Art Volume, 291–340.

Soeters, R. and van Westen, C.J. (1996) Slope instability recognition, analysis and zonation. In *Landslides: Investigation and Mitigation*, Turner, A.K. and Schuster, R.L. (eds), Transportation Research Board, Special Report 247, National Research Council, Washington, DC, 129–177.

Terzaghi, K. (1950) Mechanisms of landslides. In *Applications of Geology to Engineering Practice*, Paige, S. (ed.), Berkey Volume, American Geological Society, 83–124.

Thomas, H.S.H. and Ward, W.H. (1969) The design, construction and performance of a vibrating wire earth pressure cell. *Geotechnique*, 19, 39–51.

Tran-Duc, P.O., Ohno, M. and Mawatari, Y. (1992) An automated landslide monitoring system. *Proceedings Sixth International Symposium on Landslides*, Christchurch, Bell, D.H. (ed.), Balkema, Rotterdam, 2, 1163–1166.

Varnes, D.J. (1978) Slope movement types and processes. In *Landslides: Analysis and Control*, Schuster, R.L. and Krizek, R.J. (eds), Transportation Research Board, Special Report 176, National Academy of Sciences, Washington, DC, 11–33.

Varnes, D.J. (1984) *Landslide Hazard Zonation: A Review of Principles and Practice*. Natural Hazards 3, UNESCO, Paris.

Vaughan, P.R. and Walbanke, H.J. (1973) Pore pressure changes and delayed failure of cutting slopes in overconsolidated clay. *Geotechnique*, 23, 531–539.

Walker, B.F. and Mohen, F.J. (1987) Groundwater prediction and control, and negative pore pressure effects. *Proceedings Extension Course on Soil Slope Stability and Stabilization*, Sydney, Walker, B.F. and Fell, R. (eds), Balkema, Rotterdam, 121–181.

Walker, B.F, Blong, R.J. and MacGregor, J.P. (1987) Landslide classification, geomorphology and site investigations. *Proceedings Extension Course on Soil Slope Stability and Stabilization*, Sydney, Walker, B.F. and Fell, R. (eds), Balkema, Rotterdam, 1–52.

Wislocki, A.P. and Bentley, S.P. (1991) An expert system for landslide hazard and risk assessment. *Computers and Structures*, 40, 169–172.

Wold, R.L. and Jochim, C.L. (1989) *Landslide Loss Reduction: A Guide for State and Local Government Planning*. Colorado Geological Survey, Special Publication 33, Department of Natural Resources, Denver, Colorado.

Zaruba, Q. and Mencl. V. (1982) *Landslides and Their Control*, second edition. Elsevier, Amsterdam.

Chapter 5

Problem soils

There is a group of geohazards that exist largely because of the geotechnical properties of the material concerned, which in turn are commonly related to its fabric and mineralogy. This group of geohazards includes, for example, swelling and shrinkage of some clay soils, dispersivity in clay soils with particular chemical characteristics, and the collapse potential of certain silty soils. Other soils such as quicksands, quick clays and peat lack strength, the first two especially when disturbed, which can give rise to major problems. The effects of these hazards, while unspectacular and rarely causing loss of life, can result in considerable financial loss. Swell–shrink in clay soils, for example, has caused losses of up to £2 billion in recent years in the United Kingdom. The use of dispersive soils in southern Africa, Australia and the United States has led to the failure of dams and road embankments.

5.1 Quicksands

As water flows through silts and sands and loses head, its energy is transferred to the particles past which it is moving, which in turn creates a drag effect on the particles. If the drag effect is in the same direction as the force of gravity, then the effective pressure is increased and the soil is stable. Indeed the soil tends to become more dense. Conversely, if water flows towards the surface, then the drag effect is counter to gravity, thereby reducing the effective pressure between particles. If the velocity of upward flow is sufficient it can buoy up the particles so that the effective pressure is reduced to zero. This represents a critical condition where the weight of the submerged soils is balanced by the upward-acting seepage force. A critical condition sometimes occurs in silts and sands. If the upward velocity of flow increases beyond the critical hydraulic gradient a quick condition develops.

Quicksands, if subjected to deformation or disturbance, can undergo a spontaneous loss of strength, which causes them to flow like viscous liquids. Terzaghi (1925) explained the quicksand phenomenon in the following terms. First, the sand or silt concerned must be saturated and loosely packed. Second, on disturbance the constituent grains become more closely packed, which leads to an increase in pore water pressure, reducing the forces acting between the grains. This brings about a reduction in strength. If the pore water can escape very rapidly the loss in strength is momentary. Hence, the third condition requires that pore water cannot escape readily. This is fulfilled if the sand or silt has a low permeability and/or the seepage path is long. Casagrande (1936) demonstrated that a critical porosity existed above which a quick condition could be developed. He maintained that many coarse-grained sands, even when loosely packed, have porosities approximately equal to the critical conditions, while medium- and fine-grained sands, especially if uniformly graded, exist well above the critical porosity when loosely packed. Accordingly, fine sands tend to be potentially more unstable than coarse-grained varieties. It must also be remembered that the finer sands have lower permeabilities.

Quick conditions brought about by seepage forces are frequently encountered in excavations made in fine sands that are below the water table as, for example, in coffer dam work. As the velocity of the upward seepage force increases further from the critical gradient the soil begins to boil more and more violently (Figure 5.1). At such a point structures fail by sinking into the quicksand. Liquefaction of potential quicksands may be brought about by sudden shocks caused by the action of heavy machinery (notably pile driving), blasting and earthquakes (Figure 5.2; see also Chapter 3). Such shocks increase the stress carried by the water, the neutral stress, and give rise to a decrease in the effective stress and shear strength of the soil. There is also a possibility of a quick condition developing in a layered soil sequence where the individual beds have different permeabilities. Hydraulic conditions are particularly unfavourable where water initially flows through a very permeable horizon with little loss of head, which means that flow takes place under a great hydraulic gradient. Maps showing the degree of liquefaction hazard have been produced, for example, for the San Francisco Bay area. The zones are based on the likely response of the surface materials to seismic loading (Figure 5.3). Restriction on the location of certain types of building within particular zones can mean that severe damage due to liquefaction can be avoided.

Several methods may be employed to avoid the development of quick conditions. One of the most effective techniques is to prolong the length of the seepage path, thereby increasing the frictional losses and so reducing the seepage force. This can be accomplished by placing a clay blanket at the base of an excavation where seepage lines converge. If sheet piling is used in excavation of critical soils, then the depth to which it is sunk determines whether or not quick conditions will develop. Consequently, it should be sunk deep enough to avoid a potentially critical condition occurring at the base level of the excavation. The hydrostatic head can also be reduced by means of relief wells, and seepage can be intercepted by a wellpoint system placed around the excavation. Furthermore, a quick condition may be prevented by increasing the downward-acting force. This can be brought about by laying a load on the surface of the soil where seepage is discharging. Gravel filter beds can be used for this pur-

Figure 5.1 Sand boil near El Centro, California, formed as a result of the earthquake of 15 October 1979. Where the water table is near the surface, compaction of saturated, loosely packed sands is accompanied by ejection of water or water–sediment mixtures to form a sand boil.

Figure 5.2 Tilting of apartment buildings due to liquefaction of soil, Niigata earthquake, 1964.

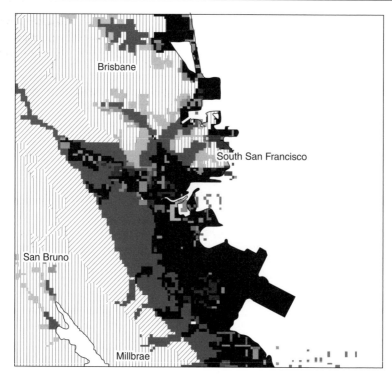

EXPLANATION

Relative liquefaction susceptibility and probability of susceptible sediment in subsurface materials

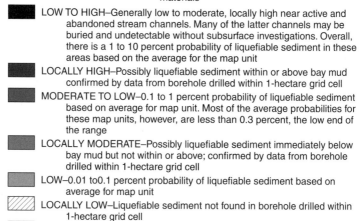

LOW TO HIGH–Generally low to moderate, locally high near active and abandoned stream channels. Many of the latter channels may be buried and undetectable without subsurface investigations. Overall, there is a 1 to 10 percent probability of liquefiable sediment in these areas based on the average for the map unit

LOCALLY HIGH–Possibly liquefiable sediment within or above bay mud confirmed by data from borehole drilled within 1-hectare grid cell

MODERATE TO LOW–0.1 to 1 percent probability of liquefiable sediment based on average for map unit. Most of the average probabilities for these map units, however, are less than 0.3 percent, the low end of the range

LOCALLY MODERATE–Possibly liquefiable sediment immediately below bay mud but not within or above; confirmed by data from borehole drilled within 1-hectare grid cell

LOW–0.01 to0.1 percent probability of liquefiable sediment based on average for map unit

LOCALLY LOW–Liquefiable sediment not found in borehole drilled within 1-hectare grid cell

VERY LOW–Less than 0.01 percent probability of liquefiable sediment based on average for map unit

BEDROCK

Figure 5.3 Part of the 1:62 500 map of San Mateo County, California, showing liquefaction susceptibility (from Youd and Perkins, 1987).

pose. Suspect soils can also be densified, treated with stabilizing grouts, or frozen.

Subsurface structures should be designed to be stable with regard to the highest groundwater level that is likely to occur. Structures below groundwater level are acted upon by uplift pressures. If the structure is weak this pressure can break it and, for example, cause a blow-out of a basement floor or collapse of a basement wall. If the structure is strong but light it may be lifted, that is, subjected to heave. Uplift can be taken care of by adequate drainage or by resisting the upward seepage force. Continuous drainage blankets are effective but should be designed with filters to function without clogging. The entire weight of structure can be mobilized to resist uplift if a raft foundation is used. Anchors, grouted into bedrock, can provide resistance to uplift.

5.2 Expansive clays

Some clay soils undergo slow volume changes that occur independently of loading and are attributable to swelling or shrinkage (see also vertisols, e.g. black cotton soils, Section 5.7). These volume changes can give rise to ground movements, which may result in damage to buildings. Low-rise buildings are particularly vulnerable to such ground movements since they generally do not have sufficient weight or strength to resist. In addition, shrinkage settling of embankments can lead to cracking and break-up of the roads they support. Maps delineating areas of expansive clay can be produced in order to facilitate land-use planning (Figure 5.4).

Problems caused by expansive soils in the United States give rise to costs of over $2 billion each year. This is frequently twice the cost of flood damage or of damage caused by landslides and more than twenty times the cost of earthquake damage. The principal cause of expansive clays is the presence of swelling clay minerals such as montmorillonite. For example, Popescu (1979) found that expansive clays in Romania contained between 40 and 80% montmorillonite. Construction damage is especially notable where expansive clay forms the surface cover in regions that experience alternating wet and dry seasons, leading to swelling and shrinkage of these soils. Again taking expansive clays in Romania as an example, Popescu noted that maximum seasonal changes in moisture content in these soils were around 20% at 0.4 m depth, 10% at 1.2 m depth and less than 5% at 1.8 m depth. The corresponding cyclic movements of the ground surface were between 100 and 200 mm.

Differences in the period and amount of precipitation and evapotranspiration are the principal factors influencing the swell–shrink response of a clay soil beneath a building. Poor surface drainage or leakage from underground pipes also can produce concentrations of moisture in clay. Trees with high water demand and uninsulated hot-process foundations may dry out clay, causing shrinkage. Cold stores may also cause desiccation of clay soil. The depth of the active zone in expansive clays (i.e. the zone in which swelling and shrinkage

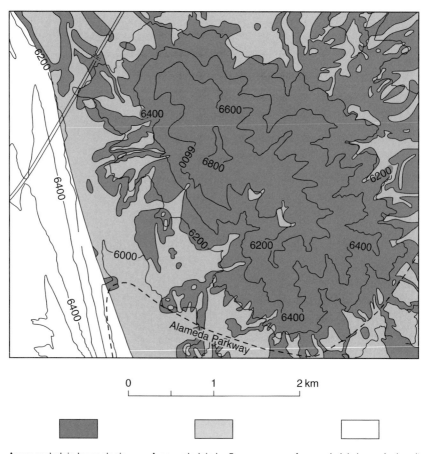

0   1   2 km

Areas underlain by geologic units that contain clays having swelling pressures higher than 2500 lb/ft^2 PVC (potential volume change). Preconstruction investigation in these areas should include engineering laboratory swell–shrink tests

Area underlain by 5 or more feet of non-swelling surficial deposits which in turn are underlain by geologic units which may have swelling pressures higher than 2500 lb/ft^2 PVC. For any construction involving foundation excavation through the surficial deposits, preconstruction investigations should include engineering laboratory swell–shrink tests

Area underlain by geologic units having no or slight swelling pressures. Preconstruction swell–shrink tests unnecessary

Figure 5.4 Map showing areas of swelling clay in the Morrison Quadrangle, Colorado (from Scott, 1972).

occur in wet and dry seasons, respectively) varies. It may extend to over 6 m depth in some semi-arid regions of South Africa, Australia and Israel. Many soils in temperate regions such as Britain, especially in southeast England, possess the potential for significant volume change due to changes in moisture content. However, owing to the damp climate in most years, volume changes are restricted to the upper 1.0 to 1.5 m in clay soils.

The potential for volume change of clay soil is governed by its initial moisture content, initial density or void ratio, microstructure and vertical stress, as well as the type and amount of clay minerals present. Cemented and undisturbed expansive clay soils often have a high resistance to deformation and may be able to absorb significant amounts of swelling pressure. Remoulded expansive clays therefore tend to swell more than their undisturbed counterparts. For example, Schmertmann (1969) maintained that some clays increase their swell behaviour when they undergo repeated large shear strains due to mechanical remoulding. He introduced the term 'swell sensitivity' for the ratio of the remoulded swelling index to the undisturbed swelling index.

Expansive clay minerals take water into their lattice structure. In less dense soils they tend to expand initially into zones of looser soil before volume increase occurs. However, in densely packed soil with low void space, the soil mass has to swell more or less immediately to accommodate the volume change. Hence clay soils with a flocculated fabric swell more than those that possess a preferred orientation. In the latter the maximum swelling occurs normal to the direction of clay particle orientation.

Because expansive clays normally possess extremely low permeabilities, moisture movement is slow and a significant period of time may be involved in the swelling–shrinking process. Accordingly, moderately expansive clays with a smaller potential to swell but with higher permeabilities than clays having a greater swell potential may swell more during a single wet season than more expansive clays.

Expansive clays are often heavily fissured due to seasonal changes in volume, which produce shrinkage cracks and shear surfaces. Consequently, near-vertical fissures are frequently found at shallow depth, with diagonal fissures at greater depth. Sometimes the soil is so desiccated that the fissures are wide open and the soil is shattered or micro-shattered. The presence of desiccation cracks enhances evaporation from the soil.

The swell–shrink behaviour of a clay soil under a given state of applied stress in the ground is controlled by changes in soil suction. The relationship between soil suction and water content depends on the proportion and type of clay minerals present, their microstructural arrangement and the chemistry of the pore water. Changes in soil suction are brought about by moisture movement through the soil due to evaporation from its surface in dry weather, by transpiration from plants, or by recharge consequent upon precipitation. The climate governs the amount of moisture available to counteract what is removed by evapotranspiration (i.e. the soil moisture deficit). In semi-arid climates, long periods of high soil moisture deficit alternate with short periods when precipitation balances or exceeds evapotranspiration.

The volume changes that occur due to evapotranspiration from clay soils can be conservatively predicted by assuming the lower limit of the soil moisture content to be the shrinkage limit. Desiccation beyond this value cannot bring about further volume change.

Transpiration from vegetative cover is a major cause of water loss from soils in semi-arid regions. Indeed, the distribution of soil suction in soil is primarily controlled by transpiration from vegetation and represents one of the most significant changes made in loading (i.e. to the state of stress in a soil). The behaviour of root systems is exceedingly complex and is a major factor in the intractability of swelling and shrinking problems. The spread of root systems depends on the type of vegetation, the soil type and groundwater conditions. The suction induced by the withdrawal of water fluctuates with the seasons, reflecting the growth of vegetation and probably varies between 100 and 1000 kPa (equivalent to pF values 3 and 4, respectively). The complete depth of active clay profiles does not usually become fully saturated during the wet season in semi-arid regions. Nonetheless, changes in soil suction may occur over a depth of 2.0 m or so between the wet and dry seasons. The suction pressure associated with the onset of cracking is approximately pF 4.6.

The extent to which the vegetation is able to increase the suction to the level associated with the shrinkage limit is obviously important. In fact, the moisture content at the wilting point exceeds that of the shrinkage limit in soils with a high content of clay and is lower in those possessing low clay contents. This explains why settlement resulting from the desiccating effects of trees is more notable in low to moderately expansive soils than in expansive ones.

When vegetation is cleared from a site, its desiccating effect is also removed. Hence the subsequent regain of moisture by clay soils leads to them swelling. Swelling movements on expansive clays in South Africa, associated with the removal of vegetation and subsequent erection of buildings, in many areas have amounted to about 150 mm, although movements over 350 mm have been recorded (Williams, 1980).

Methods of predicting volume changes in soils can be grouped into empirical methods, soil suction methods and oedometer methods. Empirical methods make use of the swelling potential as determined from void ratio, natural moisture content, liquid and plastic limits, and activity (Figure 5.5). However, because the determination of plasticity is carried out on remoulded soil, it does not consider the influence of soil texture, moisture content, soil suction or pore water chemistry, which are important factors in relation to volume change potential. Over-reliance on the results of such tests must therefore be avoided. Consequently, empirical methods should be regarded as simple swelling indicator methods and nothing more. As such, it is wise to carry out another type of test and to compare the results before drawing any conclusions.

Soil suction methods use the change in suction from initial to final condi-

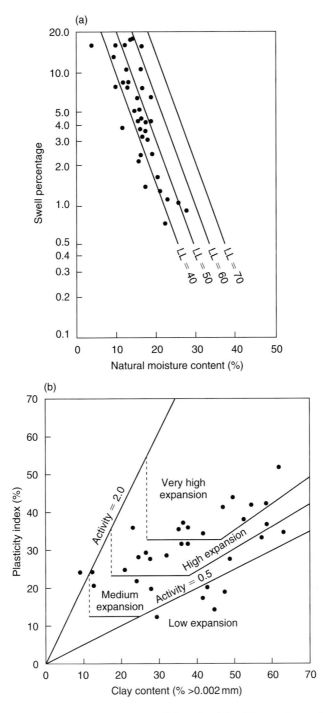

Figure 5.5 (a) Prediction of percentage swell for clay (after Vijayvergiya and Ghazzaly, 1973). Some expansive clays from Natal are shown (from Bell and Maud, 1995). LL = liquid limit. (b) Estimation of the degree of expansiveness of a clay soil (after Van der Merwe, 1964; modified by Williams and Donaldson, 1980). Some expansive clay soils from Natal are shown (from Bell and Maud, 1995).

tions to obtain the degree of volume change (Figure 5.6). Soil suction is the stress that, when removed, allows the soil to swell. In other words, the value of soil suction in a saturated, fully swollen soil is zero. O'Neill and Poormoayed (1980) quoted the United States Army Engineers Waterways Experimental Station (USAEWES) classification of potential swell (Table 5.1), which is based on the liquid limit, plasticity index and initial (*in situ*) suction. The latter is measured in the field by a psychrometer. Soil suction is not easy to measure accurately, but filter paper has been used for this purpose (McQueen and Miller, 1968). According to Chandler *et al.* (1992), measurements of soil suction obtained by the filter paper method compare favourably with measurements obtained using psychrometers or pressure plates. Nonetheless, there are a few factors that could affect the results significantly.

The oedometer methods of determining the potential expansiveness of clay soils represent more direct methods. In these methods, undisturbed samples are placed in the oedometer and a wide range of testing procedures are used to estimate the likely vertical strain due to wetting under applied vertical pres-

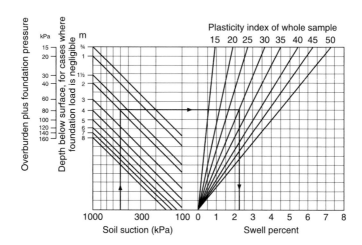

Figure 5.6 Nomogram for deriving the amount of swell of an expansive clay (after Brackley, 1980).

Table 5.1 USAEWES classification of swell potential (from O'Neill and Poormoayed, 1980).

Liquid limit (%)	Plastic limit (%)	Initial (in situ) suction (kN m^{-2})	Potential swell (%)	Classification
< 50	< 50	< 145	< 0.5	Low
50–60	25–35	145–385	0.5–1.5	Marginal
> 60	> 35	> 385	> 1.5	High

sures. The latter may be equated to overburden pressure plus that of the structure that is to be erected. In reality, most expansive clays are fissured, which means that lateral and vertical strains develop locally within the ground. Even when the soil is intact, swelling or shrinkage is not truly one-dimensional. The effect of imposing zero lateral strain in the oedometer is likely to give rise to over-predictions of heave and the greater the degree of fissuring, the greater the over-prediction. The values of heave predicted using oedometer methods correspond to specific values of natural moisture content and void ratio of the sample. Therefore any change in these affects the amount of heave predicted. Consequently, results obtained from the above-mentioned tests can show a wide variation between one test method and another. This is illustrated in Table 5.2.

Effective and economic foundations for low-rise buildings on swelling and shrinking soils have proved difficult to achieve. This is partly because the cost margins on individual buildings are low. Obviously, detailed site investigation and soil testing are out of the question for individual dwellings. Similarly, many foundation solutions that are appropriate for major structures are too costly for small buildings. Nonetheless, the choice of foundation is influenced by the subsoil and site conditions, estimates of the amount of ground movement and the cost of alternative designs (Table 5.3). In addition, different building materials have different tolerances to deflections (Burland and Wroth, 1975). Hence materials that are more flexible can be used to reduce potential damage due to differential movement of the structure.

Three methods can be adopted when choosing a design solution for building on expansive soils: namely, provide a foundation and structure that can tolerate movements without unacceptable damage; isolate the foundation and structure from the effects of the soil; or alter or control the ground conditions.

Table 5.2 Prediction of swelling potential of expansive clay from Ladysmith, South Africa (after Bell and Maud, 1995).

Method	Predicted heave (mm)	
	Maximum	Minimum
Empirical		
Van der Merwe (1964)	89	37
Brackley (1975)	215	18
Weston (1980)	59	8
Oedometer		
Swell under load	40	−23*
Frydman and Calabresi (1987)		
Double oedometer	127	33
Jennings and Knight (1957)		
Soil suction		
Brackley (1980)	57	16

*The negative sign indicates that compression rather than swelling of the soil sample was predicted.

Table 5.3 Foundation design, building procedures and precautionary measures for single-storey residential structures founded on expansive soil horizons (after Watermeyer and Tromp, 1992).

Class	Estimated total heave (mm)	Construction type	Foundation design and building procedures
H	< 5	normal	Foundations to SABS* 0400 Part H. Site drainage and service/plumbing precautions recommended
HI	5–15	modified	Lightly reinforced strip footings. Articulation joints at all internal/external doors and openings. Light reinforcement in masonry. Site drainage and plumbing/service precautions.
		soil raft	Remove all or part of expansive horizon to 10 m beyond the perimeter of the structure and replace with inert backfill compacted to 93% MOD AASHTO density at −1 to +2% of optimum moisture content. Normal construction with lightly reinforced strip footings and light reinforcement in masonry if residual movements. Site drainage and plumbing/service precautions.
H2	15–30	stiffened or cellular raft	Stiffened or cellular raft with articulation joints or solid lightly reinforced brickwork/blockwork. Site drainage and plumbing/service precautions.
		piled construction	Piled foundation with suspended floor slabs with or without ground beams. Site drainage and plumbing/service precautions.
		split construction	Combination of reinforced brickwork/blockwork and full movement joints. Suspended floors or mesh-reinforced ground slabs acting independently from the structure. Site drainage and plumbing/service precautions.
		soil raft	As for HI.
H3	> 30	stiffened or cellular raft	As for H2.
		soil raft	As for HI.
		piled construction	As for H2.

*SABS: South African Bureau of Standards.

In addition to these construction details, moisture control measures should be adopted as far as possible.

The isolation of foundation and structure has been widely adopted for 'severe' and 'very severe' ground conditions. Straight-shafted bored piles can be used in conjunction with suspended floors for severe conditions. The piles are sleeved over the upper part and provided with reinforcement. For severe conditions it may be necessary to place piles at appreciable depth (i.e. below the level of fluctuation of natural moisture content) and/or use under-reams to resist the pull-out forces.

The use of stiffened rafts is fairly commonplace (Bell and Maud, 1995). The design of the slab is dependent on assumed allowable relative deflections and it is usually necessary to incorporate certain anti-cracking features into the superstructure such as flexible joints.

Moisture control is perhaps the most important single factor in the success of foundations on shrinking and swelling clays. The aim is to maintain stable moisture conditions with minimum moisture content or suction gradients. The loss of moisture around the edges of a building, which leads to the moisture content of the soil under the centre of the building being higher, gives rise to differential heave. In order to control this an attempt should be made to maintain the same moisture content beneath a building. This can be achieved by the use of horizontal and vertical moisture barriers around the perimeter of the building, drainage systems and control of vegetation coverage.

A simple method of reducing or eliminating ground movements due to expansive soil is to replace or partially replace them with non-expansive soils. There is no requirement for the thickness of the replacement material, but a minimum of 1 m has been suggested by Chen (1988). The material should be granular, but it should not allow surface water to travel freely through the soil so that it wets any swelling soils in lower horizons. Hence the presence of a fine fraction is required to reduce permeability, or a geomembrane can surround the granular material.

If expansive soil is allowed to swell by wetting prior to construction and if the soil moisture content is then maintained, the soil volume should remain relatively constant and no heave take place. Ponding is the most common method of wetting. This may take several months to increase the water content to the required depth, notably in areas with deep groundwater surfaces. Vertical wells can be installed to facilitate flooding and thus decrease the time necessary to adjust the moisture content of the soil. Williams (1980) described a case where severe damage due to swelling was corrected by controlled wetting.

The amount of heave of expansive soils is reduced significantly when compacted to low densities at high moisture contents. Expansive soils compacted above optimum moisture content undergo negligible swell for any degree of compaction. On the other hand, compaction below optimum results in excessive swell.

Many attempts have been made to reduce the expansiveness of clay soil by chemical stabilization. For example, lime stabilization of expansive soils, prior to construction, can minimize the amount of shrinkage and swelling they undergo. In the case of light structures, lime stabilization can be applied immediately below strip footings. However, significant SO_4 content (i.e. in excess of 5000 mg kg^{-1}) in clay soils can mean that they react with CaO to form ettringite, with resultant expansion (Forster et al., 1995). The treatment is better applied as a layer beneath a raft so as to overcome differential movement. The lime-stabilized layer is formed by mixing 4 to 6% lime with the soil. A compacted layer, 150 mm in thickness, usually gives satisfactory performance. Furthermore, the lime-stabilized layer redistributes unequal moisture stresses in the subsoil, so minimizing the risk of cracking in the structure above, as well as reducing water penetration beneath the raft. Pre-mix or mix-in-place methods can be used (Bell, 1993). Alternatively, lime treatment can be used to form a vertical cut-off wall at or near the footings in order to minimize the movement of moisture.

The lime slurry pressure injection method has also been used to minimize differential movements beneath structures, although it is more expensive (Figure 5.7). The method involves pumping hydrated lime slurry under pressure into soil, the points of injection being spaced about 1.5 m apart. The lime slurry forms a network of horizontal sheets, often interconnected by vertical veins. Injection can be used to form a seam line around the perimeter of a building to provide a barrier to moisture movement, the seams extending below the critical zone of change in moisture content.

Cement stabilization has much the same effect on expansive soils as lime treatment, although the dosage of cement needs to be greater for heavy expansive clays. Alternatively, they can be pre-treated with lime, thereby reducing the amount of cement that needs to be used (Stamatopoulos *et al.*, 1992).

Figure 5.7 Lime-slurry injection method (courtesy of Hayward Baker, Fort Worth, Texas).

5.3 Dispersive soils

Dispersion occurs in soils when the repulsive forces between clay particles exceed the attractive forces, thus bringing about deflocculation so that in the presence of relatively pure water the particles repel each other. In non-dispersive soil; there is a definite threshold velocity below which flowing water causes no erosion. The individual particles cling to each other and are only removed by water flowing with a certain erosive energy. By contrast, there is no threshold velocity for dispersive soil; the colloidal clay particles go into suspension even in quiet water and are therefore highly susceptible to erosion and piping. For a given eroding fluid, the boundary between the flocculated and deflocculated states depends on the value of the sodium adsorption ratio (see below), the salt concentration, the pH value and the mineralogy.

Nonetheless, there are no significant differences in the clay fractions of dispersive and non-dispersive soils, except that soils with less than 10% clay particles may not have enough colloids to support dispersive piping. Dispersive soils contain a higher content of dissolved sodium (up to 12%) in their pore water than ordinary soils. The pH value of dispersive soils generally ranges between 6 and 8.

The sodium adsorption ratio (SAR) is used to quantify the role of sodium where free salts are present in the pore water and is defined as:

$$SAR = \frac{Na}{\sqrt{0.5 \ (Ca + Mg)}} \tag{5.1}$$

with units expressed in meq/litre of the saturated extract. There is a relationship between the electrolyte concentration of the pore water and the exchangeable ions in the adsorbed layers of clay particles. This relationship is dependent upon pH value and may also be influenced by the type of clay minerals present. Hence it is not necessarily constant. An SAR of more than 6 suggests that the soil is sensitive to leaching. However, in Australia, Aitchison and Wood (1965) regarded soils in which the SAR exceeded 2 as dispersive. As can be seen from Table 5.4, the latter value seems to be the more appropriate.

The presence of exchangeable sodium is the main chemical factor contributing towards dispersive behaviour in soil. This is expressed in terms of the exchangeable sodium percentage (ESP):

$$ESP = \frac{exchangeable \ sodium}{cation \ exchange \ capacity} \times 100 \tag{5.2}$$

where the units are given in meq/100 g of dry clay. A threshold ESP value of 10% has been recommended, above which soils that have their free salts leached by seepage of relatively pure water are prone to dispersion. Soils with ESP

Table 5.4 Some physical and chemical properties of dispersive soils from Natal (from Bell and Maud, 1994).

Location	Clay %	LL %	PI %	A	LS %	pH	K meq/l	Ca meq/l	Mg meq/l	Na meq/l	TDS meq/l	ESP %	SAR	CEC meq/100 g Clay	EC mS/m	Dispersivity %	Dispersivity potential
Makatini	30	28	14	0.47	6.0	8.6	2.6	9.0	8.0	97.6	117.2	39.4	33.4	97.9	900		HD
Makatini	29	43	24	0.83	8.0	8.6	2.6	181.9	78.9	115.9	379	30	10.1	132.86			HD
Paddock	40	31	12	0.30	5.3	8.6	1.6	95.6	147.2	23.5	268.2	89	2.1	66.3			HD
Paddock	44	33	14	0.32	6.0	8.6	0.02	0.3	0.4	4.4	5.1	16.6	7.5	60.2	52	44	HD
Winterton	31	27	13	0.42	4.0	8.6	0.03	0.8	0.4	7.7	8.9	24.4	9.9	50.9	95	71.9	HD
Winterton	44	35	10	0.23	5.3	8.95	0.04	0.8	0.5	8.3	9.6	16.1	10.2	36.6	110		HD
Makatini	30	28	14	0.47	6.0	8.6	14.7	293.4	201.5	21.3	530.9	30	1.1	132.86		53.2	HD
Makatini	12	40	14	1.17	6.0	8.85	0.1	0.6	0.5	2.7	3.9	30	3.6	137.3			HD
Heatonville	48	31	18	0.38	8.0	8.1	0.2	5.1	4.9	7.8	17.9	17.3	4.9	52.6	38		HD
Heatonville	25	22	12	0.48	3.3	8.7	0.2	2.9	3.2	5.0	11.3	35.2	4.1	45.8			HD
Weenen	43	23	10	0.23	3.5	8.1	0.3	9.9	4.6	4.1	18.9	45.8	2.2	25.0			HD
Winterton	37	28	13	0.35	5.3	7.5	0.04	1.9	1.5	4.8	8.2	38.4	3.7	53.5		35.4	HD
Rietspruit	36	47	24	0.67	8.0	6.5	2.4	71.6	46.9	13.1	134	6.3	1.7	38.8	81		D
Rietspruit	60	66	34	0.57	10.0	7.55	1.8	160.4	139.1	17.1	318.4	9.3	1.4	53.8			D
Tala	22	45	6	0.27	0.7	5.05	5.3	33.7	4.8	4.8	48.6	5.3	1.1	25.9			D
Weenen	51	34	16	0.31	7.0	8.3	0.9	12.7	5.3	2.3	21.2	8.8	1.1	25.1			D
Ramsgate	18	29	8	0.44	5	5.2	0.9	22.0	46.7	5.4	75	9.1	0.9	38	77.5		D
Ramsgate	21	30	10	0.48	5	5.1	1.2	15.8	31.7	4.6	53.5	7	0.9	30	64.8		D
Ramsgate	20	29	8	0.40	5	5.4	1.4	24.0	61.7	11.3	98.4	8	1.7	44	105.6		D
Ramsgate	23	28	4	0.17	5	5.4	1.4	15.6	63.3	14.6	94.9	11	2.3	43	98.6		D
Ramsgate	22	20	10	0.45	2	5.5	0.7	16.8	21.7	4.2	43.4	15	1.0	23	98.6		D
Ramsgate	21	32	8	0.38	6	5.6	0.9	18	55.0	9.9	83.8	9	1.7	40	70.4		D
Ramsgate	22	30	9	0.41	5	6.4	0.9	28.0	57.5	10.7	97.1	12	1.6	46	119.7		D

LL = liquid limit; PI = plasticity index; A = activity; LS = linear shrinkage; Na = sodium; TDS = total dissolved solids; ESP = exchangeable sodium percentage; SAR = sodium absorption ratio; CEC = cation exchange capacity; EC = electrical conductivity; HD = highly dispersive; D = Dispersive (according to Gerber and Harmse [1987] method).

values above 15%, are highly dispersive (Bell and Maud, 1994). Those with low cation exchange values (15 meq/100 g of clay) have been found to be completely non-dispersive at ESP values of 6% or below. Similarly, soils with high cation exchange capacity values and a plasticity index greater than 35% swell to such an extent that dispersion is not significant. High ESP values and piping potential generally exist in soils in which the clay fraction is composed largely of smectitic and other 2:1 clays. Some illites are highly dispersive. On the other hand, high values of ESP and high dispersibility are rare in clays composed largely of kaolinite.

Another property that has been claimed to govern the susceptibility of clayey soils to dispersion is the total content of dissolved salts (TDS) in the pore water. In other words, the lower the content of dissolved salts in the pore water the greater the susceptibility of sodium-saturated clays to dispersion. Sherard *et al.* (1976) regarded the total dissolved salts for this specific purpose as the total content of calcium, magnesium, sodium and potassium in milliequivalents per litre. They designed a chart in which sodium content was expressed as a percentage of TDS and was plotted against TDS to determine the dispersivity of soils (Figure 5.8). However, Craft and Acciardi (1984) showed that this chart had poor overall agreement with the results of physical tests. Furthermore, Bell and Maud (1994) showed that the use of the dispersivity chart to distinguish dispersive soils had not proved reliable in Natal, South Africa. There the determination of dispersive potential frequently involves the use of a chart designed

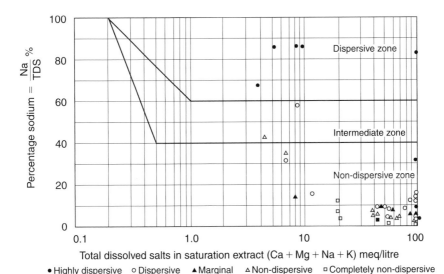

Figure 5.8 Potential dispersivity chart, based on TDS and percentage sodium, of Sherard *et al.* (1976) with results from Bell and Maud (1994).

by Gerber and Harmse (1987), which plots ESP against cation exchange capacity (CEC) (Figure 5.9).

Dispersive soils occur in semi-arid regions; for example, in South Africa they are found in areas that have less than 850 mm of rain annually. More specifically, with few exceptions, they are present in areas that experience Weinert's (1980) climatic N values between 2 and 10. Dispersive soils tend to develop in low-lying areas with gently rolling topography and smooth, relatively flat slopes where the rainfall is such that seepage water has a high SAR. In more arid regions, where N values exceed 10, the development of dispersive soils is generally inhibited by the presence of free salts, despite high SAR values.

Unfortunately, dispersive soils cannot be differentiated from non-dispersive soils by routine soil mechanics testing; for example, they cannot be identified by their plasticity or activity. Although a number of tests have been used to recognize dispersive soils, no single test can be relied on completely to identify them. For example, Craft and Acciardi (1984) found that the crumb test and the pinhole test at times yield conflicting results from the same samples of soil. Subsequently, Gerber and Harmse (1987) showed that the crumb test, the double hydrometer test and the pinhole test were unable to identify dispersive soils when free salts were present in solution in the pore water, which is frequently the case with sodium-saturated soils. Atkinson *et al.* (1990) also concluded that because internal erosion is governed by true cohesion (i.e. the strength at zero effective stress) and since true cohesion may be influenced by physico-chemical effects in the soil grains, in the pore water and in the free

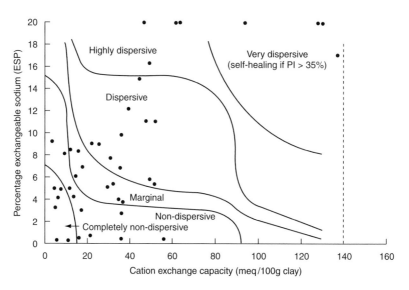

Figure 5.9 Gerber and Harmse (1987) dispersivity chart, showing results of tests carried out on soils from Natal by Bell and Maud (1994).

water, the crumb test, dispersion test and pinhole test have shortcomings. They devised the cylinder dispersion test to examine erosion and dispersion of soils from the surfaces of cracks into water.

The various tests show that the boundary between deflocculated and flocculated states varies considerably between different soils, so the transition between dispersive and non-dispersive soils is wide. Because there is no clearly defined boundary between dispersive and non-dispersive soils, it is wise to do several types of test on at least a proportion of all samples.

As no one test can be relied upon to identify dispersive soil with absolute certainty, Bell and Maud (1994) suggested a tentative rating system (Table 5.5). This included some of the tests or properties that have been used to help to recognize dispersive soils. The most important of these was the assessment based on ESP and CEC (meq/100 g clay).

Serious piping damage to embankments and failures of earth dams have occurred when dispersive soils have been used in their construction (Figure 5.10). Severe erosion damage can also form deep gullies on earth embankments after rainfall. Indications of piping take the form of small leakages of muddy-coloured water from an earth dam after initial filling of the reservoir. The pipes become enlarged rapidly and this can lead to failure of the dam. Dispersive erosion may be caused by initial seepage through an earth dam in areas of higher soil permeability, especially areas where compaction may not be so effective such as around conduits, against concrete structures and at the foundation interface; through desiccation cracks; or cracks due to differential settlement or to hydraulic fracturing (Wilson and Melis, 1991).

In many areas where dispersive soils are found there is no economic alternative to using these soils for the construction of earth dams. Experience,

Table 5.5 Suggested rating system for potentially dispersive soils (after Bell and Maud, 1994).

Crumb test	Class	Strong reaction	Moderate	Slight	No reaction
	Rating	4	2	1	0
Dispersion test	Class	Highly dispersive	Moderately	Slightly	Non-dispersive
	Rating	4	2	1	0
*ESP/CEC	Class	Highly dispersive	Dispersive	Marginal	Non-dispersive
(meq/100 g clay)	Rating	5	3	1	0
SAR	Class	> 10	2–10	< 2	< 2
	Rating	3	1	0	0
pH	Class	> 8	6–8	6–8	< 6
	Rating	2	1	1	0

Highly dispersive = 18 and above; dispersive = 9–17; cautionary zone = 5–8; marginal = 1–4; non-dispersive = 0.

* Highly dispersive includes very highly dispersive and non-dispersive includes completely non-dispersive of Gerber and Harmse (1987).

Figure 5.10 Failure of an earth dam constructed of dispersive soil near Ramsgate, Natal, South Africa.

however, indicates that if an earth dam is built with careful construction control and incorporates filters, then it should be safe enough even if it is constructed with dispersive clay. If no supply of clean pervious sand of the type used for a chimney drain is available, then a zone of silty sand or sandy silt can be considered as a line of defence against piping in dispersive soil used for a homogeneous earth dam. Such a chimney may not be pervious enough to act as a drain but it would be expected to prevent piping. Watermeyer *et al*. (1991) referred to the use of geotextiles as drainage filters in the Bloemhoek dam in South Africa. Sherard *et al*. (1977) maintained that many homogeneous dams without filters, in which dispersive clay has been properly compacted, experience no leaks and so no failure has occurred. If a leak developed under unusual conditions, such as cracking due to drying consequent upon a long period of low water level or earthquake shock, then failure could occur. In order to increase the security of such a dam, a blanket of sand filter can be placed on the downstream side covered with a weighted berm.

Alternatively, hydrated lime, pulverized fly ash, gypsum and aluminium sulphate have been used to treat dispersive clays used in earth dams. The type of stabilization undertaken depends on the properties of the soil, especially the ESP and the SAR. The quantity of hydrated lime used should be that which raises the shrinkage limit to a value near saturation moisture content based on the compaction density to be achieved in the embankment. Lime treatment is usually applied to the outer 0.3 m of the surface of the embankment. McDaniel and Decker (1979) found that the addition of 4%, by weight, of hydrated lime converted dispersive soil to non-dispersive soil. However, homogeneous mixing of small quantities of lime may not be achieved, and mixing, besides introducing brittleness, disrupts work, which can lead to shrinkage cracks developing in the dam.

Indraratna *et al.* (1991) showed that pulverized fly ash, derived from burning lignite, and with a relatively high content of CaO, can be used to stabilize dispersive soil. Around 6% is mixed with dispersive soil. However, it must be remembered that the composition and physical properties of fly ash vary depending on the coal used and the firing process, and therefore need to be determined in order to ascertain whether a particular fly ash will be suitable as a stabilizing agent.

Because of its relatively low cost and reasonable solubility in water, gypsum, when in a very fine powder form, is another stabilizing material that can be used. The rate of base exchange reaction is controlled by its solubility in water with a high pH. The gypsum is mixed with the soil during construction. The quantity of gypsum added is equivalent to the excess sodium that it is required to replace in order to bring ESP values within the desired limits. The water in a reservoir can also be dosed with gypsum so that as seepage occurs, deflocculation in the soil is prevented.

Aluminium sulphate or alum also has been added to dispersive clays. Although alum is highly soluble in water it only effects a cation exchange reaction (i.e. it is not cementitious), and it is strongly acidic. Nonetheless, Bourdeaux and Imaizumi (1977) stabilized dispersive clay at the Sobradinho dam site, Brazil, by using 0.6% aluminium sulphate, by dry weight of soil.

Dispersive soils can also present problems in earthworks, such as those required for roads on both the fill and cut slopes relating thereto. In the case of embankment fills, dispersive soil can be used provided that it is covered by an adequate depth of better-class material. Care has to be exercised in the placement and compaction of the fill layers so that no layer is left exposed during construction for such a period that it can shrink and crack, and thus weaken the fill. As in the case of the construction of earth dams, the dispersive soil material should be placed and compacted at 2% above its optimum moisture content to inhibit shrinkage and cracking therein. Where seepage areas or springs are located along the alignment of a road embankment that has to be constructed of dispersive material, special care must be exercised in the provision of adequate subsoil drainage to such areas, otherwise the long-term stability of the embankment could be jeopardized by the development of piping in the dispersive soil. In some instances, to reduce erosion, it is prudent to stabilize the outer 0.3 m or so of dispersive soil material in an embankment with lime, the soil material and lime being mixed in bulk prior to placement rather than being mixed *in situ*.

To minimize potential settlement and erosion problems when dispersive soil fill is placed against a structure, such as a bridge abutment or the wing wall or a culvert, such structures, where practically possible, are provided with sloping soil interfaces such that the soil settles on to the interface. In this way the possibility of cracks developing in the soil that could lead to piping erosion is reduced.

In the case of road and other earthwork-cut slopes in dispersive soil, unless adequately protected from surface erosion, severe runnel and gullying erosion can develop thereon. To some extent, this problem can be reduced by providing a steeper than normal slope, for instance 0.5 vertical in 1.5 horizontal, as the dispersive soil usually possesses adequate cohesion to stand in a stable condition up to a height of about 3 m (under natural conditions near-vertical slopes in erosion gullies in dispersive colluvium frequently stand satisfactorily to heights of 5 m). The steeper than normal slope means that comparatively less rainfall falls on the slope, so the amount of slope erosion is reduced. In the relatively few instances where cuts are located in more than 3 m of dispersive soil, flatter slopes of about 1 vertical in 2 horizontal have to be employed. Such slopes have to be adequately vegetated immediately on completion to limit erosion. Adequate open-channel drainage (1 to 2 m in width) has to be provided at the toes of cut slopes. These require cleaning maintenance from time to time, and such channel drains should be concrete-lined to limit erosion along their length.

Dispersive soils have low natural fertility and are frequently calcareous and have a relatively high pH. It therefore can be difficult to establish and maintain suitable vegetation on fill constructed of, and cut slopes in, dispersive soils in order to inhibit surface erosion. Apart from adequate artificial fertilization, it is usually necessary to place topsoil on such slopes to ensure satisfactory vegetative growth. The steepest slope that can be topsoiled satisfactorily is about 1 vertical in 1 horizontal, and even such a slope may have to have artificial anti-erosion measures installed on it to hold the topsoil and vegetation in place.

5.4 Collapsible soils

Soils such as loess, brickearth and certain windblown silts may possess the potential to collapse. These soils generally consist of 50–90% silt particles, and sandy, silty and clayey types have been recognized by Clevenger (1958), with most falling into the silty category (Figure 5.11). Collapsible soils possess porous textures with high void ratios and relatively low densities. They often have sufficient void space in their natural state to hold their liquid limit moisture at saturation. At their natural low moisture content these soils possess high apparent strength, but they are susceptible to large reductions in void ratio upon wetting. In other words, the metastable texture collapses as the bonds between the grains break down when the soil is wetted. Hence the collapse process represents a rearrangement of soil particles into a denser state of pack-

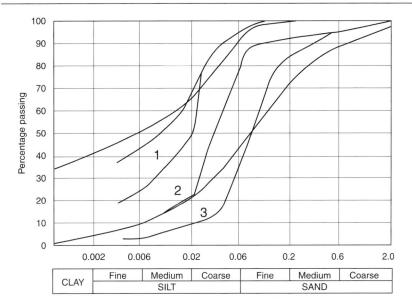

Figure 5.11 Clayey loess, silty loess and sandy loess compared with a brickearth from south Essex, England (after Northmore *et al.*, 1996).

ing. Collapse on saturation normally takes only a short period of time, although the more clay such a soil contains, the longer this period tends to be.

The fabric of collapsible soils generally takes the form of a loose skeleton of grains (generally quartz) and micro-aggregates (assemblages of clay or clay and silty clay particles). These tend to be separate from each other, being connected by bonds and bridges, with uniformly distributed pores. The bridges are formed of clay-sized minerals, consisting of clay minerals, fine quartz, feldspar or calcite. Surface coatings of clay minerals may be present on coarser grains. Silica and iron oxide may be concentrated as cement at grain contacts, and amorphous overgrowths of silica occur on grains. As grains are not in contact, mechanical behaviour is governed by the structure and quality of the bonds and bridges.

The structural stability of collapsible soils is not only related to the origin of the material, to its mode of transport and depositional environment but also to the amount of weathering undergone. For instance, Gao (1988) pointed out that the weakly weathered loess of the northwest of the loess plateau in China has a high potential for collapse, whereas the weathered loess of the southeast of the plateau is relatively stable and the features associated with collapsible loess are disappearing gradually. Moreover, in more finely textured loess deposits, high capillary potential plus high perched groundwater conditions have caused loess to collapse naturally through time, thereby reducing its porosity. This reduction in porosity, combined with a high liquid limit, makes the possibility of collapse less likely. Gao concluded that highly collapsible loess usually occurs in regions near the source of the loess, where its thickness is at a maximum, and where the landscape and/or the climatic conditions are not conducive to development of long-term saturated conditions within the soil. This view was supported by Grabowska-Olszewska (1988), who found that in Poland collapse is most frequent in the youngest loess and that it is almost exclusively restricted to loess that contains slightly more than 10% particles of clay size. Such soils are characterized by a random texture and a carbonate content of less than 5%. They are more or less unweathered and possess a pronounced pattern of vertical jointing. The size of the pores is all-important, Grabowska-Olszewska maintaining that collapse occurs as a result of pore space reduction taking place in pores greater than 1 µm in size and more especially in those exceeding 10 µm in size. Phien-wej *et al.* (1992) reported that pore size in loess in northeast Thailand frequently varied from 200 to 500 µm.

Popescu (1986) maintained that there is a limiting value of pressure, defined as the collapse pressure, beyond which deformation of soil increases appreciably. The collapse pressure varies with the degree of saturation. He defined truly collapsible soils as those in which the collapse pressure is lower than the overburden pressure. In other words, such soils collapse when saturated, since the soil fabric cannot support the weight of the overburden. When the saturation collapse pressure exceeds the overburden pressure soils are capable of supporting a certain level of stress on saturation, and Popescu defined these soils as conditionally collapsible soils. The maximum load that such soils can support is the difference between the saturation collapse and overburden pressures. Phien-wej *et al.* (1992) concluded that the critical pressure at which collapse of the soil fabric begins is greater in soils with smaller moisture content. Under the lowest natural moisture content the soils investigated posed a severe problem on wetting (the collapse potential was as high as 12.5% at 5% natural moisture content). During the wet season, when the natural moisture content could rise to 12%, there was a reduction in the collapse potential to around 4%.

Several collapse criteria have been proposed for predicting whether a soil is liable to collapse upon saturation and loading. For instance, Clevenger (1958) suggested a criterion for collapsibility based on dry density, that is, if the dry density is less than $1.28\,\mathrm{Mg\,m^{-3}}$, then the soil is liable to significant settling. On the other hand, if the dry density is greater than $1.44\,\mathrm{Mg\,m^{-3}}$, then the amount of collapse should be small, while at intermediate densities the settlement is transitional. Gibbs and Bara (1962) suggested the use of dry unit weight and liquid limit as criteria to distinguish between collapsible and non-collapsible soil types. Their method is based on the premise that a soil that has enough void space to hold its liquid limit moisture content at saturation is susceptible to collapse on wetting. This criterion applies only if the soil is uncemented and the liquid limit is above 20%. When the liquidity index in such soils approaches or exceeds 1, then collapse may be imminent. As the clay content of a collapsible soil increases, the saturation moisture content becomes less than the liquid limit, so such deposits are relatively stable. However, Northmore *et al.* (1996) concluded that this method did not provide a satisfactory means of identifying the potential metastability of brickearth (Figure 5.12). More simply, Handy (1973) suggested that collapsibility could be determined from the ratio of liquid limit to saturation moisture content. Soils in which this was less than 1 were collapsible, while if it was greater than 1 they were safe.

Collapse criteria have been proposed that depend upon the void ratio at the liquid limit (e_1), plastic limit (e_p) and natural void ratio (e_o). For example, Fookes and Best (1969) proposed a collapse index (i_c), which involved these void ratios as follows:

$$i_c = \frac{e_o - e_p}{e_1 - e_p} \tag{5.3}$$

Previously, Feda (1966) had proposed the following collapse index:

$$i_c = \frac{m/S_r - PL}{PI} \tag{5.4}$$

in which m is the natural moisture content, S_r is the degree of saturation, PL is the plastic limit and PI is the plasticity index. Feda also proposed that the

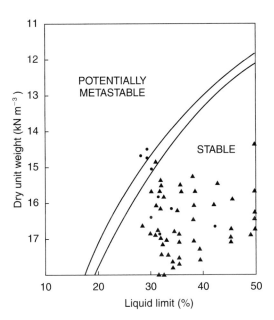

Figure 5.12 Brickearth from south Essex (triangles) compared with that from Kent (dots) on the stability diagram of Gibbs and Bara (1962) (after Northmore et al., 1996).

soil must have a critical porosity of 40% or above and that an imposed load must be sufficiently high to cause structural collapse when the soil is wetted. He suggested that if the collapse index was greater than 0.85, then this was indicative of metastable soils. However, Northmore et al. (1996) suggested that a lower critical value of collapse index, that is 0.22, was more appropriate for the brickearths of south Essex. Derbyshire and Mellors (1988) also referred to a lower collapse index for the brickearths of Kent. This may be due to the greater degree of sorting in brickearths than in loess.

The oedometer test can be used to assess the degree of collapsibility. The test involves loading an undisturbed specimen at natural moisture content in the oedometer up to a given load. At this point, the specimen is flooded and the resulting collapse strain, if any, is recorded. Then the specimen is subjected to further loading. The total consolidation upon flooding can be described in terms of the coefficient of collapsibility (C_{col}), given by Feda (1988) as:

$$C_{col} = \Delta h/h$$
$$= \frac{\Delta e}{1 + e} \tag{5.5}$$

where Δh is the change in height of the specimen after flooding, h is the height of the specimen before flooding, Δe is the change in void ratio of the specimen upon flooding and e is the void ratio of the specimen prior to flooding. Table 5.6 provides an indication of the potential severity of collapse. This table indicates that those soils that undergo more than 1% collapse can be regarded as metastable. However, in China a figure of 1.5% is taken (Lin and Wang, 1988), and in the United States values exceeding 2% are regarded as indicative of soils susceptible to collapse (Lutenegger and Hallberg, 1988).

From the above it may be concluded that significant settlements can take place beneath structures in collapsible soils after they have been wetted (in some cases in the order of metres, Feda et al., 1993). These have led to foundation failures (Clevenger, 1958; Phien-wej et al., 1992). Clemence and Finbarr (1981) recorded a number of techniques that could be used to stabilize collapsible soils. These are summarized in Table 5.7. Evstatiev (1988) also pro-

Table 5.6 Collapse percentage as an indication of potential severity (after Jennings and Knight, 1975).

Collapse (%)	Severity of problem
0–1	No problem
1–5	Moderate trouble
5–10	Trouble
10–20	Severe trouble
> 20	Very severe trouble

Table 5.7 Methods of treating collapsible foundations (based on Clemence and Finbarr, 1981).

Depth of subsoil treatment	Foundation treatment
	A. *Current and past methods*
0–1.4 m	Moistening and compaction (conventional extra heavy impact or vibratory rollers).
1.5–10 m	Over-excavation and recompaction (earth pads with or without stabilization by additives such as cement or lime). Vibro-flotation (free draining soils). Vibroreplacement (stone columns). Dynamic compaction. Compaction piles. Injection of lime. Lime piles and columns. Jet grouting. Ponding or flooding (if no impervious layer exists). Heat treatment to solidify the soils in place.
Over 10 m	Any of the aforementioned or combinations of the aforementioned, where applicable. Ponding and infiltration wells, or ponding and infiltration wells with the use of explosive.
	B. *Possible future methods* Ultrasonics to produce vibrations that will destroy the bonding mechanics of the soil. Electrochemical treatment. Grouting to fill pores.

vided a survey of methods that can be used to improve the behaviour of collapsible soils.

Another problem that may be associated with loess soils is the development of pipe systems. Extensive pipe systems, which have been referred to a loess karst, may run sub-parallel to a slope surface, and the pipes may have diameters up to 2.0 m. Pipes tend to develop by weathering and widening taking place along the joint systems in loess. The depths to which pipes develop may be inhibited by changes in permeability associated with the occurrence of palaeosols.

5.5 Quick clays

The material of which quick clays are composed is predominantly smaller than 0.002 mm, but many deposits seem to be very poor in clay minerals, containing a high proportion of ground-down, fine quartz. For instance, it has been shown that quick clay from St Jean Vienney consists of very fine quartz and plagioclase. Indeed, examination of quick clays with the scanning electron microscope has revealed that they do not possess clay-based structures, although such work has not lent unequivocal support to the view that non-clay particles govern the physical properties.

Cabrera and Smalley (1973) suggested that such deposits owe their distinctive properties to the predominance of short-range inter-particle bonding forces, which they maintained were characteristic of deposits in which there was an abundance of glacially produced, fine non-clay minerals. In other words, they contended that ice sheets had supplied abundant ground quartz in the form of rock flour for the formation of quick clays. Certainly quick clays have a restricted geographical distribution, occurring in certain parts of the northern hemisphere that were subjected to glaciation during Pleistocene times.

Gillott (1979) has shown that the fabric and mineralogical composition of sensitive soils from Canada, Alaska and Norway are qualitatively similar. He pointed out that they possess an open fabric, high moisture content and similar index properties (Table 5.8). An examination of the fabric of these soils revealed the presence of aggregations. Granular particles, whether aggregations or primary minerals, are rarely in direct contact, generally being linked by bridges of the particles. Clay minerals are usually non-oriented, and clay coatings on primary minerals tend to be uncommon, as are cemented junctions. Networks of platelets occur in some soils. Primary minerals, particularly quartz and feldspar, form a higher than normal proportion of the clay-size fraction, and illite and chlorite are the dominant phyllosilicate minerals. Gillot noted that the presence of swelling clay minerals varies from almost zero to significant amounts.

The open fabric which is characteristic of quick clays has been attributed to their initial deposition, during which time colloidal particles interacted to form loose aggregations by gelation and flocculation. Clay minerals exhibit strongly marked colloidal properties and other inorganic materials such as silica behave as colloids when sufficiently fine-grained. Gillott (*ibid.*) suggested that the open fabric may have been retained during very early consolidation because it remained a near equilibrium arrangement. Its subsequent retention to the present day may be due to mutual interference between particles and buttressing of junctions between granules by clay and other fine constituents, precipitation of cement at particle contacts, low rates of loading, and low load increment ratio.

Quick clays often exhibit little plasticity, their plasticity indices at times varying between 8 and 12%. Their liquidity index normally exceeds 1, and their liquid limit is often less than 40%. Quick clays are usually inactive, their activity frequently being less than 0.5. The most extraordinary property possessed by quick clays is their very high sensitivity (Figure 5.13). In other words, a large proportion of their undisturbed strength is permanently lost following shear. The liquidity index can be used to show sensitivity increases in clay (Figure 5.14). The small fraction of the original strength regained after remoulding quick clays may be attributable to the development of some different form of inter-particle bonding. The reason why only a small fraction of

Table 5.8 Engineering properties of sensitive soils (after Gillott, 1979).

Location[*]	Depth (m)	Natural moisture content (%)	Pre-consolidation pressure (kN m^{-2})	Undrained strength (kN m^{-2})	Sensitivity	Liquid limit (%)	Plastic limit (%)	Liquidity index	Activity
O	13.7	60	450	160	–	49	23	1.4	0.35
Q	5.2	75	150	50	–	70	26	1.1	0.64
Q	14.3	81	150	50	–	65	28	1.4	0.45
O	2.6	65	60	20	100	55	22	1.3	0.73
Q	12.2	28	590	230	–	23	16	1.7	0.18
O	5.2	78	320	120	–	65	28	1.3	0.44
BC	20.1	38	–	20	30	28	22	2.7	0.22
BC	14.0	29	–	–	4	23	16	1.9	0.33
BC	35.4	37	–	60	5	28	23	2.8	0.17
A	61.3	17	–	–	–	26	21	0.8	–
A	60.7	–	–	–	–	23	20	–	–

* O = Ontario, Q = Quebec, BC = British Columbia, A = Alaska.

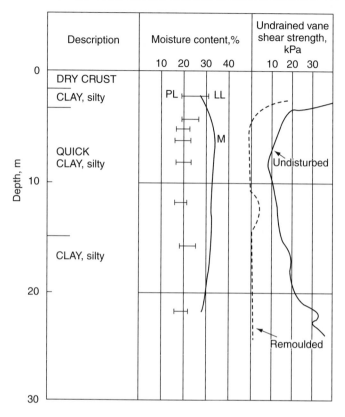

Figure 5.13 Moisture content, plastic and liquid limits, and undisturbed and remoulded shear strength (indicating sensitivity) of quick clay from near Trondheim, Norway.

the original strength can ever be recovered is because the rate at which it develops is so slow. As an example, the Leda Clay is characterized by exceptionally high sensitivity, commonly between 20 and 50, and a high natural moisture content and void ratio, the latter commonly being about 2. It has a low permeability, being around 10^{-10} m s^{-1}. The plastic limit is around 25%, with a liquid limit about 60% and an undrained shear strength of 700 kPa. When subjected to sustained load an undrained triaxial specimen of Leda Clay exhibits a steady time-dependent increase in both pore pressure and axial strain. Continuing undrained creep may often result in a collapse of the sample after long periods of time have elapsed.

In a review of the mineralogy and physical properties of sensitive clays from eastern Canada, Locat *et al.* (1984) found that plagioclase was the dominant mineral (from 25 to 48%) followed by quartz, microcline and hornblende. Small quantities of dolomite and calcite were also present. The phyllosilicates (including the clay minerals, of which illite is the most common), together with amor-

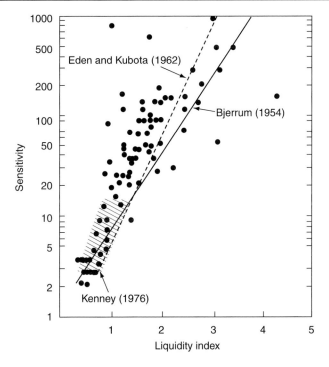

Figure 5.14 Relationship between sensitivity and liquidity index for a sensitive clay (after Lutenegger and Hallberg, 1988).

phous materials, never represented more than one-third of the minerals present in the samples analysed.

As far as the physical properties of these sensitive clays are concerned, with one exception, all the samples fall above the 'A' line on the plasticity chart, indicating inorganic material. The natural moisture contents, plastic limits, liquid limits and plasticity indices of the soils are given in Table 5.9, from which it can be seen that their natural moisture contents always exceed the plastic limits and commonly exceed the liquid limits. In such cases their liquidity indices are greater than 1. Strength decreases and sensitivity increases dramatically as the liquidity index increases. This is illustrated by the fall cone strengths quoted by Locat *et al.* (*ibid.*), the strength more or less disappearing on remoulding, giving sensitivity values varying from 10 to over 1000. Values of specific surface area, cation exchange capacity and salinity of the pore water are also provided in Table 5.9. The variation in the geotechnical properties of these soils was primarily attributed to their differences in mineralogy and texture.

Quick clays are associated with several serious engineering problems. Their bearing capacity is low and settlement is high, and predicting the consolidation of quick clays by the standard methods is unsatisfactory. Slides in quick

Table 5.9 Geotechnical properties of sensitive clays from eastern Canada (after Locat et al., 1984).

Site	c_u kPa	c_{ur} kPa	S_t	S (g/L)	w (%)	LL (%)	PL (%)	PI (%)	LI	<2μm (%)	SS (m² g⁻¹)	CEC (meq/100g)	P (%)	<2μm (%)
Grande-Baleine	32	0.1	282	0.7	58	36.0	24.0	12.0	2.8	74	26	14.4	8.5	65.5
Grande-Baleine	24	0.2	120	–	46	35.8	23.0	12.8	1.8	66	27	–	15	51.0
Grande-Baleine	23	0.2	114	0.7	58	36.9	25.0	11.9	2.7	70	24	7.5	–	–
Olga	23	1.0	19	0.3	83	73.4	26.0	47.4	1.2	94	85	44.4	34.5	59.5
Olga	23	0.9	25	0.3	85	66.7	28.0	38.7	1.5	82	48	10.0	–	–
Olga	28	1.19	24	–	88	67.6	27.8	39.8	1.5	88	55	–	31.3	56.7
St Marcel	81	0.8	24	<2	81	64.0	28.0	36.0	1.5	85	67	21.9	26.7	58.3
St Marcel	–	1.21	–	<2	82	61.8	25.8	36.0	1.6	85	58	–	24.9	60.0
St Léon	80	3.4	24	12	58	61.0	23.5	37.5	0.9	74	56	14.1	32.2	41.8
St Alban	80	2.2	37	2.0	40	36.0	21.0	15.0	1.3	43	46	11.3	33.2	9.8
St Barnabé	157	3.1	50	4	48	43.0	20.0	23.0	1.2	50	40	8.5	19.4	30.6
Shawinigan	109	1.8	62	0.6	33	34.0	18.0	16.0	0.9	36	29	6.1	12.8	23.2
Chicoutimi	245	0.5	532	–	33	28.0	18.0	10.0	1.5	55	25	22.5	14.0	41.0
Outardes	109	0.6	181	–	64	34.0	21.5	12.5	3.4	50	30	7.3	–	–
Outardes	130	<0.07	>10³	–	35	25.0	17.0	8.0	2.3	47	23	–	–	–

c_u = undrained shear strength, c_{ur} = remoulded undrained shear strength, S_t = sensitivity, S = pore water salinity, w = natural moisture content, LL = liquid limit, PL = plastic limit, PI = plasticity index, LI = liquidity index, SS = specific surface area, CEC = cation exchange capacity, P = amount of phyllosilicates and amorphous minerals.

clays have sometimes proved disastrous, and unfortunately the results of slope stability analyses are often unreliable.

Quick clays can liquefy on sudden shock. This has been explained by the fact that if quartz particles are small enough, having a very low settling velocity, and if the soil has a high water content, then the solid–liquid transition can be achieved.

5.6 Soils of arid regions

Most arid deposits consist of the products of the physical weathering of bedrock formations. The process gives rise to a variety of rock and mineral fragments, which may then be transported and deposited under the influence of gravity, wind or water. Arid conditions may give rise to a variety of evaporite deposits and cemented layers due to the precipitation of soluble minerals.

A number of engineering problems arise with Quaternary arid deposits. Some of these are common features of the materials themselves and others occur due to the climatic conditions. The shortage of surface water, together with the harsh conditions, inhibit biological and chemical activity such that in most arid regions there is only a sparse growth, or perhaps absence, of vegetation. This makes the surface regolith of upland areas and slopes highly vulnerable to intense denudation and redistribution by gravitational or aeolian processes or the action of ephemeral water. Weathering activity tends to be dominated by the physical breakdown, in places intense, of rock masses into poorly sorted assemblages of fragments ranging in size down to silts.

The rapid erosion of loose surface material by hill wash or ephemeral stream and river channels during rainfall events often leads to debris flows, especially in alluvial fans. Many of the deposits within alluvial plains and covering hillsides are poorly consolidated. As such, they may undergo large settlements, especially if subjected to vibration due to earthquakes or cyclic loading. Much of the gravel-sized material is liable to consist of relatively weak, low-durability materials.

Many arid areas are dominated by the presence of large masses of sand, usually derived from coasts, permanent or temporary streams and by direct attrition of rocks, particularly sandstones. Depending on the rate of supply of sand, the wind speed, direction, frequency and constancy and the nature of the ground surface, it may be transported and/or deposited in mobile or static dunes.

For the most part, aeolian sands are poorly graded and have smooth surfaces. In the absence of downward leaching, surface deposits become contaminated with precipitated salts, particularly sulphates and chlorides. Alluvial plain deposits often contain gypsum particles and cement, and also fragments of weak, weathered rock and clay.

Most beach materials in arid regions are carbonate sands with smooth, well-rounded, porous, chemically unstable particles. Again, contamination with evaporative salts can cause problems. Low-lying coastal zones and inland plains with shallow water tables are areas in which sabkha conditions commonly develop (see also Section 8.7). These are extensive saline flats that are underlain by sand, silt or clay and are often encrusted with salt. As remarked, sabkhas may occur along the coast (coastal sabkhas) or further inland (continental or inland sabkhas). Groundwater in coastal sabkhas is recharged directly from the sea, from inland sources or by infiltration of sea water blown inland by onshore winds. Highly developed sabkhas tend to retain a greater proportion of soil moisture than moderately developed sabkhas. In addition, the higher the salinity of the groundwater, the greater the amount of water retained by the sabkha at a particular drying temperature (Sabtan *et al.*, 1995). The groundwater surface is held at a particular level by capillary soil moisture. The height to which water can rise from the water table is a function of the size and continuity of the pore spaces in the soil. Lane and Washburn (1946) demonstrated that capillary rises of up to 3 m can occur in some fine-grained deposits. Soil above this level is subject to aeolian erosion.

Within coastal sabkhas the dominant minerals are calcite, dolomite and gypsum, with lesser amounts of anhydrite, magnesite, halite and carnalite, together with various other sulphates and chlorides (Table 5.10). For example, James and Little (1994) described the fine-grained soils of the sabkhas of Jubail on the Gulf coast of Saudi Arabia as being partially cemented with sodium chloride and calcium sulphate (i.e. gypsum). In fact, they noted that little carbonate was present. James and Little also referred to the highly saline nature of the groundwater, which at times contained up to 23% NaCl and occurred close to ground level. In such aggressively saline conditions precautions have to be taken when using concrete. James and Little recommended the use of high-density, low water/cement ratio concrete made from sulphate-resisting cement. The sodium chloride content of the water is also high enough to represent a corrosion hazard to steel reinforcement in concrete and to steel piles. Indeed, James and Little suggested that the thickness of steel may be reduced by half within 15 years. In such instances, reinforcement requires protective sheathing.

Minerals that are precipitated from groundwater in arid deposits also have high solution rates, so flowing groundwater may lead quickly to the development of solution features. Problems such as increased permeability, reduced density and settling are liable to be associated with engineering works or natural processes that result in a decrease in the salt concentration of groundwater. Hence, care needs to be exercised with irrigation, bridge, tunnel and harbour works, which may lead to the removal of soluble minerals from the ground. Changes in the hydration state of minerals, such as swelling clays and calcium sulphate, also cause significant volume changes in soils. In particular, low-density sands that are cemented with soluble salts such as sodium chloride are vulnerable to salt removal by dissolution by fresh water, leading to settling. Hence, rainstorms and burst water mains present a hazard, as does watering of grassed areas and flower beds. The latter should be controlled, and major structures should be protected by drainage measures to reduce the risks

Table 5.10 Characteristic features of zones representing incipient, slightly developed, moderately developed and well-developed sabkhas respectively in the Dahban area (from Sabtan et al., 1995).

Zone	Characteristic features	Sabkha type	Groundwater salinity	
			Description	Range (g l^{-1})
1	Uneven, ridge surface. Embryonic dunes unstable. Mottled massive sands on hard dark-coloured zones. Living green plants (zygophylum). Isolated thin salt crust lenses or layers, common after rainfall around the living zygophylum plants, animal burrows, bioturbation with mottled dark sandy zones. Some vermiform or recumbent cone-form salt ridges, common around sand dunes.	Incipient	Very saline	30 to 40
2	Stable sand accumulations. Rippled windblown sand sheets (partially wet). Spotty damp surface. Dry plants (dry zygophylum). Gypsiferous aridosols. Medium-size scattered foggy gypsum fragments or crystals. Thin layers or lenses of gypsum or halite repeated in different depths in the capillary fringe zone (gypsum mainly accretes in desert roses or concretions). Common buried plants and roots, animal burrows and bioturbation. Salt ridges in vermiform or recumbent coneform, common after rainfall or raising of water table. Mottled reddish brown to grey surface.	Slightly developed	Slightly hypersaline	45 to 70
3	Reddish brown, wet, flat surface, with windblown sand sheets trapped near the boundaries and gradually disappeared. Sediments show apparant thickness, which can be reduced by 15–25% by vehicle movement. Small lenses of light coloured (often dry) sand separating salt ridge structures. Remains of dents or small pits resulting from dissolution of salts due to rainfall or rising up of groundwater. Reddish brown stain on sand grains (weathering effect). Salt ridges in postural or brine forms (active height 10–15 cm). Isolated living and mature wrinkled algal mats common in the upper intertidal areas or during storm periods in the supratidal zone. Plants often absent. Buried carbonized plants and roots. Large elongated and flaky, partially dissolved gypsum crystals or roses in the saturated zone. Very weak rootlet form gypsum mesh band in the capillary fringe zone.	Moderately developed	Moderately hypersaline	70 to 120
4	Light grey mainly dry, or light brown mottled, partially damp crumbly flat surface. Large-scale mature salt tents (cracked crest, 20 cm high, 50 cm diameter; dip of sides 30–75°). Large euhedral flaky gypsum crystals (African combs forms). Teepee structures. Gypsum nodules. Thick (20–40 cm) lenses or layers of evaporates (mainly gypsum) in rootlet form in the upper part of the capillary fringe zone. Sticky clay (thickness variable), directly below the evaporitic lenses. No vegetation or plant evidence. Light-coloured windblown sand, small sand accumulation of abundant gypsum fragments. Scattered long flaky gypsum crystals in the saturated zone. Organic-rich layers some 70 cm below surface.	Well developed	Highly hypersaline	> 120

associated with rainstorms or burst water pipes. In the case of inland sabkhas, the minerals precipitated within the soil are much more variable than those of coastal sabkhas since they depend on the composition of the local groundwater. The same applies to inland drainage basins and salt playas that are subject to periodic desiccation.

Sabkha soils are frequently characterized by low strength (Table 5.11). Furthermore, some surface clays that are normally consolidated or lightly overconsolidated may be sensitive to highly sensitive (Hossain and Ali, 1988). The low strength is attributable to the concentrated salt solutions in sabkha brines; the severe climatic conditions under which sabkha deposits are formed (e.g. large variations in temperature and excessive wetting–drying cycles), which can give rise to instability in sabkha soils; and the ready solubility of some of the minerals that act as cements in these soils. As a consequence, the bearing capacity of sabkha soils and their compressibility frequently do not meet routine design requirements. Various ground improvement techniques have therefore been used in relation to large construction projects, such as vibro-replacement, dynamic compaction, compaction piles and under-drainage. Soil replacement and pre-loading have been used when highway embankments have been constructed. Al-Amoudi et al. (1995) suggested that sabkha soils can be improved by stabilization with cement and with lime. However, a high water to lime ratio did not prove satisfactory and lime was also not satisfactory for stabilization of sulphate-rich soils.

A number of silty deposits formed under arid conditions are liable to undergo considerable volume reduction or collapse when wetted. Such metastability

Table 5.11 Summary of geotechnical characteristics of clayey soil units (after Hossein and Ali, 1988).

Property	Ranges of values for		
	Top crust	Main body of soft clay	Stiff brown clay
Natural water content (%)	15–40	40–78	25–31
Liquid limit (%)	20–50	36–60	50–56
Plastic limit (%)	16–40	18–30	25–33
Plasticity index (%)	5–20	20–32	24–30
Undrained shear strength by field vane shear test (kPa)	20–60	12–45	> 50
Sensitivity	3.5–22	1.3–11	1–2
M-value from Mackintosh probing	15–140	3–40	50–300
Cone resistance from CPT (kg cm^{-2})	7–12	3–5	15–60
N-value from SPT	2	1	9–24
Coefficient of volume compressibility (m^2 MN^{-1})	0.5–0.7	0.52–1.47	0.17–0.22
Compression index	0.37–0.42	0.40–0.88	0.17–0.31
Overconsolidation ratio	4–37	1.3–2.4	6–11

arises due to the loss of strength of inter-particle bonds resulting from increases in water content. Thus, infiltration of surface water, including that applied in the course of irrigation, leakage from pipes and rise of the water table, may cause large settlements to occur.

Silt-sized material may be deposited in temporary lakes or by aeolian action. Lacustrine and fluvial silts that occur within alluvial plains and other areas often bear many similarities with other water-borne silts, except that they may be interbedded with evaporite deposits and have been affected by periodic desiccation. The latter process leads to the development of a stiffened crust or, where this has occurred successively, a series of hardened layers within the formation.

Various types of calcareous silty clay typically form in arid and sub-arid regions when clay is deposited in saline or lime-rich waters. These are characterized by the presence of a desiccated surface layer typically up to 2 m thick, which may be capable of supporting lightly loaded structures, although care needs to be exercised to ensure that bearing capacity failure does not occur in the underlying softer soils (Hossain and Ali, 1988). Also, the reduction in strength that typically occurs due to wetting should be borne in mind.

Apart from those derived directly by the weathering of mudrocks, clay minerals tend to be a rare constituent of arid deposits. However, aeolian clay dunes are found in situations in which the parent material has been derived as sand- or silt-sized aggregations from desiccated alluvial clays. The presence of clay minerals may significantly change the engineering behaviour of some deposits consisting predominantly of other materials.

High rates of evaporation in arid areas result in saline groundwater and the precipitation of minerals previously held in solution. Such a process may lead to ground heave due to the precipitation of minerals within the capillary fringe. For instance, Netterburg (1977) documented many cases in which the surfaces of roads in South Africa have been badly disrupted.

A common feature of arid deposits is the cementation of sediments by the precipitation of mineral matter from the groundwater. The species of salt held in solution, and also those precipitated, depend on the source of the water, as well as on the prevailing temperature and humidity conditions. The process may lead to the development of various crusts or 'cretes', in which unconsolidated deposits or bedrock, respectively, are cemented by gypsum, calcite, silica, iron oxides or other compounds. Thus cementation by gypsum would give rise to gypcrust or gypcrete, respectively. Cretes and crusts may form continuous sheets or isolated patch-like masses at the ground surface where the water table is at, or near, this level, or at some other position within the ground profile.

Well-cemented crusts and cretes may provide adequate bearing capacity for structures, but care must be taken that the underlying uncemented material is not overloaded. Also, possible changes in the engineering behaviour of the material with any changes in water conditions must be borne in mind.

5.7 Tropical soils

Engineers from temperate countries have often experienced difficulties when dealing with certain tropical soils because it has been assumed that they will behave in a similar manner to soils in the temperate zone. Consequently, methods of testing and engineering classification schemes have been used that were not designed to cope with the different conditions found when dealing with soils formed in tropical environments (Gidigasu 1988). Of course, it would be wrong to assume that all tropical soils are different from those found in other climatic zones. Vargas (1985) pointed out that many soils, such as alluvial clays and sands and organic clays, behave in the same manner and have similar geotechnical properties regardless of the climatic conditions of the area of deposition.

Water of crystallization may be present within some minerals in many tropical residual soils, as well as free water. Some of the former type of water may be lost during conventional testing for moisture content. In order to avoid this, it has been recommended (Anon., 1990) that comparative tests be carried out on duplicate samples, taking the measurement of moisture content by drying to constant mass between 105 and 110 °C on one sample, and at a temperature not exceeding 50 °C on the other. An appreciable difference in the two values indicates the presence of structural water.

Drying brings about changes in the properties of residual soils not only during sampling and testing, but also *in situ*. The latter occurs as a result of local climatic conditions, drainage and position within the soil profile. Drying initiates two important effects, namely, cementation by the sesquioxides and aggregate formation on the one hand, and loss of water from hydrated clay minerals on the other. In the case of halloysite, the latter causes an irreversible transformation to metahalloysite. Some consequences of these changes are illustrated in Table 5.12 and Figure 5.15. Drying can cause almost total aggrega-

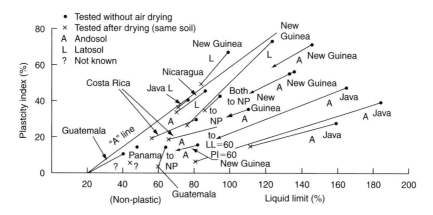

Figure 5.15 Effect of drying on the Atterberg limits of some tropical soils (from Morin and Todor, 1975).

Table 5.12 The effect of drying on index properties.
(a) Liquid (LL) and plastic limits (PL) (from Anon., 1990).

Soil type and location	Natural LL:PL	Air-dried LL:PL	Oven-dried LL:PL
laterite (Costa Rica)	81:29		56:19
laterite (Hawaii)	245:35	NP	
red clay (Kenya)	101:70	77:61	65:47
latosol (Dominica)	93:56	71:43	
andosol (New Guinea)	145:75		NP
andosol (Java)	184:146		80:74

NP indicates non-plastic.

(b) Effect of air-drying on particle size distribution of a hydrated laterite clay from Hawaii (from Gidigasu, 1974).

Index properties	Wet (at natural moisture control)	Moist (partial air-drying)	Dry (complete air-drying)
sand content (%)	30	42	86
silt content (%) (0.05–0.005 mm)	34	17	11
clay content (%) (<0.005 mm)	36	41	3

tion of clay-size particles into the silt- and sand-size ranges, and a reduction or loss of plasticity (see Table 5.10). Unit weight, shrinkage, compressibility and shear strength can also be affected. Hence, classification tests should be applied to the soil with as little drying as possible, at least until it can be established from comparative tests that drying has no effect on the results.

It is conventional to classify soils on the basis of their engineering properties. For non-tropical soils this has resulted in a number of fairly well-established classification systems that are based on geotechnical index properties, principally grading and plasticity (for example, the British Soil Classification System (Anon., 1981) and the Unified Soil Classification (Wagner, 1957)). Once a soil has been classified by means of one of these systems it may be possible to infer some of the ways in which the soil will behave from an engineering point of view. With many tropical soils this is not the case, and the use of such systems for tropical soils has been criticized by, for example, De Graft-Johnson and Bhatia (1969). For some soils the tests used are inadequate to determine the property being measured. For example, the liquid and plastic limit values obtained from 'standard' tests for some tropical red clay soils are dependent upon the precise test method employed and, in particular, the amount of energy used in mixing the soil prior to carrying out the test. Consequently, two quite different values of liquid limit may be obtained by two operators testing the same soil. Different results will also be obtained depending upon whether the soil was pre-dried prior to testing or kept close to its natural moisture content.

As far as the plasticity of tropical residual soils is concerned, disaggregation should be carried out with care so that individual particles are separated but not fragmented. In fact, disaggregation may have to be brought about in some cases by soaking in distilled water (Anon., 1990). Moreover, the sensitivity of the soil to working with water should be checked by using a range of mixing times before testing, in order to determine the shortest time for thorough mixing.

The fabric of *in situ* residual soils, particularly of the lateritic type, involves a wide range of void ratios and pore sizes. For example, the range of void ratio for lateritic soils in Hawaii, as found by Tuncer and Lohnes (1977), was 0.8 to 1.91. However, the variability of void ratio does not vary systematically with soil type, parent rock, type of weathering and state of stress. It is commonly due to differential leaching removing varying quantities of material from the soil. The void ratio at a particular state of stress may be in a metastable, stable-contractive or stable-dilatant state. The strains that a residual soil with a bonded structure experiences when it yields depend on its void ratio and degree of bonding. When the yield stress is exceeded the strains undergone are determined by the soil state. It is subjected to large contraction if it is metastable, whereas it will undergo only small strains if it is stable-dilatant. The void ratio and pore size do not change significantly with stress if this does not exceed the yield stress. Although larger pores allow the soil to de-saturate at small suctions, many small pores can retain water at high suctions.

Vaughan *et al.* (1988) introduced the concept of relative void ratio (e_R) for residual soils, defining it as

$$e_R = \frac{e - e_{opt}}{e_L - e_{opt}} \qquad (5.6)$$

where e is the natural void ratio, e_L is the void ratio at the liquid limit and e_{opt} is the void ratio at the optimum moisture content. They suggested relating engineering properties to relative void ratio rather than *in situ* void ratio.

Mineralogy and micro-fabric are the main factors that control the geotechnical properties of soil. For some tropical residual soils, macro-fabric also plays an important part in controlling the engineering properties of the soil. This factor also needs to be taken into account in deriving a classification system. Therefore, it is logical that the first level of classification should be on this basis. One of the most recent classifications of tropical residual soils (Anon., 1990) is pedologically based, and the main groupings are given in Table 5.13.

Duricrusts are not considered in detail here as they are discussed under soils of arid climates, although ferricretes (includes laterites) and alcretes (bauxites)

Table 5.13 Classification of residual soils.

Mature soils	Duricrusts
vertisols	silcrete
fersiallitic andosols	calcrete
fersiallitic (*sensu stricto*)	gypcrete
ferruginous (*sensu stricto*)	ferricrete
ferrallitic	alcrete (Alicrete)

are common in tropical regions. The term 'laterite' has been applied to soft, clay-rich horizons showing marked iron segregation or mottling and also to gravelly materials composed mainly of iron oxide concretions or pisoliths. McFarlane (1976) considered that these non-indurated materials form part of a sequence of lateritic weathering ultimately resulting in the formation of an indurated surface or near-surface sheet of duricrust.

Vertisols are characterized by the presence of clay minerals of the smectite group, which typically have a high swell and shrink potential, and possess contraction cracks and slickensides. It has been indicated (Anon., 1990) that these soils are prone to erosion and dispersion as well as swell–shrink problems. Generally, the clay fraction in these soils exceeds 50%, silty material varying between 20 and 40%, and the remainder being sand. Black cotton soils are probably the most common type and are highly plastic, silty clays. Shrinkage and swelling of these soils is a problem in many regions that experience alternating wet and dry seasons. However, the volume changes are frequently confined to an upper critical zone in the soil that is frequently less than 1.5 m thick. Below this, the moisture content remains more or less constant.

The 'red' soils of tropical regions include the fersiallitic, ferruginous and ferrallitic soils. Each type relates to a broad set of climatic conditions. Fersiallitic soils are found in subtropical or mediterranean climates. Smectite is the main clay mineral, but on older surfaces, well-drained sites or silica-poor parent rocks, kaolinite may be present. Young volcanic ashes produce fersiallitic andosols characterized by allophane, which alters to imogolite or halloysite on weathering. Ferruginous soils occur in either more humid (without dry season) or slightly hotter regions than mediterranean and are more strongly weathered than fersiallitic soils. Kaolinite is the dominant clay mineral and smectite is subordinate. Ferrallitic soils develop in the hot, humid tropics. All the primary minerals except quartz are weathered, with much silica and bases being removed in solution. Any residual feldspar is converted to kaolinite. Gibbsite may be present. Soils can be divided into ferrites and allites depending upon whether iron oxides or aluminium oxides predominate.

With this form of classification, the different groupings should be seen as part of a weathering continuum from fersiallitic soils through to ferrallitic soils. Nonetheless, this can lead to practical problems in that, for example, fersiallitic soils can be found in a climatic environment conducive to the formation of ferrallitic soils when the parent material has not been exposed to the climatic conditions for a long enough period of time. In the same way, different soils may occur within the soil profile.

Further practical problems may arise in the use of this classification because of the difficulty in identifying the variation of climate, at a particular location, with time. In other words, the present environmental conditions are not necessarily a good indicator of the soils to be expected if climate has changed since the formation of the soil.

Some values of natural moisture content for lateritic soils are given in Table 5.14, values frequently falling within the range 6 to 22%. At or near the surface, the liquid limits of laterites usually do not exceed 60% and the plasticity indices are less than 30% (Table 5.14). Consequently, such laterites are of low to medium plasticity. The activity of laterites may vary between 0.5 and 1.75. Red clays and latosols have high plastic and liquid limits (Table 5.14).

Obviously, the cementation between particles of residual soils has a significant influence on their shear strength, as does the widely variable nature of the void ratio and partial saturation, which, as noted, can occur to appreciable depth. A bonded soil structure exhibits a peak shear strength, unrelated to density and dilation, which is destroyed by yield as large strains develop. According to Mitchell and Sitar (1982), the main sources of data on strength characteristics of both lateritic soils and andosols (see below) are tests performed on remoulded specimens. The reported range of the friction angle is from 10 to 41°, with the majority of the results falling in the range 28 to 38°. The values of cohesion range from 0 to in excess of 48 kPa. Data on the undisturbed shear strength characteristics of residual soils are fairly limited. Andosols are characterized by fairly high friction angles, 27 to 57°, as opposed to 23 to 33° for lateritic soils. The cohesion of laterites can vary from 0 to over 210 kPa, while that quoted for andosols ranges from 22 to 345 kPa. Moisture content does affect the strength of andosols significantly as the degree of saturation can have an appreciable effect on cementation.

Because of the presence of a hardened crust near the surface, the strength of laterite may decrease with increasing depth. For example, Nixon and Skipp (1957) quoted values of shear strength of 90 and 25 kPa derived from undrained triaxial tests for samples of laterite taken from the surface crust and from a depth of 6 m, respectively. In addition, the variation of shear strength with depth is influenced by the mode of formation, type of parent rock, depth of water table and its movement, degree of laterization and mineral content, as well as the amount of cement precipitated.

Many tropical residual soils behave as if they are overconsolidated in that they exhibit a yield stress at which there is a discontinuity in the stress–strain behaviour and a decrease in stiffness. This is the apparent pre-consolidation pressure. The degree of overconsolidation depends on the amount of weathering. The cementation of soils formed in regions with a distinct dry season can

Table 5.14 Some engineering properties of tropical residual soils (from Bell, 1992).

Soil	Location	Natural moisture content (%)	Plastic limit (%)	Liquid limit (%)	Clay content (%)	Activity	Void ratio	Unit weight (kN m⁻³)	Dry density (Mg m⁻³)	Specific gravity	Shear strength		Compressibility			Permeability (m s⁻¹)
											ϕ'	c' (kPa)	C_c	m_v (m²MN⁻¹)	c_v(m²y⁻¹)	
1 laterite	Hawaii		135	245	36			15.2–17.3			27–57	48–345	0.0186	13	262	1.5×10^{-6}–5.4×10^{-9}
2 laterite	Nigeria	18–27	13–20	41–51							28–35	466–782				
3 laterite	Sri Lanka	16–49	28–31	33–90	15–45											
4 laterite gravel	Cameroon									2.89–3.14	37	0–40				
5 red clay	Brazil		26	42	35	0.46				2.73						
6 red clay	Ghana		24	48	31	0.77				2.8						
7 red clay	Brazil		22	28	27	0.3				2.75						
8 red clay	Hong Kong									2.86						
9 latosol	Brazil		30	47	35	0.51				2.83				0.466–0.075	2.32–232	5×10^{-6}–2×10^{-6}
10 latosol	Ghana		29	52	23	1.0				2.74	21	26	0.03–0.06	0.995–0.01		4.7–5×10^{-8}
11 latosol	Java	36–38			80		1.7–1.8			2.64	16–20	47–58				
12 black clay	Cameroon		21	75					1.27–1.75							
13 black clay	Kenya		36	103						2.72						
14 black clay	India		41	132	48			11.5			37	25				
15 andosol	Kenya	62	73	107									0.01–0.04	0.358–0.084		4×10^{-8}
16 andosol	Java	58–63			29		2.1–2.2				18	46				2.8×10^{-8}–5.6×10^{-9}

be weakened by saturation, and this leads to a collapse of the soil structure. As a consequence the apparent pre-consolidation pressure decreases and the compressibility of the soil increases.

Collapsible soils have an open-textured fabric that can withstand reasonably large stresses when partly saturated, but they undergo a decrease in volume due to collapse of the soil structure on wetting, even under low stresses. Many partially saturated tropical residual soils are of this nature (Vargas, 1990). Collapse and associated settling is normally due to loss or reduction of bonding between soil particles because of the presence of water. Intensively leached residual soils formed from quartz-rich rocks tend to be prone to collapse. Other ways in which collapse is brought about include loss of the stabilizing effect of surface tension in water menisci at particle contacts in partially saturated soil and the loss of strength of 'dry' clay bridges between particles.

The effects of leaching on lateritic soils have been investigated by Ola (1978a). As mentioned, the cementing agents in lateritic soils help to bond the finer particles together to form larger aggregates. However, as a result of leaching these aggregates break down, which is shown by the increase in liquid limit after leaching. Moreover, removal of cement by leaching gives rise to an increase in compressibility of more than 50%. Again, this is mainly due to the destruction of the aggregate structure. Conversely, there is a decrease in the coefficient of consolidation by some 20% after leaching (Table 5.15). The change in effective angle of shearing resistance and effective cohesion before and after leaching is similarly explained. Prior to leaching the larger aggregates in the soil cause it to behave as a coarse-grained, weakly bonded particulate material. The strongly curvilinear form of the Mohr failure envelopes can also be explained as a result of the breakdown of the larger aggregates.

As referred to above, some residual soils have high void ratios and large macro-pores and therefore are associated with high permeability. For instance, *in situ* permeabilities of clay-type lateritic soils may be as high as 10^{-5} to 10^{-4} m s^{-1}. Some typical values of permeability for saprolites and laterites are

Table 5.15 Engineering properties of a lateritic soil before and after leaching (after Ola, 1978a).

Property	Before leaching	After leaching
Natural moisture content (%)	14	–
Liquid limit (%)	42	53
Plastic limit (%)	25	21
Relative density	2.7	2.5
Angle of shearing resistance, ϕ'	26.5°	18.4°
Cohesion, c' (kPa)	24.1	45.5
*Coefficient of compressibility (m^2 MN^{-1})	12	15
*Coefficient of consolidation (m^2 y^{-1})	599	464

* For a pressure of 215 kPa.

as shown in Table 5.16. Both in the saprolite horizon and the underlying rock the permeability is governed by discontinuities. Brand (1985) considered that the saprolitic soils in Hong Kong were generally relatively permeable and that therefore drained conditions normally occurred. Water tables are often quite low in such soils. Blight (1990) noted that water tables are often deeper than 5 to 10 m. If evapotranspiration exceeds infiltration, then deep desiccation of the soil profile is likely. As a result, residual soils may be cracked and fissured, which further increases permeability.

Expandable clay minerals are frequently present in tropical residual clay soils of medium to high plasticity. When these clays occur above the water table, they can undergo a high degree of shrinkage on drying. By contrast, when wetted they swell, exerting high pressures. Indeed, highly expansive clay minerals may develop sufficient swelling pressure to break the bonded structure of the soil and thereby give rise to large heave movements. These can also occur when desiccated black clay soils are wetted. Alternating wet and dry seasons may produce significant vertical movements as soil suction changes. The amount of heave that takes place depends on the effective stress changes and on the swelling modulus of the soil $((\delta V/V)/\delta\sigma')$, where δV is the change in volume, V is the original volume and $\delta\sigma'$ is the change in effective stress). The swelling modulus increases with the proportion of clay in the soil and the amount of expansive clay minerals, notably montmorillonite, in the clay fraction.

The black cotton soil of Nigeria, described by Ola (1978b), is a highly plastic silty clay in which shrinkage and swelling are a problem in many regions that experience alternating wet and dry seasons (see Section 5.2). These volume changes, however, are confined to a critical upper zone of the soil, which is frequently less than 1.5 m thick. Below this the moisture content remains more or less the same, for instance around 25%. Ola (1980) noted an average linear shrinkage of 8% for some of these soils, with an average swelling pressure of 120 kPa and a maximum of about 240 kPa. He went on to state that in such situations the dead load of a building should be at least 80 kPa to counteract the swelling pressure.

Allophane-rich soils or andisols are developed from basic volcanic ashes in high-temperature–rainfall regions. The soils have very high moisture contents (60 to 250%) and liquid limits (80 to 300%), with corresponding high plasticity indices (20 to 100%). Plasticity values vary considerably if the soils are

Table 5.16 Permeability of residual soils.

Saprolite (m s^{-1})	Laterite (m s^{-1})
4×10^{-1}–5×10^{-9}	4×10^{-6}–5×10^{-9}
5×10^{-6}–1×10^{-7}	5×10^{-5}–1×10^{-6}
	3×10^{-6}–1×10^{-9}

tested with or without air-drying prior to testing, values of plasticity index and liquid limit being reduced by more than 50% in the air-dried state. The plasticity values obtained are also dependent upon the degree of working of the soil during pre-test mixing. In allophanic soils, plasticity decreases with increased working. The soils are also characterized by very low dry densities and high void ratios (sometimes as high as 6). Moore and Styles (1988) noted that the soils are very strength-sensitive, have low CBR values (around 1%) and low dry-density values on compaction when tested at natural moisture contents. Because laboratory compaction and CBR tests are normally carried out on air-dried soils, design values based on these laboratory tests will be impractical as drying of the soils in a humid tropical climate is normally very difficult. Therefore tests on samples initially at natural moisture content are recommended.

Soils containing halloysite, or its partially dehydrated form metahalloysite, have high moisture contents (30 to 65%) and liquid limits (40 to >100%). Plasticity indices can also be high (10 to 50%). Air-drying of the soils causes a reduction in plasticity, approximately parallel to the A-line but by a much smaller amount than for allophane-rich soils. Plasticity increases with mixing. Complete mixing of the soils using a grease-worker, in which both inter-ped and (probably) intra-ped bonds are broken down to produce a sticky clay paste, causes the liquid limit to increase by about 30%. The plasticity values obtained using a grease-worker can be considered to be the maximum values obtainable. Clay contents are generally high (50 to >80%), and pre-drying appears to have little effect on the particle size determination provided that complete dispersion of particles is achieved. These soils are susceptible to collapse, that is, rapid consolidation on flooding with water at constant load. However, the phenomenon does not appear to occur universally, and collapse susceptibility seems to be dependent on factors such as sample depth and disturbance, void ratio, degree of saturation and test duration prior to flooding. Again, compaction tests should be carried out without initial pre-drying. Individual samples should be used for each stage of the test, allowing partial drying from the natural moisture content as necessary. Over-compaction may increase the susceptibility of these soils to shrinkage.

5.8 Peat

Peat represents an accumulation of partially decomposed and disintegrated plant remains that has been preserved under conditions of incomplete aeration and high water content. It accumulates wherever the conditions are suitable, that is, in areas where there is an excess of rainfall and the ground is poorly drained, irrespective of latitude or altitude. Nonetheless, peat deposits tend to be most common in those regions with a comparatively cold, wet climate. Physico-chemical and biochemical processes cause this organic material to remain in a state of preservation over a long period of time. In other words, waterlogged, poorly drained conditions not only favour the growth of particular types of vegetation but they also help to preserve the plant remains. Decay is most intense at the surface, where aerobic conditions exist; it occurs at a much slower rate throughout the mass of the peat.

All present-day surface deposits of peat in northern Europe, Asia and Canada have accumulated since the last ice age and therefore have formed during the last 20 000 years. On the other hand, some buried peats may have been developed during interglacial periods. Peats have also accumulated in post-glacial lakes and marshes, where they are interbedded with silts and muds. Similarly, they may be associated with salt marshes. Fen deposits are thought to have developed in relation to the eustatic changes in sea level that occurred after the retreat of the last ice sheets. These are areas where layers of peat interdigitate with wedges of estuarine silt and clay. However, the blanket bog is a more common type of peat deposit. These deposits are found on cool, wet uplands where slopes are not excessive and drainage is impeded.

Landva and Pheeney (1980) suggested that the proper identification of peat should include a description of the constituents, since the fibres of some plant remains are much stronger than others and are consequently of interest to the engineer. The degree of humification and water content should also be taken into account. They proposed a modified form of the Von Post (1922) classification of peat (Table 5.17).

The void ratio of peat ranges from 9 for dense, amorphous, granular peat up to 25 for fibrous types with high contents of sphagnum. It usually tends to decrease with depth within a peat deposit. Such high void ratios give rise to phenomenally high water content. The water content of peats varies from a few hundreds percent dry weight (e.g. 500% in some amorphous granular peats) to over 3000% in some coarse fibrous varieties. Put another way, the water content may range from 75 to 98% by volume of peat. Moreover, changes in the amount of water content can occur over very small distances. The pH of fen peat is frequently in excess of 5, while that of bog peat is usually less than 4.5 and may be less than 3.

Gas is formed in peat as plant material decays, and this tends to take place from the centre of stems, so gas is held within stems. The volume of gas in peat varies and figures of around 5 to 7.5% have been quoted (Hanrahan, 1954). At this degree of saturation most of the gas is free and so has a significant influence on initial consolidation, rate of consolidation, pore pressure under load and permeability.

The bulk density of peat is both low and variable, being related to the organic content, mineral content, water content and degree of saturation. Peats frequently are not saturated and may be buoyant under water due to the presence of gas. Except at low water contents (less than 500%) with high mineral contents, the average bulk density of peats is slightly lower than that of water. However, the dry density is a more important engineering property of peat, influencing its behaviour under load. Hanrahan (ibid.) recorded dry densities

Table 5.17 Classification of peat (from Landva and Pheeney, 1980).

(1) Genus	(2) Designation	H	(3) Degree of humification (H)	
			Decomposition	Plant structure
Bryales (moss) = B Carex (sedge) = C Equisetum (horsetails) = Eq Eriophorum (cotton grass) = Er Hypnum (moss) = H Lignidi (wood) = W Nanolignidi (shrubs) = N Phragmites = Ph Scheuchzeria (aquatic herbs) = Sch Sphagnum (moss) = S	With few exceptions peats consist of a mixture of two or more genera. These are listed in decreasing order of content, i.e. the principal component first, e.g. ErCS.	H_1 H_2 H_3 H_4 H_5 H_6 H_7 H_8 H_9 H_{10}	None Insignificant Very slight Slight Moderate Moderately strong Strong Very strong Nearly complete Complete	Easily identified Easily identified Still identifiable Not easily identified Recognizable but vague Indistinct Faintly recognizable Very indistinct Almost unrecognizable Not discernible

(4) Water content (B)	(5) Fine fibres (F)	(6) Coarse fibres (R)	(7) Wood (W) and shrub remnant
Estimated from a scale of 1 (dry) to 5 (very high) and designated B_1, B_2, etc. Landva and Pheeney suggested the following ranges: B_2 less than 500% B_3 500–1000% B_4 1000–2000% B_5 Over 2000%	These are fibres and stems less than 1 mm in diameter, F_0 = nil; F_1 = low content; F_2 = moderate content; F_3 = high content.	These fibres have a diameter exceeding 1 mm; R_0 = nil; R_1 = low content; R_2 = moderate content; R_3 = high content.	Wood and shrub is similarly graded: W_0 = nil; W_1 = low content; W_2 = moderate content; W_3 = high content.

of drained peat within the range 65 to 120 kg m^{-3}. The dry density is influenced by the mineral content, and higher values than that quoted can be obtained when peats possess high mineral residues.

Peat undergoes significant shrinkage on drying out. Nonetheless, the volumetric shrinkage of peat increases up to a maximum and then remains constant, the volume being reduced almost to the point of complete dehydration. The amount of shrinkage that can occur ranges between 10 and 75% of the original volume. The change in peat is permanent in that it cannot recover all the water lost when wet conditions return. Hobbs (1986) noted that the more highly humified peats, even though they have lower water contents, tend to shrink more than the less humified fibrous peats.

Apart from its moisture content and dry density, the shear strength of a peat deposit appears to be influenced by its degree of humification and its mineral content. As both these factors increase so does the shear strength. Conversely, the higher the moisture content of peat, the lower is its shear strength. As the effective weight of 1 m^3 of undrained peat is approximately 45 times that of 1 m^3 drained peat, the reason for the negligible strength of the latter becomes apparent.

Consolidation of peat takes place when water is expelled from the pores and the particles undergo some structural rearrangement. Initially the two processes occur at the same time, but as the pore water pressure is reduced to a low value, the expulsion of water and structural rearrangement occur as a creep-like process (Berry and Poskitt, 1972). In other words, the initial stage of drainage can be regarded as primary consolidation, whereas the stage of continuing creep represents secondary compression. It must be remembered that primary and secondary consolidation are empirical divisions of a continuous compression process, both of which occur simultaneously during part of that process. In fact, accurate prediction of the amount and rate of settlement of peat cannot be derived directly from laboratory tests. Hence large-scale field trials seem to be essential for important projects.

Peat has a high coefficient of secondary compression, the latter being the dominant process in terms of settlement of peat, and in terms of strain, it is virtually independent of water content and degree of saturation. The short phase of primary consolidation is responsible for little distortion. Bog peats appear to possess lower values of secondary compression than fen peats. This is probably because of their non-plastic, highly frictional character.

Differential and excessive settlement is the principal problem confronting the engineer working on a peaty soil. When a load is applied to peat, settlement occurs because of the low lateral resistance offered by the adjacent unloaded peat. Serious shearing stresses are induced by even moderate loads. Worse still, should the loads exceed a given minimum, then settlement may be accompanied by creep, lateral spread, or in extreme cases by rotational slip and upheaval of adjacent ground. At any given time the total settlement in peat due to loading involves settlement with and without volume change.

Settlement without volume change is the more serious, because it can give rise to the types of failure mentioned. What is more, it does not enhance the strength of peat.

When peat is compressed the free pore water is expelled under excess hydrostatic pressure. Since the peat is initially quite pervious and the percentage of pore water is high, the magnitude of settlement is large and this period of initial settlement is short (a matter of days in the field). The magnitude of initial settlement is directly related to peat thickness and applied load. The original void ratio of a peat soil also influences the rate of initial settlement. Excess pore pressure is almost entirely dissipated during this period. Settlement subsequently continues at a much slower rate, which is approximately linear with the logarithm of time, because the permeability of the peat is significantly reduced due to the large decrease in volume. During this period the effective consolidating pressure is transferred from the pore water to the solid peat fabric. The latter is compressible and will sustain only a proportion of the total effective stress, depending on the thickness of the peat mass.

The use of pre-compression using surcharge loading in the construction of embankments across peatlands involves the removal of the surcharge after a certain period of time. This gives rise to some swelling in the compressed peat. Uplift or rebound can be quite significant, depending on the actual settlement and surcharge ratio (i.e. the mass of the surcharge in relation to the weight of the fill once the surcharge has been removed). Rebound is also influenced by the amount of secondary compression induced prior to unloading. The swelling index, C_s, is related to the compression index, C_c, in that an average C_s is around 10% of the C_c, within a range of 5 to 20%. Rebound undergoes a marked increase when surcharge ratios are greater than about 3. Normally, rebound in the field is between 2 and 4% of the thickness of the compressed layer of peat before surcharge has been removed. Hence, the compressibility of pre-consolidated peat is greatly reduced. With few exceptions, improved drainage has no beneficial effect on the rate of consolidation, because efficient drainage only accelerates the completion of primary consolidation, which anyhow is completed rapidly.

5.9 Frozen soil

Frozen ground phenomena are found in regions that experience a tundra climate, that is, in those regions where the winter temperature rarely rises above freezing point and the summer temperature is only warm enough to cause thawing in the upper metre or so of the soil. Beneath the upper or active zone the subsoil is permanently frozen and so is known as the permafrost layer. Because of this, summer meltwater cannot seep into the ground, the active zone then becoming waterlogged. Layers or lenses of unfrozen ground termed *taliks* may occur, often temporarily, in the permafrost. Permafrost is an important characteristic, although it is not essential to the definition of periglacial conditions.

It covers 20% of the Earth's land surface, and during Pleistocene times it developed over an even larger area (Figure 5.16). The temperature of perennially frozen ground below the depth of seasonal change ranges from slightly less than 0 to -12 °C. Generally, the depth of thaw is less, the higher the latitude. It is at a minimum in peat or highly organic sediments and increases in clay, silt and sand to a maximum in gravel, where it may extend to 2 m in depth.

Frost action in a soil, of course, is not restricted to tundra regions. Its occurrence is influenced by the initial temperature of the soil, as well as the air temperature, the intensity and duration of the freeze period, the depth of frost penetration, the depth of the water table, and the type of ground and exposure cover. If frost penetrates down to the capillary fringe in fine-grained soils, especially silts, then, under certain conditions, lenses of ice may be developed. The formation of such ice lenses may, in turn, cause frost heave and frost boil, which may lead to the break-up of roads, the failure of slopes, etc.

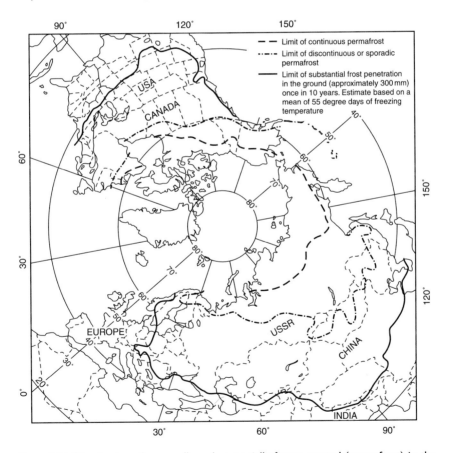

Figure 5.16 Distribution of seasonally and perennially frozen ground (permafrost) in the Northern Hemisphere.

Ice may occur in frozen soil as small disseminated crystals whose total mass exceeds that of the mineral grains, as large tabular masses that range up to several metres thick, or as ice wedges. The latter may be several metres wide and may extend to 10 m or so in depth. As a consequence, frozen soils need to be described and classified for engineering purposes. A recent method of classifying frozen soils involves the identification of the soil type and the character of the ice (Andersland and Anderson, 1978). First, the character of the actual soil is classified according to the Unified Soil Classification System. Second, the soil characteristics consequent upon freezing are added to the description. Frozen soil characteristics are divided into two basic groups based on whether or not segregated ice can be seen with the naked eye (Table 5.18). Third, the ice present in the frozen soil is classified; this refers to inclusions of ice that exceed 25 mm in thickness.

The amount of segregated ice in a frozen mass of soil depends largely upon the intensity and rate of freezing. When freezing takes place quickly no layers of ice are visible, whereas slow freezing produces visible layers of ice of various thicknesses. Ice segregation in soil also takes place under cyclic freezing and thawing conditions.

The presence of masses of ice in a soil means that, as far as behaviour is concerned, the properties of both have to be taken into account. Ice has no long-term strength, that is, it flows under very small loads. If a constant load is applied to a specimen of ice, instantaneous elastic deformation occurs. This is followed by creep, which eventually develops a steady state. Instantaneous elastic recovery takes place on removal of the load, followed by recovery of the transient creep.

The mechanical properties of frozen soil are very much influenced by the grain size distribution, the mineral content, the density, the frozen and unfrozen water contents, and the presence of ice lenses and layering. The strength of frozen ground develops from cohesion, inter-particle friction and particle interlocking, much the same as in unfrozen soils. However, cohesive forces include the adhesion between soil particles and ice in the voids, as well as the surface forces between particles. More particularly, the strength of frozen soils is sensitive to particle size distribution, particle orientation and packing, impurities (air bubbles, salts or organic matter) in the water–ice matrix, temperature, confining pressure, and rate of strain. Obviously, the difference in the strength between frozen and unfrozen soils is derived from the ice component.

The relative density influences the behaviour of frozen granular soils, especially their shearing resistance, in a manner similar to that when they are unfrozen. The cohesive effects of the ice matrix are superimposed on the latter behaviour, and the initial deformation of frozen sand is dominated by the ice matrix. Sand in which all the water is more or less frozen exhibits a brittle type of failure at low strains, for example at around 2% strain. However, the presence of unfrozen films of water around particles of soil not only means that the ice content is reduced but also that it leads to a more plastic behaviour of the

Table 5.18 Description and classification of frozen soils (from Andersland and Anderson, 1978).

I. Description of soil phase (independent of frozen state)	Classify soil phase by the Unified Soil Classification system			
	Major group		Subgroup	
	Description	Designation	Description	Designation
			Poorly bonded or friable	Nf
	Segregated ice not visible by eye	N	No excess ice	n
			Well bonded ———— Excess	Nb ———— e
II. Description of frozen soil			Individual ice crystals or inclusions	Vx
	Segregated ice visible by eye (ice 25 mm or less thick)	V	Ice coatings on particles	Ve
			Random or irregularly oriented ice formations	Vs
			Stratified or distinctly oriented ice formations	
			Ice with soil inclusions	ICE + soil type
III. Description of substantial ice strata	Ice greater than 25 mm thick	ICE	Ice without soil inclusions	ICE

soil during deformation. For instance, frozen clay, as well as often containing a lower content of ice than sand, has layers of unfrozen water (of molecular proportions) around the clay particles. These molecular layers of water contribute towards a plastic type of failure.

Lenses of ice are frequently formed in fine-grained soils frozen under a directional temperature gradient. The lenses impart a laminated appearance to the soil. In such situations the strength of the bond between soil particles and ice matrix is greater than between particles and adjacent ice lenses. Under very rapid loading the ice behaves as a brittle material, with strengths in excess of those of fine-grained frozen soils. By contrast, the ice matrix deforms continuously when subjected to long-term loading, with no limiting long-term strength. The laminated texture of the soil in rapid shear possesses the greatest strength when the shear zone runs along the contact between ice lens and frozen soil.

When loaded, stresses at the point of contact between soil particles and ice bring about pressure melting of the ice. Because of differences in the surface tension of the meltwater, it tends to move into regions of lower stress, where it refreezes. The process of ice melting and the movement of unfrozen water are accompanied by a breakdown of the ice and the bonding with the grains of soil. This leads to plastic deformation of the ice in the voids and to a rearrangement of particle fabric. The net result is time-dependent deformation of the frozen soil, namely, creep. Frozen soil undergoes appreciable deformation under sustained loading, the magnitude and rate of creep being governed by the composition of the soil, especially the amount of ice present, the temperature, the stress and the stress history.

The creep strength of frozen soils is defined as the stress level, after a given time, at which rupture, instability leading to rupture or extremely large deformations without rupture occur. Frozen fine-grained soils can suffer extremely large deformations without rupturing at temperatures near to freezing point. Hence the strength of these soils must be defined in terms of the maximum deformation that a particular structure can tolerate. As far as laboratory testing is concerned, axial strains of 20%, under compressive loading, are frequently arbitrarily considered as amounting to failure. The creep strength is then defined as the level of stress producing this strain after a given interval of time.

In fine-grained sediments the intimate bond between the water and the clay particles results in a significant proportion of soil moisture remaining unfrozen at temperatures as low as -25 °C. The more clay material in the soil, the greater is the quantity of unfrozen moisture. Nonetheless, there is a dramatic increase in structural strength with decreasing temperature. In fact, it appears to increase exponentially with the relative proportion of moisture frozen. Taking silty clay as an example, the amount of moisture frozen at -18 °C is only 1.25 times that frozen at -5 °C, but the increase in compressive strength is more than four-fold.

By contrast, the water content of granular soils is almost wholly converted into ice at a very few degrees below freezing point. Hence frozen granular soils exhibit a reasonably high compressive strength only a few degrees below freezing. The order of increase in compressive strength with decreasing temperature is shown in Figure 5.17.

Because frozen ground is more or less impermeable this increases the problems due to thaw by impeding the removal of surface water. What is more, when thaw occurs the amount of water liberated may greatly exceed that originally present in the melted-out layer of the soil (see below). As the soil thaws

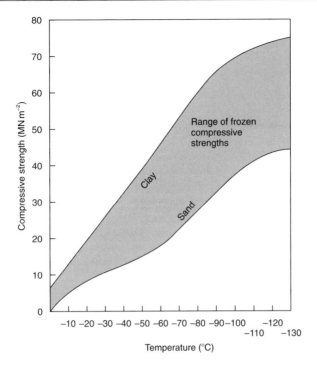

Figure 5.17 Increase in compressive strength with decreasing temperature (from Bell, 1992).

downwards the upper layers become saturated and, since water cannot drain through the frozen soil beneath, it may suffer a complete loss of strength. Indeed, under some circumstances excess water may act as a transporting agent, thereby giving rise to soil flows, which can move on gentle slopes. This movement downslope as a viscous flow of saturated debris is referred to as solifluction. It is probably the most significant process of mass wastage in tundra regions. Solifluction deposits commonly consist of gravels, which are characteristically poorly sorted, sometimes gap-graded, and poorly bedded. These gravels consist of fresh, poorly worn, locally derived material. Individual deposits are rarely more than 3 m thick and frequently display flow structures. Sheets and lobes and solifluction debris, transported by mudflow activity, are commonly found at the foot of slopes. These materials may be reactivated by changes in drainage, by stream erosion, by sediment overloading or during construction operations. Solifluction sheets may be underlain by slip surfaces, the residual strength of which controls their stability.

Settlement is associated with thawing of frozen ground. As ice melts, settlement occurs, water being squeezed from the ground by overburden pressure or by any applied loads. Excess pore pressures develop when the rate of ice melt is greater than the discharge capacity of the soil. Since excess pore pressures can lead to the failure of slopes and foundations, both the rate and amount of thaw settlement should be determined. Pore water pressures should also be monitored.

Further consolidation, due to drainage, may occur on thawing. If the soil was previously in a relatively dense state, then the amount of consolidation is small. This situation occurs only in coarse-grained frozen soils containing very little segregated ice. On the other hand, some degree of segregation of ice is always present in fine-grained frozen soils. For example, lenses and veins of ice may be formed when silts have access to capillary water. Under such conditions the moisture content of the frozen silts significantly exceeds the moisture content present in their unfrozen state. As a result, when such ice-rich soils thaw under drained conditions they undergo large settlements under their own weight.

Shrinkage, which gives rise to polygonal cracking in the ground, presents another problem when soil is subjected to freezing. The formation of these cracks is attributable to thermal contraction and desiccation. Water that accumulates in the cracks is frozen and consequently helps to increase their size. This water may also aid the development of lenses of ice. Individual cracks may be 1.2 m wide at their top, may penetrate to depths of 10 m and may be up to 12 m apart. They form when, because of exceptionally low temperatures, shrinkage of the ground occurs. When the ice in an ice wedge disappears, an ice wedge pseudomorph is formed by sediment, frequently sand, filling the crack. In addition, frozen soils may undergo notable disturbance as a result of mutual interference of growing bodies of ice or from excess pore pressures developed in confined water-bearing lenses. Involutions are plugs, pockets or tongues of highly disturbed material, generally possessing inferior geotechnical properties, which have been intruded into overlying layers. They are formed as a result of hydrostatic uplift in water trapped under a refreezing surface layer. They are usually confined to the active layer. Ice wedges, pseudomorphs and involutions usually mean that one material suddenly replaces another. This can cause problems in shallow excavations.

The following factors are necessary for the occurrence of frost heave: capillary saturation at the beginning and during freezing of the soil; a plentiful supply of subsoil water; and a soil possessing fairly high capillarity together with moderate permeability. According to Kinosita (1979), the ground surface experiences an increasingly large amount of heave, the higher the initial water table. Indeed, it has been suggested that frost heave could be prevented by lowering the water table (Andersland and Anderson, 1978).

Grain size is another important factor influencing frost heave. For example, gravels, sands and clays are not particularly susceptible to heave, while silts definitely are. The reason for this is that silty soils are associated with high capillary rises, but at the same time their voids are large enough to allow moisture to move quickly enough for them to become saturated rapidly. If ice lenses are present in clean gravels or sands, then they simply represent small pockets of moisture that have been frozen. Indeed, Taber (1930) gave an upper size limit of 0.007 mm, above which, he maintained, layers of ice do not develop. However, Casagrande (1932) suggested that the particle size critical to heave formation is 0.02 mm. If the quantity of such particles in a soil is less than 1%, no heave is to be expected, but considerable heaving may take place if this amount is over 3% in non-uniform soils and over 10% in very uniform soils. This 0.02 mm criterion has been used by the US Army Corps of Engineers (Anon., 1965), together with data from frost heave tests, to develop a frost-susceptibility system. This is outlined in Figure 5.18, which shows that soil groups exhibit a range of susceptibilities reflecting variations in particle size distribution, density and mineralogy.

Croney and Jacobs (1967) suggested that under the climatic conditions

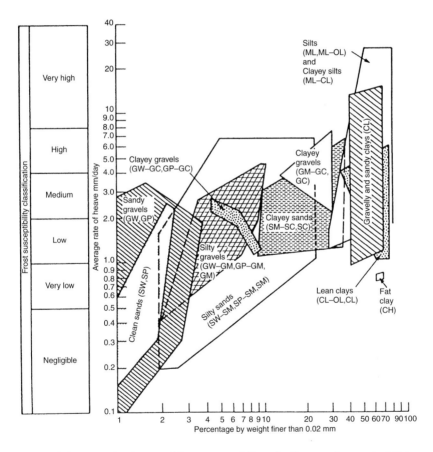

Figure 5.18 Range in the degree of frost susceptibility of soils according to the US Army Corps of Engineers (Anon., 1965).

experienced in Britain well-drained cohesive soils with a plasticity index exceeding 15% can be looked upon as non-frost-susceptible. They suggested that where the drainage is poor and the water table is within 0.6 m of formation level the limiting value of plasticity index should be increased to 20%. In addition, in experiments with sand, they noted that as the amount of silt added was increased up to 55% or the clay fraction up to 33%, the decrease in permeability in the freezing front was the overriding factor and heave tended to increase. Beyond these values, the decreasing permeability below the freezing zone became dominant and progressively reduced the heave. This indicates that the permeability below the frozen zone is principally responsible for controlling heave.

Horiguchi (1979) demonstrated, from experimental evidence, that the rate of frost heave increases as the rate of heat removal from the freezing front is increased. However, removal of heat does not increase the rate of heave indefinitely; it reaches a maximum, after which it declines. The maximum rate of heave was shown to be influenced by particle size distribution in that it increased in soils with finer grain size. Horiguchi also found that when the particles are the same size in different test specimens, then the maximum rate of heave depends upon the types of exchangeable cation present in the soil. The rate of heave is also influenced by the thickness of overburden. For instance, Penner and Walton (1979) indicated that the maximum rate of ice accumulation at lower overburden pressures occurs at temperatures nearer to 0 °C than at higher overburden pressures. However, it appears that the rate of heave for various overburden pressures tends to converge as the temperature below freezing is lowered. As the overburden pressure increases, the zone over which heaving takes place becomes greater and extends over an increasingly large range of temperature.

Maximum heaving does not necessarily occur at the time of maximum depth of penetration of the 0 °C line, there being a lag between the minimum air temperature prevailing and the maximum penetration of the freeze front. In fact, soil freezes at temperatures slightly lower than 0 °C.

As heaves amounting to 30% of the thickness of the frozen layer have frequently been recorded, moisture, other than that initially present in the frozen layer, must be drawn from below, since water increases in volume by only 9% when frozen. In fact, when a soil freezes there is an upward transfer of heat from the groundwater towards the area in which freezing is occurring. The thermal energy, in turn, initiates an upward migration of moisture within the soil. The moisture in the soil can be translocated upwards in either the vapour or liquid phase, or by a combination of both.

Before freezing, soil particles develop films of moisture around them due to capillary action. This moisture is drawn from the water table. As the ice lens grows, the suction pressure it develops exceeds that of the capillary attraction of moisture by the soil particles. Hence moisture moves from the soil to the ice lens. But the capillary force continues to draw moisture from the water table and so the process continues.

Jones (1980) suggested that if heaving is unrestrained, the heave (*H*) can be estimated as follows:

$$H = 1.09kit \tag{5.6}$$

where *k* is permeability, *i* is the suction gradient (this is difficult to derive) and *t* is time. In a discussion of frost heave, Reed *et al.* (1979) noted that predictions failed to take account of the fact that soils can exist at different states of density and therefore porosity, yet have the same grain size distribution. What is more, pore size distribution controls the migration of water in the soil and hence, to a large degree, the mechanism of frost heave. They accordingly derived expressions, based upon pore space, for predicting the amount of frost heave.

Where there is a likelihood of frost heave occurring it is necessary to estimate the depth of frost penetration. Once this has been done, provision can be made for the installation of adequate insulation or drainage within the soil and to determine the amount by which the water table may need to be lowered so that it is not affected by frost penetration. The base of footings should be placed below the estimated depth of frost penetration, as should water supply lines and other services. Frost-susceptible soils may be replaced by gravels. The addition of certain chemicals to soil can reduce its capacity for water absorption and so can influence frost susceptibility. For example, Croney and Jacobs (1967) noted that the addition of calcium lignosulphate and sodium tripolyphosphate to silty soils were both effective in reducing frost heave. The freezing point of the soil can be lowered by mixing in solutions of calcium chloride or sodium chloride, in concentrations of 0.5 to 3.0% by weight of the soil mixture. The heave of non-cohesive soils containing appreciable quantities of fines can be reduced or prevented by the addition of cement or bituminous binders. Addition of cement both reduces the permeability of a soil mass and gives it sufficient tensile strength to prevent the formation of small ice lenses as the freezing isotherm passes through.

References

Aitchison, G.D. and Wood, C.C. (1965) Some interactions of compaction, permeability and post-construction deflocculation affecting the probability of piping failures in small dams. *Proceedings Sixth International Conference on Soil Mechanics and Foundation Engineering*, Montreal, 2, 442–446.

Al-Amoudi, O.S.B., Asi, I.M. and El-Naggar, Z.R. (1995) Stabilization of arid, saline sabkha using additives. *Quarterly Journal Engineering Geology*, 28, 369–379.

Andersland, O.B. and Anderson, C.M. (eds) (1978) *Geotechnical Engineering for Cold Regions*, McGraw-Hill, New York.

Anon. (1965) *Soils and Geology – Pavement Design for Frost Conditions*. Technical Manual TM 5–818–2, US Corps of Engineers, Department of the Army, Washington, DC.

Anon. (1980) *Low-rise Buildings on Shrinkable Clay Soils, Part 3.* Building Research Establishment, Digest 240, HMSO, London, 8pp.

Anon. (1981) *Code of Practice for Site Investigations, BS 5930.* British Standards Institution, London.

Anon. (1990) Tropical residual soils. Working Party Report. *Quarterly Journal Engineering Geology*, 23, 1–101.

Atkinson, J.H., Charles, J.A. and Mhach, H.K. (1990) Examination of erosion resistance of clays in embankment dams. *Quarterly Journal Engineering Geology*, 23, 103–108.

Bell, F.G. (1992) *Engineering Properties of Soils and Rocks.* third edition. Butterworth–Heinemann, Oxford.

Bell, F.G. (1993) *Engineering Treatment of Soils.* Spon, London.

Bell, F.G. and Maud, R.R. (1994) Dispersive soils: a review from a South African perspective. *Quarterly Journal Engineering Geology*, 27, 195–210.

Bell, F.G. and Maud, R.R. (1995) Expansive clays and construction, especially of low rise structures: a view point from Natal, South Africa. *Environmental and Engineering Geoscience*, 1, 41–59.

Berry, P.L. and Poskitt, T.J. (1972) The consolidation of peat. *Geotechnique*, 22, 27–52.

Blair, M.L. and Spangle, W.E. (1979) Seismic safety and land use planning – selected examples from California. US Geological Survey Professional Paper 941B, 82pp.

Blight, G.E. (1990) Construction in tropical soils. *Proceedings Second International Conference on Geomechanics in Tropical Soils*, Singapore, Balkema, Rotterdam, 2, 449–468.

Bourdeaux, G. and Imaizumi, H. (1977) Dispersive clay at Sobrandinho dam. In *Proceedings Symposium on Dispersive Clays, Related Piping and Erosion in Geotechnical Projects*, Sherard, J.L. and Decker, R.S. (eds), American Society Testing Materials (ASTM) Publication 623, Philadelphia, 13–24.

Brackley, I.J.A. (1980) Prediction of soil heave from soil suction measurements. *Proceedings Seventh Regional Conference for Africa on Soil Mechanics and Foundation Engineering*, Accra, 159–167.

Brand, E.W. (1985) Geotechnical engineering in tropical residual soils. *Proceedings First International Conference on Geomechanics in Tropical Lateritic and Saprolitic Soils*, Brasilia, 1, 235–251.

Burland, J.B. and Wroth, C.P. (1975) Allowable and differential settlement of structures including damage and soil-structure interaction. In *Settlement of Structures*, British Geotechnical Society, Pentech Press, London, 611–654.

Cabrera, J.G. amd Smalley, I.J. (1973) Quick clays as products of glacial action: a new approach to their nature, geology, distribution and geotechnical properties. *Engineering Geology*, 7, 115–133.

Casagrande, A. (1932) Discussion on frost heaving. *Proceedings Highway Research Board*, Bulletin 12, 169.

Casagrande, A. (1936) Characteristics of cohesionless soils affecting the stability of slopes and earth fills. *Journal Boston Society Civil Engineers*, 23, 3–32.

Chandler, R.J., Crilly, M.S., and Montgomery-Smith, G. (1992) A low-cost method of assessing clay desiccation for low-rise buildings, *Proceedings Institution Civil Engineers*, 92, 82–89.

Chen, F.H. (1988) *Foundations on Expansive Soils.* Elsevier, Amsterdam.

Clemence, S.P and Finbarr, A.O. (1981) Design considerations for collapsible soils.

Proceedings American Society Civil Engineers, Journal Geotechnical Engineering Division, 107, 305–317.

Clevenger, W.A. (1958) Experience with loess as foundation material. *Proceedings American Society Civil Engineers, Journal Soil Mechanics Foundations Division*, 85, 151–180.

Craft, D.C. and Acciardi, R.G. (1984) Failure of pore water analyses for dispersion. *Proceedings American Society Civil Engineers, Journal Geotechnical Engineering Division*, 110, 459–472.

Croney, D. and Jacobs, J.C. (1967) *The Frost Susceptibility of Soils and Road Materials*. Transport and Road Research Laboratory, Report LR90, Crowthorne, Berkshire.

De Graft-Johnson, J.W.S. and Bhatia, S.H. (1969) Engineering charactertistics of lateritic soils. *Proceedings Seventh International Conference on Soil Mechanics and Foundation Engineering*, Mexico City, 2, 13–43.

Derbyshire, E. and Mellors, T.W. (1988) Geological and geotechnical characteristics of some loess and loessic soils from China and Britain: a comparison. *Engineering Geology*, 25, 135–175.

Evstatiev, D. (1988) Loess improvement methods. *Engineering Geology*, 25, 341–366.

Feda, J. (1966) Structural stability of subsidence loess from Praha-Dejvice. *Engineering Geology*, 1, 201–219.

Feda, J. (1988) Collapse of loess on wetting. *Engineering Geology*, 25, 263–269.

Feda, J., Bohac, J. and Herle, I. (1993) Compression of collapsed loess: studies on bonded and unbonded soils. *Engineering Geology*, 34, 95–103.

Fookes, P.G. and Best, R. (1969) Consolidation characteristics of some late Pleistocene periglacial metastable soils of east Kent. *Quarterly Journal Engineering Geology*, 2, 103–128.

Forster, A., Culshaw, M.G. and Bell, F.G. (1995) The regional distribution of sulphate rocks and soils of Britain. In *Engineering Geology and Construction*, Engineering Geology Special Publication No. 10, Eddleston, M., Walthall, S., Cripps, J.C. and Culshaw, M.G. (eds), The Geological Society, London, 95–104.

Frydman, S. and Calabresi, F.G. (1987) Suggested standard for one-dimensional testing. In *Proceedings of the Sixth International Conference on Expansive Soils*, New Delhi, 1, 91–100.

Gao, G. (1988) Formation and development of the structure of collapsing loess in China. *Engineering Geology*, 25, 235–245.

Gerber, A. and Harmse, H.J. von M. (1987) Proposed procedure for identification of dispersive soils by chemical testing. *The Civil Engineer in South Africa*, 29, 397–399.

Gibbs, H.H. and Bara, J.P. (1962) Predicting surface subsidence from basic soil tests. *American Society Testing Materials (ASTM), Special Technical Publication*, No. 322, 231–246.

Gidigasu, M.D. (1974) Degree of weathering and identification of laterite materials for engineering purposes – a review. *Engineering Geology*, 8, 213–266.

Gidigasu, M.D. (1988) Potential application of engineering pedology in shallow foundation engineering on tropical residual soils. *Proceedings Second International Conference on Geomechanics in Tropical Soils*, Singapore, Balkema, Rotterdam, 1, 17–24.

Gillott, J.E. (1979) Fabric, composition and properties of sensitive soils from Canada, Alaska and Norway. *Engineering Geology*, 14, 149–172.

Grabowska-Olszewska, B. (1988) Engineering geological problems of loess in Poland. *Engineering Geology*, 25, 177–199.

Handy, R.L. (1973) Collapsible loess in Iowa. *Proceedings American Society Soil Science*, 37, 281–284.

Hanrahan, E.T. (1954) An investigation of some physical properties of peat. *Geotechnique*, 4, 108–123.

Hobbs, N.B. (1986) Mire morphology and the properties and behaviour of some British and foreign peats. *Quarterly Journal Engineering Geology*, 19, 7–80.

Horiguchi, K. (1979) Effect of rate of heat removal on rate of frost heaving. *Engineering Geology*, Special Issue on Ground Freezing, 13, 63–72.

Hossain, D. and Ali, K.M. (1988) Shear strength and consolidation characteristics of Obhor sabkha, Saudi Arabia. *Quarterly Journal Engineering Geology*, 21, 347–359.

Indraratna, B., Nutalaya. P. and Kuganenthira, N. (1991) Stabilization of a dispersive soil by blending with fly ash. *Quarterly Journal Engineering Geology*, 24, 275–290.

James, A.N. and Little, A.L. (1994) Geotechnical aspects of sabkha at Jubail, Saudi Arabia. *Quarterly Journal Engineering Geology*, 27, 83–121.

Jennings, J.E. and Knight, K. (1957) The prediction of total heave from the double oedometer test. *Transactions South African Institution Civil Engineers*, 7, 285–291.

Jennings, J.E. and Knight, K. (1975) A guide to construction on or with materials exhibiting additional settlement due to collapse of grain structure. *Proceedings Sixth African Conference Soil Mechanics Foundation Engineering*, Durban, 99–105.

Jones, R.H. (1980) Frost heave of roads. *Quarterly Journal Engineering Geology*, 13, 77–86.

Kinosita, S. (1979) Effects of initial soil water conditions on frost heaving characteristics. *Engineering Geology*, Special Issue on Ground Freezing, 13, 53–62.

Landva, A.O. and Pheeney, P.E. (1980) Peat fabric and structure. *Canadian Geotechnical Journal*, 17, 416–435.

Lane, K.S. and Washburn, D.E. (1946) Capillary tests by capillarimeter and by soil filled tubes. *Proceedings Highway Research Board*, Bulletin 26, 460–473.

Lin, Z.G. and Wang, S.J. (1988) Collapsibility and deformation characteristics of deep-seated loess in China. *Engineering Geology*, 25, 271–282.

Locat, J., Lefebvre, G. and Ballivy, G. (1984) Mineralogy, chemistry and physical properties inter-relationships of some sensitive clays from eastern Canada. *Canadian Geotechnical Journal*, 21, 530–540.

Lutenegger, A.J. and Hallberg, G.R. (1988) Stability of loess. *Engineering Geology*, 25, 247–261.

McDaniel T.N. and Decker, R.S. (1979) Dispersive soil problem at Los Esteros Dam. *Proceedings American Society Civil Engineers, Journal Geotechnical Engineering Division*, 105, 1017–1030.

McFarlane, M.J. (1976) *Laterite and Landscape*. Academic Press, London.

McQueen, I.S. and Miller, R.F. (1968) Calibration and evaluation of wide ring gravimetric methods for measuring moisture stress. *Soil Science*, 106, 225–231.

Mitchell, J.K. and Sitar, N. (1982) Engineering properties of tropical residual soils. *Proceedings Specialty Conference on Engineering and Construction in Tropical and Residual Soils*, Honolulu, American Society Civil Engineers, Geotechnical Engineering Division, 30–57.

Moore, P.J. and Styles, J.R. (1988) Some characteristics of volcanic ash soil. *Proceedings Second International Conference on Geomechanics in Tropical Soils*, Singapore, Balkema, Rotterdam, 1, 161–165.

Morin, W.J. and Todor, P.C. (1975) Laterite and lateritic soils and other problem soils in the tropics. In *Engineering Evaluation and Highway Design Study*, US Agency for International Development, AID/esd 3682, Lyon Associates Inc., Baltimore, USA.

Netterburg, F. (1977) Salt damage to roads – an interim guide to its diagnosis, prevention and repair. *Institution of Municipal Engineers of South Africa*, 4, 13–17.

Nixon, I.K. and Skipp, B.O. (1957) Airfield construction on overseas soils. Part 5 – Laterite. *Proceedings Institution Civil Engineering*, 36, 253–275.

Northmore, K.J., Bell, F.G. and Culshaw, M.G. (1996) The engineering properties and behaviour of the brickearth of south Essex. *Quarterly Journal Engineering Geology*, 29, 147–161.

Ola, S.A. (1978a) Geotechnical properties and behaviour of stabilized lateritic soils. *Quarterly Journal Engineering Geology*, 11, 145–160.

Ola, S.A. (1978b) The geology and engineering properties of black cotton soils in north eastern Nigeria. *Engineering Geology*, 12, 375–391.

Ola, S.A. (1980) Mineralogical properties of some Nigerian residual soils in relation with building problems. *Engineering Geology*, 15, 1–13.

O'Neill, M.W. and Poormoayed, A.M. (1980) Methodology for foundations on expansive clays. *Proceedings American Society Civil Engineers, Journal Geotechnical Engineering Division*, 106, 1245–1267

Penner, E. and Walton, T. (1979) Effects of temperature and pressure on frost heaving. *Engineering Geology*, Special Issue on Ground Freezing, 13, 29–40.

Phien-wej, N., Pientong, T. and Balasubramanian, A.S. (1992) Collapse and strength characteristics of loess in Thailand. *Engineering Geology*, 32, 59–72.

Popescu, M.E. (1979) Engineering problems associated with expansive clays from Romania. *Engineering Geology*, 24, 43–53.

Popescu, M.E. (1986) A comparison of the behaviour of swelling and collapsing soils. *Engineering Geology*, 23, 145–163.

Radforth, N.W. (1952) Suggested classifications of muskeg for the engineer. *Engineering Journal (Canada)*, 35, 1194–1210.

Reed, M.A., Lovell, C.W., Altschaeffl, A.G. and Wood, L.E. (1979) Frost heaving rate predicted from pore size distribution. *Canadian Geotechnical Journal*, 16, 463–472.

Sabtan, A., Al-Saify, M. and Kazi, A. (1995) Moisture retention characteristics of coastal sabkhas. *Quarterly Journal Engineering Geology*, 28, 37–46.

Schmertmann, J.H. (1969) Swell sensitivity. *Geotechnique*, 19, 530–533.

Scott, G.R. (1972) Map I-790-C, US Geological Survey, Washington, DC.

Sherard, J.L., Dunnigan, L.P. and Decker, R.S. (1976) Identification and nature of dispersive soils. *Proceedings American Society Civil Engineers, Journal Geotechnical Engineering Division*, 102, 287–301.

Sherard, J.L., Dunnigan, L.P. and Decker, R.S. (1977) Some engineering problems with dispersive clays. *Proceedings Symposium on Dispersive Clays, Related Piping and Erosion in Geotechnical Projects*, Sherard, J.L. and Decker, R.S. (eds), ASTM Special Publication 623, 3–12.

Stamatopoulos, A.C., Christadoulias, J.C. and Giannaros, H.Ch. (1992) Treatment for expansive soils for reducing swell potential and increasing strength. *Quarterly Journal Engineering Geology*, 25, 301–312.

Taber, S. (1930) Mechanics of frost heaving. *Journal Geology*, 38, 303–317.

Terzaghi, K. (1925) *Erdbaumechanik auf Boden Physikalischer Grundlage*. Deuticke, Vienna.

Tuncer, R.E. and Lohnes, R.A. (1977) An engineering classification of certain basalt derived lateritic soils. *Engineering Geology*, 11, 319–339.

Van Der Merwe, D.H. (1964) The prediction of heave from the plasticity index and the percentage clay fraction, *The Civil Engineer in South Africa*, 6, 103–107.

Vargas, M. (1985) The concept of tropical soils. *Proceedings First International Conference on Geomechanics of Tropical Soils*, Brasilia, Brazilian Society for Soil Mechanics, 3, 101–134.

Vargas, M. (1990) Collapsible and expansive soils in Brazil. *Proceedings Second International Conference on Geomechanics in Tropical Soils*, Singapore, Balkema, Rotterdam, 2, 469–487.

Vaughan, P.R., Marcarini, M. and Mokhtar, S.M. (1988) Indexing the properties of residual soil. *Quarterly Journal Engineering Geology*, 21, 69–84.

Vijayvergiya, V.N. and Ghazzaly, O.I. (1973) Prediction of swelling potential for natural clays. *Proceedings Third International Conference on Expansive Soils*, Haifa, 227–231.

Von Post, L. (1922) Sveriges Geologiska Undersokings torvimventering och nogra av dess hittels vunna resultat (SGU peat inventory and some preliminary results). *Svenska Morskulltwforeningens Tidskift*, Jonkoping, Sweden, 36, 1–37.

Wagner, A.A. (1957) The use of the Unified Soil Classification for the Bureau of Reclamation. *Proceedings Fourth International Conference Soil Mechanics Foundation Engineering*, London, 1, 125–134.

Watermeyer, C.F., Botha, G.R. and Hall, B.E. (1991) Countering potential piping at an earth dam on dispersive soils. In *Geotechnics in the African Environment*, Blight, G.E., Fourie, A.B., Luker, I., Mouton, D.J. and Scheurenburg, R.J. (eds), Balkema, Rotterdam, 321–328.

Watermeyer. R.B. and Tromp, B.E. (1992) A systematic approach to the design and construction of single storey residential masonry structures on problem soils. *The Civil Engineer in South Africa*, 34, 83–96.

Weinert, H.H. (1980) *The Natural Road Construction Materials of South Africa*. Academica, Cape Town.

Weston, D.J. (1980) Expansive road treatment for southern Africa. *Proceedings Fourth International Conference on Expansive Soils*, Denver, 1, 339–360.

Williams, A.A.B. (1980) Severe heaving of a block of flats near Kimberley. *Proceedings Seventh Regional Conference for Africa on Soil Mechanics and Foundation Engineering*, Accra, 1, 301–309.

Wilson, C. and Melis, L. (1991) Breaching of an earth dam in the Western Cape by piping. In *Geotechnics in the African Environment*, Blight, G.E., Fourie, A.B., Luker, I., Mouton, D.J. and Scheurenburg, R.J. (eds), Balkema, Rotterdam, 301–312.

Youd, T.L. and Perkins, J.B. (1987) Map I-1257-G, US Geological Survey, Washington, DC.

Chapter 6

River action and control

All rivers form part of a drainage system, the form of which is influenced by rock type and structure, the nature of the vegetation cover, and the climate. An understanding of the processes that underlie river development forms the basis of proper river management.

Rivers also form part of the hydrological cycle in that they carry precipitation run-off. This run-off is the surface water that remains after evapotranspiration and infiltration into the ground have taken place. Some precipitation may be frozen, only to contribute to run-off at some other time, while any precipitation that has infiltrated into the ground may reappear as springs where the water table meets the ground surface. Although, due to heavy rainfall or in areas with few channels, the run-off may occur as a sheet, it usually becomes concentrated into channels, which become eroded by the flow of water and so eventually form valleys.

6.1 Fluvial processes

Fluvial processes comprise the full range of movement and action of surface water on sloping terrain. Unchannelled flow is termed sheet wash or overland flow. Rills and gullies are features associated with small-scale incisement. Rills possess negligible drainage areas, while gullies have steep banks with entrenched channels with steep headwalls. Streams and rivers represent larger channelled flow. Streams are the fundamental parts of a drainage basin. They develop a characteristic pattern and obey quantitative and geometrical laws in their development. The shape and size of a river basin have an important influence on water flow characteristics and the amount of erosion, transportation and deposition that occurs.

Leopold and Maddock (1953) suggested a number of simple relationships between discharge, channel dimensions and velocity as follows:

$$Q = WZv \qquad\qquad (6.1a)$$

$$W = aQ^b \qquad\qquad (6.1b)$$

$$Z = cQ^f \tag{6.1c}$$

$$v = kQ^m \tag{6.1d}$$

where Q is stream discharge, W is width, Z is depth and v is velocity; a, c and k are constants; and b, f and m are measures involving the rate of change. The discharge measurements of stream flow are usually taken at a station where differences due to rising and falling stages can be calculated and same-day discharge for downstream sites determined. The comparison of exponents is one way of contrasting streams of different regions. Typical calculations for midwestern United States have yielded exponents of b as 0.26, f as 0.4 and m as 0.34; for downstream discharge the exponents are 0.5, 0.4, and 0.1, respectively (Coates, 1987). Therefore, at a particular site, the greatest proportion of increasing discharge is absorbed by depth and the smallest width, whereas in downstream segments width is the greatest and velocity the smallest. Fahnestock (1963) showed that when width, depth and velocity exponents are about equal, streams have few constraints in their behaviour.

Subsequently, Carlston (1968) proposed that the mean annual discharge (Q_m) was related to channel slope (S) in the following manner:

$$S \propto Q_m^a \tag{6.2}$$

He found that the value of a varied significantly with the character of the channel. Where the channel was cut into an alluvial bed, and on sections that were in a steady state and a could be defined, the correlation between Q_m and S was good. In ungraded streams, however, there was no significant correlation between these two parameters. The consensus is that the discharge necessary for the formation of channels is that which approaches bankfull stage. Hence, the mean annual flood should be closely related to channel dimensions. For example, Schumm (1977) indicated that if a flood event is chiefly responsible for the dimensions of a channel, then mean annual discharge is not as closely related to channel morphology as the mean flood.

Expressions were developed by Schumm (ibid.) to relate channel hydrology and other morphological characteristics to channel width (an index of discharge) and channel width/depth ratio (an index of the type of sediment load). For example, he showed that channel width (W) and depth (Z) are related to Q_m and mass (M) as follows:

$$W = 2.3 \frac{Q_m^{0.38}}{M^{0.39}} \tag{6.3}$$

and

$$Z = 0.6M^{0.34} \times Q_m^{0.29} \tag{6.4}$$

In stable rivers with sand beds the relation between bedload (L) and channel morphology is given by:

$$L = \frac{W\lambda S}{Zp} \tag{6.5}$$

where p is sinuosity (the ratio of channel length to valley length) and λ is meander wavelength. Lane's (1955) expression

$$Ld = QS \tag{6.6}$$

indicates that as bedload and sediment size (d) increase, either water discharge (Q) or slope (S), or both, increase to compensate.

In 1957, Leopold and Wolman showed that there are three different flow patterns in river development, namely, straight, meandering and braided types, in which slope, sediment load and discharge seem to be the controlling factors. Streams in rugged terrain tend to carry coarse bedload and generally are less sinuous. Their width/depth cross-sections are large in relative terms to other streams because a wide, shallow channel is more efficient for transportation of coarse bedload since higher velocities occur nearer the channel floor. As a stream moves farther from its source it increases in sinuosity, and bedload becomes finer. Sinuous streams minimize their total work expenditure by adjusting their curvature to their slopes. By increasing their sinuosity, energy is dissipated more uniformly through each unit distance along the channel.

Straight streams possess essentially straight banks, having a sinuosity of less than 1.5, but flow between the banks is not necessarily straight. Turbulent flow does not move water in a straight line, because secondary currents develop with transverse flow producing various bedforms, pools and riffles. A pool is a deep reach, whereas a riffle is shallower with elongated sediment bars. Each succeeding riffle in a downstream direction slopes alternately towards the opposite bank. The form of the riffle can become relatively fixed in position as a node of accumulation, and particles that compose the bar are intermittently moved from one riffle to the next with each succeeding flood flow. The spacing of pools and riffles is normally 5 to 8 times channel width.

Braided stream patterns are characteristic of a main channel that is divided into a network of anastomosing and branching smaller channels, within which are small lateral and horizontal bars and islands. The network has a sinuosity that usually exceeds 1.5. Braided streams are common in glacial, periglacial, arid and semi-arid environments, and in regions where weathering produces debris that is impoverished in terms of fine-grained sediment. Rapid variations in discharge capable of creating alternating erosion and deposition sequences are also important in the formation of braided streams. Stream banks are usually low and have poor cohesiveness, so they are erodible, and the sediment load is coarser than silt. The gradient in the bifurcated reaches increases as a result of divided flow and the necessity to maintain discharge. Tributary streams supply coarser material than the main channel can transport during normal flow regimes, so bars and various bedforms develop in the lower flow regime. Braids become more elaborate in sand bed streams as particle size is reduced.

Meandering streams are characterized by a sinuosity greater than 1.5, the channel being singular and free-swinging over a plain. Empirical relationships have been developed concerning the geometry of alluvial channels; for example, meander wavelength (λ) and bankfull width (W) are related by:

$$\lambda = aW^b \tag{6.7}$$

where a varies between 7 and 10, and b is approximately 1.0. The relationship between the mean radius of curvature (R) and meander wavelength (λ) is expressed as:

$$\lambda = eR^f \tag{6.8}$$

in which e and f are approximately 5 and 1.0, respectively. The mean radius of curvature is approximately twice the bankfull width. Such relationships can only define general tendencies and so may not apply to particular situations. A more complex expression used to determine meander wavelength was proposed by Schumm (1967):

$$\lambda = \frac{1890 \, Q_m^{0.34}}{M^{0.74}} \tag{6.9}$$

in which Q_m is the mean annual discharge and M is the percentage of silt and clay in the river load.

Meander migration occurs mostly during bankfull discharge, with deposition occurring on point bars along the convex banks. Sediment is derived from the scour pool immediately upstream and on the same side of the river. Deposition results in the width of the meander bend being constricted, so causing an increase in mean velocity in order to accommodate the discharge. The increased velocity produces an increase in bed shear stress, which deepens the scour pool. As a result the concave banks become oversteepened, leading to bank caving and slumping. This keeps meander migration away from the point bar deposit. In this way, meanders migrate both laterally across and down the floodplain.

6.1.1 River flow

It is generally considered that, except at times of flood, a stream has a steady uniform flow, that is, one in which, at any given point, the depth does not vary with time. If the flow is uniform, the depth is constant over the length of the stream concerned (Figure 6.1). Although these two assumptions are not strictly true, they allow simple and satisfactory solutions for river engineering problems. Assuming a steady uniform flow, then the rate at which water passes through successive cross-sections of a stream is constant. River flow is measured as discharge of volume of water passing a given point per unit time. The height or stage of water in a channel depends on discharge, as well as on the shape and capacity of the river channel. Bankfull discharge refers to the maximum volume of water at a certain velocity of flow that a river can sustain without overbank spillage. Hydraulic geometry and bankfull discharge alter if the channel is changed by erosion.

If water flows along a smooth, straight channel at very low velocities, it moves in laminar flow, with parallel layers of water shearing one over the other. In laminar flow, the layer of maximum velocity lies below the water surface, while around the wetted perimeter it is least. Laminar flow cannot support particles in suspension and in fact is not found in natural streams except near the bed and banks. When the velocity of flow exceeds a critical value it becomes turbulent. Fluid components in turbulent flow follow a complex pattern of movement, components mixing with each other and secondary eddies

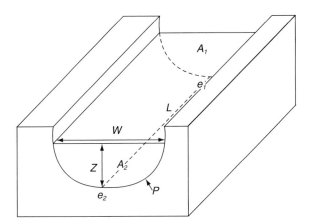

Figure 6.1 Stream channel morphometry. Stream width, W, is the actual width of water in the channel. Wetted perimeter, P, is the outline of the edge where water and channel meet. Cross-section, A, is the area of a transverse section of the river. Depth, Z, is approximately the same as the hydraulic radius, R, which is the cross-section divided by the wetted perimeter ($R = A/P$). Stream gradient, S, is the drop in elevation ($e_1 - e_2$) between two points on the bottom of a channel, divided by the projected horizontal distance between them, L.

superimposed on the main forward flow. The Reynolds number (N_R) is the starting point for calculation of erosion and transportation within a channel system and is commonly used to distinguish between laminar and turbulent flow. It is expressed as:

$$N_R = \rho\,\frac{(vR)}{(\mu)} \tag{6.10}$$

where ρ is fluid density, v is mean velocity, R is hydraulic radius (the ratio between the cross-sectional area of a river channel and the length of its wetted perimeter) and μ is viscosity. Flow is laminar for small values of Reynolds number and turbulent for higher ones. The Reynolds number for flow in streams is generally over 500, varying from 300 to 600.

There are two kinds of turbulent flow, namely, streaming and shooting flow. Streaming flow refers to the ordinary turbulence found in most streams, whereas shooting flow occurs at higher velocities such as found in rapids. Whether turbulent flow is shooting or streaming is determined by the Froude number (F_n):

$$F_n = \frac{v}{\sqrt{(gZ)}} \tag{6.11}$$

where v is the mean velocity, g is the acceleration due to gravity and Z is the depth of water. If the Froude number is less than 1, the stream is in the streaming-flow regime, while if it is greater than 1 it is in the shooting-flow regime.

Generally, the highest velocity of a stream is at its centre, below or extending below the surface. The exact location of maximum velocity depends upon channel shape, roughness and sinuosity, but it usually lies between 5 and 25% of the depth. In a symmetrical river channel the maximum water velocity is below the surface and centred. Regions of moderate velocity but high turbulence occur outward from the centre, being greatest near the bottom. Near the wetted perimeter velocities and turbulence are low. On the other hand, in an asymmetrical channel the zone of maximum velocity shifts away from the centre towards the deeper side. In such instances, the zone of maximum turbulence is raised on the shallow side and lowered on the deeper side. Consequently, channel morphology has a significant influence on erosion.

Turbulence and velocity are very closely related to the erosion, transportation and deposition. The work done by a stream is a function of the energy it possesses. Potential energy is converted by downflow to kinetic energy, which is mostly dissipated in friction. Stream energy is therefore lost owing to friction from turbulent mixing, and as such frictional losses are dependent upon channel roughness and shape. Total energy is influenced mostly by velocity, which in turn is a function of the stream gradient, volume and viscosity of water in flow, and the characteristics of the channel cross-section and bed. This relationship has been embodied in the Chezy formula, which expresses velocity as a function of hydraulic radius (R) and slope (S):

$$v = C \sqrt{(RS)} \tag{6.12}$$

where v is mean velocity and C is a constant, which depends upon gravity and other factors contributing to the friction force. The minimum bed erosion takes place when the gradient of the channel is low and the wetted perimeter is large compared with the cross-sectional area of the channel.

The Manning formula represents an attempt to refine the Chezy equation in terms of the constant C:

$$v = \frac{1.49}{n} R^{2/3} S^{1/2} \tag{6.13}$$

where the terms are the same as the Chezy equation and n is a roughness factor. The velocity of flow increases as roughness decreases for a channel of particular gradient and dimensions. The roughness factor has to be determined empirically and varies not only for different streams but also for the same stream under different conditions and at different times. In natural channels the value of n is 0.01 for smooth beds, about 0.02 for sand and 0.03 for gravel. The roughness coefficients of some natural streams are given in Table 6.1. Anything that affects the roughness of the channel changes n, including the size and shape

Table 6.1 Values of roughness coefficient n for natural streams (from Chow, 1964).

Description of stream	normal n
On a plain	
Clean straight channel, full stage, no riffles or deep pools	0.030
Same as above but with more stones and weeds	0.035
Clean winding channel, some pool and shoals	0.040
Sluggish reaches, weedy, deep pools	0.070
Mountain streams	
No vegetation, steep banks, bottom of gravel, cobbles and a few boulders	0.040
No vegetation, steep banks, bottom of cobbles and large boulders	0.050
Flood plains	
Pasture, no brush, short grass	0.030
Pasture, no brush, high grass	0.035
Brush, scattered to dense	0.050–0.10
Trees, dense to cleared, with stumps	0.150–0.04

of grains on the bed; sinuosity; and obstructions in the channel section. Variation in discharge also affects the roughness factor, since depth of water and volume influence the roughness.

The shear stress τ exerted on a fixed boundary by a turbulent fluid is a function of the fluid density and the shear velocity of the fluid. The mean boundary shear stress equation for open channels is:

$$\tau = \rho g R S \qquad (6.14)$$

where ρ is the density, g is the acceleration due to gravity, R is the hydraulic radius and S is the slope of the channel.

Since shear stress is related to the velocity gradient, as well as the energy gradient and the hydraulic radius (R), then frictional velocity (U_*) is given by:

$$U_* = gRS \qquad (6.15)$$

The hydraulics of sediment transport involves dissolved load, which offers no resistance, and suspended load, which serves to dampen turbulence thus increasing stream efficiency. In order to determine what set of flow conditions cause entrainment of a particle it is important to calculate such factors as the critical flow velocity, the critical boundary shear stress and the critical lifting force. These factors are contained in the relationship provided by the Shields entrainment function (F_s):

$$F_s = \frac{\tau}{(\rho_g - \rho_f)gd} \qquad (6.16)$$

where τ is the boundary shear stress, ρ_g is the particle density, ρ_f is the density of the water, g is the acceleration due to gravity and d is the diameter of the particle. The energy and forces of a stream are derived from its gravitational component as reflected by velocity and discharge. Thus each stream contains potential energy, which is a function of the weight and head of water. This energy is converted to kinetic energy (E_k) during downhill travel. Most kinetic energy is lost to friction but that which remains is available for erosional and transportation processes. This relationship is embodied in the kinetic equation:

$$E_k = \frac{M}{2} v^2 \qquad (6.17)$$

where M is the mass and v is the velocity, so that velocity determines energy and varies according to stream gradient and channel characteristics. When the channel cross-sectional area increases, the stream velocity decreases because water is not compressible.

6.1.2 River erosion

The work undertaken by a river is threefold: it erodes soils and rocks, and transports the products thereof, which it eventually deposits. Erosion occurs when the force provided by the river flow exceeds the resistance of the material over which it flows. Thus the erosional velocity is appreciably higher than that required to maintain stream movement.

The amount of erosion accomplished by a river in a given time depends upon the quantity of energy it possesses, which in turn is influenced by its volume and velocity of flow, and the character and size of its load. The soil, rock type and geological structure over which it flows, the infiltration capacity of the area it drains, and the vegetation cover, which directly affects the stability and permeability of the soil, also influences the rate of erosion.

Once bank erosion starts, and the channel is locally widened, the process is self-sustaining. The recession of the bank permits the current to wash more directly against the downstream side of the eroded area, and so erosion and bank recession continue. Sediment accumulates on the opposite side of the channel, or on the inside of a bend, so the channel gradually shifts in the direction of bank attack. In such a way, meandering is initiated and this process is responsible for most channel instability.

Stream meanders usually occur in series, and there is normally a downstream progression of meander loops. Meander growth is often stopped by the development of shorter chute channels across the bars formed on the inside of the beds. Chutes may develop because the resistance to flow around the lengthening bend becomes greater than that across a bar, or because changes in alignment caused by the channel shifting upstream tend to direct flow across a bar inside the bend. Meander loops may be abandoned because cut-offs (Figure 6.2) develop from adjacent bends: either the loops migrate into each other, or channel avulsions form across the necks between adjacent bends during periods of overbank flooding. Meander cut-offs or shortening by chute development reduce channel lengths and increase slopes, and hence are generally beneficial for reducing flood heights or improving drainage. But they may cause much local damage by channel shifting and bank erosion, and the resulting unstable bed conditions may interfere with navigation.

During flood the volume of a river is greatly increased, which leads to an increase in its velocity. The principal effect of flooding in the upper reaches of a river is to accelerate the rate of erosion, much of the material so produced then being transported downstream and deposited over the flood plain (Lewin, 1989). The vast increase in erosive strength during maximum flood is well illustrated by the devastating floods that occurred on Exmoor in August 1952. It was estimated that these moved 153 000 m^3 of rock debris into Lynmouth,

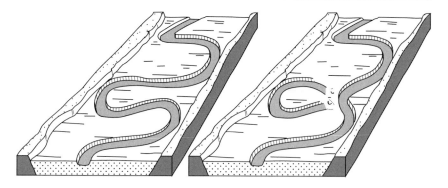

Figure 6.2 Formation of an oxbow lake.

some of the boulders weighing up to 10 tonnes. Scour and fill are characteristic of flooding. Often a river channel is filled during the early stages of flooding, but as discharge increases scour takes over. For example, streams flooding on alluvial beds normally develop an alternating series of deep and relatively narrow pools, typically formed along the concave sides of bends, together with shallow wider reaches between bends, where the main current crosses the channel diagonally from the lower end of one pool to the upper end of the next. During high flows the pools or bends tend to scour deeper, while the crossing bars are built higher by sediment deposition, although deposition does not equal rise in stage and hence water depth increases on the bars. When the stage falls, erosion takes place from the top of crossing bars leading to some filling in the pools. However, as low-stage activity is less effective, the general shape of the bed usually reflects the influence of the flood stage.

6.1.3 River transport

The load that a river carries is transported by traction, saltation, suspension or solution. The competence of a river to transport its load is demonstrated by the largest boulder it is capable of moving. This varies according to its velocity and volume, being at a maximum during flood. Generally the competence of a river varies as the sixth power of its velocity. The capacity of a river refers to the total amount of sediment that it carries, and varies according to the size of the particles that form the load on the one hand and its velocity on the other. When the load consists of fine particles the capacity is greater than when it is comprised of coarse material. Usually, the capacity of a river varies as the third power of its velocity. Both the competence and the capacity of a river are influenced by changes in the weather, and the lithology and structure of the rocks over which it flows.

The sediment discharge of a river is defined as the mass rate of transport through a given cross-section measured as mass per second per metre width, and can be divided into the bedload and suspended load. The force necessary to entrain a given particle is referred to as the critical tractive force, and the velocity at which this force operates on a given slope is the erosion velocity. The critical erosion velocity is the lowest velocity at which loose grains of a given size on the bed of a channel will move. The value of the erosion velocity varies according to the characteristics and depth of the water, the size, shape and density of particles being moved, and the slope and roughness of the floor. The Hjulstrom (1935) graph shows the threshold boundaries for erosion, transportation and deposition (Figure 6.3a). Modified versions of these curves have appeared subsequently (Figure 6.3b).

It can be seen from Figure 6.3a that fine- to medium-grained sands are more easily eroded than clays, silts or gravels. The fine particles are resistant because of the strong cohesive forces that bind them and because fine particles on the channel floor give the bed a smoother surface. There are accordingly few protruding grains to aid entrainment by giving rise to local eddies or turbulence. However, once silts and clays are entrained they can be transported at much lower velocities. For example, particles 0.01 mm in diameter are entrained at a critical velocity of about 600 mm s^{-1} but remain in motion until the velocity drops below 1 mm s^{-1}. Gravel is hard to entrain simply because of the size and weight of the particles involved. Particles in the bedload move slowly and intermittently. They generally move by rolling or sliding, or by saltation if the instantaneous hydrodynamic lift is greater than the weight of the particle. Deposition takes place wherever the local flow conditions do not re-entrain the particles.

Bedload transport in sand bed channels depends upon the regime of flow, that is, on streaming or shooting flow. When the Froude number is much smaller than 1, flow is tranquil, velocity is low, the water surface is placid, and the channel bottom is rippled. In this streaming regime resistance to flow is great and sediment transport is small, with only single grains moving along the bottom. As the Froude number increases, but remains within the streaming-flow regime, the form of the bed changes to dunes or large-scale ripples (Figure 6.4a). Turbulence is now generated at the water surface and eddies form in the lee of the dunes (Figure 6.4b). Movement of grains takes place up the lee side of the dunes to cascade down the steep front, causing the dunes to move downstream. When the Froude number exceeds 1, flow is rapid, velocity is high, resistance to flow is small, and bedload transport is great. At the transition to the upper flow regime planar beds are formed. As the Froude number increases further, standing waves form and then antidunes are developed. The particles in the suspended load have settling velocities that are lower than the buoyant velocity of the turbulence and vortices. Once particles are entrained and are part of the suspended sediment load, little energy is required to transport them. Indeed, as mentioned above, they can be carried by a current with a velocity lower than the critical erosion velocity needed for their entrainment.

(a)

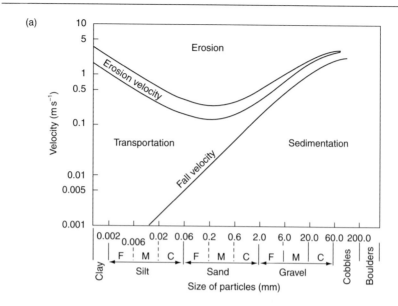

(b)

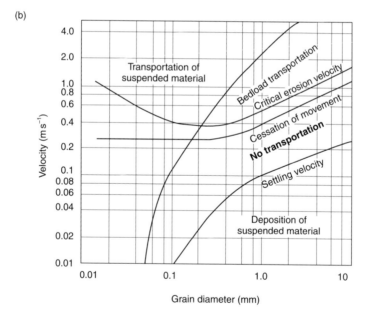

Figure 6.3 (a) Curves for erosion, transportation and deposition of uniform sediment. F, fine; M, medium; C, coarse. Note that fine sand is the most easily eroded material (after Hjulstrom, 1935). (b) Relation between flow velocity, grain size, entrainment and deposition for uniform grains with a specific gravity of 2.65. The velocities are those 1 m above the bottom of the body of water. The curves are not valid for high sediment concentrations, which increase the fluid viscosity (after Sundberg, 1956, by permission of Blackwell Publishers).

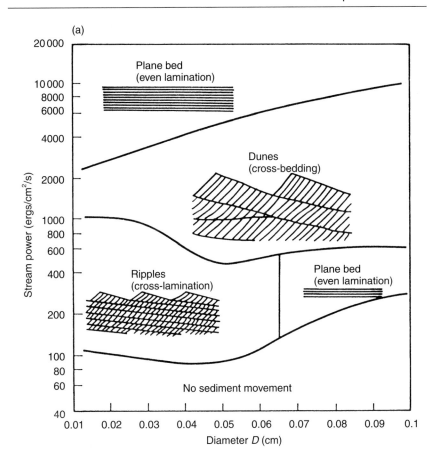

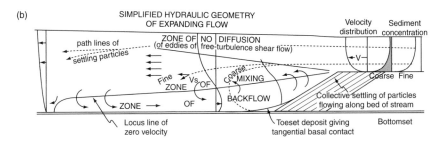

Figure 6.4 (a) Bed forms in relation to stream power and calibre of bedload material (from Simons and Richardson, 1961, by permission of American Society of Civil Engineers). (b) Hydraulic conditions at the lip of a small delta produced in a flume. The flow structure developed over the slip slope is the same as that produced over a slip slope on a dune or ripple (after Jopling, 1963, by permission of Blackwell Science Ltd).

Moreover, the suspended load decreases turbulence, which in turn reduces frictional losses of energy and makes the stream more efficient.

The distribution of suspended load increases rapidly with depth below the surface of a stream, the highest concentration generally occurring near the bed. However, there is a variation in suspended sediment concentrations at various depths of a stream for grains of different sizes. Most of the sand grains are carried in suspension near the bottom, whereas there is not very much change in silt concentration with depth.

Suspended load is commonly calculated from a sample obtained by a depth-integrating sampler, which is moved up and down along a vertical level in a stream. The weight of sediment in a sampler is determined and referred to the weight of water carrying it. This is the concentration of the suspended load and is expressed in parts per million. The suspended load concentration usually increases with an increase in stage of a stream. In large streams the peak of the sediment concentration is generally close to the peak of the discharge. In fact, during flood the amount of suspended sediment load generally increases more quickly than discharge and reaches a peak concentration maybe several hours before the floodwaters do. In such cases, the suspended load carried during the highest water flow is therefore considerably less than capacity.

Because the discharge of a river varies, sediments are not transported continuously; for instance, boulders may be moved only a few metres during a single flood. In other words, there is a threshold discharge below which no movement of bedload occurs, so there is a direct relationship between bedload movement and flood discharge. Once the rate of flow falls beneath the threshold value, the bedload remains stationary until the next flood of equal or higher magnitude occurs. Nonetheless, the contribution of bedload to total stream load is frequently high. Often the amount of bedload moved downstream during times of flood exceeds that of suspended load by several times.

As remarked, the total amount of sediment carried by a river increases significantly during flood. Part of the increase in load is derived from within the channel and part from outside the channel. Consequently, channel form changes during flood, the depth and width of the channel adapting to scour and deposition (Gupta, 1988). Scour at one point is accompanied by deposition at another. Initially during a flood, bedload may be moved into small depressions in the floor of a river, and it is not removed easily. Major floods can bring about notable changes in channel form by bank scour and slumping. However, after a flood, deposition may make good some of the areas that were removed.

Some alluvial deposits such as channel bars are transitory, existing for a matter of days or even minutes. Hence, the channels of most streams are excavated mainly in their own sedimentary deposits, which streams continually rework by eroding the banks in some places and redepositing the sediment farther downstream. Indeed, sediments that are deposited over a flood plain may be regarded as being stored there temporarily.

6.1.4 Deposition of sediments

Deposition occurs where turbulence is at a minimum or where the region of turbulence is near the surface of a river. For example, lateral accretion occurs with deposition of a point bar on the inside of a meander bend. A point bar grows as the meander moves downstream or new ones are built as the river changes course during or after floods. Old meander scars can often be seen on flood plains (Figure 6.5a). The combination of point bar and filled slough results in what is called ridge and swale topography. The ridges are composed of sand bars, and the swales are the depressions, which are subsequently filled with silt and clay.

An alluvial flood plain is the most common depositional feature of a river. Ward (1978) suggested that a flood plain can be regarded as a store of sediment across which channel flow takes place and which, in the long term, is comparatively unchanging in amount. However, dramatic changes may occur in the short term (Rahn, 1994). In fact, the morphology of a flood plain often displays an apparent adjustment to flood discharge in that different flood plain or terrace levels may be related to different frequencies of flood discharge. Obviously, the higher levels are inundated by the largest floods.

As noted above, the peak concentration of suspended load usually occurs prior to the discharge peak. Hence, much of the load of a stream is contained within its channel. The water that overflows on to the flood plain accordingly possesses less suspended sediment (Macklin *et al.*, 1992). As a consequence, the processes within the channel responsible for lateral accretion are usually the more important in the formation of a flood plain compared with overbank flow, which results in vertical accretion. In fact, lateral accretion may account for between 60 and 80% of the sediment that is deposited (Leopold *et al.*, 1964).

The alluvium of flood plains is made up of many kinds of deposit, laid down both in the channel and outside it (Marsland, 1986). Vertical accretion on a flood plain is accomplished by in-channel filling and the growth of overbank deposits during and immediately after floods. Gravels and coarse sands are moved chiefly at flood stages and deposited in the deeper parts of a river. As the river overtops its banks, its ability to transport material is lessened so that coarser particles are deposited near the banks to form levees. Levees therefore slope away from the channels into the flood basins, which are the lowest part of a flood plain. At high stages, low sections and breaks in levees may mean that there is a concentrated outflow of water from the channel into the flood plain. This outflow rapidly erodes a crevasse, leading to the deposition in the flood basin of a crevasse splay. Finer material is carried farther and laid down as backswamp deposits (Figure 6.5b). At this stage, a river sometimes aggrades its bed, eventually raising it above the level of the surrounding plain. In such a situation, when the levees are breached by flood water, hundreds of square kilometres may be inundated.

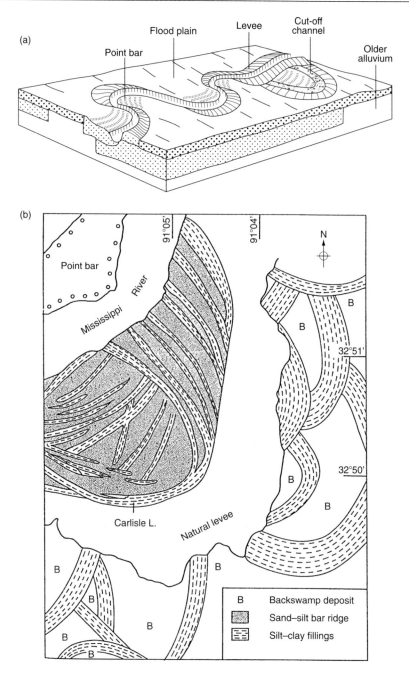

Figure 6.5 (a) The main depositional features of a meandering channel; (b) map of a portion of the Mississippi river flood plain, showing various kinds of deposit (after Fisk, 1944).

6.2 Floods

Floods represent the commonest type of geological hazard (Figure 6.6). They probably affect more individuals and their property than all the other hazards put together. However, the likelihood of flooding is more predictable than some other types of hazard such as earthquakes, volcanic eruptions and landslides.

Most disastrous floods are the result of excessive rainfall or snowmelt; that is, they are due to excessive surface run-off. In most regions floods occur more frequently in certain seasons than others. Meltwater floods are seasonal and are characterized by a substantial increase in river discharge so that there is a single flood wave (Church, 1988). The latter may have several peaks. The two most important factors that govern the severity of floods due to snowmelt are the depth of snow and the rapidity with which it melts.

Identical flood-generating mechanisms, especially those associated with climate, can generate different floods within different catchments, or within the same catchment at different times. Such differences, according to Newson (1994), are attributable to the effects of flood-intensifying factors such as certain basin, network and channel characteristics, which may increase the rate of flow through the system. For instance, flood peaks may be low and attenuated in a long narrow basin with a high bifurcation ratio, whereas in a basin with a rounded shape and low bifurcation ratio the flood peaks are higher. Generally, drainage patterns that give rise to the merging of flood flows from major tributaries in the lower part of a river basin are associated with high-magnitude

Figure 6.6 Flooding in Ladysmith, Natal, South Africa, in February 1994.

flood peaks in that part of the system. In addition, the capacity of the ground to allow water to infiltrate and the amount that can be stored influences the size and timing of a flood. Frozen ground, by inhibiting infiltration, is an important flood-intensifying factor and may extend the source area over the whole of a catchment. A catchment area with highly permeable ground conditions may have such a high infiltration capacity that it is rarely subjected to floods. Flood discharges are at a maximum per unit area in small drainage basins. This is because storms tend to be local in occurrence. High drainage density watersheds produce simultaneous flooding in tributaries, which then overload junction capacity to transmit flow. In low drainage density basins the procession of flood waves has longer delay periods, allowing the main channel to absorb incoming flows one at a time. Hence, it usually takes some time to accumulate enough run-off to cause a major disaster.

Flash floods, which are short-lived extreme events, prove the exception. They usually occur under slowly moving or stationary thunderstorms, which last for less than 24 hours. The resulting rainfall intensity exceeds infiltration capacity, so run-off takes place very rapidly. Flash floods are frequently very destructive as the high energy flow can carry much sedimentary material (Clarke, 1991).

On the other hand, long-rain floods are associated with several days or even weeks of rainfall, which may be of low intensity. They are the most common cause of major flooding (Bell, 1994). Single-event floods have a single main peak but are of notably longer duration than flash floods. They represent the commonest type of flooding. Nevertheless, some of the most troublesome floods are associated with a series of flood peaks that follow closely on each other. Such multiple-event floods are caused by more complex weather conditions than those associated with single-event floods. The effects of multiple-event floods are usually severe because of the duration over which they extend. This could be several weeks, or even months in the case of certain monsoon or equatorial regions, where seasonal floods frequently are of extended duration.

As the volume of water in a river is greatly increased during times of flood, so its erosive power increases accordingly. Thus, the river carries a much higher sediment load. Deposition of the latter where it is not wanted also represents a serious problem.

The influence of human activity can bring about changes in drainage basin characteristics; for example, the removal of forest from parts of a river basin can lead to higher peak discharges, which generate increased flood hazard. A most notable increase in flood hazard can arise as a result of urbanization; the impervious surfaces created mean that infiltration is reduced, and together with stormwater drains, they give rise to increased run-off. Not only does this produce higher discharges but lag times also are reduced. The problem of flooding is particularly acute where rapid expansion has led to the development of urban sprawl without proper planning, or worse, where informal settlements have sprung up. Heavy rainfall can prove disastrous where informal settlements are concerned.

A flood can be defined in terms of height or stage of water above some given point such as the banks of a river channel. However, rivers are generally considered in flood when the level has risen to an extent that damage occurs. The discharge rate provides the basis for most methods of predicting the magnitude of flooding, the most important factor being the peak discharge, which is responsible for maximum inundation. Not only does the size of a flood determine the depth and area of inundation but it also primarily determines the duration of a flood. These three parameters, in turn, influence the velocity of flow of flood waters, all four being responsible for the damage potential of a flood. As noted above, the physical characteristics of a river basin together with those of the stream channel affect the rate at which discharge downstream occurs. The average time between a rainstorm event and the consequent increase in river flow is referred to as the lag time. This can be measured from the commencement of rainfall to the peak discharge or from the time when actual flood conditions have been attained (e.g. bankfull discharge) to the peak discharge. This lag time is an important parameter in flood forecasting. Calculation of the lag time, however, is a complicated matter. Nonetheless, once enough data on rainfall and run-off versus time have been obtained and analysed, an estimate of where and when flooding will occur along a river system can be made.

An estimate of future flood conditions is required for either forecasting or design purposes. In terms of flood forecasting, more immediate information is needed regarding the magnitude and timing of a flood so that appropriate evasive action can be taken. In terms of design, planners and engineers require data on the magnitude and frequency of floods. Hence, there is a difference between flood forecasting for warning purposes and flood prediction for design purposes. A detailed understanding of the run-off processes involved in a catchment and stream channels is required for the development of flood forecasting but is less necessary for long-term prediction. The ability to provide enough advance warning of a flood means that it may be possible to reduce the resulting damage, because people are able to evacuate the area likely to be affected, along with some of their possessions. The most reliable forecasts are based on data from rainfall or melt events that have just taken place or are still occurring. Hence, advance warning is generally measured only in hours or sometimes in a few days. The longer-term forecasts are commonly associated with snowmelt. Basically, flood forecasting involves the determination of the amount of precipitation and resultant run-off within a given catchment area. The volume of run-off is then converted into a time-distributed hydrograph, and flood routing procedures are used, where appropriate, to estimate the changes in the shape of the hydrograph as the flood moves downstream. In other words, flood routing involves determination of the height and time of arrival of the flood wave at successive locations along a river. As far as prediction for design purposes is concerned, the design flood, namely, the maximum flood against which protection is being designed, is the most important factor to determine (see

Section 6.6). Methods of flood forecasting and prediction have been reviewed by Ward (1978).

6.3 Factors affecting run-off

The total run-off from a catchment area generally consists of four component parts, namely, direct precipitation on the stream channels, surface run-off, interflow and baseflow. Unless the catchment area contains a large number of lakes or swamps, direct precipitation on to water surfaces and into stream channels normally represents only a small percentage of the total volume of water flowing in streams. Even where the area of lakes is large, evaporation from them may equal the amount of precipitation they receive. Consequently, this component is usually ignored in run-off calculations. However, where lakes and swamps occur in the drainage basin they tend to 'absorb' high peaks of surface run-off, which is particularly beneficial in catchments with rather low infiltration capacities.

Surface run-off comprises the water that travels over the surface as sheet or channel flow. It is the first major component of flood and peak discharges during a rainstorm.

Some proportion of rainfall infiltrates into the ground, where it may meet a relatively impermeable layer that causes it to flow laterally, just below the surface, towards streams. This is referred to as interflow and, as would be expected, it moves more slowly than surface run-off. The interflow contribution to total run-off depends mainly on the soil characteristics of the catchment and the depth of the water table. In some areas, interflow may account for up to 85% of the total run-off.

The infiltration capacity is the rate at which water is absorbed by a soil. Absorption starts with an initial value, decreases rapidly, then reaches a steady value, which is taken as the infiltration capacity. Rainfall occurring after a steady rate of infiltration is reached is rainfall excess and flows off as surface run-off. Conversely, rain that falls with an intensity that is not capable of satisfying the infiltration capacity produces no rainfall excess and thus no run-off. The infiltration capacity of a particular soil is governed by soil texture: that is, the size and arrangement of grains and their state of aggregation, since they influence porosity and permeability; by vegetative cover; by biological structures such as root and worm holes; by antecedent soil moisture, that is, the moisture remaining from previous rain; and by the condition of the soil surface, for example, whether it is baked or compacted. Rain itself can reduce infiltration capacity by packing the soil, breaking down the structure of aggregates, washing down finer grains to fill the pores, and swelling colloids and clay particles by wetting them.

Often a significant proportion of total run-off is stored in the river banks, which is therefore referred to as bank storage. Bank storage takes place above the normal phreatic surface. As stream levels fall, water from bank storage is released into them. Most of the rainfall that percolates to the water table eventually reaches the main stream channels as baseflow or effluent seepage. Since water moves very slowly through the ground, the outflow of groundwater into stream channels will not only lag behind the occurrence of rainfall by several days, weeks or even years, but will also be very regular. Baseflow, therefore, normally represents the major, long-term component of total run-off, and may be particularly important during long, dry spells, when low-water flow may be entirely derived from groundwater supplies.

The flow of any stream is governed by climatic and physiographic factors. As far as the climatic factors are concerned the type, intensity, duration and distribution of precipitation contribute towards stream flow, while evapotranspiration has the opposite effect and is influenced by temperature, wind velocity and relative humidity. Indeed, the most obvious and probably the most effective influence on the total volume of run-off is the long-term balance between the amount of water gained by a catchment area in the form of precipitation and the amount of water lost in the form of evapotranspiration.

Most rivers show a seasonal variation in flow, which, although influenced by many factors, is largely a reflection of climatic variations. The pattern of seasonal variations, which tends to be repeated year after year, is frequently referred to as the river regime. Obviously, the study of river regimes plays an important part in the understanding of problems associated with flood prevention and sediment transport. The type of precipitation is important; for example, the contribution to run-off of rainfall is almost immediate, providing that its intensity and magnitude are great enough. Gentle rain that falls over a period of several hours or even days gives rise to a relatively modest flood peak with a comparatively long time base, but when the same quantity of rainfall is concentrated in a much shorter time period, then significant flooding may occur. In cold climates, in particular, a large proportion of stream flow may be derived from melting snow and ice. Where melting occurs gradually the contribution resembles that of baseflow. In other words, the snow or ice blanket acts as a store of water supply and makes a stable contribution to run-off. On the other hand, if melting occurs suddenly as a result of a rapid thaw, a large volume of water enters streams during a short period of time, giving a peak run-off. Nevertheless, only in high-latitude and high-altitude regions is the effect of accumulation and snowmelt of long-term significance.

Surface run-off does not usually become a significant feature, except in the case of intense storms, until most of the soil moisture deficit has been replenished. But once this has happened, run-off increases quite rapidly, representing an increasing proportion of the rainfall during the rest of the fall. However, the increase in stream flow does not occur at the same rate as the increase in rainfall excess because of the lag effect resulting from storage. If rainfall occurs over a frozen surface infiltration cannot take place, so that once the initial interception and depression storage has been satisfied, the remaining rainfall contributes towards run-off.

There is a critical period for an individual drainage basin for which all storms of that particular duration, irrespective of intensity, produce a period of surface run-off that is essentially the same, while for rains of longer duration the period of surface run-off is increased. The effectiveness of rainfall duration varies with the size and relief of the drainage basin. In a small catchment with steep slopes, maximum potential run-off is likely to be attained by a rainfall of shorter duration than in a large catchment with gentle slopes. The infiltration capacity is reduced during periods of extended rainfall, so the amount of run-off increases.

Storms that produce floods in large drainage basins are very rarely uniformly distributed. The highest peak flows in large basins are usually produced by storms that occur over large areas, while high peak flows in small drainage basins are commonly the result of intense thunderstorms that extend over limited areas. The amount of run-off resulting from any rainfall depends to a large extent on how the rainfall is distributed. If it is concentrated in a particular area of a basin the run-off is greater than if it is uniformly distributed throughout the basin. This is because in the former instance the infiltration capacity is exceeded quickly. The distribution coefficient provides an assessment of the run-off that results from a particular distribution of rainfall. It is expressed as:

$$\text{distribution coefficient} = \frac{\text{maximum rainfall at any point}}{\text{mean of the basin}} \qquad (6.18)$$

The peak run-off increases as the distribution coefficient increases.

The amount of surface moisture in the soil obviously influences the infiltration capacity. For instance, if the soil is saturated most rainfall will go towards run-off and flooding may occur. The run-off characteristics are influenced by the soil type since this influences the porosity and permeability of the soil, which in turn influence the infiltration capacity.

Every drainage basin or catchment area is defined by a topographic divide or watershed, which bounds the area from which the surface run-off is derived. Similarly, the groundwater contribution to a given catchment is bounded by a phreatic divide. These two divides are not necessarily coincident and accordingly inter-watershed leakage can occur. The location of the phreatic divide tends to move with fluctuations in the water table, but the higher the water table the more nearly do the two divides coincide.

The area of a basin affects the size of floods likely to occur, as well as influencing minimum flow levels. For instance, the larger the drainage basin, then the longer it takes for the total flood flow to pass a given location. What is more, the peak flow decreases relatively as the area of the basin increases, since storms become less effective and infiltration increases. Because local rains contribute to the discharge of the main stream, larger basins are likely to provide a more sustained flow than smaller ones.

One of the principal factors that governs the rate at which run-off is supplied to the main stream is the shape of the drainage basin. The outlines of large drainage basins are generally fixed, at least in part, by major geological structures, while erosional features usually form the limits of small drainage basins. The effect of shape can be demonstrated best by considering three differently shaped catchments of the same area, subjected to rainfall of the same intensity (Figure 6.7a). If each catchment is divided into concentric segments, which may be assumed to have all points within an equal distance along the stream channels from the control point, it can be seen that shape A requires 10 time units to pass before every point on the catchment is contributing to the discharge. Similarly, B requires 5 and C 8.5 time units. The shape factor also affects the run-off when a rainstorm does not cover the whole catchment at once but moves over it from one end to the other. The direction in which it moves in relation to the direction of flow can have a decided influence upon the resulting peak flow and also upon the duration of surface run-off. For example, consider catchment A to be slowly covered by a storm moving upstream and just covering the catchment after 5 time units. The flood contribution of the last segment will not arrive at the control until 15 time units from commencement. Alternatively, if the storm were moving at the same rate downstream, the flood contribution of time segment 10 would arrive at the control point simultaneously with that of all the others, so that an extremely rapid flood rise would occur. The effect of changing the direction of storm movement on the other catchments is less marked but still appreciable. The effect of storm location within a catchment is illustrated in Figure 6.7b, the steeper of the two flood hydrographs (see Section 6.4) being associated with the storm located closer to the outlet of the basin.

The index related to the shape of a drainage basin is termed the compactness coefficient. This is the ratio of the watershed perimeter to the circumference of a circle whose area is equal to that of the drainage basin. The less compact a basin is, then the less likely it is to have intense rainfall simultaneously over its entire extent. The lower the value of the coefficient, the more rapidly is water likely to be discharged from the catchment area via the main streams.

The variation in and mean elevation of a drainage basin obviously influence temperatures and precipitation, which in turn influence the amount of run-off. Generally, precipitation increases with altitude but more important is the effect of reduced evaporation and the temporary storage of precipitation in snow and ice. This affects the distribution of the mean monthly run-off, reducing it to a minimum in winter in cold climates.

Surface run-off and infiltration are related to the gradient of a drainage basin. Indeed, the slope of a drainage basin is one of the major factors controlling the time of overland flow and concentration of rainfall in stream channels, and is of especial importance as far as the magnitude of floods is concerned. Obviously, with steep slopes there is a greater chance that the water will move off the surface before it has had time to infiltrate, so that surface run-off is large.

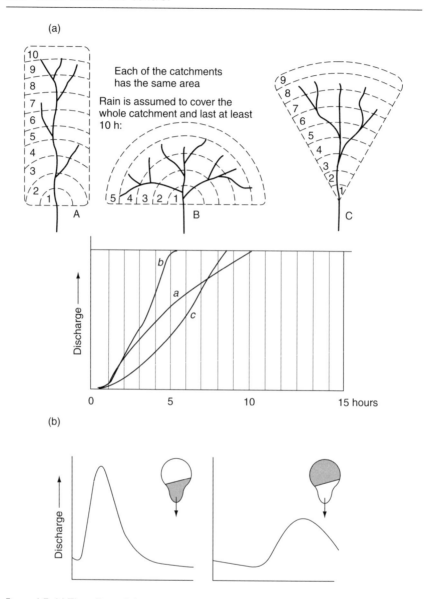

(a)

10
9
8
7
6
5
4
3
2
1
A

Each of the catchments
has the same area

Rain is assumed to cover the
whole catchment and last at least
10 h:

5 4 3 2 1
B

9
8
7
6
5
4
3
2 1
C

(b)

Figure 6.7 (a) The effect of shape on a catchment run-off (after Wilson, 1983). (b) Identical storms located in different parts of the drainage basin, giving different run-offs.

The efficiency of a drainage system is dependent upon the stream pattern. For instance, if a basin is well drained the length of overland flow is short, the surface run-off concentrates quickly, the flood peaks are high and in all probability the minimum flow is correspondingly low. The drainage pattern also generally reflects the geological conditions found within the drainage basin. A drainage network can be described in terms of stream order, length of tributaries, stream density and drainage density, and length of overland flow. First-order streams are unbranched, and when two such streams become confluent they form a second-order stream. When two of the latter types join, they form a third-order stream, and so on. It is only when streams of the same order meet that they produce one of higher rank. The frequency with which streams of a certain order flow into those of the next order above is referred to as the bifurcation ratio and is derived by dividing the number of any given order of streams by that of the next highest order. Dendritic river systems generally have bifurcation ratios in the range 3.5 to 4.0, whereas those associated with trellis patterns are much higher. The order of the main stream gives an indication of the size and extent of the drainage pattern. Such a procedure of identification provides a technique for comparing streams and basins in a numerical manner.

The length of tributaries is an indication of the steepness of the drainage basin as well as the degree of drainage. Steep, well-drained areas usually have numerous small tributaries, whereas on plains with deep, permeable soils only the relatively long tributaries are generally perennial.

The stream density or frequency can be expressed as:

$$\text{stream density} = \frac{\text{number of streams in basin}}{\text{total area of basin}} \quad (6.19)$$

Stream density does not provide a true measure of the drainage efficiency. Drainage density is the stream length per unit of area and varies inversely as the length of overland flow, thus providing some indication of the drainage efficiency of a basin. It reflects the concentration or stream frequency of an area and is inversely related to the rate of infiltration. In other words, a river basin in which the floor consists of permeable rocks tends to have fewer streams per unit area than one floored with impermeable rocks. The drainage density is also influenced by climate. The shape of drainage basins and their drainage density influence stream flow conditions (see Section 6.2).

The circularity index is the calculation of the degree of circularity of the basin. It is the ratio of the area of a circle to the area of the basin when the same circumference or perimeter length is used. Dendritic basins have circularity indices that are about 0.6 to 0.7.

The geometry of a stream channel influences run-off. A wide dish-shaped channel, for example, gives a rapid rate of increase in width with increasing discharge, while a rectangular channel has a rapid rate of increase of depth with increasing discharge.

Carlston (1963) found a highly significant relationship between mean annual flood discharge per unit area and drainage density. Peak discharge and the lag time of discharge are also influenced by drainage density, as well as by the shape and slope of the drainage basin (Gregory and Walling, 1973). Stream flow is generally most variable and flood discharges at a maximum per unit area in small basins, because storms tend to be local in occurrence. Carlston also showed that an inverse relationship existed between drainage density and baseflow. This is related to the permeability of the rocks present in a drainage basin. In other words, the greater the quantity of water that moves on the surface of the drainage system, the higher the drainage density, which in turn means that the baseflow is lower.

A permeable soil or rock allows water to percolate to the zone of saturation, from where it can be slowly discharged as springs. Open-textured sandy soils have much higher infiltration capacities than clay soils and therefore give rise to much less surface run-off. A more dramatic example is provided by the virtual disappearance of surface drainage in some areas where massive limestones are exposed. By contrast, basins on impermeable rock produce high volumes of direct run-off and very little baseflow. In flat, low-lying areas the soil type also influences the position of the water table. The water table in clay soils rises after rainfall, perhaps causing waterlogging, whereas rapid drainage through gravels allows the water table to remain below the ground surface. In the latter case, the baseflow contribution to stream flow is likely to reach the drainage channels with little delay. Although direct surface run-off is reduced by percolation the final total amount is not.

The geological structure may influence the movement of groundwater towards streams, and it generally explains the lack of correlation between the topographical and hydrological divides of adjacent catchments. The long-term relationship between groundwater and surface run-off determines the main characteristics of a stream and provides a basis for classifying streams into ephemeral, intermittent or perennial types. Ephemeral streams are those that contain only surface run-off and therefore only flow during and immediately after rainfall or snowmelt. Normally, there are no permanent or well-defined channels and the water table is always below the bed of the stream. Intermittent streams flow during a wet season, drying up during drought. Stream flow consists mainly of surface run-off, but baseflow makes some contribution during a wet season. Perennial streams flow throughout the year because the water table is always above the bed of the stream, making a continuous and significant contribution to total run-off. It is seldom possible to classify the entire length of a stream in this way.

The most important effect of the vegetation cover is to slow down the movement of water over the surface after rainfall and thus to allow more time for infiltration to take place. In this way the timing of run-off after rainfall may be considerably modified and peak stream flows may be much lower, although more prolonged.

Human factors such as agricultural practices and land use also affect run-off. In urban areas sewers, drains and paving increase run-off efficiency and reduce lag time and ground storage of precipitation, producing accelerated and high run-off peaks. Thus, stream channel networks become more efficient in collecting water quickly and may give rise to flash flooding.

6.4 Assessment of run-off

In dealing with run-off the hydrologist has to try to provide answers relating to the occurrence, size and duration of floods and droughts. Of special concern is the magnitude and duration of run-off from a particular catchment with respect to time. This can be resolved by producing graphs of the frequency and duration of individual discharges from observations over a long period of time, although if such observations are not available, estimations may be made at various probabilities. Even if measurements of rainfall and evapotranspiration were completely reliable, there would still be a need for direct measurement of stream flow. In fact, accurate stream flow data are likely to become even more important as the need to assess regional water resources grows. However, stream flow is perhaps the most difficult, and is certainly the most costly, of the hydrological parameters to measure accurately. Being a widely variable quantity there is no direct way of continuously monitoring flows in a river. Basically, however, there are three related operations in the measurement of run-off. The first of these involves the determination of the height or stage of the river; the second involves the determination of the mean velocity of the water flowing in the stream channel; and the third involves the derivation of a known relationship between stage and total volume of discharge.

6.4.1 Measurement of river stage

The measurement of the stage of a river may be made periodically or continuously, depending upon the degree of accuracy required and the hydrological characteristics of the stream. Generally, the larger a catchment area is and the more permeable the ground, the less important it is for river stage to be continuously monitored. Periodic observations of river stage are normally made by reference to a staff gauge. An alternative method is to use a surface contact gauge or wire weight gauge. It may be necessary during floods to know the height of the flood peak on a large number of tributary streams in order to ascertain which area of a drainage basin is contributing the greatest proportion of surface run-off. A peak gauge can be used in such an instance. It may consist of a hollow tube set vertically in the water, peak levels being indicated inside the tube by a non-returning float. At most major gauging stations water levels are reproduced autographically by means of continuous recorders.

6.4.2 Measurement of current velocity

The measurement of current velocity in streams is usually carried out by means of a current meter, of which there are two principal types. In the first of these the water flows against a number of cups attached to a vertical spindle, while in the second type, the flowing water acts directly on the upstream surface of a propeller, which is attached to a horizontal spindle. The speed at which the cups or propeller, are turned provides an indication of the current velocity. Due to friction between the flowing water and the wetted perimeter of the stream channel, the velocity profile of a stream is not constant. Consequently, mean velocity has to be established from the average of a number of current meter observations located in such a way as to detect the differences in velocity. Generally, the intervals between adjacent measuring points should not exceed one-fifteenth of the stream width where the bed profile is regular and one-twentieth of the width where it is irregular. Due care must be taken in the selection of the reach of stream used for current meter measurements, the ideal being a reach where the velocity profile is both regular and symmetrical. The length of channel chosen should preferably be straight for a distance of approximately three times the bankfull width of the river, the bed should be smooth so as to reduce turbulent flow to a minimum, and the direction of flow should be normal to the section of measurement.

The total volume of water or discharge flowing past a given point in a given time is the product of the cross-sectional area of the stream and its velocity. If the stream bed and banks have been accurately surveyed at the place of measurement, only data on the stage of the stream are needed to enable the cross-sectional area of the water in the stream channel to be calculated. The volume of discharge is calculated once the mean velocity of the current is known. Discharge can be measured accurately by means of a weir or flume on streams where a physical obstruction in the channel is permissible. If a continuous record of discharge is required, then it must be correlated with river stage. When discharge is plotted against corresponding stages, the curve drawn through plotted points is referred to as a rating curve (Figure 6.8). In other words, a rating curve is a graph relating the stage of a river channel at a certain cross-section to the corresponding discharge at that section. Hence, it can be used to estimate the quantity of water passing a particular location at a given time. When this ratio is greater than 1.0, flood conditions are imminent. A rating curve can be used to predict the severity of flooding as measured by the height of the flood wave that overtops bankfull height. This information can then be added to the curve developed from calculation of the frequency of flooding as related to discharge. This establishes the recurrence interval for floods of different magnitude. Using these data it can then be determined which parts of the floodplain will be affected by a storm of a given intensity and duration.

6.4.3 Peak flow

There are two specific problems related to run-off predictions. First, there is the need to forecast peak flows associated with sudden increases in surface run-off and, second, there is the prediction of minimum flow, which very much

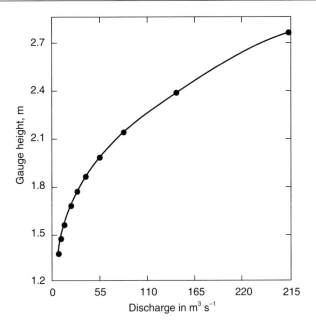

Figure 6.8 Stage discharge rating curve, showing the relationship between height of water at gauge and amount of discharge.

involves the decreasing volume of baseflow. The accuracy of run-off predictions tends to improve as the time interval is increased. Estimates of annual or even seasonal run-off totals for given catchment areas may, in some cases, be made from annual or seasonal rainfall totals, using a simple straight-line regression between the two variables.

However, correlations between rainfall and run-off may generally be expected to yield forecasts of only token accuracy, for they take no account of the contributions made by interflow and baseflow. Increases in surface run-off after rainfall or snowmelt tend to be rapid, leading to a short-lived peak, which is followed by a rather longer period of declining run-off. Since it is the peak flow that brings havoc the principal object of prediction concerns the magnitude and timing of this peak, and the frequency with which it is likely to occur.

In order to carry out a flood frequency analysis either the maximum discharge for each year or all discharges greater than a given discharge are recorded from gauging stations according to magnitude (Dalrymple, 1960; Benson, 1968; Bobee and Ashkar, 1988). The recurrence interval, that is, the period of years within which a flood of given magnitude or greater occurs, is determined from:

$$T = \frac{n + 1}{m} \qquad\qquad (6.20)$$

where T is the recurrence interval, n is the number of years of record, and m is the rank of the magnitude of the flood, with $m = 1$ as the highest discharge on record. Each flood discharge is plotted against its recurrence interval and the points are joined to form the frequency curve (Figure 6.9). The flood frequency curve can be used to determine the probability of the size of the discharge that could be expected during any given time interval. The larger the recurrence interval, then the longer is the return period and the greater the magnitude of flood flow. The probability that a given magnitude of flow will occur or be exceeded in a given time is the reciprocal of the recurrence interval. The probability, P, of a flood of recurrence interval, T years, being equalled or exceeded in x years is given by:

$$P = 1 - (1 - 1/T)^x \qquad\qquad (6.21)$$

Estimates of the recurrence intervals of floods of different sizes can be improved by using alluvial stratigraphy to date the sediments they deposited, the more widespread, thicker deposits being formed by larger floods (Jarret, 1990). The technique can be used to reconstruct past changes in climate and so to determine whether in the past floods were more or less frequent at a particular location (Costa, 1978). Clarke (1996) described how to use boulder size to estimate the probable maximum flood that has occurred in the past.

The duration and magnitude of precipitation tend to be inversely related to storm intensity. If a river basin has been saturated by precipitation, then low intensity can be compensated for by high magnitude. Consequently, the quan-

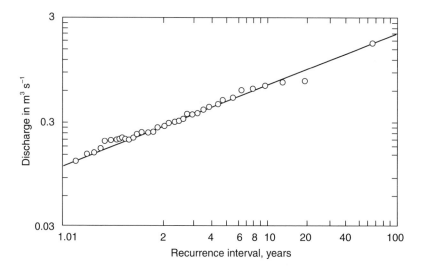

Figure 6.9 Flood frequency curve of the Licking River, Tobaso, Ohio (after Dalrymple, 1960).

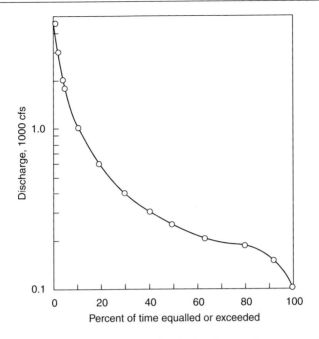

Figure 6.10 Flow duration curve (daily flow) for Bowie Creek near Hattiesburg, Mississippi, for the period 1939–48 (after Searcy, 1959).

tity of antecedent moisture in the ground is important in relation to flooding. In other words, if the ground is saturated no moisture can infiltrate and so the quantity of run-off is enhanced.

A flow-duration curve shows the percentage of time a specified discharge is equalled or exceeded. In order to prepare such a curve, all flows during a given period are listed according to their magnitude. The percentage of time that each one equalled or exceeded a given discharge is calculated and plotted (Figure 6.10). The shape of the curve affords some insight into the characteristics of the drainage basin concerned. For instance, if the curve has an overall steep slope, this means that there is a large amount of direct run-off. On the other hand, if the curve is relatively flat, there is substantial storage, either on the surface or as groundwater. This tends to stabilize stream flow.

6.4.4 Hydrograph analysis

Hydrograph analysis is commonly used in run-off prediction. The hydrograph of a river is a graph that shows how the stream flow varies with time. As such it reflects those characteristics of the watershed that influence run-off. Hydrographs are sensitive to the size of basin in that smaller basins show a quick response to fluctuations in precipitation. It may show yearly, monthly, daily or instantaneous discharges. Accordingly, the total flow, baseflow and periods of high and low flows can be determined from hydrographs. Storm hydrographs can be used to predict the passage of flood events. The rising limb of the curve is generally concave upwards and reflects the infiltration capacity of a watershed (Figure 6.11). The time before the steep climb represents the time before infiltration capacity is reached. A sudden, steeply rising limb reflects large immediate surface run-off. The peak of the curve marks the maximum run-off. Some basins may have two or more peaks for a single storm, depending upon the time distribution of the rain and basin characteristics. The recession limb represents the outflow from basin storage after inflow has ceased. Its slope is therefore dependent upon the physical characteristics that determine storage. Meanwhile, infiltration and percolation result in an elevated water table, which therefore contributes more at the end of the storm flow than at

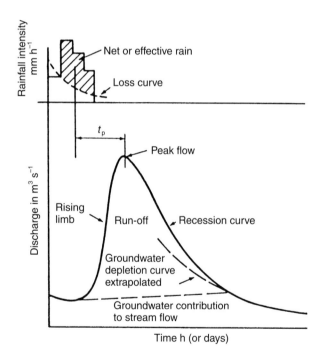

Time h (or days)

Figure 6.11 Component parts of a hydrograph. When rainfall commences there is an initial period of interception and infiltration before any measurable run-off reaches the stream channels. During the period of rainfall, these losses continue in a reduced form, so the rainfall graph must be adjusted to show effective rain. When the initial losses are met, surface run-off begins and continues to a peak value, which occurs at time t_p, measured from the centre of gravity of the effective rain on the graph. Thereafter, surface run-off declines along the recession limb until it disappears. Baseflow represents the groundwater contribution along the banks of the river.

the beginning, but thereafter declines along its depletion curve. The dividing line between run-off and baseflow on a hydrograph is indeterminate and can vary widely. If a second precipitation event follows closely upon the one preceding, then the second storm hydrograph may rise on the recession limb of the first. This complicates hydrograph analysis.

The unit hydrograph method is one of the most dependable and most frequently used techniques for predicting stream flow. The basis of the method depends on the fact that a stream hydrograph reflects many of the physical characteristics of a drainage basin, so similar hydrographs can be produced by similar rainfalls. Accordingly, once a typical or unit hydrograph has been derived for certain defined conditions, it is possible to estimate run-off from a rainfall of any duration or intensity. The unit hydrograph is the hydrograph of 25 mm (now normally taken as 10 mm) of run-off from the entire catchment area resulting from a short, uniform unit rainfall. A unit storm is defined as a rain of such duration that the period of surface run-off is not appreciably less for any rain of shorter duration. Its duration is equal to or less than the period of rise of a unit hydrograph, that is, the time from the beginning of surface run-off to the peak. For all unit storms, regardless of their intensity, the period of surface run-off is approximately the same.

It is sometimes necessary to determine unit hydrographs for catchments with few, if any, run-off records. In such instances, close correlations between the physical characteristics of the catchment area and the resulting hydrographs, are required. Snyder (1938) was one of the earlier workers to derive synthetic unit hydrographs, and he found that the shape of the catchment and the time from the centre of the mass of rainfall to the hydrograph peak were the main influencing characteristics.

6.5 Hazard zoning, warning systems and adjustments

The lower stage of a river, in particular, can be divided into a series of hazard zones based on flood stages and risk (Figure 6.12). A number of factors have to be taken into account when evaluating flood hazard, such as the loss of life and property, erosion and structural damage, disruption of socio-economic activity, including transport and communications, contamination of water, food and other materials, and damage to agricultural land and loss of livestock. A flood plain management plan involves the determination of such zones. These are based on historical evidence related to flooding, which includes the magnitude of each flood and the elevation it reached, as well as the recurrence intervals, the amount of damage involved, the effects of urbanization and any further development, and an engineering assessment of flood potential. Maps are then produced from such investigations that show, for example, the zones of most frequent flooding and the elevation of the flood waters. Flood hazard maps provide a basis for flood management schemes. Kenny (1990) recognized four

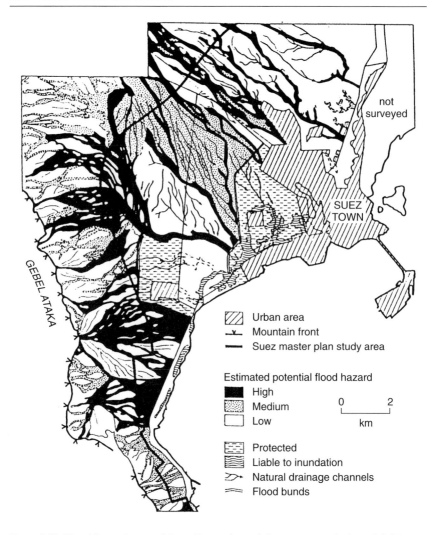

Figure 6.12 Flood hazard map of Suez, Egypt, derived from geomorphological field mapping (from Cooke *et al.*, 1982).

geomorphological flood hazard map units in central Arizona, which formed the basis of a flood management plan (Table 6.2).

Flood plain zones can be designated for specific types of land use: that is, in the channel zone water should be allowed to flow freely without obstruction; for example, bridges should allow sufficient waterway capacity. Another zone could consist of that area with recurrence intervals of 1–20 years, which could be used for parks, and agricultural and recreational purposes. Such areas act as washlands during times of flood. Buildings would be allowed in the zone

Table 6.2 Generalized flood hazard zones and management strategies (after Kenny, 1990).

Flood hazard zone I (Active flood plain area):

Prohibit development (business and residential) within flood plain
Maintain area in a natural state as an open area or for recreational uses only.

Flood hazard zone II (Alluvial fans and plains with channels less than a metre deep, bifurcating, and intricately interconnected systems subject to inundation from overbank flooding):

Flood-proofing to reduce or prevent loss to structures is highly recommended.
Residential development densities should be relatively low; development in obvious drainage channels should be prohibited.
Dry stream channels should be maintained in a natural state and/or the density of native vegetation should be increased to facilitate superior water drainage retention and infiltration capabilities.
Installation of upstream stormwater retention basins to reduce peak water discharges.
Construction should be at the highest local elevation site where possible.

Flood hazard zone III (Dissected upland and lowland slopes; drainage channels where both erosional and depositional processes are operative along gradients generally less than 5%):

Similar to flood hazard zone II
Roadways that traverse channels should be reinforced to withstand the erosive power of a channelled stream flow.

Flood hazard zone IV (Steep gradient drainages consisting of incised channels adjacent to outcrops and mountain fronts characterized by relatively coarse bedload material):

Bridges, roads and culverts should be designed to allow unrestricted flow of boulders and debris up to a metre or more in diameter.
Abandon roadways that currently occupy the wash flood plains.
Restrict residential dwelling to relatively level building sites.
Provisions for subsurface and surface drainage on residential sites should be required.
Stormwater retention basins in relatively confined upstream channels to mitigate high peak discharges.

encompassing the 20- to 100-year recurrence interval, but they would have to have some form of protection against flooding. However, a line that is drawn on a map to demarcate a flood plain zone may encourage a false sense of security and as a consequence development in the upslope area may be greater than it otherwise would be. At some point in time this line is likely to be transgressed, which will cause more damage than would have been the case without a flood plain boundary being so defined. In addition, urbanization, as seen, is a flood-intensifying land use. The replacement of permeable with impermeable surfaces and the artificial drainage systems are responsible for more rapid run-off.

Land-use regulation seeks to obtain the beneficial use of flood plains with minimum flood damage and minimum expenditure on flood protection. In other words, the purpose of land-use regulation is to maintain an adequate floodway (i.e. the channel and those adjacent areas of the flood plain that are necessary for the passage of a given flood discharge) and to regulate land-use development alongside it (i.e. in the floodway fringe, which in the United States is the land between the floodway and the maximum elevation subject to flooding by the 100-year flood). Land use in the floodway in the United States is now severely restricted to, for example, agriculture and recreation. The floodway fringe is also a restricted zone, in which land use is limited by zoning criteria based upon the degree of flood risk. However, land-use control involves the cooperation of the local population and authorities, and often central government. Any relocation of settled areas that are at high risk involves costly subsidy. Such cost must be balanced against the cost of alternative measures and the reluctance of some to move. Bell and Mason (1997) outlined such a problem at Ladysmith, South Africa. In fact, change of land use in intensely developed areas is usually so difficult and so costly that the inconvenience of occasional flooding is preferred. The purchase of land by government agencies to reduce flood damage is rare.

In some situations properties of high value that are likely to be threatened by floods can be flood-proofed. For instance, an industrial plant can be protected by a flood wall, or buildings may have no windows below high water level and possess watertight doors, with valves and cut-offs on drains and sewers to prevent backing up. Other structures may be raised above the most frequent flood levels. Most flood-proofing measures can be more readily and economically incorporated into buildings at the time of their construction than subsequently. Hence, the adoption and implementation of suitable design standards should be incorporated into building codes and planning regulations to ensure the integrity of buildings during flood events.

Reafforestation of slopes denuded of woodland tends to reduce run-off and thereby lowers the intensity of flooding. As a consequence, forests are commonly used as a watershed management technique. They are most effective in relation to small floods, where the possibility exists of reducing flood volumes and delaying flood response. Nonetheless, if the soil is saturated differences in interception and soil moisture storage capacity due to forest cover will be ineffective in terms of flood response. Agricultural practices such as contour ploughing and strip cropping are designed to reduce soil erosion by reducing the rate of run-off. Accordingly, they can influence flood response. These practices are referred to in Chapter 9.

Emergency action involves the erection of temporary flood defences and possible evacuation. The success of such measures depends on the ability to predict floods and the effectiveness of the warning systems. A flood warning system, broadcast by radio or television, or from helicopters or vehicles, can be used to alert a community to the danger of flooding. Warning systems can use rain gauges and stream sensors equipped with self-activating radio transmitters to convey data to a central computer for analysis (Gruntfest and Huber,

1989). Alternatively, a radar rainfall scan can be combined with a computer model of a flood hydrograph to produce real-time forecasts as a flood develops (Collinge and Kirby, 1987). However, widespread use of flood warning is usually only available in highly sensitive areas. The success of a flood warning system depends largely on the hydrological characteristics of the river. Warning systems often work well in large catchment areas which allow enough time between the rainfall or snowmelt event and the resultant flood peak to allow evacuation and any other measures to be put into affect. By contrast, in small tributary areas, especially those with steep slopes or appreciable urban development, the lag time may be so short that, although prompt action may save lives, it is seldom possible to remove or protect property. In the United States, flash flood warning systems have been installed in certain areas, where sensors at an upstream station detect critical water levels and relay them to an alarm station in the community about to be affected. The flood warning is issued from the station.

Financial assistance in the form of government relief or insurance payouts do nothing to reduce flood hazard. Indeed, by attempting to reduce the economic and social impact of a flood, they encourage repair and rebuilding of damaged property, which may lead to the next flood of similar size giving rise to more damage. What is more, the expectation that financial aid will be made available in such an emergency may result in further development of flood-prone areas. Hence, it would be more realistic to adjust insurance premiums for flooding in relation to the degree of risk. Be that as it may, the basic principle of insurance needs to be modified in the case of floods so that premiums paid over many years cover the large losses that will be encountered in a few years. In the United States, the Federal Flood Insurance Program is administered by the Federal Emergency Management Agency (FEMA), from which interested parties can purchase subsidized flood insurance. The programme provides incentives to local governments to plan and regulate land use in flood hazard areas. The boundary of the 100-year flood defines the area that is subject to flood damage compensation, and if a community wishes to qualify for federal aid it must join the Federal Flood Insurance Program. All property owners within the 100-year flood boundary must then purchase flood insurance. The hope is to achieve flood damage abatement and efficient use of the flood plain.

6.6 River control and flood regulation

River control refers to projects designed to hasten the run-off of flood waters or confine them within restricted limits, to improve drainage of adjacent lands, to check stream bank erosion or to provide deeper water for navigation (Brammer, 1990). What has to be borne in mind, however, is that a river in an alluvial channel is continually changing its position due to hydraulic forces acting on its banks and bed. As a consequence, any major modifications to the river system that are imposed without consideration of the channel will give rise to a prolonged and costly struggle to maintain the change.

6.6.1 River training

River training works have evolved from practices centuries old. Today many of these methods are still largely unsophisticated and vary from area to area depending on the availability of suitable materials. They have two objectives, namely, to prevent erosion of river banks, and to prevent the deterioration or improve the discharge capacity of a river channel. A careful study of the current pattern in a river or estuary should be made before deciding the location of training works, as they may cause changes in the behaviour of river flow that may be undesirable. In estuaries, it may be desirable to check the proposed line of training works with the aid of hydraulic models.

Willow piling is one of the most simple methods of controlling erosion, although revetments composed of fascines of willows (faggots) have been more widely used in Britain (Figure 6.13). When stone is readily available it is used in preference to faggotting as it is more durable. Stone has frequently been used as pitching to counteract erosion and for training walls to stabilize river channels. Pre-cast concrete blocks are used more extensively as revetment than stone. Blockwork revetment is more inflexible than stone and requires a rigid toe beam to maintain its alignment. Gabions consist of steel-mesh, rectangular boxes filled with stone (Figure 6.14). Along rivers where there is an ample

Figure 6.13 Placing fascines along the west bank of the River Ouse south of King's Lynn, Lincolnshire, England.

Figure 6.14 Gabions protecting a cut slope in the valley of the River Taff, South Wales.

supply of boulders gabions have some advantages. In urban areas channels may be lined with concrete.

Channel regulation can be brought about by training dykes, jetties or wing dams, which are used to deflect channels into more desirable alignments or confine them to lesser widths. Dykes and dams can be used to close secondary channels and thus divert or concentrate a river into a preferred course. Permeable pile dykes are the principal means of training and contracting the lower Mississippi, restricting the low-water channel to about its normal average width, thereby eliminating local sections of excessive width where shoaling is most troublesome. Sand-fill dams and dykes, built by hydraulic dredging, are used in many places to direct the current or close off secondary channels. The sand-fill dams are not expected to be permanent. In some cases, ground sills or weirs need to be constructed to prevent undesirable deepening of the bed by erosion. Bank revetment by pavement, rip-rap or protective mattresses to retard erosion is usually carried out along with channel regulation.

6.6.2 Dredging

River channels may be improved by dredging. When a river is dredged its floor should not be lowered to such a degree that the water level is appreciably lowered. In addition, the nature of the materials occupying the floor should be investigated. First, this gives an indication of which plant may be suitably employed. Indeed removal of unconsolidated material usually revolves around the selection of suitable floating equipment for the dredging work. Of special importance is the possible presence of boulders. Suction dredgers should always be fitted with a simple trap device in the suction line for catching boulders and rock fragments. Underwater rock excavation can be carried out by underwater drilling and blasting or from a floating rock breaker. Second, it provides information relating to the stability of the slopes of the channel. The rate at which sedimentation takes place provides some indication of the regularity with which dredging should be carried out. Dredging of river mouths and depositing the sediment at sea, however, may lead to erosion of neighbouring beaches subjected to reduced sediment supply.

6.6.3 The design flood and flood control

No structure of any importance, either in or adjacent to a river, should ever be planned or built without due consideration being given to the damage it may cause by its influence on flood waters or the damage to which it may be subjected by those same waters. To avoid disaster, bridges must have the required waterway opening; flood walls and embankments must be high enough for overtopping not to occur; reservoirs must have sufficient capacity; and dams must have sufficient spillway capacity as well as adequate protection against scour at the toe.

The maximum flood that any such structure can safely pass is called the design flood. If a flood of a given magnitude occurs on average once in 100 years there is a 1% chance that such a flood will occur during any one year. The important factor to be determined for any design flood is not simply its magnitude but the probability of its occurrence. In other words, is the structure safe against the 2, 1 or the 0.1% chance flood or against the maximum flood that may ever be anticipated? Once this has been answered the magnitude of the flood that may be expected to occur with that particular average frequency has to be determined. The design of any flood protection works must also consider:

1   The extent to which human life will be endangered. Any structure whose failure would seriously endanger human lives should be designed to pass the greatest flood that will probably ever occur at that point, with safety. However, this point of view has been questioned by Cordery and Pilgrim (1976).
2   The value of any property that would be destroyed by any particular flood. This can be weighed against the estimated cost of the necessary flood protection works.
3   The inconvenience resulting from failure of a structure. A flood control system has a psychological effect in that it provides a sense of security against floods.

In the United States, the Flood Disaster and Protection Act (1973) specified the 100-year flood as the limit of the flood plain for flood insurance purposes, and this became widely accepted as the standard of risk. However, flood mitigation projects should not be constrained by the design standard being linked to a particular flood event. When assessing the future affectiveness of a specific flood mitigation project, its likely reduction of damage and safety afforded to the community over the entire spectrum of possible floods should be evaluated. The design flood should therefore be chosen in relation to all relevant economic, social and environmental factors.

A relatively simple solution to the problem of flooding is to build flood defences consisting of either earth embankments or masonry or concrete walls around the area to be protected (Thampapillai and Musgrave, 1985). Levees have been used extensively in the United States to protect flood plains from overflow. Their use along the lower Mississippi River is given in an account of river engineering activities, along this most engineered of rivers, by Smith and Winkley (1996). Levees are earth embankments, the slopes of which should be protected against erosion by planting trees and shrubs, by paving or with riprap. Without protection against bank erosion, a river will probably begin to meander again. The rate at which river channels revert to their former condition depends upon many factors. Nevertheless, this means that there is a need for maintenance. Levees reduce the storage of flood water by eliminating the natural overflow basins of a river on the flood plain. Furthermore, they contract the channel and so increase flood stages within, above and below the leveed reach. Accordingly, wherever possible levees should be located away from river channels and ideally outside the meander belt. In rural areas this normally poses no great problems, but it is impractical in urban areas. On the other hand, because levees confine a river to its channel this means that its efficiency is increased, hence they expedite run-off. But then consideration must be given to the fact that more rapid movement of water through a section of river that has been hydraulically improved can enhance flood peaks further downstream and be responsible for accelerated erosion. Levees often encourage new development at lower levels where previously no one was inclined to build. Consequently, when the exceptional flood occurs and overtops the levees, the hazard may be reduced by building fuseplugs into the levees, that is, by making certain sections deliberately weaker than the standard levee section, thereby determining that if breaks occur, they do so at locations where they cause minimum damage.

Levees or embankments for flood waters along a river inevitably come in contact with tributaries. In such instances, the protective structures must be continued along the tributary until high ground is reached. Alternatively, the tributary channel can be blocked off during times of flood, for example by sluice gates, but this causes a problem of interior drainage (i.e. water collecting behind the embankment). This problem may be solved by collecting the water during a flood and pumping it over the embankment; by conveying the water in an open channel on the landward side of the embankment to some point downstream where it can be discharged; or by collecting water in a storage basin for subsequent disposal. Embankment stability is another consideration (Gilvear *et al.*, 1994). Excessive seepage of flood water into an embankment may bring about instability or induce piping. The latter can cause an embankment to collapse. If breached, an embankment can accentuate the flood problem in the 'protected' area by inhibiting drainage of floodwater downstream of the breach back into the channel.

Flood walls may be constructed in urban areas where there is not enough space for embankments. They should be designed to withstand the hydrostatic pressure (including uplift pressure) exerted by the water when at design flood level. If the wall is backed by an earthfill, then it must also act as a retaining wall.

Diversion is another method used to control flooding. This involves opening a new exit for part of the river water. Diversion schemes may be temporary or permanent. In the former case, the river channel is supplemented or duplicated by a flood relief channel, a bypass channel or a floodway. These operate during times of flood. Flood water is diverted into a diversion channel via a sluice or fixed-crest spillway. A permanent diversion acts as an intercepting or cut-off channel that replaces the existing river channel, diverting either all or a substantial part of the flow away from a flood-prone reach or river. Such schemes are normally used to protect intensively developed areas. Temporary diversion channels are most effective in reducing flood water levels when they do not return further downstream into the river concerned, that is, they exit into the sea, a lake or another river system. Old river channels can be used as diversions to relieve the main channel of part of its flood water, the Atchafalaya floodway on the lower Mississippi providing an example. Usually, however, it is necessary to return the diverted water into the river at a point downstream. Therefore the diversion channel should be long enough to minimize backwater effects in the stretch of river being protected. Any diversion must be designed in such a way that it does not cause excessive deposition to occur in the main channel, otherwise it defeats its purpose. Bypasses have particular application in highly developed areas, where it is impossible to increase the size of existing rivers.

The flood hazard can often be lessened by stage reduction by improving the hydraulic capacity of the channel without affecting the rate of discharge. This can be accomplished by straightening, and widening and deepening a river channel. Inasmuch as the quantity of discharge through any given cross-section of a river during a given time depends upon its velocity, the stage can be reduced by increasing the velocity (Brookes, 1988). However, the extent of the benefits that can be obtained by straightening depend upon the initial conditions of the channel. For example, even though a river course is extremely sinuous, if the fall is slight, then the amount of stage reduction that can be accomplished by straightening is usually quite limited. Nevertheless, the velocity and therefore the efficiency of a river is generally increased by cutting through constricted meander loops. River channels may be enlarged to carry the maximum flood discharges within their banks without overspill. However, over-widened channels eventually revert to their natural sizes unless dredged continuously. Over-large channels also mean that during periods of low flow the depths of water are shallow and that riparian land may become over-drained.

Flood routing is a procedure by which the variation of discharge with time at a point on a stream channel can be determined by consideration of similar data for a point upstream (Wilson, 1983). In other words, it is a process that shows how a flood wave may be reduced in magnitude and lengthened in time by the use of storage in a reach of the river between the two points. Flood routing therefore depends on a knowledge of storage in the reach. This can be evaluated either by making a detailed topographical and hydrographical survey of the river reach and the riparian land, thereby determining the storage capacity of the channel at different levels, or by using records of past levels of flood waves at the limits of the reach and hence deducing its storage capacity.

Peak discharges can be reduced by temporarily storing a part of the surface run-off until after the crest of the flood has passed. This is done by inundating areas where flood damage is not important, such as water meadows or waste land. If, however, storage areas are located near or in towns and cities they sterilize large areas of land that, if useable for other purposes, could be extremely valuable. On the other hand, it may be feasible to develop these storage areas as recreational centres. This method is seldom sufficient in itself and should be used to complement other measures.

Reservoirs help to regulate run-off, so helping to control floods and improve the utility of a river (Hager and Sinniger, 1985). There are two types of storage, regardless of the size of the reservoir, controlled and uncontrolled. In controlled storage, gates in the impounding structure may regulate the outflow. Only in unusual cases does such a reservoir have sufficient capacity to completely eliminate the peak of a major flood. As a result the regulation of the outflow must be planned carefully. This necessitates an estimation of how much of the early portion of a flood can be safely impounded, which in turn requires an assessment of the danger which can arise if the reservoir is filled before the peak of a flood is reached. Where reservoirs exist on several tributaries, the additional problem of the timing of the release of the stored waters becomes a matter of great importance, since to release these waters in such a way that the peak flows combine at a downstream point can bring disaster. In uncontrolled storage there is no regulation of the outflow capacity, and the only flood benefits result from the modifying and delaying effects of storage above the spillway crest.

A significant although largely qualitative criterion for evaluating a flood-mitigation reservoir is the percentage of the total drainage area controlled by the reservoir. Generally, one-third or more of the total drainage area should be controlled by the reservoir if flood control is to be affective (Ward, 1978). The storage capacity of the reservoir is another criterion. An approximate idea of the effectiveness of a reservoir can be derived by comparing its storage capacity with potential storm rainfall over the area. The maximum capacity that is needed is represented by the difference in volume between the safe release from the reservoir and the design flood inflow.

Reservoirs for flood control should be so operated that the capacity required for storing flood water is available when needed (Bell and Mason, 1998). This can generally be accomplished by lowering the water level of the reservoir as soon as is practicable after the flood passes. On the other hand, the greatest effectiveness of a reservoir for increasing the value of a river for utilization is realized by keeping the reservoir as near full as possible. Hence, there must be some compromise in operation between these two purposes. Furthermore, after a flood has occurred, a part of the storage capacity will be occupied by the flood-water and will not be available until this water is released. However, another part of the storage capacity should be reserved just in case a second flood follows before floodwater release is effected. In other words, the full capacity of a reservoir cannot be assumed to be available for a single flood event. Accordingly, the effectiveness of a reservoir in controlling the regimen of a river increases as the reservoir capacity is augmented and is measured by the ratio of capacity to total run-off. A reservoir must generally be designed and situated so that the quantity of inflow that has to be stored does not ordinarily exceed reservoir capacity. The effect of a reservoir on the regimen of flow in any part of a river varies inversely with the distance from the reservoir because of the time involved in transit, natural losses, fluctuations in the flow of intervening tributaries and decreasing relative effect as river flow increases with increase in drainage area.

The economic aspect of controlling rivers by means of reservoirs is largely affected by the availability of possible reservoir sites. Reservoir sites are principally located in the middle reaches of a river course. Thus limitations are placed upon the use of reservoirs for river control. The value of small reservoirs in the headwaters of a river as a method of flood control is questionable, because such reservoirs are likely to be full or partly full at the time of a flood-producing rain. Moreover, there is little point in inundating a large area of land to protect other areas if they are only slightly more valuable. Reservoirs reduce the flow velocity of streams as they enter it, thereby causing siltation. In time, the growth of the delta may reach upstream and can effect the economy and installations at upstream localities. Water discharged from the reservoir below the dam has renewed ability to erode and entrain sediment from the channel immediately below the dam. Such channel lowering causes lowering of tributary channels. The sediment eroded from positions near the dam is transported further downstream, where it is deposited at slack water sites. These new deposits upset the normal channel system and can lead to renewed flooding.

Retarding basins are much less common than reservoirs. They are provided with fixed, ungated outlets, which regulate outflow in relation to the volume of water in storage. An ungated sluiceway acting as an orifice tends to be preferable to a spillway functioning as a weir. The discharge from the outlet at full reservoir capacity ideally should be equal to the maximum flow that the downstream channel can accept without causing serious flood damage. One of the advantages of retarding basins is that only a small area of land is permanently removed from use after their construction. Although land will be inundated at times of flood, this will occur infrequently, so the land can be used for farming but obviously not for permanent habitation. As with reservoirs, the planning of a system of retarding basins must avoid making a flood event worse by synchronizing the increased flow during drawdown with flood peaks from tributaries. Consequently, retarding basins are better suited to small than large catchment areas.

References

Bell, F.G. (1994) Floods and landslides in Natal and notably the greater Durban area, September 1987, a retrospective view. *Bulletin Association Engineering Geologists*, 31, 59–74.

Bell, F.G. and Mason, T.R. (1997) The problems of flooding in Ladysmith, Natal, South Africa. In *Geohazards and Engineering Geology*, Engineering Geology Special Publication No. 14, Eddleston, M. and Maund, J.G. (eds), Geological Society, London.

Benson, M.A. (1968) Uniform flood frequency estimating methods for federal agencies. *Water Resources Research*, 4, 891–908.

Bobee, B. and Ashkar, F. (1988) Review of statistical methods for estimating flood risk with special emphasis on the log Pearson type 3 distribution. In *Natural and Man-Made Hazards,* El-Sabh, M.I. and Murty, T.S. (eds), Reidel, Dordrecht, 357–368.

Brammer, H. (1990) Floods in Bangladesh: flood mitigation and environmental aspects. *Geographical Journal*, 156, 158–165.

Brookes, A. (1988) *Channelized Rivers: Perspectives for Environmental Management.* Wiley, Chichester.

Carlston, C.W. (1963) Drainage density and streamflow. US Geological Survey, Professional Paper 422-C.

Carlston, C.W. (1968) Slope discharge relations for eight rivers in the United States. US Geological Survey, Professional Paper 600–D, 45–47.

Chow, C.T. (ed.) (1964) *Handbook of Applied Hydrology.* McGraw-Hill, New York.

Church, M. (1988) Floods in cold climates. In *Flood Geomorphology*, Baker, V.R., Kochel, R.C. and Patton, P.C. (eds), Wiley, New York, 205–229.

Clarke, A.O. (1991) A boulder approach to estimating flash flood peaks. *Bulletin Association Engineering Geologists*, 28, 45–54.

Clarke, A.O. (1996) Estimating probable maximum floods in the upper Santa Ana Basin, southern California, from stream boulder size. *Environmental and Engineering Geoscience*, 2, 165–182.

Coates, D.R. (1987) Engineering aspects of geomorphology. In *Ground Engineer's Reference Book*, Bell, F.G. (ed.), Butterworths, London, 2/1–2/37.

Collinge, V. and Kirby, C. (eds) (1987) *Weather Radar and Flood Forecasting.* Wiley, Chichester.

Cooke, R.U., Brunsden, D., Doornkamp, J.C. and Jones, D.K.C. (1982) *Urban Geomorphology in Drylands*. Oxford University Press.

Cordery, I. and Pilgrim, D.H. (1976) Engineering attitudes to flood risk. *Water Services*, 673–676.

Costa, J.E. (1978) Holocene stratigraphy and flood frequency analysis. *Water Resources Research*, 14, 626–632.

Dalrymple, T. (1960) Flood frequency analysis. *US Geological Survey Water Supply Paper* 1543-A, 1–80.

Fahnestock, R.K. (1963) Morphology and hydrology of a glacial stream – White River, Mount Rainier, Washington. US Geological Survey, Professional Paper 424–A.

Fisk, H.N. (1944) *Geological Investigation of the Alluvial Valley of the Lower Mississippi River*. US Army Corps of Engineers, Mississippi River Commission, Vicksburg, Mississippi.

Gilvear, D.J., Davies, J.R. and Winterbottom, S.J. (1994) Mechanisms of floodbank failure during large flood events on the rivers Tay and Earn, Scotland. *Quarterly Journal Engineering Geology*, 27, 319–332.

Gregory, K.J. and Walling, D.E. (1973) *Drainage Basin Forms and Process: A Geomorphological Approach*. Edward Arnold, London.

Gruntfest, E.C. and Huber, C. (1989) Status report on flood warning systems in the United States. *Environmental Management*, 13, 357–368.

Gupta, A. (1988) Large floods as geomorphic events in the humid tropics. In *Flood Geomorphology*, Baker, V.R., Kochel, R.C. and Patton, P.C., (eds), Wiley, New York, 301–315.

Hager, W.H. and Sinniger, R. (1985) Flood storage in reservoirs. *Proceedings American Society Civil Engineers, Journal Irrigation and Drainage Engineering Division*, 111, 76–85.

Hjulstrom, F. (1935) Studies of the morphological activity of rivers, as illustrated by the river Fynis, *Uppsala University Geological Institute*, Bulletin, 25.

Jarret, R.D. (1990) Palaeohydrologic techniques used to define the special occurrence of floods. *Geomorphology*, 3, 181–195.

Jopling, A.V. (1963) Hydraulic studies on the origin of bedding. *Sedimentology*, 2, 115–121.

Kenny, R. (1990) Hydrogeomorphic flood hazard evaluation for semi-arid environments. *Quarterly Journal Engineering Geology*, 23, 333–336.

Lane, E.W. (1955) The importance of fluvial morphology in hydraulic engineering. *Proceedings American Society Civil Engineers, Journal Hydraulics Division*, 81, 1–17.

Leopold. L.B. and Maddock, T. (1953) The hydraulic geometry of stream channels and some physiographic implications. US Geological Survey, Professional Paper 252.

Leopold, L.B. and Wolman, M.G. (1957) River channel patterns, braided, meandering and straight. US Geological Survey, Professional Paper 282–B.

Leopold, L.B., Wolman, M.G. and Miller, J.P. (1964) *Fluvial Processes in Geomorphology*. Freeman, San Francisco.

Lewin, J. (1989) Floods in fluvial geomorphology. In *Floods: Hydrological, Sedimentological and Geomorphological Implications*, Beven, K. and Carling, F. (eds), Wiley, Chichester, 265–284.

Macklin, M.G., Rumsby, M.T. and Newson, M.D. (1992) Historic overbank floods and floodplain sedimentation in the lower Tyne valley, north east England. In *Gravel Bed Rivers*, Hey, R.D. (ed.), Wiley, Chichester.

Marsland, A. (1986) The flood plain deposits of the Lower Thames. *Quarterly Journal Engineering Geology*, 19, 223–247.

Newson, M.D. (1994) *Hydrology and the River Environment*. Clarendon Press, Oxford.

Rahn, P.H. (1994) Flood plains. *Bulletin Association Engineering Geology*, 31, 171–183.

Schumm, S.A. (1967) Meander wavelength of alluvial rivers. *Science*, 157, 1549–1550.

Schumm, S.A. (1969) River metamorphosis. *Proceedings American Society Civil Engineers, Journal Hydraulics Division*, 95, 255–273.

Schumm, S.A. (1977) Applied fluvial geomorphology. In *Applied Geomorphology*, Hails, J.R. (ed.), Elsevier, Amsterdam, 119–156.

Searcy, J.M. (1959) Flow-duration curves. US Geological Survey, Water Supply Paper 1542A.

Simons, D.B. and Richardson, E.V. (1961) Forms of bed roughness in alluvial channels. *Proceedings American Society Civil Engineers, Journal Hydraulics Division*, 87, 87–105.

Smith, L.M. and Winkley, B.R. (1996) The response of the Lower Mississippi River to river engineering. *Engineering Geology*, 45, 433–455.

Snyder, F.F. (1938) Synthetic unit graphs, *Transactions American Geophysical Union*, 19, 447–463.

Sundberg, A. (1956) The river Klarelren, a study of fluvial processes. *Geografiska Annaler*, 38, 127–316.

Thampapillai, D.J. and Musgrave, W.F. (1985) Flood damage and mitigation: a review of structural and non-structural measures and alternative decision frameworks. *Water Resources Research*, 21, 411–424.

Ward, R.C. (1978) *Floods: A Geographical Perspective*. Macmillan, London.

Wilson, E.M. (1983) *Engineering Hydrology*. Macmillan, London.

Chapter 7

Marine action and control

Johnson (1919) distinguished three elements in a shoreline: the coast, the shore and the offshore. The coast was defined as the land immediately behind the cliffs, while the shore was regarded as that area between the base of the cliffs and low-water mark; the area that extended seawards from the low-water mark was termed the offshore. The shore was further divided into foreshore and back-shore, the former embracing the intertidal zone, while the latter extended from the foreshore to the cliffs. Those deposits that cover the shore are usually regarded as constituting the beach.

Marine activity in coastal environments acts within a narrow and often varying vertical zone. Significant changes in sea level, associated with the Pleistocene glaciation, have taken place in the recent geological past that have influenced the character of present-day coastlines. When ice sheets expanded the sea level fell, rising when the ice melted. In addition, as the ice sheets retreated the land beneath began to rise in an attempt to regain isostatic equilibrium, so complicating the situation. Hence, coasts can be characterized by features developed in response to submergence on the one hand or emergence on the other. And such changes in the relative level of the sea are not only a feature of the past; they are still occurring at present. Tides also have an influence upon the coast, the tidal range governing the vertical interval over which the sea can act. In addition, tidal currents can influence the distribution of sediments on the sea floor.

Waves acting on beach material are a varying force. They vary with time and place due to changes in wind force and direction over a wide area of sea, and with changes in coastal configuration and offshore relief. This variability means that the beach is rarely in equilibrium with the waves, in spite of the fact that it may take only a few hours for equilibrium to be attained under new conditions. Such a more or less constant state of disequilibrium occurs most frequently where the tidal range is considerable, as waves are continually acting at a different level on the beach.

7.1 Waves

When wind blows across the surface of deep water it causes an orbital motion in those water particles in the plane normal to the wind direction (Figure 7.1). The motion decreases in significance with increasing depth, dying out at a depth equal to that of the wavelength. Because adjacent particles are at different stages in their circular course a wave is produced. However, there is no progressive forward motion of the water particles in such a wave, although the form of the wave profile moves rapidly in the direction in which the wind is blowing. Such waves are described as oscillatory waves.

Forced waves are those formed by the wind in the generating area; they are usually irregular. On moving out of the area of generation, these waves become long and regular. They are then referred to as free waves. As these waves approach a shoreline they feel bottom, which disrupts their pattern of motion, changing them from oscillation to translation waves. Where the depth is approximately half the wavelength the water particle orbits become ellipses with their major axes horizontal.

The forward movement of the water particles, as a whole, is not entirely compensated for by the backward movement. As a result, there is a general movement of the water in the direction in which the waves are travelling. This is known as mass transport. The time required for any one particle to complete its orbital revolution is the same as the period of the waveform. The orbital velocity, u, is equal to the length of orbital travel divided by the wave period, T. Hence:

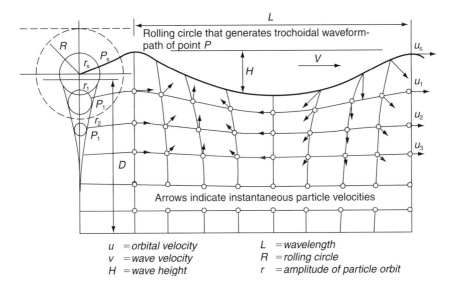

Figure 7.1 Trochoidal water waves.

$$u = \frac{2\pi r}{T} \tag{7.1}$$

Similarly, the wave velocity or celerity, c, is

$$c = \frac{2\pi R}{T} \tag{7.2}$$

where R is the rolling circle required to generate a trochoidal waveform and r is the amplitude of particle orbit. The celerity of waves can also be derived from the following expression:

$$c = \frac{gL}{2\pi} \left(\tanh \frac{2\pi Z}{L} \right) \tag{7.3}$$

where g is the acceleration due to the gravity, Z is the still water depth and L is the wavelength. If Z exceeds L, then tanh $2Z/L$ becomes equal to unity, so that in deep water:

$$c = \frac{gL}{2\pi}$$

$$= 2.26\sqrt{L} \tag{7.4}$$

When the depth is less than one-tenth of the wavelength, tanh $2\pi Z/L$ approaches $2\pi Z/L$ and then c equals $\sqrt{(gZ)}$. As waves move into shallowing water their velocity is reduced and so their wavelength decreases.

7.1.1 Force and height of waves

The forces exerted by waves include jet impulse, viscous drag and hydrostatic pressure. The source of dynamic wave action lies in the inertia of the moving particles. Each particle can be considered as having a tangential velocity due to rotation about the centre of its orbit, and to the velocity of translation corresponding to mass transport. The vectorial sum of these two components is the actual velocity of the particle at any instant; when the particle is at a wave crest this resultant velocity is horizontal.

The effectiveness of wave impact on the shoreline or marine structure depends on the depth of water and the size of the wave; it drops sharply with increasing depth. If deep water occurs alongside cliffs or sea walls, then waves may be reflected without breaking and in so doing they may interfere with incoming waves. In this way standing waves, which do not migrate, are formed, in which the water surges back and forth between the obstruction and a distance equal to half the wavelength away. The crests are much higher than in the original wave. This form of standing wave is known as a clapotis. It is claimed that the oscillation of standing waves causes an alternating increase and decrease of pressure along any discontinuities in rocks or cracks in marine structures that occur below the water line. It is assumed that such action gradually dislodges blocks of material. It has been estimated that translation waves reflected from a vertical face exert six times as much pressure on the wall as oscillating waves of equal dimensions. When waves break, jets of water are thrown at approximately twice the wave velocity, which also causes increases in the pressure in discontinuities and cracks, thereby causing damage.

It has been shown experimentally that the maximum pressure exerted by breaking waves on a sea wall is 1.3 $(c + u_c)^2$, where c is the wave celerity and u_c is the orbital velocity at the crest. Alternatively, Little (1975) gave the velocity of waves breaking in shallow water as 5.8 H_c, where H_c is the height of the crest above the sea bed. He derived the simple expression $p = H/10$ for the pressure, in tons per sq. ft, exerted by breaking waves on sea works. This is applicable to most short waves, and beach and wall slopes exceeding 1 in 50. In the case of long waves, he suggested $p = H/8$, this being the wave height out at sea. These expressions overestimate the pressure, but in design it is always better to overestimate rather than to underestimate. Bascom (1964) estimated that a storm that has wave heights averaging 3 m can develop a maximum force equal to 145 kPa, that a 6.1 m swell could develop 115 kPa and that strong gales could develop waves with forces in excess of 290 kPa.

Fetch is the distance that a wind blows over a body of water and is the most important factor determining wave size and efficiency of transport. For instance, winds of moderate force that blow over a wide stretch of water generate larger waves than do strong winds that blow across a short reach. For a given fetch, the stronger the wind, the higher are the waves (Figure 7.2a). Their period also increases with increasing fetch. The maximum wave height (H_{max}) for a given fetch can be derived from the expression:

$$H_{max} = \frac{0.3v_w}{g} \qquad (7.5)$$

where v_w is the velocity of a uniform wind in cm s^{-1} and g is the acceleration due to gravity. Wind duration also influences waves: the longer it blows, the faster the waves move, while their lengths and periods increase (Figure 7.2b). Usually, wavelengths in the open sea are less than 100 m and the speed of propagation is approximately 48 km h^{-1}. They do not normally exceed 8.5 m in height. Waves that are developed in storm centres in the middle of an ocean may journey outwards to the surrounding land masses. This explains why large waves may occur along a coast during fine weather. Waves frequently approach

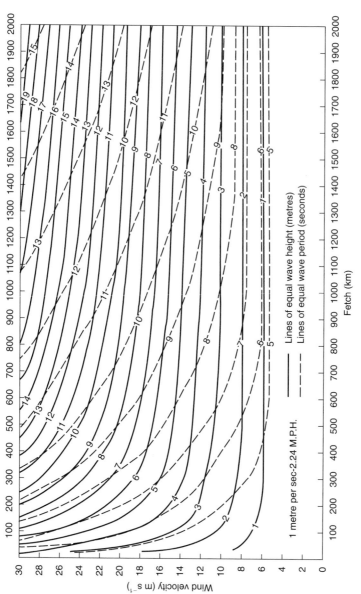

Figure 7.2a Wave height and period in relation to fetch and wind. Caution must be exercised in using these curves, since the quantities are rarely accurately known, so the results read from the curves will be correspondingly open to doubt. For instance, it is rarely possible to say with certainty of a particular storm what its duration or the wind velocity has been; and it is certain that during the storm the wind velocity was not constant.

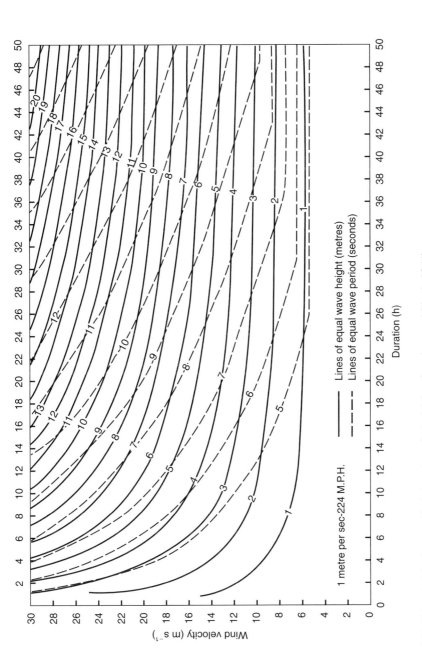

Figure 7.2b Wave height for specified duration of wind (after Sverdrup and Munk, 1946).

a coastline from different areas of generation; if they are in opposition, then their height is decreased, while their height is increased if they are in phase. Moreover, when the wind shifts in direction or intensity the new generation of waves differs from the older waves, which still persist in their original course. The more recent waves may overtake or, if the storm centre has migrated meanwhile, may intercept the earlier waves, so complicating the wave pattern.

7.1.2 Wave refraction

Wave refraction is the process whereby the direction of wave travel changes because of changes in the topography of the nearshore sea floor (Figure 7.3). When waves approach a straight beach at an angle they tend to swing parallel to the shore due to the retarding effect of the shallowing water. At the break point, such waves seldom approach the coast at an angle exceeding $20°$ irrespective of the offshore angle to the beach. As waves approach an irregular shoreline, refraction causes them to turn into shallower water so that the wave crests run roughly parallel to the depth contours. Along an indented coast shallower water is first met off headlands. This results in wave convergence and an increase in wave height, with wave crests becoming concave towards the headlands. Conversely, where waves move towards a depression in the sea floor they diverge, decrease in height and become convex towards the shoreline. In both cases the wave period remains constant. The concentration of erosion on headlands leads to the coast being gradually smoothed.

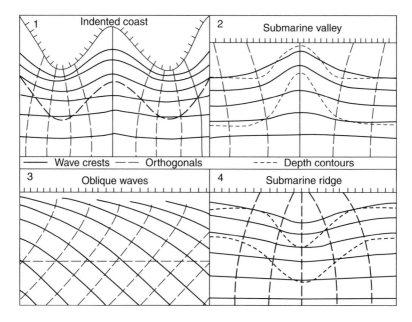

Figure 7.3 Diagrammatic wave refraction patterns.

Refraction diagrams often form part of a shoreline study, graphic or computer methods being used to prepare them from hydrographic charts or aerial photographs. These diagrams indicate the direction in which waves flow, and the spacing between the orthogonal lines is inversely proportional to the energy delivered per unit length of shore. Because of refraction it is possible by constructing one or more raised areas on the sea bed to deflect waves passing over them so that calm water occurs at harbour entrances.

Waves are smaller in the lee of a promontory or off a partial breakwater. The nature of the wave pattern within this sheltered area is determined by diffraction as well as refraction phenomena. That portion of the advancing wave crest that is not intercepted by the barrier immediately spreads out into the sheltered area and the wave height shrinks correspondingly. This lateral dissipation of wave energy is termed diffraction. Diffraction results from interference with the horizontal components of wave motion. The depth of water is not a relevant factor.

7.2 Tides

Tides are the regular periodic rise and fall of the surface of the sea, observable along shorelines. They are formed under the combined attraction of the Moon and the Sun, chiefly the Moon. Normal tides have a dual cycle of about 24 h and 50 min. The tide height is at a maximum or minimum when Sun, Moon and Earth are in line, so that the attractive force of both upon the ocean waters is combined and thus creates spring tides. Neap tides are generated when the attractive forces of the Sun and Moon are aligned in opposing directions. In this instance, the range between high and low tide is lowest. The tidal range is also affected by the configuration of a coastline. For example, in the funnel-shaped inlet of the Bay of Fundy a range of 16 m is not uncommon. By contrast, along relatively straight coastlines or in enclosed seas like the Mediterranean the range is small.

The tidal current that flows into bays and estuaries along the coast is called the flood current. The current that returns to the sea is named the ebb current. These are of unequal duration as the higher flood tide travels more quickly than the lower ebb tide.

The inshore current all along the coast need not necessarily be in the same direction as the offshore tidal current; for example, the effect of a headland may be to cause an eddy, giving an inshore current locally in the opposite direction to the tidal current. Information with regard to the directions and velocities of tidal currents at definite points, and at tabulated periods in hours before and after high water at a nearby port, can be obtained in Britain from the relevant Admiralty charts. What is more, data on local tidal conditions are available in the Admiralty tide tables. Nonetheless, local conditions affecting tidal estuaries are often so variable that a detailed study of each case is usually required. Scale models are frequently used in such studies.

7.3 Beach zones

Four dynamic zones have been recognized within the nearshore current system of the beach environment: they are the breaker zone, the surf zone, the transition zone and the swash zone (Figure 7.4). The breaker zone is that in which waves break. The surf zone refers to that region between the breaker zone and the effective seaward limit of backwash. The presence and width of a surf zone is primarily a function of the beach slope and tidal phase. Those beaches that have gentle foreshore slopes are often characterized by wide surf zones during all tidal phases, whereas this zone may not be represented on steep beaches. The transition zone includes that region where backwash interferes with the water at the leading edge of the surf zone and it is characterized by high turbulence. The region where water moves up and down the beach is termed the swash zone.

The breaking of a wave is influenced by its steepness, the slope of the sea floor and the presence of an opposing or supplementary wind. When waves enter water equal in depth to approximately half their wavelength they begin to feel bottom, their velocity and length decrease, while their height, after an initial decrease, increases. The wave period remains constant. As a result, the wave grows steeper until it eventually breaks.

Three types of breaking wave can be distinguished: plunging breakers, spilling breakers and surging breakers (Figure 7.5a). The beach slope influences the type of breaking waves that develop; for example, spilling waves are commonest on gentle beaches, whereas surging breakers are associated with steeper beaches (Figure 7.5b). In fact, Galvin (1968) showed that the type of breaker that occurs can be predicted from $H_w/L_w\beta$, where H_w is the wave height in deep water, L_w is the wavelength in deep water and β is the angle of the beach slope. Spilling breakers would appear to have values in excess of 4.8, plunging breakers values between 0.09 and 4.8, and those of surging waves are below 0.09.

Plunging breakers collapse suddenly when their wave height is approximately equal to the depth of the water. At the plunge point, wave energy is transformed into energy of turbulence and kinetic energy, and a tongue of water rushes up the beach. The greater part of the energy released by the waves is used in overcoming frictional forces or in generating turbulence, and lesser amounts are utilized in shifting bottom materials or in developing longshore currents. The plunge point is defined as the final breaking point of a wave just before water rushes up the beach. For any given wave there is a plunge line along which it breaks, but the variations in positions where waves strike the shore usually result in a plunge zone of limited width. At the plunge line, the distribution of sand is almost uniform from the bottom to the water surface during the breaking of a wave, but seaward of the plunge line the sand content of the surface water rapidly decreases. Accordingly, the maximum disturbance of the sand due to turbulence occurs in the vicinity of the plunge line, it tending to migrate from the plunge line towards less turbulent water on both sides.

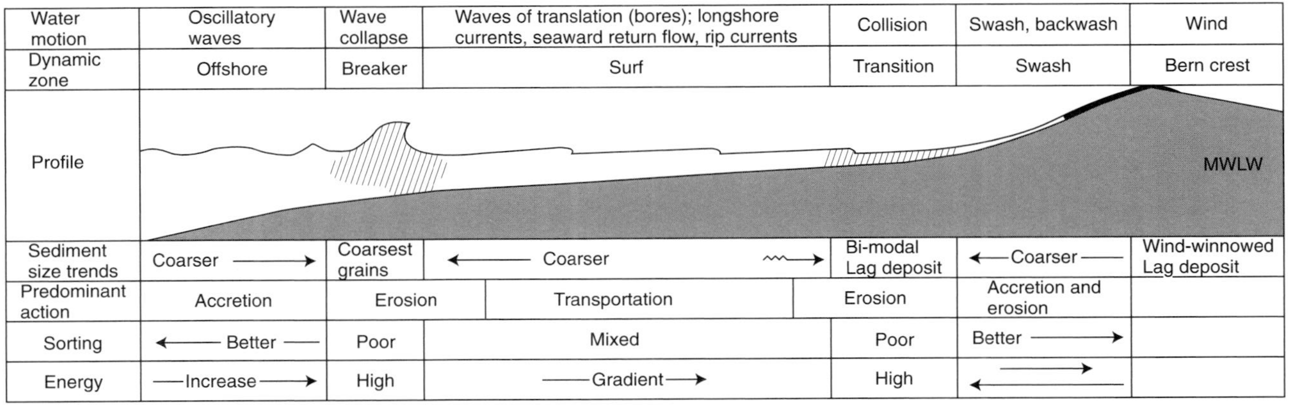

Water motion	Oscillatory waves	Wave collapse	Waves of translation (bores); longshore currents, seaward return flow, rip currents	Collision	Swash, backwash	Wind
Dynamic zone	Offshore	Breaker	Surf	Transition	Swash	Bern crest
Profile						MWLW
Sediment size trends	← Coarser →	Coarsest grains	← Coarser →	Bi-modal Lag deposit	← Coarser →	Wind-winnowed Lag deposit
Predominant action	Accretion	Erosion	Transportation	Erosion	Accretion and erosion	
Sorting	← Better →	Poor	Mixed	Poor	Better →	
Energy	← Increase →	High	Gradient →	High	High	

Figure 7.4 Summary diagram schematically illustrating the effect of the four main dynamic zones in the beach environment. Hatched areas represent zones of high concentrations of suspended grains. Dispersion of fluorescent sand and electro-mechanical measurements indicate that the surf zone is bounded by two high-energy zones: the breaker zone and the transition zone. MWLW = mean water low water (after Ingle, 1966).

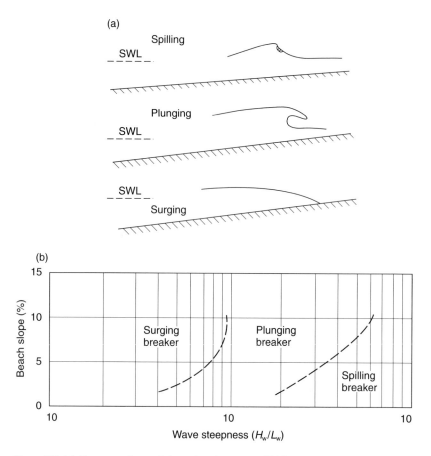

Figure 7.5 (a) Cross-sections of three breaker types (SWL = still water line); (b) type of breaking wave in relation to wave steepness and beach gradient.

This may result in a trough along the plunge line. Seaward of the plunge line shingle is not generally moved by waves in depths much greater than the wave height, but sand is moved for a considerable distance offshore.

Spilling breakers begin to break when the wave height is just over one half the water depth. They do so gradually over some distance.

Surging breakers or swash rush up the beach and are usually encountered on beaches with a steep profile. The height to which the swash rises determines the height of shingle crests and whether or not sea walls or embankments are overtopped. The more impermeable and steep the slope, the higher is the swash height. The term 'backwash' is used to describe the water that subsequently descends the beach slope.

Swash tends to pile water against the shore and thereby gives rise to currents that move along it, which are termed longshore currents. After flowing parallel to the beach the water runs back to the sea in narrow flows called rip currents (Figure 7.6). In the neighbourhood of the breaker zone, rip currents extend from the surface to the floor, while in the deeper reaches they override the bottom water, which still maintains an overall onshore motion. The positions of rip currents are governed by submarine topography, coastal configuration, and the height and period of the waves. They frequently occur on the up-current sides of points and on either side of convergence, where the water moves away from the centre of the convergence and turns seawards.

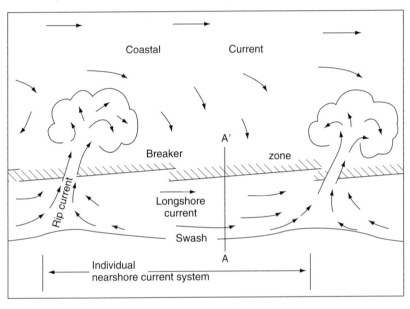

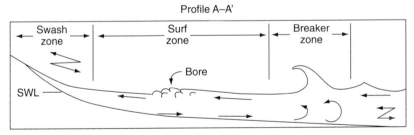

Figure 7.6 Terminology of nearshore current systems. Each individual system begins and terminates with a seaward-flowing rip current. Arrows indicate direction of water movement in plan and profile. Existence of the controversial seaward return flow along the foreshore–inshore bottom (profile A–A') has recently been confirmed by electro-mechanical measurements in the surf zone. These measurements indicated that a seaward bottom flow exceeding $0.36 \, \mathrm{m \, s^{-1}}$ often occurs at the same time that surface flow is shorewards. The surf zone is here defined as the area between the seaward edge of the swash zone and the breaker zone. SWL = still water line (from Ingle, 1966).

Material is sorted according to its size, shape and density, and the variations in the energy of the transporting medium. During the swash, material of various sizes may be swept along in traction or suspension, whereas during the backwash the lower degree of turbulence results in a lessened lifting effect so that most of the movement of grains is by rolling along the bottom. This can mean that the maximum size of particles thrown onto a beach is larger than the maximum size washed back to the surf zone. However, grains of larger diameter may roll down slope farther than smaller particles. The continued operation of waves on a beach, accompanied by the winnowing action of wind on dry sand, tends to develop patterns of variation in average particle size, sorting, firmness, porosity, permeability, moisture content, mineral composition and other attributes of the beach.

7.4 Coastal erosion

Coastal erosion is not necessarily a disastrous phenomenon; however, problems arise when erosion and human activity come into conflict. Increases in the scale and density of development along coastlines has led to increased vulnerability, which means that better coastal planning is necessary.

Those waves with a period of approximately 4 s are usually destructive, while those with a lower frequency, that is, a period of about 7 s or over, are constructive. When high-frequency waves collapse they form plunging breakers and the mass of water is accordingly directed downwards at the beach. In such instances, swash action is weak and, because of the high frequency of the waves, is impeded by the backwash. As a consequence, material is removed from the top of the beach. The motion within waves that have a lower frequency is more elliptical and produces a strong swash, which drives material up the beach. In this case the backwash is reduced in strength because water percolates into the beach deposits and therefore little material is returned down the beach. Although large waves may throw material above the high-water level and thus act as constructive agents, they nevertheless have an overall tendency to erode the beach, while small waves are constructive. For example, when steep storm waves attack a sand beach they are usually entirely destructive, and the coarser the sand the greater the quantity that is removed to form a submarine bar offshore. In some instances a vertical scarp is left on the beach at the high-tide limit. It is by no means a rarity for the whole beach to be removed by storm waves: witness the disappearance of sand from some of the beaches along the Lincolnshire coast of England after the storm flood of January–February 1953, which exposed their clay base. Storm waves breaking on shingle throw some shingle to the backshore to form a storm beach ridge, which may extend far above high-tide level. Chesil Beach in southern England provides an example, extending 13 m above high-tide level at its eastern end. Storm waves may also bring shingle down the foreshore, so that a step is developed at their break point.

The rate at which coastal erosion proceeds is influenced by the nature of the coast itself. Marine erosion is most rapid where the sea attacks soft, unconsolidated sediments (Figure 7.7). Some stretches of such coasts may disappear at a rate of a metre or more annually. In addition, beaches that are starved of sediment supply due to, for example, natural or artificial barriers inhibiting movement of longshore drift (see Section 7.5) are exposed to erosion.

Steep cliffs made of unconsolidated materials are prone to landsliding (Hutchinson *et al.*, 1991). For example, clay may be weakened by wetting due to waves or spray, or the toe of the cliffs may be removed by wave action. Thus rotational slips and mudflows may develop (Figure 7.8), and they may carry away protective structures. For erosion to continue the debris produced must be removed by the sea. This is usually accomplished by longshore currents. On the other hand, if material is deposited to form extensive beaches and the detritus is reduced in size, then the submarine slope, because of its small angle of rest, is very wide. Wave energy is therefore dissipated as the water moves over the beach, and cliff erosion ceases.

Obviously, marine erosion is concentrated in areas along a coast where the rocks offer less resistance. In most cases a retreating coast is characterized by steep cliffs, at the base of which a beach is excavated by wave action. Erosive forms of local relief include such features as wave-cut notches, caves, blowholes, marine arches and stacks. Debris from the cliff is washed seawards to form a terrace, which marks the outer limit of the beach. The degree to which rocks are traversed by discontinuities affects the rate at which they are removed.

Figure 7.7 Marine erosion of till near Bridlington, England, undermining houses.

Figure 7.8 Slides along the Palisades coast in southern California, northwest of Santa Monica.

In particular, the attitude of joints and bedding planes is important. Where the bedding planes are vertical or dip inland, then the cliff recedes vertically under marine attack. But if beds dip seawards blocks of rock are more readily dislodged since the removal of material from the cliff means that the rock above lacks support and tends to slide into the sea. Marine erosion also occurs along fault planes. The height of a cliff also influences the rate at which erosion takes place. The higher the cliff the more material falls when its base is undermined by wave attack. This in turn means that a greater amount of debris has to be broken down and removed before the cliff is once more attacked with the same vigour.

7.5 Beaches and longshore drift

Beaches are supplied with sand, which is usually derived entirely from the adjacent sea floor, although in some areas a larger proportion is produced by cliff erosion. During periods of low waves the differential velocity between onshore and offshore motion is sufficient to move sand onshore, except where rip currents are operational. Onshore movement is particularly notable when long-period waves approach a coast, whereas sand is removed from the foreshore during high waves of short period. Of course, all beaches are continually changing. Some develop during periods of small waves and disappear during periods of high waves, whereas other beaches change in height and width during stormy seasons. If seasonal changes bring about changes in the wave approach, then sand is shifted along the beach, beaches tending to form at right angles to the direction of the wave approach. If sand forms dunes and these migrate landwards, then sand is lost to the beach.

The beach slope is produced by the interaction of swash and backwash. Beaches undergoing erosion tend to have steeper slopes than prograding beaches. The beach slope is also related to grain size; in general, the finer the material the gentler the slope and the lower the permeability of the beach. For example, the loss of swash due to percolation into beaches composed of grains of 4 mm in median diameter is 10 times greater than into beaches where the grains average 1 mm. As a result, there is almost as much water in the backwash on fine beaches as there is in the swash, so the beach profile is gentle and the sand is hard-packed. The grain size of beach material is continually changing, however, because of breakdown and the removal or addition of material.

Storm waves produce the most conspicuous constructional features on a shingle beach, but they remove material from a sandy beach. A small foreshore ridge develops on a shingle beach at the limit of the swash when constructive waves are operative. Similar ridges or berms may form on a beach composed of coarse sand, these being built by long, low swells. Berms represent a marked change in slope and usually occur at a small distance above high-water mark. They are not conspicuous features on beaches composed of fine sand. A fill on the upper part of a beach may be balanced by a cut lower down the foreshore.

Dunes are formed by onshore winds carrying sand-sized material landwards from the beach along low-lying stretches of coast where there is an abundance of sand on the foreshore. Leathermann (1979) maintained that dunes act as barriers, energy dissipators and sand reservoirs during storm conditions; for example, the broad sandy beaches and high dunes along the coast of the Netherlands present a natural defence against inundation during storm surges (Figure 7.9). Because dunes provide a natural defence against erosion, once they are breached the ensuing coastal changes may be long-lasting (Gares, 1990). On the other hand, along parts of North Carolina, where dune protection is limited, washovers associated with storm tides have been responsible for high rates of erosion (Cleary and Hosier, 1987).

In spite of the fact that dunes inhibit erosion, Leathermann (1979) stressed that without beach nourishment they cannot be relied upon to provide protection, in the long term, along rapidly eroding shorelines. Beach nourishment widens the beach and maintains the proper functioning of the beach–dune system during normal and storm conditions.

When waves approach a coast at an angle, material is moved up the beach by the swash, in a direction normal to that of wave approach, and then it is rolled down the beach slope by the backwash. In this manner material is moved in a zig-zag path along the beach, the phenomenon being referred to as longshore or littoral drift. Currents, as opposed to waves, seem incapable of moving material coarser than sand grade. Since the angle of the waves to the shore

Figure 7.9 Planting of marram grass to protect the base of coastal dunes in the Netherlands.

affects the rate of drift, there is a tendency for erosion or accretion to occur where the shoreline changes direction. The determining factors are the rate of arrival and the rate of departure of beach material over the length of foreshore concerned. Moreover, the projection of any solid structure below mean tide level results in the build-up of drift material on the updrift side of the structure and perhaps in erosion of material from the other side. If the drift is of any magnitude, the effects can be serious, especially in the case of works protecting harbour entrances. On the other hand, sand can be trapped by deliberately stopping longshore drift in order to build up a good beach section. In normal longshore drift the bulk of the sand is moved in a relatively shallow zone, with perhaps 80% of shore drift being moved within depths of 2 m or less.

The amount of longshore drift along a coast is influenced by coastal outline and wavelength. Short waves can approach the shore at a considerable angle and generate consistent downdrift currents. This is particularly the case on straight or gently curving shores and can result in serious erosion where the supply of beach material reaching the coast from updrift is inadequate. Conversely, long waves suffer appreciable refraction before they reach the coast.

An indication of the direction of longshore drift is provided by the orientation of spits along a coast. Spits are deposits that grow from the coast and are supplied chiefly by longshore drift. Their growth is spasmodic and alternates with episodes of retreat. The distal end of a spit is frequently curved. A spit that extends from the mainland to link up with an island is referred to as a tombola. A bay bar is constructed across the entrance to a bay by the growth of a spit being continued from one headland to the other. A cuspate bar arises where a change in the direction of spit growth takes place so that it eventually joins the mainland again, or where two spits coalesce. If progradation occurs, then bars give rise to cuspate forelands. Bay-head beaches are one of the commonest types of coastal deposit and tend to straighten a coastline. Wave refraction causes longshore drift from headlands to bays, sweeping sediments with it. Since the waves are not so steep in the innermost reaches of bays they tend to be constructive. Marine deposition also helps to straighten coastlines by building beach plains.

Offshore bars consist of ridges of shingle or sand, which may extend for several kilometres and may be located up to a few kilometres offshore. They usually project above sea level and may cut off a lagoon on their landward side. The distance from the shore at which a bar forms is a function of the deep water wave height, the water depth and the deep water wave steepness. If water depth and wave steepness remain constant, and wave height is reduced, the bar moves shorewards. If wave height and wavelength stay the same, and the depth of water increases, again the bar moves shorewards. If water depth and wave height are unchanged but steepness is reduced, then the bar moves seawards. During storms sand is shifted outwards to form an offshore bar, and during the ensuing quieter conditions this sand is wholly or in large part moved back to the beach.

7.6 Storm surges and marine inundation

Except where caused by failure of protection works, marine inundation is almost always attributable to severe meteorological conditions giving rise to abnormally high sea levels, referred to as storm surges. For example, Murty (1987) described storm surges as oscillations of coastal waters in the period range from a few minutes to a few days that were brought about by weather systems such as hurricanes, cyclones and deep depressions. Low pressure and driving winds during a storm may lead to marine inundation of low-lying coastal areas, particularly if this coincides with high spring tides. This is especially the case when the coast is unprotected. Floods may be frequent, as well as extensive where flood plains are wide and the coastal area is flat. Coastal areas that have been reclaimed from the sea and are below high-tide level are particularly suspect if coastal defences are breached. A storm surge can be regarded as the magnitude of sea level along the shoreline that is above the normal seasonally adjusted high-tide level. Storm surge risk is often associated with a particular season. The height and location of storm damage along a coast over a period of time, when analysed, provides some idea of the maximum likely elevation of surge effects. The seriousness of the damage caused by storm surge tends to be related to its height and the velocity of water movement.

Factors that influence storm surges include the intensity in the fall in atmospheric pressure, the length of water over which the wind blows, the storm motion and offshore topography. Obviously, the principal factor influencing storm surge is the intensity of the causative storm, the speed of the wind piling up the sea against the coastline. For instance, threshold wind speeds of approximately 120 km h^{-1} tend to be associated with central pressure drops of around 34 mb. Normally, the level of the sea rises with reductions in atmospheric pressure associated with intense low-pressure systems. In addition, the severity of a surge is influenced by the size and track of a storm and, especially in the case of open coastline surges, by the nearness of the storm track to the coastline. The wind direction and the length of fetch are also important, both determining the size and energy of waves. Because of the influence of the topography of the sea floor, wide shallow areas on the continental shelf are more susceptible to damaging surges than those where the shelf slopes steeply. Surges are intensified by converging coastlines, which exert a funnel effect as the sea moves into such inlets.

Ward (1978) distinguished two types of major storm surge. First, these are storm surges on open coastlines, which travel as running waves over large areas of the sea. Such surges are associated with tropical cyclones (typhoons and hurricanes). Second, surges occur in enclosed or partially enclosed seas where the area of sea is small compared with the size of the atmospheric disturbance. Thus the surges more or less affect the whole sea at any one time. This type of surge may be frequent as well as damaging.

Most deaths have occurred as a result of storm surges generated by hurricanes. The height of a storm surge associated with a hurricane is influenced by the distance from the centre of the storm to the point at which maximum wind speeds occur; the barometric pressure in the eye of the hurricane; the rate of forward movement of the storm; the angle at which the centre of the storm crosses the coastline; and the depth of sea water offshore. In addition to the storm surge, enclosed bays may experience seiching. Seiches involve the oscillatory movement of waves generated by a hurricane. Seiching may also occur in open bays if the storm moves forward very quickly, and in low-lying areas this can be highly destructive.

The east coast of India and Bangladesh experience serious inundation due to storm surges once in a decade or so (Sharma and Murty, 1987). Several major rivers flow into the Bay of Bengal, and excessive siltation of their estuaries, together with the formation of sand bars, provides ideal conditions for the propagation of surges. The deltaic area of the River Ganges is especially prone to storm surges. The width of the continental shelf, in places up to 300 km, also favours storm surge development. Furthermore, the orientation of the track of the cyclone in relation to the coast is of importance where the width of the continental shelf is significant. For example, a surge induced at an angle of incidence of 90° may be almost twice as large as one that is produced by a cyclone of the same intensity when the angle of incidence is 135°. The winds associated with tropical cyclones can blow at speeds between 50 and 120 km h^{-1}, and storm surges of over 5 m in height have been recorded. The duration of marine inundation may be related to the duration of the surge, or in relatively slow-draining areas such as tidal marshes to the time it takes for drainage to occur.

Obviously, prediction of the magnitude of a storm surge in advance of its arrival can help to save lives. Storm tide warning services have been developed in various parts of the world. Warnings are usually based upon comparisons between predicted and observed levels at a network of tidal gauges such as that along the coast of the North Sea, where coastal surges tend to move progressively along the coast. To a large extent, the adequacy of a coastal flood forecasting and warning system depends on the accuracy with which the path of the storm responsible can be determined. Storms can be tracked by satellite, and satellite and ground data used for forecasting. Numerous attempts have been made to produce models that predict storm surges (Jelesnianski, 1978; Johns and Ali, 1980).

Because of the increased urbanization of many coastal areas, the damaging effects of storm surges has increased in the last 30 years. For instance, this is the case with several large coastal cities in Japan, Osaka being a notable example (Tsuchiya and Kawata, 1987). Typhoons frequently move northeastwards before striking the Japanese coast, and unfortunately Osaka Bay is elliptical in shape, with its major axis running northeast to southwest. This means that water can pile up at the northeast part of the bay during storm surges. During high spring tides, the tidal range is approximately 1.85 m. Hence a meteorological tide due to a storm surge combined with such an astronomical tide can give rise to an exceptional increase in sea level. The characteristics of three storm surges and associated typhoons are given in Table 7.1. After the Muroto typhoon of 1934, embankments were constructed in low-lying areas. Dykes of wooden piles, concrete or soil, together with locks and pumping stations, were constructed along rivers and canals, and in harbour areas. Tens of kilometres of breakwaters were constructed. Also an early warning system was introduced so that people could be evacuated from areas that could be affected by storm surges. Historical data relating to storm surges are available for over 1000 years for this area and suggest a mean recurrence interval of 150 years for severe storm surges, that is, a surge having a height of 3.5 m or more above astronomical tide level.

Protective measures, such as noted above, are not likely to be totally effective against every storm surge. Nevertheless, flood protection structures are used extensively and often prove substantially effective. Sand dunes and high beaches offer natural protection against floods and can be maintained by artificial nourishment with sand. Dunes can also be constructed by tipping and bulldozing sand or by aiding the deposition of sand by the use of screens and

Table 7.1 Characteristics of the Muroto, Jane and Daini–Muroto typhoons and their accompanying storm surges (after Tsuchiya and Kawata, 1987).

Items	Muroto Typhoon	Typhoon Jane	Daini–Muroto Typhoon
Date	9 Sept 1934	3 Sept 1950	16 Sept 1961
Lowest atmospheric pressure (mb)	954.3	970.3	937.3
Moving velocity of typhoon (km h^{-1})	60	58	50
Maximum mean wind velocity for 10 min (m s^{-1})	42.0	28.1	33.3
Instantaneous maximum wind velocity (m s^{-1})	>60	44.7	50.6
Highest tidal level above O.P. (m)	4.2	3.85	4.12
Total precipitation (mm)	22.3	62.2	44.2

O.P. = Osaka Port.

fences. Coastal embankments are usually constructed of local materials. Earth embankments (see Section 7.8) are susceptible to erosion by wave action and if overtopped can be scoured on the landward side. Slip failures may also occur. Sea walls may be constructed of concrete or sometimes of steel sheet piling. Permeable seawalls may be formed of rip-rap or concrete tripods. The roughness of their surfaces reduces swash height and the scouring effect of backwash on the foreshore.

Barrages can be used to shorten the coastline along highly indented coastlines and at major estuaries. Although expensive, barrages can serve more than one function, not just flood protection. They may produce hydroelectricity or carry communications, and freshwater lakes can be created behind them. The Delta scheme in the Netherlands provides one of the best examples (Figure 7.10). This was a consequence of the disastrous flooding that resulted from the storm surge of February 1953. This inundated 150 000 ha of polders and damaged over 400 000 buildings. The death toll amounted to 1835, and 72 000 people were evacuated. Some 47 000 cattle were killed. The scheme involved the construction of four massive dams to close off sea inlets and a series of secondary dams, with a flood barrier northwest of Rotterdam. The length of the coastline was reduced by 700 km and coastal defences were improved substantially.

Estuaries that have major ports require protection against sea flooding. The use of embankments may be out of the question in urbanized areas as space is not available without expensive compulsory purchase of waterfront land, and barrages with locks would impede sea-going traffic. In such situations movable flood barriers, which can be emplaced when storm surges threaten, may be used. The Thames Barrier in London offers an example (Figure 7.11).

(a)

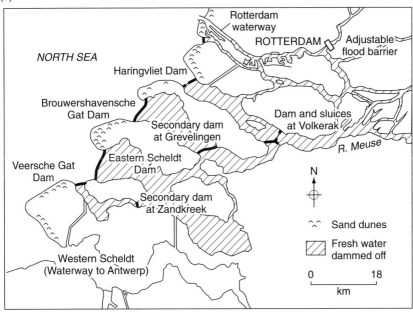

(b)

Figure 7.10 (a) The Netherlands Delta project; (b) the main Oosterschelde flood gates of the Delta scheme.

Figure 7.11 The Thames Barrier (photograph courtesy of The Steel Construction Institute Library).

7.7 Shoreline investigation

Protective works or littoral barriers have to be planned and built to prevent the destruction of a shoreline by marine action. Before any project or beach planning and management scheme can be started a complete study of the beach must be made. The preliminary investigation of the area concerned should therefore first consider the landforms and rock formations along the beach and adjacent rivers, paying particular attention to their durability and stability. In addition, consideration must be given to the width, slope, composition, and state of accretion or erosion of the beach, the presence of bluffs, dunes, marshy areas or vegetation in the backshore area, and the presence of beach structures such as groynes.

The stage of development of the beach area in terms of the cycle of shore erosion, the sources of supply of beach material, probable rates of growth, and probable future history merit considerable discussion. Estimates of the rate of erosion and the proportion and size of eroded material contributed to the beach must be made, as well as whether they are influenced by seasonal effects (Martinez *et al.*, 1990). Not only must the character of the rocks be investigated but the presence of joints and fractures must also be studied. The latter afford some estimate of the average size of fragments resulting from erosion. The rapidity of the breakdown of rock fragments into individual particles should be noted. The behaviour of many unconsolidated materials that form cliffs, when weathered, together with their slope stability and likelihood of sliding has to be taken into account. Samples of the beach and underwater material have to be collected and analysed for such factors as their particle size distribution and mineral content. Mechanical analysis may prove useful in helping to determine the amount of material that is likely to remain on the beach, for beach sand is seldom finer than 0.1 mm in diameter. The amount of material moving along the shore must be investigated, because the effectiveness of the structures erected may depend upon the quantity of drift available.

A characteristic of sand and shingle foreshores is their permeability, which is an important factor in foreshore stability. Any impermeable surface introduced tends to alter stability and hence it is often advisable to use permeable defences. In some instances, erosion problems may arise as the result of artificial structures along the shore.

7.7.1 Recording devices

Valuable data may be obtained from a wave gauge, which can be a graduated pole erected on the foreshore from which the height of waves may be estimated. The wave period can be determined by timing the rise and fall of the surface of the sea at the wave gauge. The tide level is derived from gauge board readings in a tidal basin or from recorders on nearby maritime structures. A number of different types of wave-measuring instrument have been developed. A step-resistance wave gauge consists of a series of graduated resistors connected to a common source of electric potential. The circuit is arranged so that the rising water surface furnishes an electrical path that short-circuits successive resistors and causes an increase in the current. Pressure gauges may also be used to measure wave height and, like electrical resistance gauges, are generally open-framed structures. Wave data can also be obtained by using wave-recording buoys or by ship-borne wave recorders. Analysis of wave records is accomplished by measuring a number of wave heights and periods. Aerial photography can be used to study refraction patterns of long swell.

Surface floats are used to obtain information on the pattern of tidal streams over an appreciable area. In a float test, the tidal stream at any particular locality has to be estimated throughout the tidal cycle. Velocity profiles determined by current meter will usually be required to translate the results of float tests into overall transport. The direct current meter is used extensively for measuring currents at sea, the measurement being obtained from the speed of rotation of a propeller when the meter is suspended in the sea. The depth of the instrument is usually recorded automatically, and current profiles are obtained by raising and lowering the instrument in stages, and taking readings at predetermined levels. Recording current meters measure the speed and direction of currents in the same way, but the information is recorded on magnetic tape.

The depth of water can be recorded by an echo-ranging device, the technique being referred to as transit sonar or oblique asdic. A narrow transverse beam of sound energy pulses is transmitted to one side of a vessel following a steady course. Echoes are continuously recorded and the surface topography can be deduced from their interpretation, augmented as required by visual, radio-wave or laser instrument sights. Seismic refraction can also be used to measure the depth of water to the sea floor, the energy pulses being generated by an air gun or sparker. A survey vessel traverses the area concerned, releasing the pulses and recording the refracted shock waves. In addition, it may be possible to estimate the depth of sediment on the sea floor by using this method.

The most reliable method for measuring suspended load at some distance away from the bed is to use oceanographers' sampling bottles, suspended at given depth intervals. Sampling is done by using an automatic recorder that counts the sediment particles. The Delft bottle can be used to measure the rate of transport of material in suspension or saltation near the bed. The velocity of the water entering the orifice of this bottle is reduced and thus sediment down to about 0.1 mm in size falls from suspension and is retained in the bottle. There is no satisfactory method of direct measurement of bed load transport at sea, although a trap like the bottom transport meter can be used. However, this is unsuitable for measuring bed transport where the sediments are soft. There are a number of techniques for obtaining samples from the sea bed, each being appropriate to a fairly narrow range of sediments. The drop sampler falls freely into sediments, while the piston sampler or explosive sampler is forced into the sea bed.

Where physiographical or structural features provide a complete barrier, it is only necessary to measure the rate at which material is accumulating on the updrift side or diminishing on the downdrift side. If sand is involved, the survey will have to extend some distance offshore and inshore. On the other hand, shingle drift is usually confined to the littoral zone, the exceptions occurring in those localities where rapid currents transport the shingle into deep water. Submarine investigation at shallow depth is sometimes of special importance. In such cases divers may be employed.

Tracer techniques can be used to assess the movement of sedimentary particles. For example, radioactive tracers can be attached to the particles concerned, or alternatively irradiated artificial material of similar size may be introduced into the sedimentary material under study. The movement and distribution of the tracer is surveyed subsequently at intervals by means of a Geiger counter or scintillation counter lowered by cable to the sea bed. While radioactive tracers have proved of the greatest value offshore, fluorescent tracers have been most widely adopted for determination of longshore drift, with particular success on shingle beaches. The usual procedure is to deposit the tracer (this, on a shingle beach, is often crushed concrete) at a steady rate at predetermined positions on the foreshore and to detect its presence at regular intervals of time and distance along the foreshore. Calculations of longshore drift are based on recording the count of visible particles and assuming uniform mixing of the tracer with the natural beach.

7.7.2 Topographic and hydrographic surveys

Topographic and hydrographic surveys of an area allow the compilation of maps and charts from which a study of the changes along a coast can be made. Topographic maps extending back to the earliest editions, remote-sensing imagery and aerial photographs taken at successive intervals of time prove of value in evaluating changes in coastlines over the recent past. Historical evidence has also been used to determine coastal changes, the study of the retreat of the Holderness coast in England by Steers (1964) providing an example (Figure 7.12). Bathymetric charts provide information on variations in offshore topography.

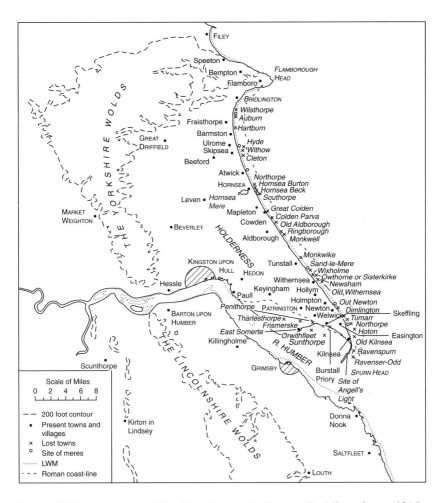

Figure 7.12 The lost towns of East Yorkshire and Holderness 'Bay' (from Steers, 1964).

Observations should be taken of winds, waves and currents, and information gathered on streams that enter the sea in or near the area concerned. The wind rose must differentiate between periods of light and strong winds. It is the distribution and direction of the latter that concern the coastal engineer.

Where there is an appreciable tidal range, it is necessary to consider the most severe combination of water depth and wave height, since the breaking or surging wave gives rise to very much greater forces and water velocities than does the wave that strikes the structure without breaking. The maximum wave is determined by a statistical method similar to that used to find the exceptional flood. It may likewise be defined in terms of a 100-year period. Where structural failure could lead to loss of life or other unacceptable risk, then the maximum wave must be considered. The selection of the 'design wave' in deep water is, according to the nature and importance of the particular structure, normally contained between the limits of the significant wave height (H_s) and the maximum wave height (H_{max}) for the period corresponding to the aggregate length of storms of such intensity as is anticipated during the life of the structure. For example, for a revetment that can be readily repaired following storm damage without risk of major resultant costs, it may be economic to design the structure to withstand waves of height, say, ($H_{1/10}$) (the average height of the highest 10% of all waves).

Local currents will affect the layout and siting of marine works, and when determined by hydrographic surveying, they usually must be integrated into the design. The contributions made by large streams may vary; for example, material brought down by large floods may cause a temporary, but nevertheless appreciable, increase in the beach width around the mouths of rivers. Inlets across a beach need particular evaluation. During normal times there may be relatively little longshore drift, but if upbeach breakthroughs occur in any bars off the inlet mouth, then sand is moved down-beach and is subjected to longshore drift.

Selection of the measures to be taken necessitates consideration of whether the beach conditions represent a long-term trend or whether they are cyclical phenomena that may recur at some time in the future. The recent history of a beach and the marine processes operating on it may have to be evaluated. Observations must be extended over at least one complete storm cycle, because beach slopes and other features may change rapidly during such times. The effects of any likely changes in sea level should also be taken into consideration (Clayton, 1990).

7.7.3 Modelling

Scale models of particular stretches of coastline can be constructed in wave tanks in laboratories in an attempt to simulate coastal processes and so evaluate the results that individual projects may have if developed along a coastline. Such small-scale modelling allows variation of various factors in the coastal regime and represents a three-dimensional reproduction of the situation. Computers can also be used to model changes in coastlines. Again, the consequences of varying parameters can be evaluated, as can the influence that various protection works will have on a coastline.

7.8 Protective barriers

7.8.1 Sea walls

Training walls are designed to protect inlets and harbours as well as to prevent serious wave action in the area they protect. They also impound any longshore drift material upbeach of an inlet and thereby prevent sanding of the channel. However, this may cause or accelerate erosion downdrift of the walls. Training walls are usually built at right angles to the shore, although their outer segments may be set at an angle. Two parallel training walls may extend from each side of a river mouth for some distance out to sea and because of this confinement the velocity of river flow is increased, which lessens the amount of deposition that takes place between the walls. Training walls may be built on rubble mounds.

Sea walls may rise vertically or they may be curved or stepped in cross-section (Figure 7.13). They are designed to prevent wave erosion and overtopping by

Figure 7.13 Sea wall at Newbiggin-by-the-Sea, Northumberland. Note the stepped run-up in front of the wall and the curved design with bullnosed coping to reflect waves.

high seas by dissipating or absorbing destructive wave energy. Sea walls are expensive, so they tend to be used where property requires protection. They must be stable, and this is usually synonymous with weight. They should also be impermeable and resist marine abrasion. Sea walls can be divided into two main classes: those from which waves are reflected and those on which waves break (Thorn and Simmons, 1971). The second category can be subdivided into those types where the depth of the water in front of the wall is such that waves break on the structure and those types where the bigger waves break on the foreshore in front of the wall.

It is generally agreed that a wall that combines reflection and breaking sets up a very severe erosive action immediately in front of it. Consequently, in the layout of a sea wall, particular attention should be given to wave reflection and the possibility of the crests of the waves increasing in height as they travel along the wall due to their high angle of obliquity. The toe of the wall should not be allowed to become exposed. If base erosion becomes serious, a row of sheet piling may be driven along the seaward edge of the foot of the wall. Unless the eroding foreshore in front of a sea wall is stabilized, the apron of the wall must be repeatedly extended seawards as the foreshore falls. Generally speaking, on a deteriorating foreshore shingle drops much more rapidly than sand, and the estimated rate of fall determines the extent to which an apron should be extended. The shingle fill must extend slightly above swash height to achieve stability, and the seaward profile must have a stable slope.

On a shingle beach a sea wall prevents percolation of the upper part of the swash into the beach. If a wall can be positioned near the top of the swash it has a negligible effect on the beach profile. Also, if the wave energy in front of a sea wall can be dampened the deleterious effects are reduced. This can be accomplished by means of a permeable revetment or armouring. However, a stepped profile absorbs only a small fraction of the energy of storm waves, since the steps are usually too small by comparison with the wave height. A curved profile helps to avoid a violent reflection, and a bullnosed coping at the top of the wall tends to deflect the plume from a breaking wave towards the sea.

A simple sloping wall is not subject to the full force of waves. As they travel up the wall their break is accelerated, if they have not already broken, and much of their energy is expended in their run up the slope, so the impact is a glancing rather than a direct blow. The effect of roughness is greater on a gentle slope, where velocities are greater and the total distance travelled by the swash is also greater. An inclined berm in a swash wall provides an effective means of reducing wave run-up. Providing the width of the berm represents a significant part, say 20%, of the wavelength, then its effect is approximately the same as if its slope were continued to the crest of the wall.

A sea wall is subjected to earth pressure on its landward side. The position of the water table is therefore important and is complicated by the probable existence of appreciable fluctuations due to the rise and fall of the tides. For example, water can saturate the backfill of a sea wall during the flood tide and descends during the ebb. As the water level behind the wall does not fall as quickly as the ebb water, the resultant lag leads to an increase of pore water pressure against the wall. In the case of high tides this lag should be carefully estimated in relation to the permeability of the material.

7.8.2 Breakwaters

Breakwaters disperse the waves of heavy seas and provide shelter for harbours. These commonly run parallel to the shore or at slight angles to it and are attached to the coast at one end, their orientation being chosen with respect to the direction which storm waves approach the coast. The detached offshore breakwater is constructed as a barrier parallel to the shoreline and is not connected to the coast. Attached breakwaters cause sand or shingle of longshore drift, on shorelines so affected, to accumulate against them and so rob downdrift beaches of sediment. Consequently, severe erosion may take place in the downdrift area after the construction of a breakwater. Furthermore, accumulation of sand can eventually move around the breakwater to be deposited in the protected area. Komar (1976) described this as happening at Santa Barbara, California, after a breakwater had been built to protect the harbour. As a result, the harbour entrance has to be dredged and the sand is deposited on the downdrift shoreline. Although long offshore breakwaters shelter their lee side, they cause wave refraction and may generate currents in opposite directions along the shore towards the sheltered area, with resultant impounding of sand.

The character of the sea bed at the site of a proposed breakwater must be carefully investigated. Not only must the sea bed be investigated fully with respect to its stability, bearing capacity, and ease of removal if dredging is involved, but the adjacent stretches of coast must also be studied in order to determine local currents, longshore drift and any features that may in any way be affected by the proposed structure.

The oldest form of breakwater is a simple pile of dumped rock, in other words, a rubble mound. Rubble-mound breakwaters dissipate storm waves by turbulence, as sea water penetrates the voids between the blocks of rock (Silvester, 1974). Most of the attenuation occurs near the surface of the water, with oscillatory motions of water at depth being reduced somewhat. Rubble-mound breakwaters are adaptable to any depth of water, are suitable for nearly all types of foundation from solid rock to soft mud and can be repaired readily. The stability of such a structure is not determined by the integrity of the entire mass in the presence of wave action but upon the ability of the individual stones to resist displacement. The quality and durability of the stone is also important (Latham, 1991). The structural design of a rubble-mound breakwater is therefore mainly concerned with specifications for rock size, density and durability, and the selection of an appropriate slope angle (Fookes and Poole, 1981). In general, the wave force acting upon an individual stone is proportional to the exposed area of the stone, while the resistance of the stone is proportional to its volume. Hence, the larger the stone, the greater is its stability. Tetrapods are four-legged concrete blocks that weigh up to 40 t (Figure 7.14). They have been used mainly in harbour works. Their design enables them to interlock and so give relatively steep slopes, while their roughness reduces swash height. Vertical wall breakwaters include masonry walls, timber cribs, caissons, and sheet piling, alone or in combination. This type of breakwater is not very much used, but a combined type, consisting of a wall placed on top of a rubble-mound structure, has many applications in practice.

A pneumatic breakwater provides protection against relatively short waves by releasing bubbles from a compressed air pipe (Silvester, 1974). Pipelines supplying air for pneumatic breakwaters can soon become silted up if placed too close to the sea bed. The release of air bubbles into a wave system reduces wave height and therefore wave energy. However, the pneumatic breakwater is inefficient as far as protection against long waves is concerned.

Sea walls and breakwaters are often made up of interconnected earth-filled cylinders of steel sheet piling. Each constituent cell is a stable unit, deriving its stability from the composite strength of both the earth fill and the steel shell. It is undesirable to carry clay backfill above the water line, because poor drainage allows the development of excess pore pressures, hence sand fill should be used above the low-water level. The sand fill should be adequately drained so that groundwater levels do not lag much behind water levels in, say, a harbour. With a weak foundation the backfill should be all sand.

Figure 7.14 Tetrapods used to construct a breakwater near Port Elizabeth, South Africa.

7.8.3 Embankments

Embankments have been used for centuries as a means of coastal protection. The weight of an embankment must be sufficient to withstand the pressure of the water. The material composing the embankment should be impervious or at least it should be provided with a clay core to ensure that seepage does not occur. It requires a protective stone apron to withstand the destructive power of the waves. Clay is the usual material for embankment construction.

The top of a clay embankment should preferably be at least 1 m above the highest known tide level so that the risk of water flowing through surface fissures is reduced to a minimum. This fissures zone is caused by the drying out of the surface. If the nature of the soil and the space available is such that the desirable level for the top of an embankment cannot be attained without the possibility of rotational slip occurring, a common cause of embankment failure, then two alternatives are possible. The first is to construct a crest wall and the second is to provide adequate protection to the landward side of the embankment so that it may be overtopped during severe storms without being damaged. Usually, the period of overtopping is short and the quantity of water involved is often small, and so can be easily dealt with by a system of dykes whereby it can be discharged back into the sea.

The general effect of flattening the slope of an embankment is to lessen the swash height, lessen foreshore erosion and reduce the wave pressure on the upper part of the apron. The slope therefore may vary from 1 in 2 for an estuary embankment well protected by a wide mud flat at a high level, to 1 in 5 or even flatter for one in deep water that permits large waves to approach and break on it. A berm is generally constructed on the seaward side of an embankment at about the level of high water to reduce swash height.

Clay embankments may fail due to direct frontal erosion by wave action; flow through the fissured zone causing a shallow slump to occur on the landward face of the wall; scour of the back of the wall by overtopping; or rotational slip. Scouring of the back of the wall, if it is too severe for turf to provide sufficient protection, generally can only be combated by constructing a protective revetment. Delayed slips sometimes occur up to nine months or more after construction, and are usually due to a redistribution of pore water pressure, excessive pore water pressures resulting from changes in tidal level. If a permeable stratum lies beneath a clay embankment uplift pressures could develop under it at high tides. These can reduce stability by lessening the shear strength of the soil and even by causing the landward toe to heave. Suitable remedies are to provide more weight on the landward slope and to provide relief filter drains near the landward toe to reduce the uplift pressures. Even where failure does not occur, slow settlement of embankments is very common, particularly where the underlying material is very soft.

7.8.4 Revetments

A revetment affords an embankment protection against wave erosion (Thorn and Simmons, 1971). Stone is used most commonly for revetment work, although concrete is also employed. The chief factor involved in the design of a stone revetment is the selection of stone size, it being important to guard against erosion between stones (Figure 7.15). Consequently, coarse rip-rap must be isolated from the earth embankment by one or more courses of filter stone. Stone pitching is an ancient form of embankment protection, consisting of stone properly placed on the clay face of the wall and keyed firmly into place. Revetments of this type are flexible and have proved very satisfactory in the past. Flexibility is an important requirement of revetment since slow settlement is likely to occur in a clay embankment. As an alternative to keying, stones may be grouted or asphalt-jointed.

Figure 7.15 Rip-rap used as revetment, Fremantle, Western Australia.

Individual blockwork units may be used so that in the event of a wash-out the spread of damage is limited. Blockwork formed of igneous rock is one of the finest forms of revetment for resisting severe abrasion. On walls where wave action is not severe, interlocking concrete blocks provide a very satisfactory form of revetment. The effectiveness of a revetment lies more in its water-tightness than in the weight of the stones.

7.8.5 Bulkheads

Bulkheads are vertical walls either of timber or of steel sheet piling that support an earth bank. A bulkhead may be cantilevered out from the foundation soil or it may be additionally restrained by tie rods. Except for installations of moderate height and stable foundation, a cantilever bulkhead will be rejected in favour of a design having one or two anchorages. The tie rod on an anchored bulkhead must be carried to a secure anchorage; generally this is provided by a deadman, the placement of which must be such that the potential slip surface behind it does not invade the zone influencing bulkhead pressures.

Foundation conditions for bulkheads must be given careful attention, and due consideration must be given to the likelihood of scour occurring at the foot of the wall and to changes in beach conditions. Cut-off walls of steel sheet piling below reinforced concrete superstructures provide an effective method of construction ensuring protection against scouring. Providing the wall structure does not encroach far below the high-tide mark, the existence of this wall should not affect the stability of the beach. If a bulkhead wall has to be constructed either within the tidal range on the beach or below the low-water mark, then study must be made of the possible consequences of construction.

7.9 Stabilization of longshore drift

Before any scheme for beach stabilization is put into operation it is necessary to determine the prevailing direction of longshore drift, its magnitude and whether there is any seasonal variation. It is also necessary to know whether the foreshore is undergoing a net gain or loss of beach material and what is the annual rate of change.

7.9.1 Groynes

The groyne is the most frequently used structure to stabilize or increase the width of a beach by arresting longshore drift (Fig. 7.16). Groynes are used to limit the movement of beach material, and to stabilize the foreshore by encouraging accretion. However, groynes do not usually halt all drift. Groynes should be constructed transverse to the mean direction of the breaking crest of storm waves, which usually means that they should be approximately at right angles to the coastline. Standard types usually slope at about the same angle as the beach. With abundant longshore drift and relatively mild storm conditions almost any type of groyne appears satisfactory, but when the longshore drift is lean, the choice is much more difficult.

By arresting longshore drift, groynes cause the beach line to be reoriented between successive groynes so that the prevailing waves arrive more nearly parallel to the beach. If the angle of wave incidence is greater than that for maximum rate of drift, groynes serve no useful purpose, since accumulation of beach against groynes orients the beach immediately updrift of them in such a way as to lead initially to increased longshore drift (Figure 7.17). The beach crest rapidly swings around until the original rate of drift is restored, resulting in loss of beach material (Muir-Wood and Fleming, 1983).

The height of a groyne determines the maximum beach profile updrift of it. In general, there is little advantage in building groynes that are higher than the highest level normally reached by the sea. Low groynes are simpler to construct. The stability must take account of the storm profile and local scour. On a sandy foreshore the height of groyne should not exceed 0.5 to 1.0 m above the beach if it is to minimize scour from wave and tidal currents, particularly at the seaward end. It is therefore advisable to provide low groynes at relatively close spacing to achieve the desired value Q_g/Q_o (see below). Closely spaced groynes on a sand beach have the additional advantage that they tend to control rip currents so that these occur at correspondingly more frequent intervals, and consequently the longshore drift in the surf zone is reduced. Where long

Figure 7.16 Groynes used to prevent longshore drift at Ventnor, Isle of Wight, England. Note that the beach is starved of sand beyond the far groyne.

groynes are too close, a proportion of the beach material will not be captured by each groyne bay. The effect is more marked for sand than shingle, because of beach gradient and the mode of transport.

The length of groynes is often determined by the tidal range and beach slope. Limiting groynes to the beach above the level of low spring tides is utterly defenceless in terms of efficiency, since the total rate of sand movement in the coastwise direction is greater below the level of low spring tides than above it. Length is also related to the desired effectiveness of the groyne system. In other words, as the ratio of the quantity of longshore drift with groynes (Q_g) to the quantity without (Q_o) becomes smaller, the necessary length of the groyne becomes greater until $Q_g/Q_o = 0$, when the groyne must extend to the limit of longshore drift. Intermediate groynes have sometimes been used to form littoral cells capable of retaining virtually all the shingle on the foreshore. Groynes should extend landwards to the cliff or protective structure or on shingle shores to somewhat higher than the highest swash height at high water. This is to protect them from being outflanked by storm waves.

The maximum groyne spacing is frequently governed by the resulting variation in beach level on either side (Flemming, 1990). For instance, in bays where the direction of wave attack is confined, groynes may be more widely spaced than on exposed promontories. Indeed, it may be necessary to provide

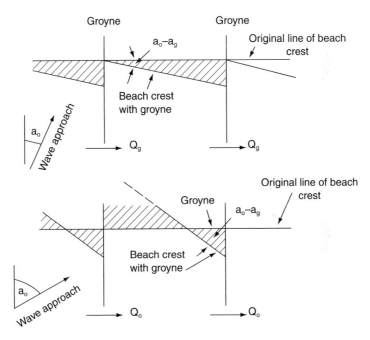

Figure 7.17 Diagram of groyned shore (after Muir-Wood and Fleming, 1983).

a more substantial groyne where a change in the direction of the coastline occurs to counter the accentuated wave attack and change in the rate of drift. The common spacing rule for groynes is to arrange them at intervals of one to three groyne lengths. More specifically, when groyne spacing is being considered, estimation of the desired value of Q_g/Q_o, the beach profile, the direction and strength of the prevailing or storm waves, the amount and direction of longshore drift, the relative exposure of the shore, and the angle of the beach crest between groynes, should be taken into account.

Permeable groynes have openings that increase in size seawards and thereby allow some drift material to pass through them. The use of permeable groynes for a sand foreshore may help to reduce local scour but may be considered only where Q_g/Q_o is to remain fairly high, since they cause no more than a slight reduction in the rate of longshore drift. On a shingle foreshore, groyne-induced scour need not be a serious problem, and the permeable groyne is not an economic expedient.

Groynes reduce the amount of material passing downdrift and therefore can prove detrimental to those areas of the coastline (Silvester, 1974). Their effect on the coastal system should therefore be considered before installation. To prevent such damage to adjacent areas of beach, groynes can be filled artificially by beach nourishment. In this way, natural longshore drift will continue to reach the downdrift beaches.

7.9.2 Beach replenishment

Artificial replenishment of the shore (Figure 7.17) by building beach fills is used either if it is economically preferable or if artificial barriers fail to defend the shore adequately from erosion (Whitcombe, 1996). In fact, beach nourishment represents the only form of coastal protection that does not adversely affect other sectors of the coast. Unfortunately, it often is difficult to predict how frequently a beach should be renourished. Ideally, the beach fill used for renourishment should have a similar particle size distribution to the natural beach material.

The material may be obtained from a borrow pit away from the coastline, in which case it is necessary to consider the effects of using a material of different grain size and sorting on the stability of the beach, or it may be obtained from points of accretion on the foreshore. A common method of beach replenishment is to take the material from the downdrift end of a beach and return it at the updrift end. This is usually the case when beach feeding is combined with groynes. However, investigation of the foreshore may indicate that drift increases in the downdrift direction and therefore that recharge is required at intermediate points to make good such a deficiency. Special attention should be given to areas where there is a change in the direction of coastline. The immediate advantage from beach feeding is experienced updrift rather than downdrift of the supply point. A variant on such a system is to bypass obstructions that cause unwanted accretions and scour, respectively, on their updrift and downdrift sides.

Sediment bypassing involves the mechanical or hydraulic transfer of sediment across a shoreline structure, for instance, a harbour entrance towards the downdrift zone, where there is the potential for sediment starvation. The purpose of bypasses is to prevent the detention of too much sediment by the structure or the diversion of sediment into deeper water out of reach of the beach regime. Sediment bypassing systems, however, are expensive to build and maintain.

Sand dunes can be built up by the construction of permeable screens. As the screens become buried more screens are constructed above them to allow sand to accumulate continuously. Marram grass is generally planted to stabilize the dunes.

7.9.3 Coastal land-use management

Because of the large cost of engineering works and the expense of rehabilitation, coastal management represents an alternative way of tackling the problem of coastal erosion and involves prohibiting or even removing developments that are likely to destabilize the coastal regime (Gares and Sherman, 1985; Ricketts, 1986). The location, type and intensity of development must be planned and controlled. Public safety is the primary consideration. Generally, the level at which risk becomes unacceptable is governed by the socio-economic cost. Conservation areas and open space zones may need to be established. However, in areas which have been developed the existing properties along a coastline may be expensive, so that public purchase may be out of the question. What is more, owners of private property may resist acquisition attempts.

7.10 Tsunamis

One of the most terrifying phenomena that occur along coastal regions is inundation by large masses of water called tsunamis (Figure 7.18). Most tsunamis originate as a result of fault movement on the sea floor, although they can also be developed by submarine landslides or volcanic activity. However, even the effects of large earthquakes are relatively localized compared with the impact of tsunamis. Seismic tsunamis are most common in the Pacific Ocean and are usually formed when submarine faults have a significant vertical movement. Faults of this type occur along the coasts of South America, the Aleutian Islands and Japan. Resulting displacement of the water surface generates a series of tsunami waves, which travel in all directions across the ocean. As with other forms of wave, it is the energy of tsunamis that is transported, not the mass. Oscillatory waves are developed with periods of 10–60 minutes that affect the whole column of water from the bottom of the ocean to the surface. Together with the magnitude of an earthquake and its depth of focus, the amount of

Figure 7.18 The northern end of Resurrection Bay at Seward, Alaska, after it had been affected by a tsunami. The earthquake epicentre was 75 km distant.

vertical crustal displacement determines the size, orientation and destructiveness of a tsunami. Horizontal fault movements, such as occur along the Californian coast, do not result in tsunamis.

In the open ocean, tsunamis normally have a very long wavelength and their amplitude is hardly noticeable. Their wavelength (L) is equal to:

$$L = v.T \tag{7.6}$$

in which v is the velocity and T is the period. Successive waves may be from five minutes to an hour apart. They travel rapidly across the ocean at speeds of around 650 km h^{-1}. In fact, the velocity (v) of propagation of tsunamis in deep water is given by:

$$v = \sqrt{gZ} \tag{7.7}$$

where g is the acceleration due to gravity and Z is the depth of water. This allows prediction of tsunami arrival times along coasts. Unfortunately, however, the vertical distance between the maximum height reached by the water along a shoreline and mean sea level, that is, the tsunami run-up, is not possible to predict. Because their speed is proportional to the depth of water, this means that the wave fronts always move towards shallower water. It also means that in coastal areas the waves slow down and increase in height, rushing onshore as highly destructive breakers. Waves have been recorded up to nearly 20 m in height above normal sea level. Tsunamis, like other waves, are refracted by offshore topography and by the differences in the configuration of the coastline.

Due to the long period of tsunamis, the waves are of great length (e.g. 200–700 km in the open ocean; 50–150 km on the continental shelf). It is therefore almost impossible to detect tsunamis in the open ocean because their amplitudes (0.1–1.0 m) are extremely small in relation to their length. They can only be detected near the shore.

The size of a wave as it arrives at the shore depends upon the magnitude of the original displacement of water at the source and the distance it has travelled, as well as underwater topography and coastal configuration. Soloviev (1978) devised a classification of tsunami intensity, which is given in Table 7.2.

Free oscillations develop when tsunamis reach the continental shelf, which modify their form. Usually the largest oscillation level is not the first but one of the subsequent oscillations. However, to the observer on the coast, a tsunami appears not as a sequence of waves but as a quick succession of floods and ebbs (i.e. a rise and fall of the ocean as a whole) because of the great wavelength involved. Shallow water permits tsunamis to increase in amplitude without significant reduction in velocity and energy. On the other hand, where the water is relatively deep off a shoreline the growth in the size of the wave is restricted. Large waves, several metres in height, are most likely when tsunamis move into narrowing inlets. Such waves cause terrible devastation and can wreck settlements. For example, if the wave is breaking as it crosses the shore, it can destroy houses merely by the weight of water. In fact, water flow exerts a force on obstacles in its path that is proportional to the product of the depth of water and the square of its velocity. Wiegel (1970) suggested that the maximum pressure (P_{max}) on an obstacle can be derived from:

$$P_{max} = 0.5 \ C_D \rho \ v_c^2 \ F_n^2 \tag{7.8}$$

where C_D is a factor including the shape of the obstacle, ρ is the density of water, $v_c = 2\sqrt{gZ_c}$, Z_c is the depth of the inundation zone and F_n is the Froude number. The subsequent backwash may carry many buildings out to sea and may remove several metres depth of sand from dune coasts. Damage is also caused when the resultant debris smashes into buildings and structures.

Usually the first wave is like a very rapid change in tide; for example, the sea level may change 7 or 8 m in 10 minutes. A bore occurs where there is a concentration of wave energy by funnelling, as in bays, or by convergence, as on points. A steep front rolls over relatively quiet water. Behind the front the crest of such a wave is broad and flat, and the wave velocity is about 30 km h^{-1}. Along rocky coasts large blocks of material may be dislodged and moved shorewards.

Table 7.2 Scale of tsunami intensity (after Soloviev, 1978).

Intensity	Run-up height (m)	Description of tsunami	Frequency in Pacific Ocean
I	0.5	*Very slight.* Wave so weak as to be perceptible only on tide gauge records.	one per hour
II	1	*Slight.* Waves noticed by people living along the shore and familiar with the sea. On very flat shores waves generally noticed.	one per month
III	1	*Rather large.* Generally noticed. Flooding of gently sloping coasts. Light sailing vessels carried away onshore. Slight damage to light structures situated near the coast. In estuaries, reversal of river flow for some distance upstream.	one per eight months
IV	4	*Large.* Flooding of the shore to some depth. Light scouring on made ground. Embankments and dykes damaged. Light structures near the coast damaged. Solid structures on coast lightly damaged. Large sailing vessels and small ships swept inland or carried out to sea. Coasts littered with floating debris.	one per year
V	8	*Very large.* General flooding of the shore to some depth. Quays and other heavy structures near the sea damaged. Light structures destroyed. Severe scouring of cultivated land and littering of the coast with floating objects, fish and other sea animals. With the exception of large ships, all vessels carried inland or out to sea. Large bores in estuaries. Harbour works damaged. People drowned, waves accompanied by a strong roar.	once in three years
≥VI	16	*Disastrous.* Partial or complete destruction of man-made structures for some distance from the shore. Flooding of the coasts to great depths. Large ships severely damaged. Trees uprooted or broken by the waves. Many casualties.	once in ten years

Because of development in coastal areas within the last 30 or more years, the damaging effects of future tsunamis will probably be much more severe than in the past. Hence it is increasingly important that the hazard is evaluated accurately and that the potential threat is estimated correctly. This involves an analysis of the risk for the purpose of planning and putting into place mitigation measures.

Nonetheless, the tsunami hazard is not frequent, and when it does affect a coastal area its destructiveness varies with both location and time. Accordingly, analysis of the historical record of tsunamis is required in any risk assessment. This involves a study of the seismicity of a region to establish the potential threat from earthquakes of local origin. In addition, tsunamis generated by distant earthquakes must be evaluated. The data gathered may highlight spatial differences in the distribution of the destructiveness of tsunamis that may form the basis for zonation of the hazard. If the historical record provides sufficient data, then it may be possible to establish the frequency of recurrence of tsunami events, together with the area that would be inundated by a 50-, 100- or even 500-year tsunami event. On the other hand, if insufficient information is available, then tsunami modelling can be resorted to using either physical or computer models. The latter are now much more commonly used and provide reasonably accurate predictions of potential tsunami inundation that can be used in the management of tsunami hazard. Such models permit the extent of damage to be estimated and the limits for evacuation to be established. The ultimate aim is to produce maps that indicate the degree of tsunami risk, which in turn aid the planning process, thereby allowing high-risk areas to be avoided or used for low-intensity development (Pararas-Carayannis, 1987). Models also facilitate the design of defence works.

Various instruments are used to detect and monitor the passage of tsunamis. These include sensitive seismographs that can record waves with long-period oscillations, pressure recorders placed on the sea floor in shallow water, and buoys anchored to the sea floor and used to measure changes in the level of the sea surface. The locations of places along a coast affected by tsunamis can be hazard-mapped to show, for example, the predicted heights of tsunami at a certain location for given return intervals (e.g. 25, 50 or 100 years). Homes and other buildings can be removed to higher ground and new construction prohibited in the areas of highest risk. Resettlement of coastal communities and prohibition of development in high-risk areas has occurred at Hilo, Hawaii. However, the resettlement of all coastal populations away from possible danger zones is not a feasible economic proposition. Hence, there are occasions when evacuation is necessary. This depends on estimating just how destructive any tsunami will be when it arrives on a particular coast. Furthermore, evacuation requires that the warning system is effective and that there is an adequate transport system to convey the public to safe areas.

Breakwaters, coastal embankments and groves of trees tend to weaken a tsunami wave, reducing its height and the width of the inundation zone. Sea walls may offer protection against some tsunamis. Buildings that need to be located at the coast can be constructed with reinforced concrete frames and elevated on reinforced concrete piles with open spaces at ground level (e.g. for car parks). Consequently, the tsunami can flow through the ground floor without adversely affecting the building. Buildings are usually oriented at right angles to the direction of approach of the waves, that is, perpendicular to the shore. It is, however, more or less impossible to protect a coastline fully from the destructive effects of tsunamis.

Ninety percent of destructive tsunamis occur within the Pacific Ocean, averaging more than two each year. For example, in the past 200 years the Hawaiian Islands have been subjected to over 150 tsunamis. Hence, a long record of historical data provides the basis for prediction of tsunamis in Hawaii and allows an estimation of the events following a tsunamigenic earthquake to be made.

The Pacific Tsunami Warning System (PTWS) is a communications network covering the countries bordering the Pacific Ocean and is designed to give advance warning of dangerous tsunamis (Dohler, 1988). The system uses 69 seismic stations and 65 tide stations in major harbours around the Pacific Ocean. Earthquakes with a magnitude of 6.5 or over cause alarms to sound and those over 7.5 give rise to around-the-clock tsunami watch. Nevertheless, it is difficult to predict the size of waves that will be generated and to avoid false alarms. Clearly, the PTWS cannot provide a warning of an impending tsunami to those areas that are very close to the earthquake epicentre that is responsible for the generation of the tsunami. In fact, 99% of the deaths and much of the damage due to tsunamis occur within 400 km of the area where it was generated. Considering the speed of travel of a tsunami wave, the warning afforded in such instances is less than 30 minutes. Recently, however, a system has been developed whereby an accelerometer transmits a signal, via a satellite over the eastern Pacific Ocean, to computers when an earthquake of magnitude 7 or more occurs within 100 km of the coast. The computers decode the signal, cause water-level sensors to start monitoring and transmit messages to those responsible for carrying out evacuation plans. On the other hand, waves generated off the coast of Japan take 10 hours to reach Hawaii. In such instances, the PTWS can provide a few hours for evacuation to take place if it appears that it is necessary (Bernard, 1991).

References

Bascom, W. (1964) *Waves and Beaches.* Doubleday, New York.

Bernard, E.N. (1991) Assessment of Project THRUST: past, present and future. *Natural Hazard*, 4, 285–292.

Clayton, K.M. (1990) Sea level rise and coastal defences in the United Kingdom. *Quarterly Journal Engineering Geology*, 23, 283–288.

Cleary, W.J. and Hosier, P.E. (1987) North Carolina coastal geologic hazards, an overview. *Bulletin Association Engineering Geologists*, 24, 469–488.

Dohler, G.C. (1988) A general outline of the ITSU Master Plan for the tsunami warning system in the Pacific. *Natural Hazards*, 1, 295–302.

Flemming, C.A. (1990) Principles and effectiveness of groynes. In *Coastal Protection*, Pilarczyk, K.W. (ed.), Balkema, Rotterdam, 121–156.

Fookes, P.G. and Poole, A.B. (1981) Some preliminary considerations on the selection and durability of rock and concrete materials for breakwaters and coastal protection works. *Quarterly Journal Engineering Geology*, 14, 97–128.

Galvin, C.J. (1968) Breaker type classification on three laboratory beaches. *Journal Geophysical Research*, 73, 3651–3659.

Gares, P.A. (1990) Predicting flooding probability for beach–dunes systems. *Environmental Management*, 14, 115–123.

Gares, P.A. and Sherman, D.J. (1985) Protecting an eroding shoreline: the evolution of management response. *Applied Geography*, 5, 55–69.

Hutchinson, J.N., Bromhead, E.N. and Chandler, M.P. (1991) Investigation of the coastal landslides at St Catherine's Point, Isle of Wight. *Proceedings Conference on Slope Stability Engineering: Developments and Applications,* Thomas Telford Press, London, 151–161.

Ingle, J.G. (1966) *The Movement of Beach Sand*. Elsevier, Amsterdam.

Jelesnianski, C.P. (1978) Storm surges. In *Geophysical Predictions*, National Academy of Sciences, Washington, DC, 185–192.

Johns, B. and Ali, A. (1980) The numerical modelling of storm surges in the Bay of Bengal. *Quarterly Journal Royal Meteorological Society*, 106, 1–18.

Johnson, D.W. (1919) *Shoreline Processes and Shoreline Development*. Wiley, New York.

Komar, P.D. (1976) *Beach Processes and Sedimentation*. Prentice-Hall, Englewood Cliffs, NJ.

Latham, J.-P. (1991) Degradation model for rock armour in coastal engineering. *Quarterly Journal Engineering Geology*, 24, 101–118.

Leathermann, S.P. (1979) Beach and dune interactions during storm conditions. *Quarterly Journal Engineering Geology*, 12, 281–290.

Little, D.H. (1975) Harbours and docks. In *Civil Engineer's Reference Book*, Blake, L.S. (ed.), Newnes-Butterworths, London, Section 24, 2–40.

Martinez, M.J., Espejo, R.A., Bilbao, I.A. and Cabrera, M.D. del R. (1990) Analysis of sedimentary processes on the Las Canteras beach (Las Palmas, Spain) for its planning and management. *Engineering Geology*, 29, 377–386.

Muir-Wood, A.M. and Fleming, C.A. (1983) *Coastal Hydraulics*, second edition. Macmillan, London.

Murty, T.S. (1987) Mathematical modelling of global storm surge problems. In *Natural and Man-Made Hazards*, El-Sabh, M.I. and Murty, T.S. (eds), Riedel, Dordrecht, 183–192.

Pararas-Carayannis, G. (1987) Risk assessment of the tsunami hazard. In *Natural and Man-Made Hazards*, El-Sabh, M.I. and Murty, T.S. (eds), Riedel, Dordrecht, 183–192.

Pilarczyk, K.W. (1990) Design of sea wall and dikes – including an overview of revetments. In *Coastal Protection*, Pilarczyk, K.W. (eds) A.A. Balkema, Rotterdam, 197–288.

Ricketts, P.J. (1986) National policy and management responses to the hazard of coastal erosion in Britain and the United States. *Applied Geography*, 6, 197–221.

Sharma, G.S. and Murty, A.J. (1987) Storm surges along the east coast of India. In *Natural and Man-Made Hazards*, El-Sabh, M.I. and Murty, T.S. (eds), Riedel, Dordrecht, 257–278.

Silvester, R. (1974) *Coastal Engineering*, Volumes 1 and 2. Elsevier, Amsterdam.

Soloviev, S.L. (1978) Tsunamis. In *The Assessment and Mitigation of Earthquake Risk*, UNESCO, Paris, 91–143.

Steers, J.A. (1964) *The Coastline of England and Wales*. Cambridge University Press, Cambridge.

Sverdrup, H.U. and Munk, W.H. (1946) Empirical and theoretical relations. In *Forcasting Breakers and Surf*, American Geophysical Union, 27, 823–827.

Thorn, R.B. and Simmons, J.C.F. (1971) *Sea Defence Works*. Butterworths, London.

Tsuchiya, Y. and Kawata, K. (1987) Historical changes of storm-surge disasters in Osaka. In *Natural and Man-Made Hazards*, El-Sabh, M.I. and Murty, T.S. (eds), Riedel, Dordrecht, 279–304.

Ward, R.C. (1978) *Floods: A Geographical Perspective*. Macmillan, London.

Whitcombe, L.J. (1996) Behaviour of an artificially replenished shingle beach at Hayling Island, U.K. *Quarterly Journal Engineering Geology*, 29, 265–272.

Wiegel, R.L. (1970) Tsunamis. In *Earthquake Engineering*, Wiegel, R.L. (ed.), Prentice-Hall, Englewood Cliffs, NJ.

Chapter 8

Wind action and arid regions

8.1 Introduction

Although winds are frequently violent in temperate humid regions, the part they play in sculpting landscapes is negligible compared with that of weathering, mass movement and rivers. This is largely because vegetation acts as a protective cover against wind erosion. In arid regions, however, because there is little vegetation and the ground surface may be dry, wind action is much more significant and sediment yield may be high. The most serious problem attributable to wind action is soil erosion, which is dealt with in Chapter 9. This makes the greatest impact in relation to agriculture. Dust or sandstorms and migrating sand dunes also give rise to problems. As mentioned in Chapter 9, hazards such as dust storms and migrating dunes are frequently made more acute because of the intervention of man. Commonly, the problems are most severe on desert margins, where rainfall is uncertain and there may be significant human pressure on land.

In temperate humid regions the work of wind is most significant along low-lying stretches of sandy coasts. In such localities, if the prevailing winds are onshore, sand is heaped into dunes behind the beach. Successive waves of dunes frequently migrate landwards and spoil land that lies in their path. Wind erosion also occurs in temperate humid regions, notably on agricultural lands that may be subject to occasional droughts. The Fenlands, parts of East Anglia and the Vale of York in England provide examples.

By itself, wind can only remove uncemented debris up to a certain size, which it can perform more effectively if the debris is dry rather than wet. But once armed with rock particles, the wind becomes a noteworthy agent of abrasion. The size of individual particles that the wind can transport depends on the strength of the wind, and particle shape and density. The distance that the wind, given that its velocity remains constant, can carry particles depends principally upon their size. Fine-grained particles can be transported for hundreds or even thousands of kilometres.

Arid and semi-arid regions account for one-third of the land surface of the Earth (Figure 8.1). These are regions that experience a deficit of precipitation and where the range of climatic conditions, particularly the magnitude,

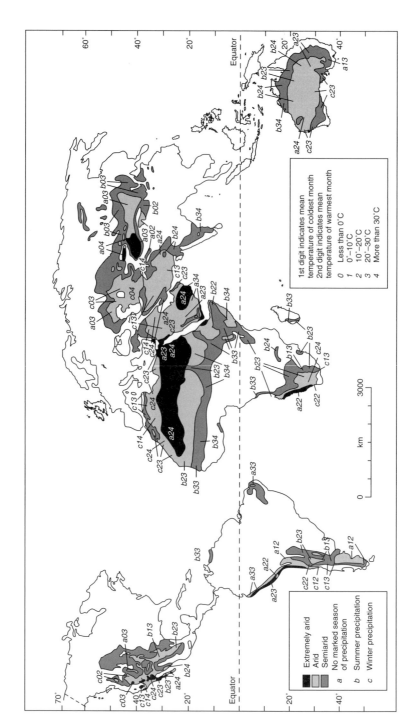

Figure 8.1 World distribution of arid lands (after Meigs, 1953).

frequency and duration of precipitation and wind activity, vary appreciably. Consequently, run-off is usually ephemeral, varying between rainfall events and type of flow, and aeolian processes are commonly dominant. Nonetheless, rainfall can be torrential at times and can generate flash floods. Because of the lack of vegetation, especially in very arid regions, the landscape is vulnerable to erosion.

Although the overall distribution of arid or desert regions is governed by the global pattern of atmospheric circulation, in that they tend to coincide with high-pressure belts, the position is much more complicated. Additional influences affecting their distribution include the distribution of continents and oceans, the size and shape of water and land masses, the influence of warm and cold ocean currents, the topography of the land masses, the seasonal migration of the climatic belts and the seasonal presence of high- and low-pressure cells. Hence, there are interior deserts far from ocean supplies of moisture and mountain barrier deserts on the leeward side of high mountains, which rob winds of their moisture, as well as low-latitude deserts, which occur within the high-pressure belts. Some of the most arid regions occur in low latitudes on the western sides of continents, where the winds are offshore and the coasts are washed by cold currents.

The semi-arid transition lands on the borders of deserts are hazardous because of the vagaries of their climate. Periods of drought are common in these regions and can last for weeks or months. A number of dry years often alternates with a wetter cycle of years. Precipitation is unreliable, and the only consistent factor about the climate is its inconsistency.

8.2 Wind action

Bagnold (1941) maintained that the wind velocity, V, at any height z above the ground surface is given by:

$$V_z = 5.75 \, V_* \, \log \frac{z}{k} \qquad (8.1)$$

where V_* is drag velocity and k is a roughness constant. This indicates that the velocity of the wind increases with height above the ground surface. However, as particles are picked up and moved by wind, this reduces the surface wind velocity, the mobile particles exerting a drag effect on airflow. The drag velocity is related to the drag exerted on the wind by friction at the surface, τ, and to the density of air, ρ_a, and can be expressed as follows:

$$V_* = \sqrt{\frac{\tau}{\rho_a}} \qquad (8.2)$$

As the shear strength of the wind increases, loose particles on the ground surface are subjected to increasing stress so that they ultimately begin to move.

According to Bagnold (*ibid.*), the critical velocity (which is sometimes referred to as the fluid threshold velocity) at which particle movement is initiated is given by:

$$V_{*t} = A \sqrt{\left(\frac{\rho_g - \rho_a}{\rho_a}\right) gD} \qquad (8.3)$$

where V_{*t} is the threshold velocity, ρ_g is the relative density of the grains, ρ_a is the relative density of air, D is the diameter of the grain (cm) and A is a coefficient, which is 0.1 for particles larger than 0.1 mm in diameter. Particles smaller than this do not conform to this law, and the velocities increase with decreasing grain size (Figure 8.2). This is probably due to the increased cohesion between fine particles, to their greater water retention and to lower values of surface roughness. Bagnold proposed a figure of 16 km h^{-1} as the threshold velocity for most desert sands. Chepil (1945) subsequently showed that few particles with actual diameters exceeding 0.84 mm are eroded.

Once particle movement has been initiated, then the ground is impacted by moving grains, thereby causing further movement. In this situation sediment movement can be maintained at lower velocities than are required to start movement. Bagnold (1941) therefore introduced another threshold value, which he termed the impact threshold velocity. The impact threshold velocity is expressed as:

$$V_{*t} = 680 \sqrt{D} \log 30/D \qquad (8.4)$$

where V_{*t} is the threshold wind velocity and D is the diameter of the grain. The fluid and impact threshold velocities are similar for most particles, although with increasing particle size the impact threshold velocity becomes less (Figure 8.2).

Wind erosion takes place when air pressure overcomes the force of gravity on surface particles. Three types of pressure are exerted on the surface of a particle at the threshold of grain movement, namely, impact pressure, viscosity pressure and static pressure. Impact or velocity pressure occurs on the windward side of the particle, whereas viscosity pressure occurs on the leeward side. Static pressure actually represents a reduction in pressure on the top of the grain as a result of increased air velocity. Accordingly, drag is exerted on the top of the particle as a result of the pressure difference between the windward and leeward sides and the lift due to the reduction in static pressure. However, once a particle has been entrained, drag and lift change rapidly.

At first, particles are moved by saltation, that is, in a series of jumps. The impact of saltating particles on others may cause them to move by creep, saltation or suspension. Saltation accounts for up to three-quarters of the grains transported by wind, most of the remainder being carried in suspension; the rest are moved by creep or traction. Saltating grains may rise to heights of up

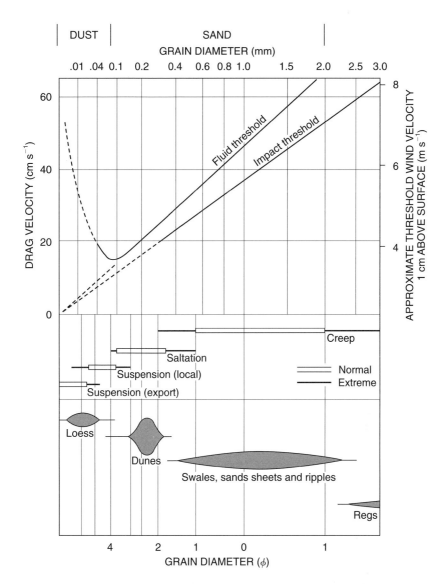

Figure 8.2 Relationship between grain size, fluid and impact threshold wind velocities, characteristic modes of aeolian transport and resulting size grading of aeolian sand formations (after Kerr and Nigra, 1952).

to 2 m, their trajectory then being flattened by more quickly moving air and tailing off as the grains fall to the ground. Bagnold (*ibid.*) maintained that the angle at which particles strike the ground varies between 10 and 16° from the horizontal, irrespective of the height reached. He suggested that this was due to the balance between the force of gravity acting downwards and the maximum forward velocity. It has been found that the saltation height generally is inversely related to particle size and directly related to roughness. The length of the trajectory is roughly ten times the height. If a hard surface crust exists, then before saltation can begin this crust must be broken. Consequently, this means that the initial fluid threshold velocity has to be higher than that which removes erodible particles. Particles are moved in suspension when the terminal velocity of fall is lower than the mean upward eddy currents in the airflow.

One of the most important factors in wind erosion is its velocity. Its turbulence, frequency, duration and direction are also important. As far as the mobility of particles is concerned, the important factors are their size, shape and density. It would appear that particles less than 0.1 mm in diameter are usually transported in suspension, those between 0.1 and 0.5 mm are normally transported by saltation and those larger than 0.5 mm tend to be moved by traction or creep. Grains with a relative density of 2.65, such as quartz sand, are most suspect to wind erosion in the size range 0.1 to 0.15 mm. A wind blowing at 12 km h^{-1} will move grains 0.2 mm diameter – a lower velocity will keep the grains moving.

Wind can only remove particles of limited size range, so if erosion is to proceed beyond the removal of existing loose particles, then the remaining material must be sufficiently reduced in size by other agents of erosion or weathering, or it will seriously inhibit further wind erosion. Removal of fine material leads to a proportionate increase in particles of larger size that cannot be removed. The latter affords increasing protection against continuing erosion, and eventually a wind-stable surface is created. This stage is defined by the critical surface barrier ratio (i.e. the ratio of the height of non-erodible surface projections to the distance between projections that will barely prevent movement of erodible fractions by wind). Desert pavements therefore consist of stones left behind after the finer sediment has been winnowed out. The surfaces of the stones may be flattened and polished by sand-blasting to provide a smooth surface.

Binding agents, such as silt, clay and organic matter, hold particles together and so make wind erosion more difficult. Soil moisture also contributes to cohesion between particles. Indeed, according to Woodruff and Siddoway (1965), the rate of soil movement varies approximately inversely as the square of effective surface soil moisture. The structural units in the soil may represent non-erodible or less erodible obstacles to wind erosion. These structural units or particle aggregates may be reduced in size by weathering or erosive processes. Their resistance to wind erosion varies inversely according to their mechanical stability, which depends on cohesion between particles, and therefore varies with the type of soil.

Generally, a rough surface tends to reduce the velocity of the wind immediately above it. Consequently, particles of a certain size are not as likely to be blown away as they would on a smooth surface. Even so, Bagnold (1941) found that grains of sand less than 0.03 mm in diameter are not lifted by the wind if the surface on which they lie is smooth. On the other hand, particles of this size can easily remain suspended by the wind. Vegetation affords surface cover and increases surface roughness. The taller and denser the vegetation, the more effective it is in affording protection against wind erosion. The characteristics of the wind, the organic content in the soil, the soil moisture content and the vegetation cover can change over short periods of time, particularly according to seasonal variation. The longer the surface distance over which a wind can blow without being interrupted, the more likely it is to attain optimum efficiency.

Particles removed by wind are primarily of sand and dust size. Bagnold (*ibid.*) vividly described the difference between the two in terms of their transportation. He indicated that in dust storms dust rises in dense clouds to great height and that dust particles, because of their very low terminal velocity of fall, are kept in suspension by the turbulence of the wind. By contrast, sand is moved principally by saltation and moves across the surface as a thick, low-flying cloud, with the bulk of the sand movement taking place near the ground. This can produce sand-blast effects on obstacles in the path of movement.

Bagnold (*ibid.*) noted that sand flow can increase downwind until it becomes saturated; that is, the amount of material in motion is the maximum sustainable by the wind. Beyond this point no further removal downwind takes place. The downwind increase in quantity of material transported is referred to as avalanching and leads to increased surface abrasion. However, particles that are too large to be removed inhibit wind erosion. Hence, as wind erosion continues the proportion of non-erodible particles increases until eventually they protect the erodible particles from wind erosion. At this point a wind-stable surface is produced. The area of removal accordingly migrates downwind. Hence, particles may move across a surface that may not suffer a net loss of material. In fact, the transport of particles by wind is related to supply sources and sediment stores, which may be permanent or temporary (see Section 8.6).

The downwind increase in the amount of material transferred was referred to as avalanching by Chepil (1957). He suggested that this was due to a progressive increase in grain impacts consequent upon the entrainment of more particles. This leads to more abrasion, which reduces soil aggregates in size, thereby increasing the supply of erodible particles. Then as particles come to rest in depressions in the surface, its roughness is gradually reduced. This process is termed 'detrusion'. Subsequently, Chepil (1959) pointed out that saturated flow was independent of soil type; however, the distance from the point of initiation to that where saturated flow begins varies with soil erodibility. This distance is approximately 65 m for most erodible soils and around 1900 m for the least erodible soils.

Bagnold (1941) proposed the following expression for sand movement, q:

$$q = C\sqrt{\frac{d}{D}}\ \frac{\rho}{g}\ V_*^3 \tag{8.5}$$

where d is the mean diameter of the grains, D is a standard grain diameter of 0.25 mm, ρ/g is the mass density in air, V_* is the drag velocity above the eroding surface and C is a coefficient that varies primarily with particle size distribution. Bagnold quoted values for C of 1.5 for nearly uniform sand, 1.8 for naturally graded sand and 2.8 for sand with a wide range of particle size distribution. The expression indicates that the rate of movement depends upon the mean particle diameter, the degree of uniformity of the particle size and the velocity gradient of the wind above the erosion surface. Subsequently, Chepil and Woodruff (1963) suggested a simplified form of Equation 8.5 whereby the quantity of material, X (in tons per acre), that can be removed from a particular area can be determined as follows:

$$X = a(V_*)^3 \tag{8.6}$$

where a is a coefficient that similarly depends on the particle size distribution of eroding particles, the position in an eroding zone and the moisture content of the soil, and V_* is the drag velocity.

There are three types of wind erosion, namely, deflation, attrition and abrasion (Chapil and Woodruff, 1963). Deflation results in the lowering of land surfaces by loose, unconsolidated rock waste being blown away by the wind. The effects of deflation are seen most acutely in arid and semi-arid regions. The presence of small deflation hollows has been referred to by Cooke and Warren (1973). They explained the formation of such hollows as a result of strong winds blowing over bare, dry surfaces of unconsolidated sediment, giving rise to differential erosion. The latter may be aided by the sediment being less well bound in the areas where hollows form. Basin-like depressions are formed by deflation in the Sahara and Kalahari deserts. However, downward lowering is almost invariably arrested when the water table is reached since the wind cannot readily remove moist rock particles. What is more, deflation of sedimentary material, particularly alluvium, creates a protective covering if the material contains pebbles. The fine particles are removed by the wind, leaving a surface formed of pebbles, which are too large to be blown away.

The suspended load carried by the wind is further comminuted by attrition. Turbulence causes the particles to collide vigorously with one another and so to break down.

When the wind is armed with grains of sand it possesses great erosive force, the effects of which are best displayed in rock deserts. Wind abrasion occurs as a result of windblown grains impacting against a surface and involves the transfer of kinetic energy (Greeley and Iversen, 1985). The collision may damage either the particle or the surface, or both. Basically, the greater the kinetic energy involved (depending on grain size and/or wind velocity), the larger is the amount of abrasion. Accordingly, any surface subjected to prolonged attack by windblown sand is polished, etched or fluted. Abrasion has a selective action, picking out the weaknesses in rocks. For example, discontinuities are opened and rock pinnacles developed. Since the heaviest rock particles are transported near to the ground, abrasion is there at its maximum and rock pedestals are formed.

The differential effects of wind erosion are well illustrated in areas where alternating beds of hard and soft rock are exposed. If strata are steeply tilted then, because soft rocks are more readily worn away than hard, a ridge-and-furrow relief develops. Such ridges are called yardangs. Yardangs are elongated in the direction of the wind, with rounded upwind faces and long, pointed downwind projections (Cooke and Warren, 1973). Conversely, when an alternating series of hard and soft rocks is more or less horizontally bedded, features known as zeugens are formed. In such cases, the beds of hard rock act as resistant caps affording protection to the soft rocks beneath. Nevertheless, any weaknesses in the hard caps are picked out by weathering and the caps are eventually breached, exposing the underlying soft rocks. Wind erosion rapidly eats into the latter and in the process the hard cap is undermined. As the action continues tabular masses, known as mesas and buttes, are left in isolation.

8.3 Desert dunes

About one-fifth of the land surface of the Earth is desert (see Figure 8.1). Approximately four-fifths of this desert area consists of exposed bedrock or weathered rock waste. The rest is mainly covered with deposits of sand (Glennie, 1970). Desert regions may have very little sand; for example, only about one-ninth of the Sahara Desert is covered by sand. Most of the sand that occurs in deserts does so in large masses referred to as sand seas or ergs. In fact, Wilson (1971) estimated that 99.8% of aeolian sand is found in ergs that exceed 125 km^2 in area, and 85% is in ergs greater than $32\,000 \text{ km}^2$. The smaller ergs, according to Cooke and Warren (1973), include dunes of one type only, but most ergs have different patterns of dunes in different areas.

Bagnold (1941) recognized five main types of sand accumulation, namely, sand drifts and sand shadows, whalebacks, low-scale undulations, sand sheets, and true dunes. He further distinguished two kinds of true dune, the barkhan and the seif. Several factors control the form that an accumulation of sand adopts. First, there is the rate at which sand is supplied; second, there is wind speed, frequency and constancy of direction; third, there is the size and shape of the sand grains; and fourth, there is the nature of the surface across which the sand is moved. Sand drifts accumulate at the exits of the gaps through which wind is channelled and are extended downwind. However, such drifts, unlike true dunes, are dispersed if they are moved downwind. Whalebacks are large mounds of comparatively coarse sand that are thought to represent the relics of seif dunes. Presumably, the coarse sand is derived from the lower parts of seifs, where accumulations of coarse sand are known to exist. These features develop in regions devoid of vegetation. By contrast, undulating mounds are found in the peripheral areas of deserts, where the patchy cover of vegetation slows the wind and creates sand traps. Large undulating mounds are composed of fine sand. Sand sheets are also developed in the marginal areas of deserts. These sheets consist of fine sand that is well sorted; indeed, they often present a smooth surface that is capable of resisting wind erosion. A barkhan is crescentic in outline (Figure 8.3) and is oriented at right angles to the prevailing wind direction, while a seif is a long ridge-shaped dune running parallel to the direction of the wind. Seif dunes are much larger than barkhans; they may extend lengthwise for up to 90 km and reach heights up to 100 m (Figure 8.4). Barkhans are rarely more than 30 m in height and their width is usually about twelve times their height. Generally, seifs occur in great numbers running approximately equidistant from each other, the individual crests being separated from one another by anything from 30 to 500 m or more.

It is commonly believed that sand dunes come into being where some obstacle prevents the free flow of sand, sand piling up on the windward side of the obstacle to form a dune. But in areas where there is an exceptionally low rainfall and therefore little vegetation to impede the movement of sand, observation has revealed that dunes develop most readily on flat surfaces, devoid of large obstacles. It would seem that where the size of the sand grains varies or

Figure 8.3 Barkhan dunes in Death Valley, California.

Figure 8.4 Seif dune near Sossusvlei, Namibia.

where a rocky surface is covered with pebbles, dunes grow over areas of greater width than 5 m. Such patches exert a frictional drag on the wind, causing eddies to blow sand towards them. Sand is trapped between the larger grains or pebbles, an accumulation results. If a surface is strewn with patches of sand and pebbles, deposition takes place over the pebbles. However, patches of sand exert a greater frictional drag on strong winds than do patches of pebbles, so deposition under such conditions takes place over the sand. When strong winds sweep over a rough surface they become transversely unstable, and barkhans may develop.

It would appear that barkhans do not form unless a pile of sand exceeds 0.3 m in height. At this critical height the heap of sand develops a shallow windward and a steep leeward slope. Sand is driven up the windward slope and emptied down the leeward, but eddying of the wind occurs along the leeward slope, impeding the fall of sand and imparting a concave outline to it. With a smaller height than 0.3 m, the sand deposit cannot maintain its form, since individual grains, when moved by the wind, fall beyond its boundaries. It is then dispersed to form sand ripples. Windblown sand is commonly moved by saltation, that is, in a series of jumps. The distance that an individual particle can leap depends on its size, shape and weight, the wind velocity and the angle of lift, which in turn is influenced by the nature of the surface. If a deposit of sand reaches a height of 0.3 m, then the grains that leap from its windward face usually land on the leeward slope. As the infant barkhan grows, an increasing amount of sand is piled at the top of the leeward slope since the trajectories of sand grains cannot carry them further. The leeward slope is increasingly steepened until the angle of rest, which is approximately 35°, is reached. Subsequent deposition leads to sand slumping down the face. The dune advances in this manner.

The question that now has to be answered is, how does the sand become piled to the critical height? It has been asserted that if the wind increased in velocity it does not immediately become fully loaded with sand. Hence, until this happens the wind is capable of moving a greater quantity of sand than normal. Although downwind the sand flow is increasing this also means that frictional drag and turbulence are also increasing. The velocity of the wind is therefore reduced and so more sand is deposited than is removed.

As there is more sand to move in the centre of a barkhan than at its tails the latter advance at a faster rate and the deposit gradually assumes an arcuate shape. The tails are drawn out until they reach a length where their obstructive power is the same as that of the centre of the dune. At this point the dune adopts a stable form, which is maintained as long as the factors involved in dune development do not radically alter. Barkhans tend to migrate downwind in waves, their advance varying from about 6 to 16 m annually, depending on their size (the rate of advance decreases rapidly with increasing size). If there is a steady supply of material the dune tends to advance at a constant speed and to maintain the same shape, whereas if the supply is increased the dune grows and its advance decelerates. The converse happens if the supply of sand decreases.

Longitudinal dunes may develop from barkhans. Suppose that the tails of a barkhan for some reason becomes fixed, for example, by vegetation or by the water table rising to the surface, then the wind continues to move the central part until the barkhan eventually loses its convex shape, becoming concave towards the prevailing wind. As the central area becomes further extended, the barkhan may split. The two separated halves are gradually rotated by the eddying action of the wind until they run parallel to one another, in line with the prevailing wind direction. Dunes that develop in this manner are often referred to as 'blow-outs'.

Seif dunes appear to form where winds blow in two directions, that is, where the prevailing winds are gentle and carry sand into an area, the sand then being driven by occasional strong winds into seif-like forms. Seifs may also develop along the bisectrix between two diagonally opposed winds of roughly equal strength.

Because of their size, seif dunes can trap coarse sand much more easily than can barkhans. This material collects along the lower flanks of the dune. Indeed, barkhans sometimes occur in the troughs of seif dunes. On the other hand, the trough may be floored by bare rock. The wind is strongest along the centre of the troughs as the flanks of the seifs slow the wind by frictional drag, so creating eddies. The wind blowing along the flanks of seifs is accordingly diverted up their slopes. For a further discussion of dunes and their behaviour, the reader is referred to Mabbutt (1977), and Greeley and Iversen (1985).

8.4 Stream action in arid and semi-arid regions

The processes of weathering, erosion and sedimentation in arid regions are similar to those in humid regions but differ in relative importance, intensity and superficial effects. Weathering, although widespread, is dominantly physical because of the deficiency of moisture and vegetation. Although intermittent, streams are the most important agents of erosion on barren slopes because of the intense run-off.

In fact, stream activity plays a significant role in the evolution of landscape in arid and semi-arid regions. Admittedly, the amount of rainfall occurring in arid regions is small and falls irregularly, while that of semi-arid regions is markedly seasonal. Consequently, many drainage courses in arid lands may be dry for a good part of the year, and run-off events do not necessarily activate whole drainage systems. The extent to which river courses are occupied after rainfall events depends on the magnitude, duration and frequency of the events. What is more, river channels on alluvial ground frequently migrate during flow events so that they flow along different courses during successive events. However, large rivers may enter arid lands from outside, such as the Fish River in Namibia and, perhaps more notably, the Nile. The Nile, with its annual floods and sediment increment, has had an influence upon human settlement in Egypt for centuries. Be that as it may, rain often falls as heavy and sometimes violent showers. The result is that the river channels cannot cope with the amount of rainwater, and extensive flooding takes place. These floods develop with remarkable suddenness and either form raging torrents, which tear their way down slopes excavating gullies as they go, or they may assume the form of sheet floods. Dry wadis are filled rapidly with swirling water and therefore are enlarged. However, such floods are short-lived, since the water soon becomes choked with sediment and the consistency of the resultant mudflow eventually reaches a point when further movement is checked. Much water is lost by percolation, and mudflows are also checked where there is an appreciable slackening in gradient. Gully development frequently is much in evidence.

Some of the most notable features produced by stream action in arid and semi-arid regions are found in intermontane basins, that is, where mountains circumscribe basins of inland drainage (Figure 8.5), as for example, in the southwest United States. The mountain catchment area has an appreciable influence upon the quantity and duration of run-off, and so upon the floods and associated erosion that develop. Mechanical weathering also plays a significant role in the mountain zone (Fookes, 1978). Channels in mountain catchments may be largely devoid of debris as it has been swept away by floods; or of bedrock overlain by thin deposits of sediment of mixed origin. They may

Figure 8.5 View across the playa of Death Valley, California. Note the alluvial fans merging into bahadas at the foot of the far mountain range.

contain braided stream courses or debris flow deposits. The types of channel that occur affect the nature of a flood as the character of deposits influence sediment transport, baseflow, recharge and the initiation of debris flows. Short-term changes in stream channels may be brought about by changes in the catchment area such as overgrazing or construction operations, as well as by flooding. Run-off is generally very rapid in arid regions, so the nature of slopes can have a significant affect on flood and sediment problems. For instance, run-off is especially rapid from the bare rock slopes found in the upper reaches of mountain catchments. Accumulations of debris occur further downslope. Fluvial processes are only important on the lower parts of these slopes, where the materials are finer and may have been cemented to form duricrusts. Rills and gullies develop on the lowermost slopes, which are formed of fine-grained sediment or stone pavements. They may also be washed over by sheet flows. Rockfalls and rock slides occur from the upper slopes. After rainfall or snowmelt, debris flows can originate in small mountain catchments that have a dense network of gullies carved into relatively loose sediment. The flows build levees along their channels. They consist mainly of sandy material and tend to spread out on the lower slopes to form lobate deposits. Even on gently sloping fans the run-out can be up to 30 km, but if the fan deposits are highly permeable, then rapid drainage from the flow restricts its movement. In addition, the rapid infiltration of water into highly permeable fan deposits can give rise to undulating irregular topography formed of sieved deposits, that is, the fine sediments are washed out of the coarser sediments. This frequently means that such deposits have a variable load-carrying capacity. Debris flows represent a hazard on the debris slopes, and as well as descending from the adjoining mountains, develop especially on the finer material, which can become saturated by rainfall or snowmelt. Furthermore, ancient landslides, which formed when the climate was wetter, where present, can be reactivated by human interference, notably construction operations.

Alluvial cones or fans, which consist of irregularly assorted sediment, are found along the foot of the mountain belt where it borders the pediment, the marked change in gradient accounting for the decrease in flow velocity and the rapid deposition. The rapid accumulation of debris is attendant upon rapid run-off in the mountain catchment bringing down copious quantities of sediment. The particles composing the cones are almost all angular in shape, boulders and cobbles being more frequent upslope, grading downslope into fine gravels, sands and silts. The cones have a fairly high permeability. A decrease occurs in channel depth downslope, with a tendency for flow to be directed into unstable, shifting channels.

Alluvial fans are commonly made up of distinctive features such as contemporary deposition areas, abandoned surfaces and an array of channels. The contemporary areas of deposition occur at the ends of active channels. Abandoned surfaces may be distinguished by desert varnish or by duricrusts. Small localized stream channels are present and are supplied with water and sediment from the catchment areas of these old surfaces. Consequently, flood risk in these areas is low. Channels include entrenched feeder channels, distributary channels and small dendritic systems, which develop on the fan itself. These have a relatively high flood risk and tend to migrate laterally. Although stream flow on alluvial cones is ephemeral, nevertheless flooding can constitute a serious problem, occurring along the margins of the main channels and in the zone of deposition beyond the ends of supply channels (Figure 8.6). The flood waters are problematic because of their high velocities, their variable sediment content and their tendency to change locations with successive floods, abandoning and creating channels in a relatively short time. Hydrocompaction may occur on alluvial cones, particularly if they are irrigated. The dried surface layer of these cones may contain many voids. Percolating water frequently reduces the strength of this material, which in turn causes collapse of the void space, giving rise to settlement. The streams that descend from the mountains rarely reach the centre of the basin, since they become choked by their own deposits and split into numerous distributaries, which spread a thin veneer of gravels over the pediment. When alluvial cones merge into one another they form a bahada.

Pediments are graded plains, which are formed by the lateral erosion of mainly ephemeral streams. They occur at the foot of mountain slopes and may be covered with a thin veneer of alluvium. Pediments are adjusted to dispose

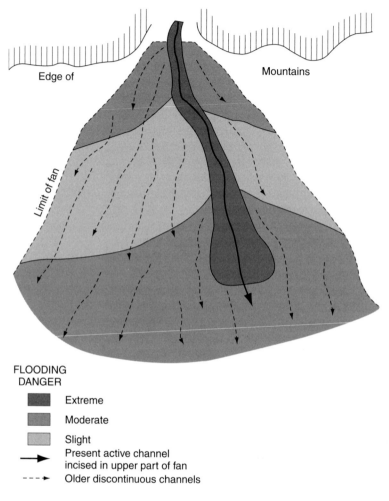

Edge of

Mountains

Limit of fan

FLOODING
DANGER

Extreme

Moderate

Slight

Present active channel
incised in upper part of fan

Older discontinuous channels

Figure 8.6 The pattern of flood hazard on a typical alluvial fan in the western United
States (from Cooke *et al.*, 1982).

of water in the most efficient way and when heavy rain falls, this often means
that it takes the form of sheet wash. Although true laminar flow occurs during sheet wash, as the flowing water deepens laminar flow yields to turbulent
flow. The latter possesses much greater erosive power and occurs during and
immediately after heavy rainfall. This, it is argued, is why these pediments
carry only a thin veneer of rock debris. With a lesser amount of rainfall there
is insufficient water to form sheets and it is confined to rills and gullies.
Distributary channels may develop in the upper parts of pediments. Streams
on pediments occasionally are active and liable to flood, to scour and fill, and

tend to migrate laterally. According to King (1963), the rock waste transported across the pediment is relatively fine and is deposited in hollows, thereby smoothing the slope. The abrupt change in the slope at the top of the pediment is caused by a change in the principal processes of earth sculpture, the nature of the pediment being governed by sheet erosion, while that of the steep hillsides is controlled by the downward movement of rock debris.

The extension of pediments on opposing sides of a mountain mass means that the mountains are slowly reduced until a pediplain is formed. The pediments are first connected through the mountain mass by way of pediment passes. The latter become progressively enlarged, forming pediment gaps. Finally, opposing pediments meet to form a pediplain, on which there are residual hills. Such isolated, steep-sided residual hills have been termed inselbergs or bornhardts. They are characteristically developed in the semi-arid regions of Africa, where they are usually composed of granite or gneiss, that is, of more resistant rock than that which forms the surrounding pediplain. The pediplains represent extensive planation surfaces that have been subjected to long periods of denudation, where tectonic activity has frequently been of little consequence or absent.

The central and lowest area of a basin is referred to as the playa and it sometimes contains a lake. This area may be subjected to seasonal flooding and be covered with deposits of sand, silt, clay and evaporites. The silts and clays often contain crystals of salt, whose development further comminutes their host. Silts usually exhibit ripple marks, while clays are frequently laminated. Desiccation structures such as mud cracks are developed on an extensive scale in these fine-grained sediments. Lister and Secrest (1985) described giant polygonal desiccation cracks in Red Lake playa in Arizona. They also referred to fissures that may have been formed as a result of flowage in an underlying bed of salt. If the playa lake has contracted to leave a highly saline tract, then this area is termed a 'salina'. The capillary rise generally extends to the surface, leading to the formation of a salt crust. Where the capillary rise is near to, but does not normally reach the surface, desiccation ground patterns (some cracks are extremely large) provide an indication of its closeness. Aeolian and fluvial deposits, notably sand, may also be laid down in the intermediate zone between the pediment and the playa. However, if deflation is active this zone may be barren of sediments. Sands are commonly swept into dunes, and the resultant deposits are cross-bedded.

Yet another type of fluvial landscape developed in arid regions is found where there are thick surface formations of more or less horizontally bedded sedimentary rocks. Large rivers coming from outside the region frequently flow in canyons, such as the Colorado River in the southwest of the United States and the Fish River in Namibia. The plateau lands that characterize such areas are interrupted by buttes and mesas, and by escarpments or cuestas. Such landscapes are typified by the Colorado Plateau, and are found in the semi-arid and arid areas of southern Africa.

Duricrusts are surface or near-surface hardened accumulations or encrusting layers, formed by precipitation of salts on evaporation of saline groundwater. When describing duricrusts those terms ending in 'crete' refer to hardened surfaces usually occurring on hard rock; those ending in 'crust' represent softer accumulations, which are usually found in salt playas, salinas or sabkhas. Duricrusts may be composed of calcium or magnesium carbonate, gypsum, silica, aluminium or iron oxide, or even halite, in varying proportions, (Goudie, 1973). However, enrichment of silica or iron to form silcrete or ferricrete, respectively, occurs very occasionally in arid regions. Duricrusts may occur in a variety of forms, ranging from a few millimetres in thickness to over a metre. A leached cavernous, porous or friable zone is frequently found beneath the duricrust.

The most commonly precipitated material is calcium carbonate. These caliche deposits are referred to as 'calcrete' (Braithwaite, 1983; Netterberg, 1980). Calcrete occurs where soil drainage is reduced due to long and frequent periods of deficient precipitation and high evapotranspiration. Nevertheless, the development of calcrete is inhibited beyond a certain aridity, since the low precipitation is unable to dissolve and drain calcium carbonate towards the water table. Consequently, in arid climates gypcrete may take the place of calcrete. Climatic fluctuations that, for example, took place in North Africa during Pleistocene times, therefore led to alternating calcification and gypsification of soils. Certain calcretes were partially gypsified and elsewhere gypsum formations were covered with calcrete hardpans (Horta, 1980). Similarly, the cemented sands, known as gatch, in Kuwait also owe their origin to past conditions. These occur at depth and bear little relationship to present-day groundwater levels, their cementation having occurred sometime in the past. They are probably being continually modified by dissolution and cementation (Al Sanad *et al.*, 1990). Cementation provides a high proportion of the shear strength of many of these soils. However, only small amounts (around 2%) of such salts as calcite and gypsum are required to give this enhanced strength. In the case of cementation with halite, which is frequently on a seasonal basis, the strength of soils close to the surface may be increased by three to four times. Nonetheless, soaking and prolonged throughflow of fresh water can lead to a loss of soil strength. This can be brought about by over-irrigation or leaking drains.

The hardened calcrete crust may contain nodules of limestone or be more or less completely cemented (this cement may, of course, have been subjected to differential leaching). In the initial stages of formation, calcrete contains less than 40% calcium carbonate, which is distributed throughout the soil in a discontinuous manner. At around 40% carbonate content, the original colour of the soil is masked by a transition to a whitish colour. As the carbonate content increases it first occurs as scattered concentrations of flaky habit, then as hard concretions. Once it exceeds 60%, the concentration becomes continuous. The calcium carbonate in calcrete profiles decreases from top to base, as generally does the hardness.

Gypcrete is developed in arid zones, that is, where there is less than 100 mm precipitation annually. In the Sahara, aeolian sands and gravels are often encrusted with gypsum deposited from selenitic groundwaters. A gypcrete profile may contain three horizons. The upper horizon is rich in gypsified roots and has a banded and/or nodular structure. Beneath this occur massive gypcrete–gypsum cemented sands. Massive gypcrete forms above the water table during evaporation from the capillary fringe (newly formed gypcrete is hard, but it softens with age). At the water table gypsum develops as aggregates of crystals; this is the sand rose horizon.

8.5 Flooding and sediment problems

Desert rainfall is erratic both in time and spatially. In any one year, the annual amount of rainfall may be four times the average or more and may occur in a single storm. Conversely, in other years very little rain may fall. A high proportion of rainfall in arid regions is lost by evaporation or infiltrates into the ground. In fact, with rare exceptions such as the Fish, Nile, Niger and Colorado, drainage does not escape from arid regions. Streams fade away as a result of evaporation or infiltration into the ground. Those that persist have sufficient volume to do so because their headwaters are located in regions outside deserts. Furthermore, the boundaries of arid and semi-arid regions migrate periodically with changing distribution of rainfall. In wet periods semi-arid regions expand, to contract back in dry periods. The proportion of rain that remains for run-off in extremely arid regions may be as little as 10%. In fact, not all low-intensity storms give rise to run-off, so intensity becomes a more important factor in this regard than total rainfall. Intensities of probably at least 1 mm min^{-1} over a minimum of 10 min may be needed to generate channelled run-off. On the mountain or fan catchments, run-off may either occur as sheet flow or be concentrated into rills and gullies. Overland flow may be concentrated quickly to give rise to floods where rills and gullies occur. Such rapid flow from hillslopes governs the major characteristics of flow generated in ephemeral stream channels. Ephemeral run-off in small channels rapidly attains peak discharge, and wave-like rapid advance of a flood is typical. Hence, rapid flash floods of relatively small size are associated with small to medium-sized streams, and these may have catastrophic consequences.

Run-off is rapid from mountain catchments because of the steep slopes, which frequently are bare or covered with the deposits of sediment and lacking vegetation. However, extensive deep deposits of coarse-grained alluvium may occur in many ephemeral stream channels, and such material possesses a high infiltration capacity and a potential for significant storage of water. The amount of water that infiltrates such alluvium depends on a number of factors, which include the permeability, moisture content and position of the water table within the alluvium on the one hand and the size of the channel, flow duration and peak discharge on the other. At times, all the water involved in run-off may eventually infiltrate into the bed of the stream system. Most infiltration occurs in the early stages of a flood, and the distance downstream achieved by floods is obviously very much influenced by the size of the rainfall event and the amount of run-off generated. Because of high infiltration in the upper parts of stream systems, groundwater recharge downstream is affected adversely, which in turn may mean that less water is available for any irrigation schemes.

Movement of sediment on slopes is caused primarily by run-off, supplemented by raindrop impact and splash erosion. High run-off associated with torrential rainfall can flush sediment from gullies. Small fans may develop at the exits from gullies, spreading into stream channels. This, in turn, may lead to the formation of gravel sheets, lobes and bars on the one hand and/or bank erosion on the other. Major flows can transport massive quantities of sediment, which when deposited raises valley floor levels. Channels are frequently flanked by terraces, which provide evidence of previous flood events. Sediment load increases downstream as slopes and tributaries contribute more load, and as water is lost by infiltration. Hence, the lower reaches of streams are dominated by deposition. The deposits exhibit some lateral grading in that the larger material, that is, the bedload, is deposited first, the suspended sediment being carried further downstream. Individual flood deposits also tend to show an upward fining sequence. In fact, the surface of channel deposits may consist of a thin layer of sandy or silty material, which can affect infiltration. An important factor to be borne in mind is that the annual sediment yield in such regions is highly variable.

Piping refers to the development of subsurface channels, sometimes up to several metres in diameter. Pipes are typically associated with dispersive soils in semi-arid regions (see Chapter 5). They can develop very quickly and can discharge large quantities of water and sediment through a slope. Once a pipe has developed it provides a new base for the local hydraulic gradient. Water flow is then concentrated towards the pipe and the rate of piping accelerates. Eventually, the roof of a pipe becomes unstable and it collapses, leading to the formation of a gully. Undermining and collapse associated with piping can prove a subsidence hazard and has threatened and damaged roads, bridges, railways, canals and other structures. Piping has led to the failure of dams constructed with dispersive soils in semi-arid regions (Bell and Maud, 1994).

Desert floods, although not common, are of inevitable occurrence and can take place rapidly after rainfall. Avoidance of flood-prone areas is much more important in arid areas than elsewhere because of the unpredictability of flash floods. Although flood warning systems that have been developed for one purpose or another do exist in certain arid regions, particularly in the United States, it cannot be expected that they could be established in most arid regions. Indeed, in many arid regions the lack of data militates against the setting up of flood warning systems. However, in certain areas such as Suez, Cooke *et al.* (1982) were able to produce a hazard map of flooding indicating areas of low, medium and high risk (see Figure 6.12). Maps such as this can be used for planning purposes, including land zonation, which in this case involved flood control. Any development that occurs should be restricted to those areas of low risk. Furthermore, conventional flood control engineering solutions are not always appropriate in arid regions, and design should be more in keeping with the landscape. For instance, disturbance should be kept to a minimum. As an illustration, Cooke *et al.* (*ibid.*) suggested that on low-cost roads bridgeless crossings of stream channels are more effective, and less expensive, than the construction of bridges and associated structures. In such an instance, the road should be slightly below channel grade so that during flooding it will be covered by a protective veneer of sediment. This can be removed after the flood has abated.

Although run-off events are infrequent in arid regions, when they occur they, and the sediment they carry, can cause extensive damage in areas that have been urbanized. Furthermore, urbanization is increasing in these regions, so the problem is being accentuated. Unfortunately, scant data are usually available on extreme rainfall events and associated flood frequency in arid regions. This obviously means that it is difficult to assess the degree of flood risk and probable damage from such rainfall events. Hence, it is difficult to institute measures to mitigate flooding and sediment movement.

8.6 Movement of dust and sand

As noted previously, dust and sand movement by wind are different. Nonetheless, dust and sand movement are governed by the same factors, namely, the velocity and turbulence of the wind, the particle size distribution of the grains involved, and the roughness and cohesion of the surface over which the wind blows. In other words, movement is concerned with the erosivity of the wind on the one hand and with the erodibility of the surface on the other. As far as the erosivity of the wind is concerned, the most important factors are its mean velocity, its direction and frequency, the period and intensity of gusting, and the vertical turbulence exchange. The two main factors governing erodibility are the nature of the surface and its vegetation cover. As dust particles are bound, to varying degrees, by cohesive forces, high-velocity winds may be needed to entrain particles, although saltating sand particles can entrain dust as a result of impact. Once entrained, dust can be carried aloft by fairly gentle winds. Dust storms are common in arid and semi-arid regions, and dust from deserts is often transported hundreds of kilometres. These storms at times cover huge areas; for example, the dust storm that affected North America on 12 November 1933 covered an area greater than France, Italy and Hungary combined (Goudie, 1978). As mentioned above, sand moves principally by saltation and as such is concentrated near the ground. The problems of sand and dust movement are most acute in areas of active sand dunes or where dunes have been destabilized by human interference disturbing the surface cover and vegetation of fixed dunes (Jones *et al.*, 1986).

The rate of movement of sand dunes depends on the velocity and persistence of the winds, the constancy of wind direction, the roughness of the surface over which they move, the presence and density of vegetation, the size of the dunes and the particle size distribution of the sand grains. Large dunes may move up to 6 m per year and small ones may exceed 23 m per year. For instance, Watson (1985) recorded that the barkhans, up to 25 m in height, that occur in the Eastern Province of Saudi Arabia have an average rate of movement of 15 m per annum and that drift rates reach 30 $m^3 m^{-1}$ width annually. In fact, the largest amounts of material in motion occur in the sand deserts, because the grains tend to be loose and not aggregated together. Some of the major types of soil surface in arid areas are given in Table 8.1.

Deflation leads to the removal of finer material, that is, silt and clay, together with organic matter from soil, leaving behind coarser particles, which are less capable of retaining moisture. Once lost, soil in arid and semi-arid regions does not re-form quickly. Deflation may lead to scour and undermining of railway lines, roads and structures, which in turn can lead to their collapse. However, deflation can give rise to a wind-stable surface such as a stone pavement or surface crust. But human interference can initiate deflation on a stable surface by causing its destruction, for example, by the construction of unpaved roads and other forms of urbanization.

Sand and dust storms can reduce visibility, bringing traffic and airports to a halt. During severe sandstorms the visibility can be reduced to less than 10 m. In addition, dust storms may cause respiratory problems and even lead to suffocation of animals, disrupt communications, and spread disease by transport of pathogens. The abrasive effect of moving sand is most notable near the ground (i.e. up to a height of 250 mm). However, over hard man-made surfaces, the abrasion height may be higher because the velocity of sand movement and saltation increase. Structures may be pitted, fluted or grooved depending on their orientation in relation to the prevailing wind direction.

Movement of sand in the form of dunes can bury obstacles in its path such as roads and railways or accumulate against large structures. Such moving sand necessitates continuous and often costly maintenance activities. Indeed, areas may be abandoned as a result of sand encroachment (Figure 8.7). Stipho (1992) pointed out that only a few centimetres on a road surface can constitute a major driving hazard. Deep burial of pipelines by sand makes their inspection and maintenance difficult, and unsupported pipes on active dunes can be left high above the ground as dunes move on, causing the pipes to move and possibly fracture (Cooke et al., 1982).

The deposition of dust may bury young plants, contaminate food and drinking water, clog equipment so that it has to be maintained more frequently and make roads impassable. Jones et al. (1986) pointed out that the mobility of dust, along with the variability of the stability of the ground surface near towns in deserts due to human impact, means that it is especially difficult to predict

Table 8.1 Soil-surface types and their relative erodibilities (after Petrov, 1976).

Soil-surface type	Arabic name	Local names	Amount of sand and/or dust movement
Sand deserts	Erg	tamakhak (Arabic) kum, barchan (Turkish) edeien (Berber) sha-mo (Chinese) elisun (Mongolian)	Large amounts of sand and dust in motion when unvegetated. Fixed sand sheets and dunes become unstable when vegetation cover is degraded
Sand–pebble deserts	Serirs	ergs, azrirs (Arabic) gobis (Mongolian) regs (Persian)	Quantity of sand and dust in motion depends on the development of the pebble pavement. Intermediate amounts moved when undisturbed. Large quantities moved when pebbles and/or vegetation removed from surface
Pebble deserts	Regs	hamada (Arabic)	Little or no sand and dust movement in undisturbed state; intermediate amounts when pavement disturbed or removed
Gravelly hamada			Intermediate between pebble desert and rocky hamada
Rocky hamada			Little or no movement of sand and dust
Solonchak	Sabkha	sabkha, chott (Arabic) sor, shor (Turkish) kevir, kebir (Persian)	Dust movement increases as surface becomes desiccated when water table drops. Breakdown of crusts results in large quantities of dust being suspended. Floodouts and playas are the most important source of dust

Soil-surface type descriptions:

Sand deserts: Sand sheets and dune fields developed on thick loose arenaceous sediments

Sand–pebble deserts: Poorly developed pebble pavement with variable quantities of sand filling the space between pebbles

Pebble deserts: Well-developed multi-layered pebble pavement. Weakly cemented, poorly sorted gravel. Plant cover very sparse

Rocky hamada: Weak removal of weathering products. In lower horizons the weathering products are coarsely fragmented, while upper horizons contain poorly rounded gravel. Virtually devoid of fine material

Solonchak: Saline depression with water table close to surface. Sometimes saline crusts on surface

Figure 8.7 Abandoned house partially filled with sand, Kolmanskop, Namibia.

the magnitude, frequency or extent of movements or to control the creation of new sources of supply.

Aerial photographs and remote sensing imagery of the same area and taken at successive time intervals can be used to study dune movement, rate of land degradation and erodibility of surfaces (Jones *et al.*, 1986). The recognition of these various features of arid landscapes, by the use of aerial photographs, should be checked in the field. Field mapping involves the identification of erosional and depositional evidence of sediment movement (e.g. sand drifts, dunes, surface grooving, ventifacts and yardangs), which can provide useful information on the direction of sand movement but less on dust transport. For instance, indirect evidence of the wind regime can be derived from trends of active dunes. However, care needs to be taken when assessing the data obtained from the field in that the features identified should represent forms that are developing under present-day conditions. Fixed dunes that have been developed in the past are not necessarily related to the contemporary wind regime. As their destabilization by human activity should be avoided, it is important that they should be identified and monitored. The methods available for monitoring of sand have been reviewed by Cooke *et al.* (1982). They include surveys, sequential photography, tracing with fluorescent sand grains and the use of trenches and collectors to monitor movement. In the latter case, the measurement of sand flow is not easy, since the sample collector interferes with the airflow. The variability of dust movement makes its monitoring difficult, but settling jars and sediment collectors have been used.

Meteorological investigations revolve around recording wind speed and direction, and the visibility conditions associated with sand and dust movements, together with analysis of patterns of aeolian transport relating to prevailing synoptic conditions. However, analysis of meteorological data provides only an indication of the wind potential for sand and dust movement. The actual patterns of sand and dust drift also reflect availability of mobile material, near-ground wind speed and turbulence characteristics, and topographical conditions. The measurement of near-surface wind velocities can be accomplished with an anemometer or vane. Wind tunnels can be used to examine velocity profiles and turbulence of wind. The aerodynamic effects of different building configurations can also be studied in a wind tunnel so that areas of high velocity and shelter can be identified. In the former areas flying sand and dust leads to abrasion, as well as causing discomfort to individuals, and deposition usually occurs in the sheltered areas. The use of predictive equations, notably the Wind Erosion Equation, to assess the removal of particles by wind is referred to in Chapter 9.

Of particular importance in an assessment of the sand and dust hazard is the recognition of sources and stores of sediment. Jones *et al.* (1986) defined sources as sediments that contributed to the contemporary aeolian transport system and included weathered rock outcrops, fans, dunes, playa/sabkha deposits and alluvium. Stores are unconsolidated or very weakly consolidated surface deposits that have been moved by contemporary aeolian processes and have accumulated temporarily in surface spreads, dunes, sabkha and stone pavements.

Erodibility can be assessed in the field by the use of experimental plots or by soil classification. The environmental conditions can be controlled to a large extent when using an experimental plot. Soil classification according to erodibility takes into consideration the depth and structure of the surface horizon, as well as the depth of non-erodible subsurface material.

Examples of hazard maps of geomorphology, dune morphology and migrating dunes have been provided by Doornkamp *et al.* (1979). For example, they produced a number of such maps for two proposed sites for a new airport in Dubai (Figure 8.8). Field survey revealed that dune characteristics, notably their form and vegetation density, could be related to their mobility, and these characteristics could also be recognized on aerial photographs. High angular fresh dunes lacking vegetation were regarded as potentially or actively mobile. By contrast, broad, rounded, low, well-vegetated dunes were regarded as posing a low hazard risk. It was suggested that the small unstable dunes could be removed, while the larger dunes required extensive stabilization. The low-hazard dunes required moderate control measures to check their migration. Jones *et al.* (1986) assessed the relative importance of geomorphological units as potential sources of aeolian sand and dust, recognizing high, medium and low classes of potential (Figure 8.9).

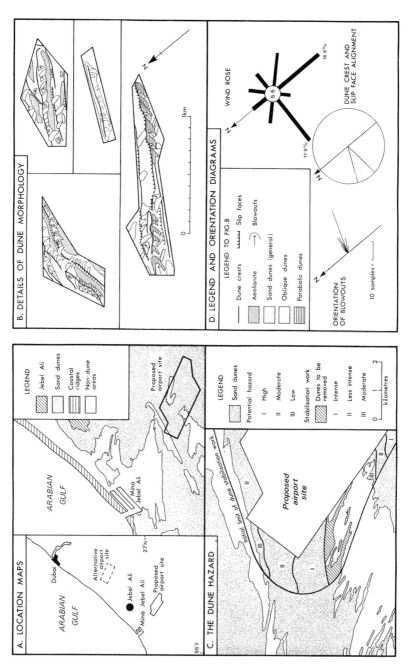

Figure 8.8 Geomorphological analysis of a proposed airport site in Dubai with respect to the threat from mobile sand dunes (after Doornkamp et al., 1979).

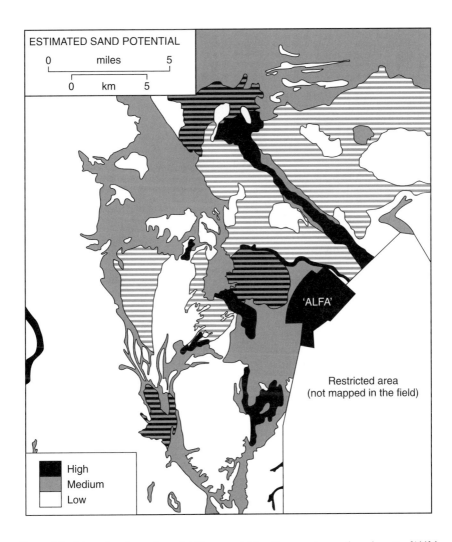

ESTIMATED SAND POTENTIAL

0   miles   5

0   km   5

'ALFA'

Restricted area
(not mapped in the field)

High
Medium
Low

Figure 8.9a Map of estimated sand drift potential for the area to north and west of 'Alfa'.

As Cooke *et al.* (1982) very sensibly pointed out, the best way to deal with a hazard is to avoid it, this being both more effective and cheaper than resorting to control measures. This is particularly relevant in the case of aeolian problems, notably those associated with large, active dune fields, where even extensive control measures are likely to prove only temporary. Hence, sensible site selection, based on thorough site investigation, is necessary prior to any development.

It is extremely unlikely that all moving sand can be removed. This can happen only where the quantities of sand involved are small and so can apply only

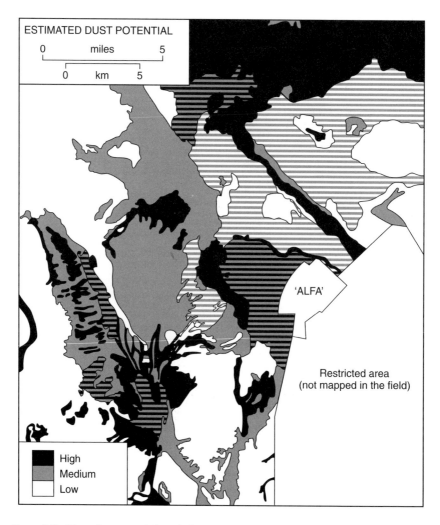

ESTIMATED DUST POTENTIAL

'ALFA'

Restricted area
(not mapped in the field)

High
Medium
Low

Figure 8.9b Map of estimated dust drift potential for the area to north and west of 'Alfa' (after Jones *et al.*, 1986).

to small dunes (a dune 6 m in height may incorporate 20 000–26 000 m^3 of sand with a mass of between 30 000 and 45 000 tonnes). Even so, this is expensive, and the excavated material has to be disposed of. Often the removal of dunes is practical only when the sand can be used as fill or ballast; however, the difficulty of compacting aeolian sand frequently precludes its use (Watson, 1985). In addition, flattening dunes does not represent a solution, since the wind will develop a new system of dunes on the flattened surface within a short period of time.

Accordingly, a means of stabilizing mobile sand must be employed (Table 8.2). One of the best ways to bring this about is by the establishment of a vegetative cover. Plants used for the stabilization of dunes must be able to exist on moving sand and either survive temporary burial or keep pace with deposition. The plants normally used have coarse, stiff stems to resist sand-blasting and are unpalatable to livestock, as well as being fast-growing. Cooke *et al.* (1982) noted that if successful stabilization is going to be achieved, then the character of the substrate, the thickness of the sand deposit, the quantity and quality of the water in the sand and the substrate, the position of the water table, and the rate of sand movement should be investigated. Obviously, vegetation is more difficult to establish in arid than semi-arid regions. In arid regions, young plants have to be protected from burial by windblown sand or from being blown out of the soil. Generally, the more degraded the vegetation is, the longer it takes for natural recovery, recovery being confined to areas where groundwater occurs at shallow depth and the threat of deflation is small. When a planting programme is undertaken, it is usually necessary to apply fertilizers liberally. Mulches can be used to check erosion and to provide organic material. Brush matting has been used to check sand movement temporarily, especially in deflation hollows. The brush is laid in rows so that succeeding rows overlap. Trees may be planted in the brush mat. However, brush mats are generally impractical where wind velocities exceed 65 km h^{-1}.

Natural geotextiles can be placed over sand surfaces after seeding to protect the seeds, to help to retain moisture and to provide organic matter eventually. These geotextiles can contain seeds within them (see Chapter 9).

Gravel or coarse aggregate can be placed over a sand surface to prevent its deflation. A minimum particle diameter of around 20 mm is needed for the gravels to remain unaffected by strong winds. In addition, the gravel layer should be at least 50 mm in thickness, since if it is disrupted the underlying sand can be subjected to scouring action by the wind. The gravel material should not be susceptible to abrasion.

Artificial stabilization, which provides a protective coating or bonds grains together, may be necessary on loose sand. In such cases, rapid-curing cutback asphalt and rapid-setting asphalt emulsion may be used. Asphalts and oils are prone to oxidation over a period of one or two years, and once this has occurred the treatment rapidly becomes ineffective. Cutback asphalts are less prone to oxidation and their tendency to become brittle is reduced by additives such as liquid latex. Nonetheless, the performance of cutbacks is limited by their low degree of penetration. Chemical sprays have been used to stabilize loose sand surfaces, and the thin protective layer that develops at the surface helps to reduce water loss. Many chemical stabilizers are only temporary, breaking down in a year or so, and therefore they tend to be used together with other methods of stabilization, such as the use of vegetation. Seeds are sometimes sprayed together with the chemical spray and fertilizers in a type of hydro-seeding. For example, wood cellulose fibres can be used to make a suspension with water,

Table 8.2 Objectives and methods of aeolian sand control (after Kerr and Nigra, 1952).

Objectives

1 The destruction or stabilization of sand accumulations in order to prevent their further migration and encroachment.
2 The diversion of windblown sand around features requiring protection.
3 The direct and permanent stoppage or impounding of sand before the location or object to be protected.
4 The rendition of deliberate aid to sand movement in order to avoid deposition over a specific location, especially by augmenting the saltation coefficient through surface smoothing and obstacle removal.

Methods
The above objectives are achieved by the use of one or more types of surface modification:

1 Transposing. Removal of material (using anything from shovels to bucket cranes) – rarely economical or successful, and does not normally feature in long-term plans.
2 Trenching. Cutting of transverse or longitudinal trenches across dunes destroys their symmetry and may lead to dune destruction. Excavation of pits in the lee of sand mounds or on the windward side of features to be protected will provide temporary loci for accumulation.
3 Planting of appropriate vegetation is designed to stop or reduce sand movement, bind surface sand, and provide surface protection. Early stages of control may require planting of sand-stilling plants (e.g. *Ammophila arenaria*, beach grass), protection of surface (e.g. mulching), seeding, and systematic creation of surface organic matter. Planting is permanent and attractive, but expensive to install and maintain.
4 Paving is designed to increase the saltation coefficient of wind-transported material by smoothing or hard-surfacing a relatively level area, thus promoting sand migration and preventing its accumulation at undesirable sites. Often used to leeward of fencing, where wind is unladen of sediment, and paving prevents its recharge. Paving may be with concrete, asphalt or wind-stable aggregates (e.g. crushed rock).
5 Panelling, in which solid barriers are erected to the windward of areas to be protected, is designed either to stop or to deflect sand movement (depending largely on the angle of the barrier to wind direction). In general, this method is inadequate, unsatisfactory and expensive, although it may be suitable for short-term emergency action.
6 Fencing. The use of relatively porous barriers to stop or divert sand movement, or destroy or stabilize dunes. Cheap, portable and expendable structures are desirable (using, for example, palm fronds or chicken wire).
7 Oiling involves the covering of aeolian material with a suitable oil product (e.g. high-gravity oil), which stabilizes the treated surface and may destroy dune forms. It is, in many deserts, a quick, cheap and effective method.

seed and fertilizer, together with an asphalt or resin emulsion, and sprayed on to sand. Resin-in-water emulsion, latex-in-water emulsion and gelatinized starch solutions have been used as stabilizers. Large aggregates are formed when emulsions are mixed with sand. This helps to increase the rate of infiltration, thereby decreasing run-off and enhancing the moisture content in the soil. Large aggregates are also more resistant to wind erosion. Polyacrylamide solutions when applied to soil also give rise to high rates of infiltration, irrespective of whether large or small aggregates are formed (Gabriels et al., 1979).

A dune will not migrate if its windward side is stabilized. Polyurea polymers have been used to stabilize dune sands successfully. These polymers contain a mixture of hydrophilic ethylene oxide and hydrophobic propylene oxide (hydrophilic conditioners, if used in the right concentrations, make the surface less permeable). These can be injected to appreciable depth and may extend the life expectancy of the treatment to more than five years. The amounts of each component used in the solution depends on whether hydrophilicity or hydrophobicity is required. When hydrophobic conditioners such as asphalt or latex emulsions are applied as surface treatments in concentrated form (i.e. not highly diluted), they form an impermeable surface on the sand (Gabriels et al., 1979). Nonetheless, sand will be blown over the stabilized surface. Barkhans, for example, become oval-shaped deposits of sand that are several times larger than the original dune. The technique can only be used when the migrating dune is three times its width from the structure requiring protection. Otherwise, an impounding fence is positioned on the windward side of the dune to reduce sand supply to the leeward side, the windward side again having been stabilized.

Gravels, cobbles and crushed rock provide a stable cover for dunes and can be used where environmental considerations preclude the use of oil or chemicals. However, spreading such material over unstable sand may present a problem.

Windbreaks are frequently used to control wind erosion and obviously are best developed where groundwater is near enough the surface for trees and shrubs to have access to it. It is essential to select species of trees with growth rates that exceed the rate of sand accumulation and that have a bushy shape. *Tamarix* and *Eucalyptus* species have proved most successful (Watson, 1985). Windbreaks can act as effective dust and sand traps. Their shape, width, height and permeability influence their trap efficiency and the amount of turbulence generated as the wind blows over them. For example, low permeability enhances trap efficiency but gives rise to eddying on the leeward side of a windbreak.

Fences can be used to impound or divert moving sand. Most sand accumulation occurs on the lee side of the fence, chiefly in a zone four times as wide as the fence is high. They can be constructed of various materials from brushwood or palm fronds, wood or metal panels, stone walls or earthworks. The surfaces around fences should be stabilized in order to avoid erosion and undermining, with consequent collapse. Individual fence designs have different sand-trapping abilities at different wind velocities, so fence design must be tailored to local environmental conditions. The alignment of fences relative to the direction of sand drift, and the number and spacing of rows of fences, is critical (Watson, 1985). The optimum spacing of rows of fences is about four times the height of the fences, so that 1 m high fences have a maximum capacity to trap approximately $16 \text{ m}^3 \text{ m}^{-1}$ width, assuming that there are four rows of fences 4 m apart. Multiple rows of fencing can trap more than 80% of wind-borne sand (Figure 8.10), even under variable wind conditions, if optimal alignment and porosity for the specific area are employed. The practical number of rows of fences is about four. If more are installed, then the increase in trap efficiency is negligible.

Fences must be located in areas where the development of a large artificial dune will not pose problems. Obviously, a fence must be able to resist the strength of the wind. Estimates of average wind velocity, size of sand particles, and type and distance of the protected object are required for design purposes, as are the fence type, arrangement of spaces, height, alignment and whether single or multiple rows should be used.

Like a windbreak, the most effective fences are also semi-permeable. Although such fences reduce wind velocity to a lesser extent than impermeable fences, they restrict diffusion and eddying effects and their influence extends further downwind. Manohar and Bruun (1970) found that the amount of sand trapped by a row of 40 mm wide slot-type fence varied from 60% when the wind speed was 10 m s^{-1} to 10% when it was 18 m s^{-1}. In areas that experience a wide range of wind speeds, the optimum fence porosity is around 40%. If the wind speed exceeds 18 m s^{-1}, then no sand is trapped by a single fence with a porosity of 40%. Under such conditions, a double row will trap about 30% of the sand being moved.

Sand accumulates against an impounding fence that is built normal to the prevailing wind direction until it reaches the top of the fence, when a further

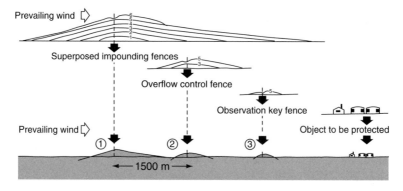

Figure 8.10 Three-fence system to protect extensive areas, such as shops, villages, yards, industrial plants, etc. (after Kerr and Nigra, 1952).

fence needs to be installed or the fence heightened (see Figure 8.10). The initial height and times at which the heights have to be increased are governed by the rate of sand flow. The next fence installed on top of the sand mound will he a greater sand-trapping capacity, since the volume impounded varies as the square of the height of the fence. Hence, the use of fences over a long period of time represents a commitment to a sand control policy based on dune building. Alternatively, a number of fences can be positioned at an acute angle (usually around $45°$) to the prevailing wind, or a V-shaped barrier pointing into the sand stream can be used. Multiple rows are more effective than single rows, especially when the wind speed exceeds 18 m s^{-1}. Such fences not only trap sand but they also deflect the wind, thereby removing some of the accumulated sand. These fences can be relatively low, since most sand moves near the ground.

Fences erected on the windward side of a dune bring about local deposition of sand. However, when the shelter effect of the fence declines, the wind will begin to erode more sand from the windward side of the dune than if the fence had not led to deposition on its leeward side. This causes the dune to migrate more quickly, so that it diminishes in size. Indeed, the dune may ultimately disappear if the area between the fence and the dune is stabilised to prevent the removal of sand.

Panels can be used to direct airflow over a road or around a building. They are only used where the rate of deposition of sand is low. However, although panels can be arranged to increase wind velocity and so keep surfaces sand-free, when the wind leaves the panelling it decreases in speed, so some deposition of sand occurs.

Stipho (1992) mentioned the use of trenches and pits to reduce the amount of blown sand. These must be wider than the horizontal leap of saltating particles (up to 3–4 m) and sufficiently deep to prevent scouring. As regular removal of accumulated sand is necessary, pits and trenches require continual removal of this sand and so are expensive to maintain. In effect, trenches are useful only for short-term protection. Stipho also referred to the use of sand traps to collect sand to prevent it accumulating on highways. He suggested that the sand could be removed subsequently. Again in relation to highways, Stipho suggested that in some situations it may be necessary to elevate a carriageway well above the surrounding dunes in order to allow the wind to accelerate over the road, while reducing cutting in the sand to a minimum.

The deposition of sand can be reduced either by removing areas of low wind energy or by increasing the velocity of the wind over the surface by shaping the land surface (Watson, 1985). For example, flat slopes in the range of 1:5 to 1:6 with rounded shoulders are necessary for small to medium embankments, such slopes helping to streamline airflow. Cuttings require flatter slopes of perhaps 1:10 to allow for free sand transport and are usually accompanied by means to collect blown sand such as a wide ditch at the base of a slope. Bridges should be constructed as open structures with the minimum size of columns so as to maximize the free flow of the wind. The upwind slope of road and rail cuttings must be graded at a shallower angle than the windward slopes if sand accumulation is to be avoided. The surface can be treated in order to enhance saltation and traction of sand grains across it. A variety of materials have been used in this respect, including asphalt, synthetic latex, polyvinyl polymers, sodium silicate and gelatine. Most have short effective life spans, generally around 1 to 5 years. Curing resins, although more expensive than non-curing oil-based substances, have a life expectancy of 20–30 years. They are also colourless, as opposed to asphalts, which give a blanketed black surface and form hard crusts.

8.7 Sabkha soil conditions

Sabkhas are fairly level salt-encrusted surfaces and are common in arid regions (Table 8.3). They occur in both coastal (Figure 8.11) and inland areas, but the mode of development and the properties of the sabkhas in these areas are different. Inland sabkhas have also been referred to as playas and salinas. Coastal sabkhas may consist of cemented and uncemented layers of reworked aeolian sand and muddy sand of varying thickness. These sabkhas are highly variable in both horizontal and vertical extent. The horizontal variation may be related to the changing position of the shoreline, while the vertical variation is connected with the depositional sequence and subsequent diagenesis. The groundwater is saline, containing calcium, sodium, chloride and sulphate ions. Evaporative pumping, whereby brine moves upwards from the water table under capillary action, appears to be the most effective mechanism for the concentration of brine in groundwater and precipitation of minerals in sabkhas. The aragonite and calcite in a sabkha, according to Akili and Torrance (1981), originate primarily during the subaqueous state, when the lagoon waters are being concentrated by evaporation. However, some diagenetic aragonite forms during the early stages of brine concentration in the sabkha and may give rise to a thin, lithified crust. All the same, gypsum is the commonest of the

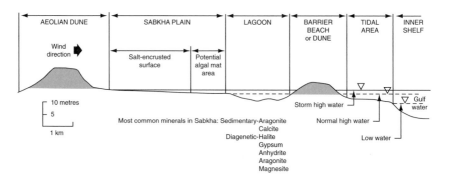

Figure 8.11 Generalized cross-section across a typical coastal sabkha with typical surface features (after Akili and Torrance, 1981).

Table 8.3 Ground conditions associated with sabkhas, playas and salinas (after Fookes and Collis, 1975).

Name	Terrain	Water table	Salts	Special significance	Construction notes
Sabkha	Coastal flat, inundated by sea water during exceptional floods.	Very near the surface.	Thick surface salt crusts from evaporating brines. Salts usually include carbonates, sulphates, chlorides and others.	Generally aggressive to all types of foundation by salt weathering of stone and concrete and/or sulphate attack on cement-bound materials. Evaluate bearing capability carefully.	Carefully investigate; consider tanking concrete foundations; use dense concrete. For surfaced roads consider using inert aggregate, raised embankment or positive cut-off below sub-bases. Use as fill may be suspect. May not be deleterious to unsurfaced roads.
Playa	Inland, shallow, centrally draining basin – of any size.	Too deep for the capillary moisture zone to reach the ground surface, but area will be a temporary lake during floods.	None if temporary lake is of salt-free water.	Non-special. Ground surface may be silt/clay or covered by windblown sands. Evaluate bearing capability.	Non-special.
Salt playa	As playa, but sometimes smaller.	As above, but lake of salty water.	Surface salt deposits from evaporating temporary salty lake water. Salts usually include chlorides and sometimes sulphates and carbonates.	Can be slightly to moderately aggressive to all types of foundation by salt weathering and sulphate attack. More severe near water table.	As sabkha.
Salina	Inland basin of any size.	Near surface; capillary moisture zone from salty groundwater can reach the surface.	Surface crusts from evaporating salty groundwater. Salts include sulphates and others.	Can be slightly to exceptionally aggressive to all types of foundation by salt weathering and sulphate attack.	As sabkha.

diagenetic minerals formed during this stage. Anhydrite is the dominant sulphate mineral above the standing level of the brines in the sabkha. Halite forms as a crust at or near the surface or sometimes as massive layers. However, it is removed by rain, floods, inundating high tides and even strong winds. In fact, wind may move surface salt crystals to form drifts or even dunes. In those sabkhas that are only intermittently invaded by the sea, halite forms a surface crust that may be in the shape of polygonal ridges. The salt may be deposited for some decimetres below the surface, decreasing downwards in amount (Fookes *et al.*, 1985). Dolomite is another diagenetic mineral, and where the concentration of magnesium in the brine is very high, magnesite may make its appearance. The formation of these diagenetic minerals and their positions within a sabkha depend mainly on its hydrogeology and the way in which groundwater at a given location is replenished, from fresh water from flash floods or infrequent rains, or from marine inundation.

One of the main problems with sabkhas is the decrease in strength that occurs, particularly in the uppermost layers, after rainfall, flash floods or marine inundation due to the dissolution of soluble salts, which act as cementing materials. This decrease in strength can render normally stable surface crusts impassable to traffic. Hence when wetted, sabkhas may not be capable of supporting heavy equipment. There is a possibility of differential settlement occurring on loading due to the different compressibility characteristics resulting from differential cementation of sediments. Excessive settlement can also occur due to the removal of soluble salts by flowing groundwater. This can cause severe disruption to structures within months or a few years. Movement of groundwater can also lead to the dissolution of minerals to the extent that small caverns, channels and surface holes can be formed. In addition, gypsum can undergo dehydration in sabkha environments, leading to volume changes in the soil. Heave resulting from the precipitation and growth of crystals, according to Bathurst (1971), can elevate the surface of a sabkha in places by as much as 1 m.

Continental sabkhas, playas and salinas are saline deposits found in inland areas with no hydrological connection that allows groundwater replenishment by the sea. Akili and Torrance (1981) indicated that in Arabia they tend to occur in areas that are dominated by dune sand. Groundwater brines near the surface are concentrated by evaporation, from which minerals such as gypsum and anhydrite are precipitated. Gypsum tends to occur below the water table, whereas anhydrite tends to occur above. Some of these sabkhas, however, contain little carbonate. Salts are precipitated at the ground surface when the capillary fringe extends from the water table to the surface. Although halite frequently forms at the ground surface, it is commonly dispersed by wind. The sediments may consist of layers of fine- to medium-grained quartz sand alternating with muddy sand, in which the grains are partially cemented by evaporating minerals. Salt-bearing soils tend to be strongly hygroscopic, depending on the type of salt.

Coarse-grained soils formed in arid regions differ from their counterparts developed in temperate and humid climates. Such soils are frequently characterized by easy transportation by wind and lack of vegetation, and where the water table occurs at depth are often unsaturated and highly cemented. On the other hand, where the water table is at shallow depth the soils may possess a salty crust and are chemically aggressive due to the precipitation of salts from saline groundwater. Windblown silt- to clay-sized particles of carbonate generally occur between coarser grains. Fookes and Gahir (1995) indicated that these coarse-grained soils exhibit large variations in grain size distribution, are free-draining and achieve high dry densities upon compaction. Occasional wetting and subsequent evaporation have been responsible for a patchy development of weak, mainly carbonate and occasionally gypsum cement, often with clay deposited between and around the coarser particles. This has given rise to a metastable structure. These soils may therefore undergo collapse, especially where localized changes in the soil water regime are brought about by construction activity. Collapse is attributed to a loss of strength in the binding agent, and the amount of collapse undergone depends upon the initial void ratio. Loosely packed aeolian sandy soils with a density of less than 1.6 Mg m^{-3} commonly exhibit a tendency to collapse according to Jennings and Knight (1975).

Stipho (1985) found that increased salt content increased the shear strength of sandy soils as a result of increased cohesion but actually led to a reduction in the angle of friction. Remoulding gave rise to a significant decrease in soil strength. He also found that the residual strength of salt-cemented sand was close to that of uncemented sand.

An excess of evaporation over infiltration gives rise to negative pore water pressure (soil suction) as water is removed form the soil by evaporation. If the soil suction is large enough, then air enters the soil. Under these circumstances, soil behaviour has to be described in terms of total applied stress and soil water suction. Changes in total applied stress and soil water suction separately bring about changes in volume and strength. Blight (1994) maintained that soil water suction has physical (matrix) and osmotic components and that the former governs the mechanical behaviour of the soil. If the soil possesses an unstable structure, then collapse may take place once the soil water suction falls below a critical value.

The occurrence of calcareous expansive clays in eastern Saudi Arabia has been described by Abduljauwad (1994). These clays have a moderate to very high swelling potential, depending on their smectite content. They have been formed by weathering of limestone and marl in an alkaline environment with an abscence of leaching. In some cases, the natural moisture content of the clays is higher than that of the plasticity index.

8.8 Salt weathering

Because of the high rate of evaporation in hot, arid regions the capillary rise of near-surface groundwater is normally very pronounced (Fookes *et al.*, 1985). The type of soil governs the height of capillary rise, but in clay soils it may be a few metres (see Table 9.21). Normally, however, the capillary rise in desert conditions does not exceed 2 to 3 m. Although the capillary fringe can be located at depth within the ground, obviously depending on the position of the water table, it can also be very near or at the ground surface. For instance, Al-Sanad *et al.* (1990) referred to the occurrence of saline groundwater (chloride content 1000–3000 mg l^{-1}; sulphate content 2000–5500 mg l^{-1}) within 1 metre of the ground surface at the coast in Kuwait. In the latter case, evidence is provided in the form of efflorescences on the ground surface due to the precipitation of salts. Salts are also precipitated in the upper layers of the soil. Among the most frequently occurring salts are calcium carbonate, gypsum, anhydrite and halite. Nonetheless, Netterberg (1970) showed that the occurrence of salts is extremely variable from place to place. Salt concentrations are highest in areas where saline water, either surface or subsurface, is evaporated, such as low-lying areas of internal drainage, especially with salinas and saline bodies of water, or along coasts particularly where sabkhas are present. Concentrations of salt in coastal areas may occur due to the transport inland of salt by onshore winds. Conversely, free-draining deep deposits of sand and upland soils generally have a low salt contents.

Groundwater flow causes dissolution of soluble phases in older sediments beneath the water table. The rate of dissolution or precipitation varies with the individual properties of the rock–soil water system and usually is rapid enough to be of engineering significance. New layers of minerals can be formed within months, and thin layers can be dissolved just as quickly (Fookes *et al.*, 1985). The groundwater regime can be changed by construction operation and so may lead to changes in the positions at which mineral solution or precipitation occurs. Groundwaters are high in sodium, chloride, potassium, sulphate and magnesium but generally low in calcium, but normally are not saturated in these ions (Table 8.4).

Salt weathering leading to rock disintegration is brought about as a result of the stresses set up in the pores, joints and fissures in rock masses due to the growth of salts, the hydration of particular salts and the volumetric expansion that occurs due to the high diurnal range of temperature (Evans, 1970). The aggressiveness of the ground depends on the position of the water table and the capillary fringe above in relation to the ground surface, the chemical composition of the groundwater and the concentration of salts within it, the type of soil and the soil temperature. The pressures produced by the crystallization of salts in small pores are appreciable: for instance, gypsum ($CaSO_4.H_2O$) exerts a pressure of 100 MPa; anhydrite ($CaSO_4$), 120 MPa; kieosrite ($MgSO_4.H_2O$), 100 MPa; and halite (NaCl), 200 MPa. Some salts are more effective in the

Table 8.4 Typical compositions of sea water and sabkha water (after Fookes et al., 1985).

Ion*	Open sea	Coastal seawater Arabian Gulf	Sabkha water
Ca^{++}	420	420	1250
Mg^{++}	1320	1550	4000
Na^{+}	10 700	20 650	30 000
K^{+}	380	650	1300
SO_4^{--}	2700	3300	9950
Cl^{-}	19 300	35 000	56 600
HCO_3^{-}	75	170	150

* In $mg\,l^{-1}$.

breakdown process than others. For example, Goudie et al. (1970) showed, from a series of experiments, that Na_2SO_4, followed closely by $MgSO_4$, were much more effective in bringing about the disintegration of cubes of sandstone than other salts and that NaCl brought about little change. Limestone also proved susceptible to this form of breakdown but not igneous rocks. Also, it appears that when some salts are combined in solution their effectiveness as agents of breakdown is increased, such as a combination of $CaSO_4$ and NaCl.

Ground surface temperatures in hot deserts can have a diurnal range exceeding 50 °C. This, together with the fact that the coefficients of thermal expansion of the common salts found in desert regimes are greater than those of most common rock-forming minerals, would suggest that the potential for disruptive disintegration is significant. The hydration, dehydration and rehydration of hydrous salts may occur several times throughout a year and depends upon the temperature and relative humidity conditions on the one hand and dissociation vapour pressures of the salts on the other. In fact, Obika et al. (1989) indicated that the crystallization and hydration–dehydration thresholds of the more soluble salts such as sodium chloride, sodium carbonate, sodium sulphate and magnesium sulphate may be crossed at least once daily.

Salt weathering also attacks structures and buildings, leading to cracking, spalling and disintegration of concrete, brick and stone. The extend to which damage due to salt weathering occurs depends upon climate, soil type and groundwater conditions, since they determine the type of salt, its concentration and mobility in the ground, and the type of building materials used. In addition, faulty design or poor workmanship can increase the susceptibility of structures to salt attack by leaving cracks or hollows in which salts can accumulate (Fookes and Collis, 1975). One of the most notable forms of damage to buildings and structures is that brought about by salt weathering that is attributable to sulphate attack. For example, Robinson (1995) described damage done to concrete house slabs due to corrosion and heaving. The deterioration of the concrete arose because the capillary fringe, and so aggressive groundwater, extended to the surface, the Portland cement reacting with calcium sulphate in solution. Heaving was due to the precipitation of salts beneath the slabs. The doming of concrete slabs as a result of heaving leads to them cracking and hence to the concrete becoming more susceptible to corrosion. Where slabs have suffered moderate to severe damage, they may be removed and 0.65 m of underlying soil removed. This is replaced with gravel over which a vapour barrier (e.g. 6 mm vizqueen or polythene) is placed. The latter should be contained within a layer of sand. A new slab of reinforced concrete is then laid.

The most serious damage caused to brickwork, and limestone and sandstone building stone occurs in low-lying salinas, playas and sabkhas, where saline groundwater occurs at shallow depth, giving rise to aggressive ground conditions. In such instances, the capillary rise may continue into the lower parts of buildings, giving rise to dampness, subflorescence, efflorescence, surface disintegration, spalling, and the development of cavities in the building materials. Salt attack on brick foundations has sometimes led to their disintegration to a powdery material, which in turn results in the building undergoing settlement. If this is to be avoided, the position of the upper surface of the capillary fringe in relation to the ground surface should be determined, since this will indicate whether or not the foundations of buildings and structures will be subjected to saline attack. If the potential capillary rise is above the ground surface, then the buildings and structures themselves will be affected. Obviously, any such damage can be avoided by not undertaking construction in low-lying areas with aggressive ground conditions. Alternatively, foundations can be provided with a protective coverings, together with damp courses to prevent capillary rise in buildings. In addition, it is advisable to use building materials that offer a high degree of resistance to salt weathering.

Damage, especially in the form of cracking, to concrete buildings and structures is caused by salt accumulation either as a result of the salt content in the aggregate or mix water, or by introduction consequent upon capillary rise from the groundwater. Indeed, Fookes and Collis (1975) suggested that concrete appears to deteriorate more quickly in arid regimes than elsewhere. Crack development is probably enhanced by the high temperatures, which lead to surface drying and shrinkage. The use of reactive alkali rocks is also responsible for cracking. As the resultant silicate gels develop they generate osmotic pressures, which first produce hairline cracking, the gels moving into the cracks and causing them to expand. This can mean that deterioration ultimately reaches a point where the concrete becomes unsafe. Use of hydratable aggregates can also give rise to expansion within concrete, resulting in 'pop-outs'. Salt weathering tend to be concentrated on cracks that are formed in the concrete.

Most of the problems associated with salts in concrete are consequent upon chlorides and sulphates being introduced in the mix water or in the aggregates.

Chlorides, generally sodium chloride, are responsible for the corrosion of steel reinforcement in concrete. Fookes and Collis (1976) proposed that total acid-soluble chloride levels (as NaCl) should not exceed 0.5% by weight of cement in general-purpose reinforced concrete. They also recommended a limit of 0.1% by weight for fine aggregate and 0.05% by weight for coarse aggregate. Consequently, the production of chemically sound aggregates dictates that material selection and quarrying processes must be undertaken with care. Fookes and Collis went on to suggest that the total sulphate content of concrete should not exceed 4% by weight of cement, thereby implying that the limit for both coarse and fine aggregates should be 0.4% by weight of aggregate.

Both the scale and rate of sulphate attack in arid regions exceeds that encountered in other climatic regimes. Again capillary rise, this time of sulphate-bearing groundwater, is a major problem, the sulphate reacting with calcium aluminium hydrate in cement to give rise to ettringite (hydrated calcium sulphate), which is responsible for an increase in volume. One of the consequences of such a reaction is heave in concrete ground slabs above damp fill containing sulphate. Again, protective coverings can prevent sulphate attack of foundations, sulphate-resisting cement can be used and concrete can be densified (Table 8.5).

Salt weathering of bituminous paved roads built over areas where saline groundwater is at or near the surface is likely to result in notable signs of damage such as heaving, cracking, blistering, stripping, potholing, doming and disintegration. For example, Blight (1976) mentioned that the surface layer can be damaged in damp soils as a result of upward capillary migration and concentration of calcium, magnesium and sulphate salts due to evaporation from the surface. Such movements are probably encouraged in the immediate vicinity of the road by the fact that the road has a higher temperature than the surrounding ground. The physical and chemical consequences of the attack on roads mainly depend on the salinity of the groundwater, the type of aggregate used and the design of the road (Table 8.6). Assessment of local groundwater regimes and salt profiles in relation to pavement location is necessary before construction commences in areas where the water table is high. Aggregates used for base and sub-base courses can be attacked and this can mean that roads undergo settlement and cracking (Fookes and French, 1977). Once cracks appear, the intensity and extent of salt damage tends to increase with time. As a result of upward migration of capillary moisture, salts may be precipitated beneath the bituminous surface, leading to its degradation and gradually to it being heaved to form ripples in the road, with the ripples containing tension cracks.

Obika *et al.* (1989) reviewed the work done in various semi-arid and arid regions to determine salt limits in relation to the construction of bituminous roads. Their findings are summarized in Table 8.7. However, they did point out that the salt content depends on the method of testing and that there is a

Table 8.5 Some engineering consequences of salt water systems (after Fookes et al., 1985).

Effect	Cause	Time scale	Significance	Techniques to investigate and confirm occurrence	Preventive and remedial measures
Chemical sulphate attack on concrete.	Calcium compounds in cement react with sulphates to give expansion and disintegration. Flowing water helps to remove products and expose fresh concrete to attack. Can also occur in wet conditions if there are high initial sulphate levels in mix.	Years.	Varies from disintegration of surface and loss of reinforcement cover to total disintegration.	Observe locations of attack with respect to groundwater level. Chemical testing of groundwater. Testing of concrete possible but complicated.	Existing concrete – attack might be slowed down by keeping as dry as possible. New concrete – ensure concrete is dense and waterproof. Provide tanking. Maximum total acid soluble sulphates in concrete (as SO_3) – 4% by weight of cement.
Physical sulphate attack on concrete masonry, rip-rap and other stone structures.	Expansive forces set up as salts crystallize in pores of concrete and stone.	Half to many years.	Surface deterioration. Loss of reinforcement cover.	Observe locations of attack and local environment.	New concrete – ensure concrete is dense. Use clean sound stone and aggregates. Tanking.
Physical sulphate attack on road pavements.	As above.	As above.	Surface deterioration. Weakening and susceptibility to mechanical damage. Potholes. Salt blisters.	As above.	Use clean, sound aggregates. Surfacing impermeable to water vapour (min. 40 mm DBWC). Elevate pavement above capillary rise.
Chloride contamination of concrete.	Salts in mix constituents. Migration of chloride ions into concrete, especially if exposed to wet/dry conditions. Ingress may be increased where concrete is cracked due to thermal or shrinkage movements.	Months for ingress. Months to years for signs of reinforcement corrosion.	Mass concrete – possible slight weakening of surface layers. Reinforced concrete – continuing corrosion of reinforcement.	Chloride testing of concrete samples.	New concrete – ensure concrete is dense. Use sound aggregates. Use cover of at least 75 mm in aggressive (salty, wet/dry) conditions. Tanking. Maximum total allowable acid soluble chloride in concrete (as NaCl) – varies 0.1 to 1% by weight of cement.* Use O.P.C.
Heave.	Accumulation of soluble salts from groundwater. Hydration of massive or dispersed salt deposits (e.g. anhydrite).	Months to years. Indeterminate, but perhaps months to years.	Blisters in thin blacktop. Movement of lightly loaded floor slabs. May vary from small movements to complete destruction of structures.	Level surveys, observe local environment, e.g. groundwater level and changes since construction. Chemical testing of soil samples to check changes with time.	Minimize changes in groundwater equilibrium caused by construction. Provide thick floor and pavement structures.
Dissolution of soluble deposits.	Solutions of salts, especially under conditions of rapid flow.	Depends on flow path length, concentration and solubility of salts.	Settlement and perhaps failure of foundations.	Level surveys, observe water flow paths. Chemical testing of water and soil to check changes with time.	Install cut-off, impermeable membrane or grout to reduce water flows.

* Varies depending on cement type and whether concrete is mass or reinforced.

Table 8.6 Outline of potential physical and chemical problems encountered in road construction (after Fookes and French, 1977).

	Bituminous wearing course/base course		Unbound base e.g. wet or dry bound macadam, wet mix or all-in granular material	Unbound sub-base e.g. gravel, crushed stone
	Thick/dense	Thin/porous		
Potential migration of water	Very low	Moderate	High (varies with aggregate, etc.)	High (varies with aggregate, etc.)
Potential migration of salt	Very low	Moderate to high	Moderate (varies with aggregate, etc.)	High (varies with aggregate, etc.)
Physical changes in presence of groundwater – permanent, intermittent or capillary	Unlikely in short or medium term	Probable	Probable	Probable
Distress	Possible long-term aggregate disintegration, surface erosion and stripping	Short-term aggregate disintegration, surface erosion scabbing, blisters, potholes	Short- to medium-term disintegration, stripping settlement	Medium-term disintegration, settlement
Physical changes in presence of transient water (rain, dew, etc.) – depends on aggregate and salt content	Unlikely unless salt in aggregates high	Unlikely unless high salt in aggregates	Unlikely unless salt in aggregates	Unlikely unless salt in aggregate
Distress	Very slow aggregate disintegration	Slow aggregate disintegration	May be slight disintegration, settlement	May be slight disintegration, settlement
Chemical changes in presence of groundwater – movement intermittent, capillary, depending on salt levels and type	Unlikely	Possible bitumen and aggregate may decompose and disintegrate if salt high	Possible if salt content high	Possible if salt content high
Distress	Unlikely	Potholes, scabbing, stripping	Volume changes and loss of strength	Volume changes and loss of strength
Chemical changes in presence of transient water (rain, dew, etc.) – depends on aggregate and salts present	Unlikely	Unlikely bitumen may decompose if salt high	Possible with some aggregate and some salts	Possible with some aggregate and some salts
Distress	Unlikely	Unlikely	Small volume change or loss of strength	Small volume change or loss of strength

Table 8.7 Some recommended salt limits (from Obika et al., 1989).

Author	Salt	Maximum limit[1]	Material location	Criteria/remarks
Cole and Lewis (1960)	Chloride as NaCl	0.2% (0.5% may be safe)	Sandy clay soils and lateritic gravel base semi-arid. Western Australia.	Observations of soluble salt-damaged base courses and laboratory tests with compacted gravels to which various amounts of salt solution had been added. No damage was encountered with materials containing up to 0.5% but authors suggest 0.2% and warn against the limited scope of their tests. Method of salt analysis not given. Soluble salt damage in Western Australia results mainly from capillary rise of chloride in groundwater.
Weinert and Clauss (1967)	Chloride, sulphate as SO_3 (mostly $MgSO_4$)	0.2% 0.05%	For road foundations materials generally but principally for quartzite mine waste. South Africa.	Chloride limit adopted from Cole and Lewis (1960). Sulphate limit suggested from considerations of the critical limit (i.e. 0.05%) for building stones and sulphate content of base material where damage had occurred. Salt contents as obtained by analysis of 5:1 water:soil extracts. Analysis was carried out on material obtained from the top few inches of base, probably after upward migration of salt took place. Damage in this case resulted from the use of saline construction material.
Netterberg (1970)	As an indication of total sulphate	Reject material if electrical conductivity is >1.5 mmhos/cm at 15.8°C. If 0.6–1.5 mmhos/cm test for sulphates and reject if >0.05%. If 0.06 mmhos/cm accept material.	For any base and sub-base materials. South Africa.	Conductivity limits correspond to salt limits suggested by Cole and Lewis (1960) and Weiner and Clauss (1967). Electrical conductivity can be used as a rapid indication of salinity; however, it has a major drawback in that it does not identify the type of salt present. Applicable only to quartzite mine waste construction material as conductivity/salinity relationship varies depending on the material.
Netterberg et al. (1974)	Total soluble salt Total soluble sulphate	0.2% 0.15%	Base and sub-base materials. Witwatersrand quartzite mine waste – see Weinert and Clauss. South Africa.	Authors stress the peculiar nature of the experience (mine waste material) upon which the suggested limits are based. Sulphate analysis carried out on water extract. Note: some authors (e.g. Fookes and French, 1977; Netterberg, 1979) refer to acid-soluble sulphate. The salts considered are from quartzite mine waste construction materials.

Source	Parameter	Limit	Application	Comments
Blight et al. (1974)	Total soluble salt (mostly sulphate)	2% (3% may be safe)	For sand used for asphaltic mixes. South Africa	Observation of asphaltic surfaces of various ages made with sulphate-bearing sands. Also laboratory tests on specimens containing various quantities of added soluble salt.
SABS1083–1976 (South Africa Bureau of Standards)	Total soluble salt	0.5%	South Africa.	Quoted in Netterberg (1979). Based on an understanding that total soluble salts when determined will include other harmless salts in the material. Therefore 0.2% of Netterberg et al. (1974) considered too stringent. Soluble salt content determined on water extracts.
Blight (1976)	Total soluble salt	0.2%	Base and sub-base materials.	The salts discussed are sulphates of predominantly magnesium and iron. Author suggests that 0.2% can also be applied where the salt is predominantly chloride. The limit does not acknowledge that sulphates are more deleterious than chlorides.
Fookes and French (1977)		See Table 8.6, Fookes and French (1977)		Primarily on field experience in the Middle East and on assessment of the type of material and conditions found in that region. Authors point to the possible effect of salt mixtures on the limits specified and take into account the predominant type of salts in the region.
Netterberg (1979)	Sulphate as SO_3	0.25%	Lime and cement treated materials if cohesive.	Criteria for suggested limits not discussed. Determinations are carried out on water extracts. Material similar to that discussed in Netterberg et al. (1974).
Sir William Halcrow and partners (date unknown)	Total soluble salt	1.0%	If not cohesive	Experience of road construction in the Middle East.
	Total soluble salt	0.2%	Untreated material	
	Total soluble salt	0.25%	Untreated material-fines	
	Acid-soluble sulphate	0.3%	Wearing course and base course material	
	Acid-soluble sulphate	0.5%	Roadbase and hard shoulder	
	Acid-soluble sulphate	2.0%	Sub-base	

[1] Figures refer to % by weight of dry material.

need to standardize methods and they cautioned that extreme care has to be taken when adopting a total soluble salt content, as the latter does not take account of the types and proportions of salts present.

Wherever possible, roads should avoid areas in which saline groundwater occurs at or near the surface, and aggregate sources should be investigated for salt content prior to use, those with high salt content being rejected. As it is not always possible to avoid the construction of roads on saline ground or the use of materials containing salts, a number of precautions should be taken. Usually pavement damage attributable to soluble salts is confined to thin bituminous surfaces. Consequently, Blight (1976) suggested that this could be avoided by laying a surface layer of at least 30 mm of dense asphalt concrete to prevent evaporation and thereby migration and crystallization of salt at or near the surface. Even so, salt may accumulate beneath such surfaces, leading to the degradation of base and sub-base material. It appears that cutback bitumen and emulsion bitumen primers perform better than tar primers in relation to the reduction of surface degradation. This may be due to emulsion primers resting on the surface rather than penetrating the pavement layer and thereby providing lower permeability. Januszke and Booth (1984) suggested that evaporation and therefore the accumulation of salt at the surface could be avoided by placing a bituminous surfacing immediately after compaction. Alternatively, French *et al.* (1982) have suggested placement of an impermeable membrane in the base course. A granular layer in the base course may also help to reduce the capillary rise.

If the problems associated with aggressive salty ground are to be avoided, then the first object must be to identify the limits of the hazard zone and the spatial variability of the hazard within it. Fookes and French (1977) recognized five moisture zones (Figure 8.12; Table 8.8) of which zones C, D and E can generally be identified and mapped (Figure 8.13), thereby enabling a thematic hazard map of aggressive salty ground conditions to be produced. Such a map indicates those areas that should be avoided as far as construction is concerned or where precautionary measures must be adopted. Cooke *et al.* (1982) provided an alternative zonation scheme that was concerned primarily with foundations, as opposed to the scheme offered by Fookes and French, which was devised in relation to road construction. Both are compared in Table 8.9. The concentration of salts in the groundwater can allow hazard zones to be subdivided into those of high, moderate or low salinity, thereby expressing the relative hazard between different areas. Cooke *et al.* described the production of groundwater level, electrical conductivity and saline hazard intensity maps for northern Bahrain. The position where the capillary fringe intersected the ground surface was also marked on the groundwater level map, as was the position of the 10 m contour. The latter marked the inland limit of the capillary fringe hazard. Groundwater salinity was assessed, at the same sampling points where water level was determined, by means of a conductivity meter, the specific conductance being directly related to the concentration of salts present. In addition,

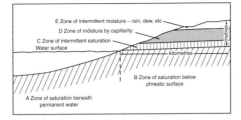

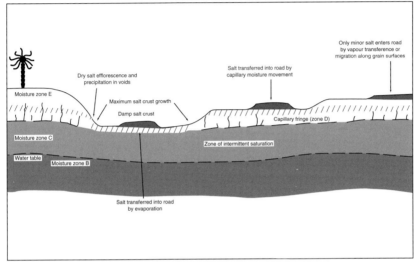

Figure 8.12 Soil moisture zones and road construction in drylands (after Fookes and French, 1977).

the ionic concentrations of chloride, sulphate, sodium, potassium, calcium and magnesium in groundwater samples were determined. The areas of highest ionic concentrations corresponded to those of highest electrical conductivity. The map of hazard intensity was based on the assumption that it is related to the depth of the water table on the one hand and the conductivity (salinity) of the groundwater on the other (Figure 8.14). Similar hazard maps have been produced for aggressive ground conditions encounted in other areas of the Middle East and, although like any other type of hazard map these represent broad generalizations of the ground conditions, they nevertheless can be used for preliminary planning purposes.

Table 8.8 Summary of the influence of moisture zones A to E on behaviour of roads (after Fookes and French, 1977).

Zone	Moisture conditions	Salt conditions	Possible damage
Moisture zone E	Transient water from rain, dew, etc.	Salts may be removed in solution and may accumulate by subsequent evaporation or by vapour transfer, etc.	Damage not serious unless aggregate is rich in salt or road is of thin construction with unsound aggregates, in long term.
Moisture zone D	Water present by capillary moisture movement.	Salts may be precipitated at all levels of road construction and in large quantities.	Aggregate and bitumen may decompose, blisters may develop, small holes and cracks likely. Serious damage only in thin construction.
Moisture zone C	Water present by capillary movement or ground may be saturated at times of high water table.	Salts precipitated and may be redissolved.	Large potholes develop, aggregates and bitumen decompose rapidly. Irregular surface develops. Maximum damage in thin construction.
Moisture zone B	Permanently saturated zone below capillary fringe.	Soil and rock properties may be changed in long term.	Damage by long-term deformation possible.
Moisture zone A	Saturated zone below water.	May create sabkha conditions in reclaimed ground or embankments.	Damage as moisture zones E, D, and C depending on elevation of construction.

Note: For explanation of moisture zones see Figure 8.12

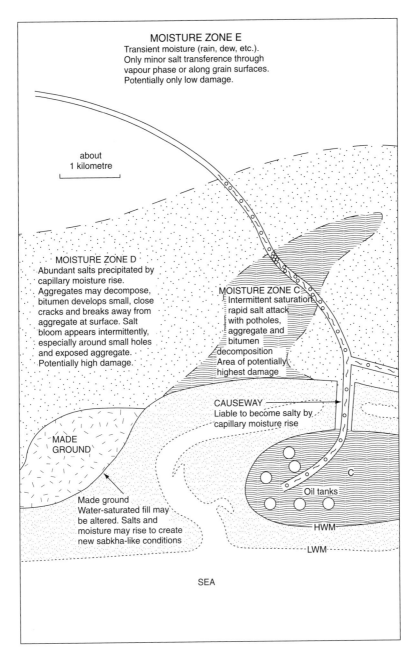

MOISTURE ZONE E
Transient moisture (rain, dew, etc.).
Only minor salt transference through
vapour phase or along grain surfaces.
Potentially only low damage.

about
1 kilometre

MOISTURE ZONE D
Abundant salts precipitated by
capillary moisture rise.
Aggregates may decompose,
bitumen develops small, close
cracks and breaks away from
aggregate at surface. Salt
bloom appears intermittently,
especially around small holes
and exposed aggregate.
Potentially high damage.

MOISTURE ZONE C
Intermittent saturation
rapid salt attack
with potholes,
aggregate and
bitumen
decomposition
Area of potentially
highest damage

CAUSEWAY
Liable to become salty by
capillary moisture rise

MADE
GROUND

Made ground
Water-saturated fill may
be altered. Salts and
moisture may rise to create
new sabkha-like conditions

Oil tanks

C

HWM

LWM

SEA

Figure 8.13 Hypothetical map to show how identification of soil moisture zones can assist in road construction and the determination of maintenance priority areas (after Fookes and French, 1977).

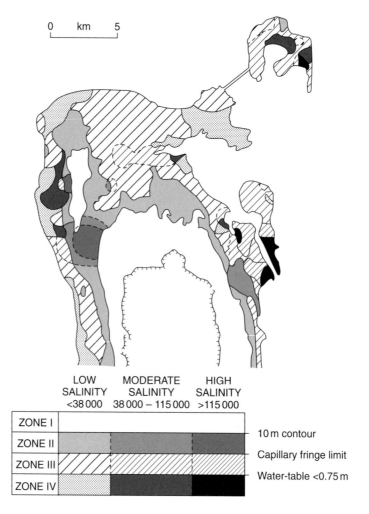

Figure 8.14 Aggressive ground conditions in northern Bahrain: predicted hazard intensity (from Cooke *et al.*, 1982).

References

Abduljauwad, S.N. (1994) Swelling behaviour of calcareous clays from the Eastern Province of Saudi Arabia. *Quarterly Journal Engineering Geology*, 27, 333–351.

Akili, W. and Torrance, J.K. (1981) The development and geotechnical problems of sabkha, with preliminary experiments on the static penetration resistance of cemented sands. *Quarterly Journal Engineering Geology*, 14, 59–73.

Al-Sanad, H.A., Shagour, F.M., Hencher, S.R. and Lumsden, A.C. (1990) The influence of changing groundwater levels on the geotechnical behaviour of desert sands. *Quarterly Journal Engineering Geology*, 23, 357–364.

Anon. (1990) Tropical residual soils. Geological Society, Engineering Group Working Party Report. *Quarterly Journal Engineering Geology*, 23, 1–101.

Bagnold, R.A. (1941) *The Physics of Wind Blown Sand and Desert Dunes*. Methuen, London.

Bathurst, R.G.C. (1971) *Carbonate Sediments and their Diagenesis*. Elsevier, Amsterdam.

Bell, F.G. and Maud, R.R. (1994) Dispersive soils and earth dams with some experiences from South Africa. *Bulletin Association Engineering Geologists*, 31, 433–446.

Blight, G.E. (1976) Migration of subgrade salts damages thin pavements. *Proceedings American Society Civil Engineers, Transportation Engineering Journal*, 102. 779–791.

Blight, G.E. (1994) The geotechnical behaviour of arid and semi-arid zone soils – South African experience. *Proceedings First International Symposium on Engineering Characteristics of Arid Soils*, London, Fookes P.G. and Parry, R.H.G. (eds), Balkema, Rotterdam, 221–235.

Braithwaite, C.J.R. (1983) Calcrete and other soils in Quaternary limestones: structures, processes and applications. *Journal Geological Society*, 140, 351–364.

Chepil, W.S. 1945. Dynamics of wind erosion: Initiation of soil movement. *Soil Science*, 60, 397–411.

Chepil, W.S. (1957) Sedimentary characteristics of dust storms. *American Journal Science*, 255, 12–22; 206–213.

Chepil, W.S. (1959) Equilibrium of soil grains at the threshold of movement by wind. *Proceedings Soil Science Society America*, 23, 422–428.

Chepil, W.S. and Woodruff, N.P. (1963) The physics of sand erosion and its control. *Advances in Agronomy*, 15, 211–302.

Cooke, R.U. and Warren, A. (1973) *Geomorphology in Deserts*. Batsford, London.

Cooke, R.U., Brunsden, D., Doornkamp, J.C. and Jones, D.K.C. (1982) *Urban Geomorphology in Drylands*. Oxford University Press, Oxford.

Doornkamp, J.C., Brunsden, D., Jones, D.K.C., Cooke, R.U. and Bush, P.R. (1979) Rapid geomorphological assessments for engineering. *Quarterly Journal Engineering Geology*, 12, 189–204.

Evans, I.S. (1970) Salt crystallization and rock weathering: a review. *Revue Geomorphique Dynamique*, 19, 153–177.

Fookes, P.G. (1978) Engineering problems associated with ground conditions in the Middle East: inherent ground problems. *Quarterly Journal Engineering Geology*, 11, 33–50.

Fookes, P.G. and Collis, L. (1975) Problems in the Middle East. *Concrete*, 9, No. 7, 12–17.

Fookes, P.G. and Collis, L. (1976) Cracking and the Middle East. *Concrete,* 10, No. 2, 14–19.

Fookes, P.G. and French, W.J. (1977) Soluble salt damage to surfaced roads in the Middle East. *Journal Institution Highway Engineers*, 24, No. 12, 10–20.

Fookes, P.G., French, W.J. and Rice, S.M.M. (1985) The influence of ground and groundwater chemistry on construction in the Middle East. *Quarterly Journal Engineering Geology*, 18, 101–128.

Fookes, P.G. and Gahir, J.S. (1995) Engineering performance of some coarse grained arid soils in the Libyan Fezzan. *Quarterly Journal Engineering Geology*, 28, 105–130.

French, W.J., Poole, A.B., Ravenscroft, P. and Khiabani, M. (1982) Results from preliminary experiments on the influence of fabrics on the migration of groundwater and water soluble minerals in the capillary fringe. *Quarterly Journal Engineering Geology*, 15, 187–199.

Gabriels, D. Maene, L.J., Leavin, J. and De Boodt, M. 1979. Possibilities of using soil conditioners for soil erosion control. In *Soil Conservation and Management in the Humid Tropics*, Greenland, D.J. and Lal, R.(eds), Wiley, Chichester, 99–108.

Glennie, K.N. (1970) *Desert Sedimentary Environments*. Elsevier, Amsterdam.

Goudie, A.S. (1973) *Duricrusts in Tropical and Subtropical Latitudes*. Oxford University Press, Oxford.

Goudie, A.S. (1978) Dust storms and their geomorphological implications. *Journal Arid Environments,* 1, 291–310.

Goudie, A.S.., Cooke, R.U. and Evans, I.S. (1970) Experimental investigation of rock weathering by salts. *Area*, 4, 42–48.

Greeley, R. and Iverson, J.D. (1985) *Wind as a Geological Process*. Cambridge University Press, Cambridge.

Horta, J.D. de S.O. (1980) Calcrete, gypcrete and soil classification. *Engineering Geology*, 15, 15–52.

Januszke, R.M. and Booth, E.H.S. (1984) Soluble salt damage to sprayed seals on the Stuart highway. *Proceedings Twelfth Australian Road Research Board Conference*, Hobart, 3, 18–30.

Jennings, J.E. and Knight, K. (1975) A guide to construction on or with materials exhibiting additional settlements due to collapse of grain structure. *Proceedings Sixth Regional Conference on Soil Mechanics and Foundation Engineering*, Durban, 99–105.

Jones, D.K., Cooke, R.U. and Warren, A. (1986) Geomorphological investigation, for engineering purposes, of blowing sand and dust hazard. *Quarterly Journal Engineering Geology*, 19, 251–270.

Kerr, R.C. and Nigra, J.O. (1952) Eolian sand control. *Bulletin American Association Petroleum Geologists*, 36, 1541–1573.

King, L.C. 1963. *South African Scenery.* Third Edition, Oliver and Boyd, Edinburgh.

Lister, L.A. and Secrest, C.D. (1985) Giant desiccation cracks and differential surface subsidence, Red Lake playa, Mohave County, Arizona. *Bulletin Association Engineering Geologists*, 22, 299–314.

Mabbutt, J.A. (1977) *Desert Landforms*. Massachussetts Institute of Technology Press, Cambridge, Mass.

Manohar, M. and Bruun, P. (1970) Mechanics of dune growth by sand fences. *Dock and Harbour Authority*, 51, 243–252.

Meigs, P. (1953) World distribution of arid and semi-arid homoclimates. In *Arid Zone Hydrology*, UNESCO Paris, 203–209.

Netterberg, F. (1970) Occurrence and testing for deleterious salts in road construction materials with particular reference to calcretes. *Proceedings Symposium on Soils and Earth Structures in Arid Climates*, Adelaide, 87–92.

Netterberg, F. (1980) Geology of South African calcretes: terminology, description and classification. *Transactions Geological Society South Africa*, 83, 255–283.

Obika, B., Freer-Hewish, R.J. and Fookes, P.F. (1989) Soluble salt damage to bituminous road and runway surfaces. *Quarterly Journal Engineering Geology*, 22, 59–73.

Petrov, M.P. (1976) *Deserts of the World*. Wiley, New York.

Robinson, D.M. (1995) Concrete corrosion and slab heaving in a sabkha environment: Long Beach–Newport Beach, California. *Environmental and Engineering Geoscience*, 1, 35–40.

Stipho, A.S. (1985) On the engineering properties of salina soil. *Quarterly Journal Engineering Geology*, 18, 129–137.

Stipho, A.S. (1992) Aeolian sand hazards and engineering design for desert regions. *Quarterly Journal Engineering Geology*, 25, 83–92.

Watson, A. (1985) The control of wind blown sand and moving dunes: a review of methods of sand control in deserts with observations from Saudi Arabia. *Quarterly Journal Engineering Geology*, 18, 237–252.

Wilson, I.G. (1971) Desert sandflow basins and a model for the development of ergs. *Geographical Journal*, 137, 180–197.

Woodruff, N.P. and Siddoway, F.H. (1965) A wind erosion equation. *Proceedings Soil Science Society America*, 29, 602–608.

Chapter 9

Soil erosion and desertification

9.1 Introduction

Loss of soil due to erosion by water or wind is a natural process. Soil erosion removes the topsoil, which contains a high proportion of the organic matter and the finer mineral fractions in soil that provide nutrient supplies for plant growth.

Soil erosion is part of the process of denudation along with the other agents that are responsible for changing the landscape. According to Kirby (1980), soil erosion becomes a more effective agent of denudation once slopes are reduced to gradients that are relatively stable as far as mass movements are concerned. Unfortunately, however, this process can be accelerated by the activity of man: witness the disasters in the Dust Bowl and Tennessee Valley of the United States prior to the Second World War. It is difficult however, to separate natural from human-induced changes in erosion rates. Although accelerated, human-induced soil erosion does not figure as prominently among environmental issues now as it did then, this does not mean that the problem no longer exists. It most certainly does, and in some countries it is as serious as it ever was, if not more so.

In the case of soil erosion by water, it is most active where most of the rainfall finds it difficult to infiltrate into the ground so that most of it flows over the surface and in so doing removes soil. The intensity of rainfall is at least as important in terms of soil erosion as the total amount of rainfall, and as rainfall becomes more seasonal, the total amount of erosion tends to increase. This run-off can take the form of sheet flow or can be concentrated into rills and gullies. Such conditions are met most frequently in semi-arid and arid regions. Nevertheless, soil erosion occurs in many different climatic regions where vegetation has been removed, it being at a maximum when intense rainfall and the vegetation cover are out of phase, that is, when the surface is bare.

Sheet flow may cover up to 50% of the surface of a slope during heavy rainfall, but erosion does not take place uniformly across a slope. The depth of sheet flow, up to 3 mm, and the velocity of flow are such that both laminar and turbulent flow take place. Erosion due to turbulent flow occurs only where flow is confined in linear concentrations within sheet flow. Hence, the flow elsewhere is laminar and non-erosive. The velocity of sheet flow ranges between 15 and 300 mm s^{-1}. Velocities of 160 mm s^{-1} are required to erode soil particles of 0.3 mm diameter, but velocities as low as 20 mm s^{-1} will keep these particles in suspension.

Rills and gullies begin to form when the velocity of flow increases to rates in excess of 300 mm s^{-1} and flow is turbulent. Whether they form depends on soil factors as well as velocity and depth of water flow. Rills and gullies remove much larger volumes of soil per unit area than sheet flow. Severe soil erosion, associated with the formation of gullies (Figure 9.1), can give rise to mass movements on the steepened slopes at the sides of these gullies.

Soil erosion by wind, like erosion by water, depends upon the force that the wind can exert upon soil particles. This is influenced by the roughness of the surface over which wind blows. Where the surface is rough, for instance due to the presence of boulders, plants or other obstacles, this reduces the wind speed immediately above the ground. Obviously particle size is important, for wind can only remove particles of a certain size, silt-sized particles being especially prone to wind erosion. Nonetheless, wind erosion would appear to be effective in lowering landscapes; for example, it is estimated that during the period when parts of Kansas formed the 'Dust Bowl', wind erosion accounted for soil losses up to 10 mm y^{-1} at certain locations. By comparison, in some semi-arid regions, water may remove around 1 mm y^{-1}.

Figure 9.1 Gully erosion in the Molteno Formation, northern Lesotho (Coppin and Richards, 1990).

The removal of soil by wind can damage crops by exposing the roots of plants or by subjecting the leaves to wind blasting. It can also remove seeds from the soil. The smaller, more easily removed grains of soil and organic matter are readily subject to wind erosion. This action leads to a reduction in the fertility of the soil, as well as lowering its water-retention capacity. Erosion of soil by wind can give rise to dust storms and sandstorms, which may choke small water courses, accumulate against buildings, bury small obstacles, block highways and even strip paint from cars. People and animals affected may suffer from respiratory problems.

Wind erosion is most effective in arid and semi-arid regions, where the ground surface is relatively dry and vegetation is absent or sparse. The problem is most acute in those regions where land-use practices are inappropriate and rainfall is unreliable. This means that the ground surface may be left exposed. Nonetheless, soil erosion by wind is not restricted to drylands. It also occurs, though on a smaller scale, in humid areas.

Of paramount importance as far as soil erosion is concerned is vegetation cover. Generally, an increase in erosion occurs with increasing rainfall and erosion decreases with increasing vegetation cover. However, the growth of natural vegetation depends on rainfall, so producing a rather complex variation of erosion with rainfall. Although vegetation cover depends primarily upon rainfall, agriculture, especially where irrigation occurs, can mean that vegetation becomes more or less independent of rainfall. Hence, farming practices have had and do have an influence on soil erosion, since they alter the nature of the vegetation cover, the total rainfall and its intensity during times of low cover being most important. Consequently, land is most vulnerable to erosion during periods when ploughing and harvesting take place. In addition, overgrazing reduces the amount of vegetation cover and so can lead to an increase in the rate of soil erosion. Semi-arid regions appear to be the most sensitive in terms of the rate of soil erosion to changes in the amount of rainfall and therefore vegetation cover (Figure 9.2).

The overall influence of vegetation as far as soil erosion is concerned depends upon the relationship between beneficial and adverse influences (Table 9.1). The properties of vegetation that influence its engineering function behaviour and value may vary with the stage of vegetation growth, and so may change with the seasons. Although vegetation can control the amount of erosion that may occur, conversely erosion may produce an environment that cannot sustain vegetation. Most vegetation is self-generating, but interference by man and animals may adversely affect the natural cycles of plant growth.

The influence of vegetation on evapotranspiration can be expressed in terms of the E_t/E_o ratio, in which E_t is the evapotranspiration rate for vegetation cover and E_o is the evaporation rate from open water. Some typical values of the E_t/E_o ratio are provided in Table 9.2. Evapotranspiration is not limited by the supply of water, in that it takes place at the potential rate (E_p). Nonetheless, where the rate of evapotranspiration is high, the top layers of the soil dry out rapidly,

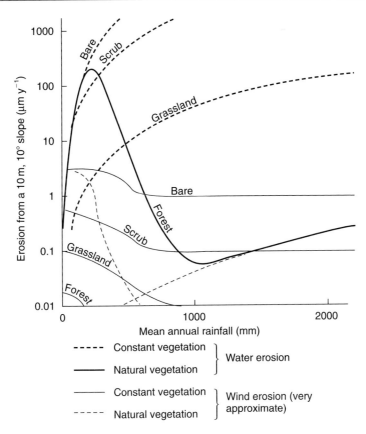

Figure 9.2 Estimated rates of soil erosion by wind and water as a function of rainfall and vegetation cover (after Kirkby, 1980).

so that plants find it increasingly difficult to abstract water from the soil. Consequently, in order to avoid dehydrating, plants reduce their transpiration. At this point, actual evapotranspiration (E_a) becomes less than potential evapotranspiration. The E_a/E_p ratio is governed by the soil moisture deficit, that is, the difference between the reduced level of soil moisture and that at field capacity.

Rainfall that is interrupted by vegetation is referred to as being intercepted. Some intercepted rainfall remains on the plants and subsequently is evaporated; the rest (i.e. temporarily intercepted throughfall) reaches the ground either as leaf drainage or stem flow. Interception storage varies widely but it may account for 15 to 25% of annual rainfall in temperate deciduous forests, and for 20 to 25% of annual rainfall in tropical rainforests. The interception storage therefore reduces the quantity of rainfall reaching the ground surface. Not only does vegetation affect the volume of rain reaching the ground, it also affects its local

Table 9.1 Beneficial and adverse effects of vegetation (from Coppin and Richards, 1990).

Hydrological effects		Mechanical effects	
Foliage intercepts rainfall causing:		Roots bind soil particles and permeate the soil, resulting in:	
1 Absorptive and evaporative losses, reducing rainfall available for infiltration	B	1 Restraint of soil movement, reducing erodibility	B
2 Reduction in kinetic energy of raindrops and thus erosivity	B	2 Increase in shear strength through a matrix of tensile fibres	B
3 Increase in drop size through leaf drip, thus increasing localized rainfall intensity	A	3 Network of surface fibres creates a tensile mat effect, restraining underlying strata	B
Stems and leaves interact with flow at the ground surface, resulting in:		Roots penetrate deep strata, giving:	
1 Higher depression storage and higher volume of water for infiltration	A/B	1 Anchorage into firm strata, bonding soil mantle to stable subsoil or bedrock	B
2 Greater roughness on the flow of air and water, reducing its velocity, but	B	2 Support to upslope soil mantle through buttressing and arching	B
3 Tussocky vegetation may give high localized drag, concentrating flow and increasing velocity	A	Tall growth of trees, so that:	
Roots permeate the soil, leading to:		1 Weight may surcharge the slope, increasing normal and downslope force components	A/B
1 Opening up of the surface and increasing infiltration	A	2 When exposed to wind, dynamic forces are transmitted into the ground	B
2 Extraction of moisture, which is lost to the atmosphere in transpiration, lowering pore water pressure and increasing soil suction, both increasing soil strength	B	Stems and leaves cover the ground surface, so that:	
3 Accentuation of desiccation cracks, resulting in higher infiltration	A	1 Impact of traffic is absorbed, protecting soil surface from damage	B
		2 Foliage is flattened in high-velocity flows, covering the soil surface and providing protection against erosive flows	B

A = Adverse effect; B = Beneficial effect.

Table 9.2 E_t/E_0 ratios for selected plant covers (from Morgan, 1995).

Plant (crop) cover	E_t/E_0 ratio
Wet (paddy) rice	1.35
Wheat	0.59–0.61
Maize	0.67–0.70
Barley	0.56–0.60
Millet/sorghum	0.62
Potato	0.70–0.80
Beans	0.62–0.69
Groundnut	0.50–0.87
Cabbage/Brussels sprouts	0.45–0.70
Banana	0.70–0.77
Tea	0.85–1.00
Coffee	0.50–1.00
Cocoa	1.00
Sugar cane	0.68–0.80
Sugar beet	0.73–0.75
Rubber	0.90
Oil palm	1.20
Cotton	0.63–0.69
Cultivated grass	0.85–0.87
Prairie/savanna grass	0.80–0.95
Forest/woodland	0.90–1.00

intensity and drop size distribution. Hence, the energy of rainfall that is available for splash erosion under a vegetation cover is governed by the proportion of rain falling as direct throughfall and that of leaf drainage, as well as the height of the canopy, which determines the energy of leaf drainage.

The presence of vegetation can increase the infiltration rate of the soil by maintaining a continuous pore system due to root growth, by the presence of organic matter and by enhancing biological activity. This also leads to an increase in the moisture storage capacity of the soil and a decrease in run-off. Vegetation also increases the time taken for run-off to occur, so heavier rainfall is required to produce a critical amount of run-off.

As commented above, soil erosion is a natural process. Hence, ideally as soil is being removed it should be replaced by newly formed soil. If this does not occur, then soil cover is gradually removed. This is accompanied by an increase in run-off by overland flow, which in turn reduces the amount of water that infiltrates into the ground. This means that chemical breakdown of rock material becomes less effective, so adversely affecting soil formation.

The exhaustion of the organic matter and soil nutrients in soil is closely allied to soil erosion. Organic matter in the soil fulfils a similar role to clays in holding water, and both organic and inorganic nutrients. The organic matter in the soil is important in terms of aggregation of soil particles, and a good

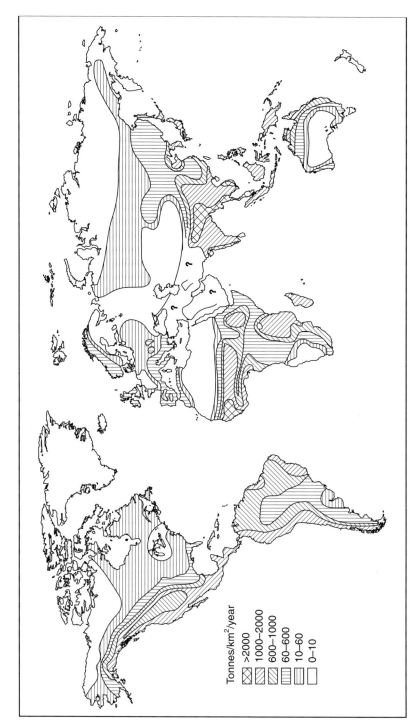

Tonnes/km²/year

>2000
1000–2000
600–1000
60–600
10–60
0–10

Figure 9.3 World distribution of erosion rates (after Fournier, 1960).

vegetation cover tends to increase biological activity and the rate of aggregate formation. Furthermore, increasing aggregate stability leads to increasing permeability and infiltration, as well as a moister soil. High permeability and aggregate strength minimize the risk of overland flow. Loss of organic matter depends largely on the vegetation cover and its management. Partial removal of vegetation or wholesale clearance prevents the addition of plant debris as a source of new organic material for the soil. Over a period of years, this results in a loss of plant nutrients and in a dry climate there is a significant reduction in soil moisture. The process can turn a semi-arid area into a desert in less than a decade. This organic depletion leads to lower infiltration capacity and increased overland flow, with consequent erosion on slopes of more than a few degrees.

The strength of the soil is increased by the presence of plant roots. Grasses and small shrubs can have a notable strengthening effect to depths of up to 1.5 m below the surface, whereas trees can enhance soil strength to depths of 3 m or more. Roots have their greatest effects in terms of soil reinforcement near to the soil surface where the root density is highest. The reinforcing effect is limited with shallow rooting vegetation, where roots fail by pullout before their peak tensile strength is attained. In the case of trees, because their roots penetrate soil to greater depths, then roots may be ruptured when the forces placed upon them exceed their tensile strength.

Soil erosion is obviously not restricted to the arid and semi-arid areas of the world. However, the latter areas face a greater risk of soil erosion (Figure 9.3). Humid areas are most at risk when the land lies fallow during ploughing or after harvesting. It is in these areas where conservation practices have been used most frequently.

9.2 Estimation of soil loss

Soil loss refers to the amount of soil moved from a particular area, and it is transported elsewhere in the form of sediment yield. Techniques used to predict the amount of soil loss have been developed over many years as the processes of soil erosion have become better understood. The most widely used method for predicting soil loss is the Universal Soil Loss Equation (USLE). Several empirical methods are available for estimating sediment yield. These relate sediment concentrations to flow stage, or sediment yield to watershed or hydrological parameters.

Soil erosion by water involves the separation of soil particles from the soil mass and their subsequent transportation. There are two agents responsible for this, namely, raindrops and flowing water. Erosion by raindrops involves the detachment of particles from the soil by impact and their movement by splashing. Run-off removes soil by the action of turbulent water flowing either as sheets or in rills or gullies. The nature of soil erosion depends on the relationship between the erosivity of raindrops and run-off, and the erodibility of the soil. Splash and sheet erosion combined are referred to as inter-rill erosion (Watson and Laflen 1986; Liebenow *et al.*, 1990).

The results of research by several workers have shown that the amount of soil in run-off increases rapidly with raindrop energy (Ellison, 1947; Hudson, 1981). Indeed, many have regarded raindrop impact as a fundamentally important and frequently initial phase of soil erosion, claiming that as much as 90% of erosion of agricultural land may be the result of this process. However, Kirby (1980), although not disputing the role that raindrop impact plays in compacting the soil surface, thereby reducing infiltration and increasing run-off, questioned the significance of raindrop splash in terms of soil particle movement, especially when compared with flow. The success with which soil particles can be separated from a soil mass depends upon the character of the rainfall on the one hand and the type of soil and ground cover on the other. In this context, the principal factors as far as a raindrop is concerned are its mass, size, direction, rainfall intensity and terminal velocity of the drop. According to Hudson (1981) the median size of raindrops increases with the intensity of rainfall (the median drop size of rainfall no longer increases when the intensity exceeds 76.2 mm h^{-1}). The maximum size of raindrop in high-intensity rainfall is approximately 5–6 mm in diameter.

The impact velocity of raindrops is related to drop size. The relationship between drop size, terminal velocity and the fall distance needed to develop 95% of terminal velocity as derived in the laboratory can be seen in Figure 9.4. Obviously, winds and air turbulence affect the terminal velocity of rain; nonetheless, Figure 9.4 suggests that the fall distances needed to obtain maximum terminal velocity are small.

As Hudson (1981) pointed out, with the size, terminal velocity and intensity data of the raindrops, the momentum and kinetic energy of rainfall can be determined by summation of these values for individual drops. The intensity of rainfall can be derived from rain-gauge data, and this can be compared with kinetic energy (Figure 9.5). As the kinetic energy of rain obviously varies from one fall to the next, the frequency and duration of rainfall must also be taken into account if a worthwhile annual value is to be obtained.

Raindrop impact may give rise to surface crusting in soils containing high proportions of silt or fine sand or in those with collapsible structures. When detached, these finer particles tend to clog pores and fill cracks in the soil, thereby reducing the infiltration rate. For example, Brakensiek and Rawls (1983) suggested that on average crusted soils can reduce the infiltration rate by between 15 and 20 mm h^{-1}.

Run-off on slopes occurs primarily when the intensity of the rainfall exceeds the infiltration capacity of the soil. The latter varies with soil type (e.g. 0.25–0.5 mm h^{-1} in clayey loams, 25–50 mm h^{-1} in sandy loams) but tends to be highest when a soil is dry. Hence, the infiltration capacity can change during a rainstorm in that it decreases as the storm continues as a result of the soil becoming wetter and being slightly compacted. Once a soil is fully saturated, it can take

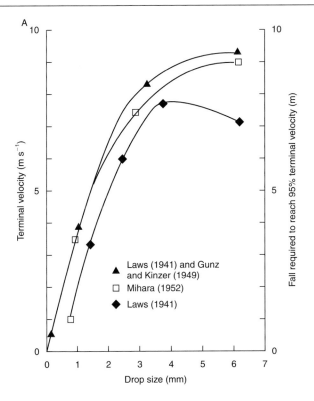

Figure 9.4 Relations between drop size, terminal velocity and fall required to attain 95% of terminal velocity (after Smith and Wischmeier, 1962, by permission of Academic Press, Inc.).

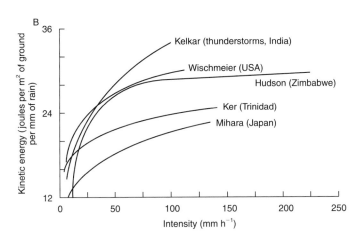

Figure 9.5 The relation between kinetic energy and intensity of rainfall, from studies by various workers in different localities (after Hudson, 1981).

up no further rainfall. Infiltration is also affected by vegetation in that not only does vegetation impede surface flow but also the plant root system can aid penetration of water into the soil. Once the infiltration capacity is satisfied, overland flow begins in the form of sheet wash. As remarked above, the laminar movement of water, as occurs at the beginning of sheet wash, does not cause erosion. Flow becomes turbulent when a critical depth and velocity are reached as water proceeds downslope. Erosion begins at this juncture; therefore, upslope of this critical distance a zone exists where only raindrop erosion can take place. Rills and gullies are formed downslope of this critical distance. To summarize, the factors that govern run-off erosion include rainfall intensity, infiltration capacity, length of overland flow, slope angle and surface cover or roughness.

Erosion by flowing water begins above a certain critical velocity, which varies primarily with particle size. Particles of fine to medium sand are the most easily eroded. Once entrained, the velocity required to keep a particle in suspension is less than that required for its removal. This is especially the case in fine-grained soils.

The depth and velocity of water increase as it is concentrated into rills, with erosive power being enhanced accordingly. The initial system of rills develops rapidly and as they are enlarged and join, so gullies are formed. Flow in rills takes place only after rainfall, that is, they are ephemeral. Smaller gullies may also be empheral, and in semi-arid and especially arid regions large gullies are occupied by water only when it rains. Gullies may also form independently where, for example, breaks in vegetation cover allow erosion to be initiated. Their growth and development involves landward erosion, bank collapse and downward scouring. Frequently, the accelerated erosion has been attributable to an intensification of cultivation of upland watersheds, where run-off volumes have trebled in some instances. The presence of pseudo-karst features in loessic soils can aid the development of gullies.

Vegetation can affect the capacity of flowing water to erode and transport soil by reducing the flow velocity as a result of increased surface roughness. The greatest reductions in flow velocity are brought about by dense uniform vegetation covers. Open tussocky vegetation is less effective and in fact can give rise to localized increases in flow velocity between tussocks, this generating higher erosion. In addition to retarding flow, vegetation also tends to filter soil particles being transported by the flow: the denser the vegetation, the more sediment can be trapped.

9.3 Sediment yield

The movement of sediment through a river system is fundamentally important in relation to the environmental management of the drainage basin. Once sedimentary particles have been removed from slopes they are normally incorporated into the river system. Their transport is generally intermittent in that they are stored within the system. Wolman (1967) showed that in a particular drainage basin in Maryland, 52% of eroded sediment was stored on slopes, 14% was stored in flood plains and 34% of total upland erosion was delivered to the outflow point as sediment yield. Hence, only a proportion of the sediment generated within a drainage basin is moved out of it and the details of movement within the system are complex.

Sediment yield or net erosion is measured at a particular point in a drainage basin. The movement of sediment on slopes tends to be sporadic and uneven, whereas storage occurs in flood plains and valley fills. Therefore, it is usually difficult to obtain an accurate measure of the overall rate at which soils are being removed from slopes in a catchment area. Thus sediment yield is a measure of the erosion that takes place in a drainage basin, minus storage, and quantifies the rate of sediment exported out of the local erosional system. Sediment yield can be divided by the total production of sediment to give the sediment delivery ratio, where individual slopes can be monitored to obtain the rate of actual erosion that occurs. Moreover, the average areal rate of denudation decreases with increasing size of drainage basin.

The sediment yield is defined as the total sediment outflow from a watershed or drainage basin in a given period of time. As noted, all the sediment loss is not delivered to the river system, since some of it is deposited at various positions within the drainage basin. The usual method of obtaining values of sediment yield is by measuring the suspended sediment and bedload. However, as this can be an expensive operation, it may be necessary to use alternative methods to estimate sediment yield. These include predictive equations, gross erosion and sediment delivery ratio computations, and suspended sediment load or reservoir sediment deposition measurements. According to Mitchell and Bubenzer (1980), predictive equations can normally only be applied regionally. They are based on watershed parameters such as amount and intensity of rainfall, amount or peak rate of run-off, temperature, size of drainage area, slope, soil descriptions, and land-use descriptions. Geological and time factors may also be incorporated into equations.

The gross erosion in a drainage basin includes erosion arising from inter-rill, rill, gully and stream action. The sediment delivery ratio is dependent upon the size of the drainage area, the relief of the drainage area, the stream length, the bifurcation ratio, the texture of the eroded material and the proximity of the sediment source to the stream. This ratio shows what proportion of the gross erosion is moved beyond a given point and in so doing yields the amount of sediment storage within a drainage area. Estimates of the sediment delivery ratio are given in Table 9.3. However, the use of such estimates should take into consideration other factors that may affect the ratio at a particular location. In particular, a higher sediment delivery ratio should be used where silt or clay soils are present and a lower one for coarse-grained soils.

Reservoirs of known age and sedimentation history provide good data sources for determining sediment yields. However, reservoir deposition and sediment

Table 9.3 General sediment delivery ratio estimates.

Drainage area (square kilometres)	Sediment delivery ratio
0.05	0.58
0.1	0.52
0.5	0.39
1	0.35
5	0.25
10	0.22
50	0.153
100	0.127
500	0.079
1000	0.059

Source: US Soil Conservation Service (1971) *Sediment Sources, Yields and Delivery Ratios*. SCS National Engineering Handbook, US Department of Agriculture, Washington, DC.

yield are not synonymous and the reservoir trap efficiency must be used to account for the difference. Nonetheless, accumulation of sediment in a reservoir over a given period of time can be used to derive an average annual sediment yield once the trap efficiency has been taken into account (Rausch and Heinemann, 1984). The trap efficiency of a reservoir is the portion of the total sediment delivered to the reservoir that is retained in the reservoir. Trap efficiency has been related to the reservoir storage capacity–mean annual inflow ratio.

The suspended sediment load is assessed by measuring the suspended sediment transported by a stream. Stream discharge is determined by stream gauging. The measured sediment concentration is converted to sediment load, in mass per time period, using the average concentration and the volume of flow for a time period. The sediment load versus stream discharge rates for many sample periods can then be plotted on logarithmic paper to provide the basis for a sediment rating curve for a stream. A flow duration curve (see Figure 6.9) can be used with a sediment rating curve (Figure 9.6) to determine the average annual suspended sediment yield. This determination does not include the bedload. The sediment rating curve–flow duration method provides a useful means of predicting sediment yield, allowing extrapolation from a short period of sediment records to much longer periods.

Brune (1951) summarized the empirical relationships between sediment yield, drainage area, proportion of cultivated land and run-off for certain areas of the United States. He also produced a nomogram for computing sediment yield in the upper Mississippi region (Figure 9.7).

In a discussion of the many empirical methods that have been used to estimate sediment yield, Cooke and Doornkamp (1990) concluded that they provide a useful means of examining erosion rates worldwide. However, they went on to state that they only define those major regions that are exceptionally

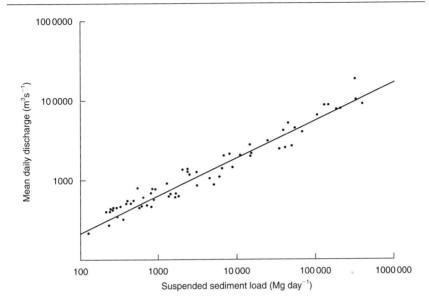

Figure 9.6 Sediment rating curve for the Elkhorn River, Waterloo, Nebraska (after Livesey, 1974).

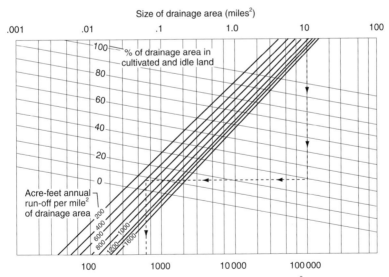

Figure 9.7 Nomogram for calculating sediment yield (for the Upper Mississippi region). The example is for a 10-square-mile basin – follow the arrow on drainage area, to percentage of cultivated land, to run-off and then to sediment yield (after Brune, 1951).

prone to erosion and where man has to take the greatest care when interfering with nature.

The bedload transport is generally determined separately. Bedload may vary from zero to nearly all of the total load, depending on the sediment sources and transport capability of the stream. Bedload transport dominates when the ratio between the lifting forces and the stabilizing forces on a particle is less than 0.2, that is, when particles are large in relation to the force of flow.

9.4 The Universal Soil Loss Equation

As noted above, the most frequently used method to determine the soil loss is the Universal Soil Loss Equation (USLE) developed by the United States Soil Conservation Service:

$$A = RKLSCP \tag{9.1}$$

where A is the average annual loss of soil, R is rainfall erosivity, K is the soil erodibility factor, L is the slope length factor, S is the slope gradient factor, C is the cropping management factor and P is the erosion control practice factor. The method was developed as a means of predicting the average annual soil loss from inter-rill and rill erosion. However, the equation is not universally applicable. Its primary purpose is to predict losses from arable or non-agricultural land by sheet and rill erosion and to provide guides for the selection of adequate erosion control practices (Wischmeier and Smith, 1978; Mitchell and Bubenzer, 1980). As such, the USLE can be used to determine how conservation practices can be applied or altered to permit more intensive cultivation, to predict the change in soil loss that results from a change in cropping or conservation practices, and to provide soil loss estimates for conservationists to use for determining conservation needs. Lastly, the USLE provides a means of estimating the yield of sediments from watersheds.

The rainfall erosivity factor in the USLE was developed by Wischmeier (1959), who showed that storm losses from fallow plots were highly correlated with the product of total kinetic energy and the maximum 30 minute rainfall intensity. This is expressed in terms of the EI_{30} index, which shows how much soil is lost due to rainfall, where E is the kinetic energy and I_{30} is the greatest intensity of rainfall in a 30 minute period (Smith and Wischmeier, 1962). The rainfall erosivity factor, R, is obtained by dividing EI_{30} by 173.6. The rainfall erosivity indices can be summed for any time period to provide the erosivity of rainfall for that period. Long-term records of rainfall provide average annual values of erosivity index, and these can be used to produce maps of rainfall erosivity (Figure 9.8). Unfortunately, this index does not always correlate well with soil erosion outside the United States. An alternative erosivity index, developed by Fournier (1960), included suspended sediment yield, month with highest precipitation, mean annual precipitation, and mean height and slope of

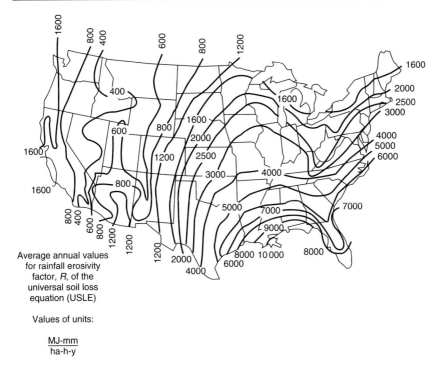

Average annual values
for rainfall erosivity
factor, R, of the
universal soil loss
equation (USLE)

Values of units:

$$\frac{\text{MJ-mm}}{\text{ha-h-y}}$$

Figure 9.8 Rainfall and run-off erosivity index, R, by geographic location (after Foster *et al.*, 1981).

basin. Although the correlation is not perfect, high erosivity is frequently associated with very high values of drainage density developed in areas of weak sediments.

Soil splash varies according to how easily a soil particle can be separated and transported away on the one hand and to the vulnerability of the surface to the splash on the other. In this regard, the type and density of soil cover is all-important, as is the type of soil, its infiltration capacity and the slope of the surface. For example, splash erosion, E_s, has been represented in terms of:

$$E_s = C(D/AkS) \tag{9.2}$$

where C is a constant of proportionality, D is soil dispersion by rain (or its cohesiveness), A is the infiltration capacity of the soil, k is the hydraulic conductivity of the soil and S is the mean grain size of the soil. Although individual particles of soil may be ejected up to 600 mm into the air by splashing raindrops and moved as much as 1.5 m away, they need to be on a sloping surface to move out of the area. For example, Ellison (1947) showed that there can be a net loss of soil from a slope of $6°$ as a result of raindrop splash, 75%

of soil splash being downslope. Splash erosion is most effective shortly after the surface has been wetted and decreases soon afterwards. This is due to the soil being slightly compacted, the surface film of water increasing in thickness and the finer soil particles having been removed. In addition, splash action can further facilitate soil erosion by destroying the soil structure and so making it more susceptible to run-off, it can remove nutrients from the soil and can lead to the formation of a dry crust at the surface, which reduces infiltration capacity.

Additional methods of defining an erosion factor have been introduced with the intention of simplifying the original method and deriving values that are more applicable to regional conditions. For example, attempts have been made to relate the rainfall erosion index to rainfall intensity, duration and frequency. Mitchell and Bubenzer (1980) discussed these various methods, noting their advantages or disadvantages. More recently, thaw and snowmelt run-off have been taken into account, and Wischmeier and Smith (1978) recommended that a rainfall erosivity factor equal to 0.0591 times the December to March precipitation be added to the R factor value normally determined.

The soil erodibility factor, K, in the USLE is a quantitative description of the inherent erodibility of a particular soil. It reflects the fact that different types of soil are eroded at different rates. Many of these factors have been discussed above, such as infiltration capacity, saturation, permeability, dispersion, splash and transport of soil particles. Other factors of importance include the texture or crumb structure of the soil, the stability of the mineral aggregates, the strength of the soil, and its chemical and organic components. A nomogram can be used to derive the soil erodibility factor (Figure 9.9). Five soil parameters are needed to use the nomogram, namely, percentage of silt and fine sand; percentage of sand (0.1–2.0 mm); organic matter content; structure; and permeability. Some examples of the value of K are given in Table 9.4.

The effects of slope length and gradient are represented separately in the USLE, but they are frequently evaluated as a single topographic factor, LS. In this instance, slope is defined as the distance from the point at which overland flow begins to the position where there is a sufficient decrease in the gradient of the slope for deposition to occur or where run-off enters a defined channel. The gradient of the slope is usually expressed as a percentage and is the field or segment slope. The development of the USLE was based on a standard plot length of 22 m. Consequently, the slope length factor, L, was expressed as

$$L = x^m/22 \tag{9.3}$$

where x is the slope length and m is an exponent. The values of m for slopes from 1 to 5% increase from 0.2 to 0.5, respectively. The LS factor is therefore used to predict the soil loss per unit area of a given field slope to that from a standard plot length. It can be obtained from Figure 9.10. This figure should be used only for single uniform slopes. The topographic factor usually overestimates the loss of soil from concave slopes and underestimates that from

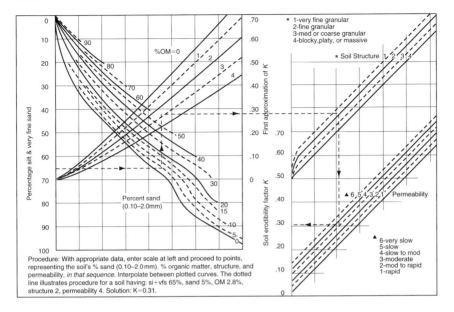

Figure 9.9 Nomogram for determining the soil erodibility factor, K, for US mainland soils (Anon., 1975).

Table 9.4 Indications of the general magnitude of the soil erodibility factor K*.

Texture class	Organic matter content		
	<0.5 percent	2 percent	4 percent
Sand	0.05	0.03	0.02
Fine sand	0.16	0.14	0.10
Very fine sand	0.42	0.36	0.28
Loamy sand	0.12	0.10	0.08
Loamy fine sand	0.24	0.20	0.16
Loamy very fine sand	0.44	0.38	0.30
Sandy loam	0.27	0.24	0.19
Fine sandy loam	0.35	0.30	0.24
Very fine sandy loam	0.47	0.41	0.33
Loam	0.38	0.34	0.29
Silt loam	0.48	0.42	0.33
Silt	0.60	0.52	0.42
Sandy clay loam	0.27	0.25	0.21
Clay loam	0.28	0.25	0.21
Silty clay loam	0.37	0.32	0.26
Sandy clay	0.14	0.13	0.12
Silty clay	0.25	0.23	0.19
Clay		0.13–0.29	

* The values shown are estimated averages of broad ranges of specific soil values. When a texture is near the borderline of two texture classes, use the average of the two K values. For specific soils, use of Figure 9.9 or Soil Conservation Service K-value tables will provide much greater accuracy.

Source: Agricultural Research Service (1975) *Control of Water Pollution from Cropland Volume 1, A Manual for Guideline Development*. Report ARS-H-5-1, US Department of Agriculture, Washington, DC.

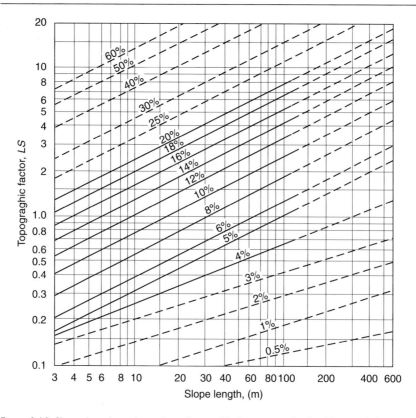

Figure 9.10 Slope length and gradient factor, LS, for use with the Universal Soil Loss Equation (after Wischmeier and Smith, 1978).

convex slopes. Irregular slopes have to be divided into a series of segments, with each segment being uniform in gradient and soil type. This allows the soil loss for the whole slope to be determined (Forster and Wischmeier, 1974).

The cropping management factor, C, represents the ratio of soil loss from a specific cropping cover to the soil loss from a fallow condition in the same soil type, same slope and same rainfall. As mentioned above, vegetation inhibits soil erosion by reducing the effectiveness of raindrop impact and run-off, as well as improving infiltration capacity. This factor takes account of the nature of the cover, crop sequence, productivity level, length of growing season, tillage practices, residue management, and expected time distribution of erosion events. Hence the evaluation of this factor frequently proves difficult. Table 9.5 gives the cropping and residue management variables as percentages of soil loss for crops to that for continuous fallow. The annual distribution of the rainfall erosivity index varies with location, and examples are provided in Figure 9.11. The C factor for a particular crop rotation is derived by multiplying the soil

Table 9.5 Percentage of soil loss from crops to corresponding loss from continuous fallow (after Wischmeier and Smith, 1978).

Cover sequence and management[a]	Spring planting		Soil loss ratio (%)[d] for crop stage period and canopy cover[e]							
	Residue[b] (kg)	Cover[c] (%)	F	SB	1	2	3:80	3:90	3:96	4L
Continuous										
Corn, rdl., sprg, TP	5000		36	60	52	41		24	20	30
Small grain	5000	60		16	14	12	7	4	2	
Meadow							1			
Rotation										
Rowcrop after meadow			12	27	23	20		14	12	21
Corn after beans, sprg,			47	78	65	51		30	25	37
Beans after corn, sprg, TP			39	64	56	41		21	18	
Conservation tillage										
Beans or corn after corn	5000	60		13	11	10		10	8	20
Corn after beans		30		33	29	25	22	18	14	33
Small grain after corn	5000	60		16	14	13	7	4	2	

a   rdl, crop residue left in field; sprg, spring tillage; TP, ploughed with mouldboard.
b   Dry mass per hectare, after winter loss and reductions by grazing or partial removal; 5000 kg ha^{-1} represents a yield of 6 to 8 Mg ha^{-1}.
c   Percentage of soil surface covered by plant residue mulch after crop seeding. The difference between spring residue and that on the surface after crop seeding is reflected in the soil loss ratios as residues mixed with the topsoil.
d   The soil loss ratios assume that the indicated crop sequence and practices are followed consistently. One-year deviations from normal practices do not have the effect of a permanent change.
e   Crop stage periods: F, rough fallow; SB, seedbed until 10% canopy cover; 1, establishment until 50% canopy cover; 2, development until 75% canopy cover; 3, maturing until harvest for three different levels of canopy cover; 4L, residue or stubble.

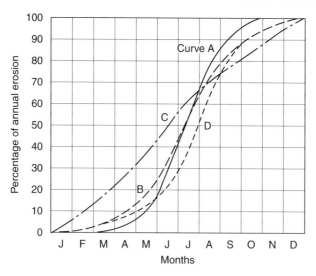

Figure 9.11 Monthly distribution of the rainfall and run-off erosivity index. Curve A: northwestern Iowa, northern Nebraska and southeastern South Dakota. Curve B: northern Missouri, central Illinois, Indiana and Ohio. Curve C: Louisiana, Mississippi, western Tennessee and eastern Arkansas. Curve D: Atlantic coastal plains of Georgia and the Carolinas (after Smith and Wischmeier, 1962).

loss ratios for each growth period in the crop rotation (Table 9.5) by the percentage of annual erosion during each period (Figure 9.11). These products are then summed to give the *C* factor. In addition, each segment of the cropping and management sequence must be evaluated along with the rainfall erosivity distribution for the region. Some examples of *C* values are given in Table 9.6.

Wischmeier (1975) subdivided the *C* factor into three classes. Thus C_1 refers to the effect of a plant canopy at some elevation above the soil; C_{11} refers to the effects of a mulch or close-growing vegetation in direct contact with the soil; and C_{111} refers to the residual effects of land use on the soil structure, porosity and density, organic content, and the effects of roots and biological activity in the soil, as well as the effects of tillage or lack of it on surface roughness. However, Styczen and Morgan (1995) pointed out the limitations in doing this since at least two of the subfactors may influence more than one erosion process. It is therefore difficult to assign them a precise physical meaning. They went on to argue that if the role of vegetation in protecting the soil against erosion is to be understood, then the erosion processes and how each of the processes may be affected by vegetation must be understood. The salient properties of the vegetation which most affect these processes should be determined in an attempt to quantify the combined effect of vegetation on the processes acting together in different situations.

Table 9.6 C-factor values for the Universal Soil Loss Equation (from Morgan, 1995).

Practice	Average annual C-factor
Bare soil	1.00
Forest or dense shrub, high-mulch crops	0.001
Savanna or prairie grass in good condition	0.01
Overgrazed savanna or prairie grass	0.10
Maize, sorghum or millet: high productivity, conventional tillage	0.20–0.55
Maize, sorghum or millet: low productivity, conventional tillage	0.50–0.90
Maize, sorghum or millet: low productivity, no or minimum tillage	0.02–0.10
Maize, sorghum or millet: high productivity, chisel ploughing into residue	0.12–0.20
Maize, sorghum or millet: low productivity, chisel ploughing into residue	0.30–0.45
Cotton	0.40–0.70
Meadow grass	0.01–0.025
Soya beans	0.20–0.50
Wheat	0.10–0.40
Rice	0.10–0.20
Groundnuts	0.30–0.80
Palm trees, coffee, cocoa with crop cover	0.10–0.30
Pineapple on contour: residue removed	0.10–0.40
Pineapple on contour: with surface residue	0.01
Potatoes: rows downslope	0.20–0.50
Potatoes: rows across-slope	0.10–0.40
Cowpeas	0.30–0.40
Strawberries: with weed cover	0.27
Pomegranate: with weed cover	0.08
Pomegranate: clean-weeded	0.56
Ethiopian tef	0.25
Sugar cane	0.13–0.40
Yams	0.40–0.50
Pigeon peas	0.60–0.70
Mungbeans	0.04
Chilli	0.33
Coffee: after first harvest	0.05
Plantains: after establishment	0.05–0.10
Papaya	0.21

The erosion control practice factor, P, is the ratio of soil loss using a specific practice compared with the soil loss that takes place when ploughing up and down a slope. This factor therefore takes into account erosion control practices such as contouring, contour-strip cropping and terracing. The P factor is therefore derived as follows:

$$P = P_c \times P_s \times P_t \tag{9.4}$$

where P_c is the contouring factor based on slope; P_s is the strip cropping factor for given strip widths; and P_t is the terracing factor. Table 9.7 gives values of the erosion control practice factor for these three types of control practice, according to the slope of the land.

Kirkby (1980) pointed out that there were a number of difficulties inherent in the USLE. For instance, the method does not allow for any non-linear interactions between the factors involved. In addition, the overland flow produced in a small quadrat (and added to flow accumulated from upslope) is equal to rainfall intensity minus infiltration rate. As the maximum infiltration rate is dependent on soil rather than rainfall properties, neither overland flow nor the soil transported by it can be expressed as a rainfall factor multiplied by a soil factor, as occurs in the USLE. Hence, the USLE does not represent a totally satisfactory means of evaluating the soil erosion system. Because the USLE was developed in the United States, attempts have been made to refine it so that it can be applied to other regions of the world. For example, Elwell (1984) developed a soil loss estimation system for southern Africa.

The aim of soil erosion control is to reduce soil loss in order that soil productivity can be maintained economically. In this context, the soil loss tolerance is defined as the maximum rate of soil erosion that allows a high level of productivity to be sustained. Tolerances vary from place to place because soil type and rate of erosion vary. In the United States, deep, medium-textured, moderately permeable soils have been assigned tolerance losses of around 1.1 kg m^{-2} y^{-1} (Mitchell and Bubenzer, 1980). Soil with a shallow root zone or other detrimental characteristics were given lower tolerance values. Kirkby (1980) suggested that tolerances of 0.251 to 1.255 kg m^{-2} y^{-1} generally may be suitable.

In order to predict the effect that soil erosion has on soil productivity, Williams (1985) developed a computer model named EPIC (Erosion

Table 9.7 Erosion control practice factor, P, with maximum slope lengths and maximum strip crop widths for different slopes[a] (after Wischmeier and Smith, 1978).

Land slope percentage	Contouring	Maximum slope length[b] (m)	Maximum strip-crop width (m)	Contour strip-cropping and irrigated furrows	Terracing[c]
1–2	0.60	120	40	0.30	0.12
3–8	0.50	90	30	0.25	0.10
9–12	0.60	36	24	0.30	0.12
13–16	0.70	24	24	0.35	0.14
17–20	0.80	18	18	0.40	0.16
21–25	0.90	15	15	0.45	0.18

[a] Factor for farming upslope and downslope is 1.0.
[b] Maximum slope length for strip-cropping can be twice that for contouring only.
[c] For prediction of contribution to off-field sediment load.

Productivity Impact Calculator). The model includes various factors that take account of erosion on the one hand and plant growth on the other, such as hydrology, weather, water and wind erosion, nutrients, soil temperature, and tillage. Economics and crop budgets also figure in the model.

9.5 Soil erosion by wind

Wind erosion is most effective in those regions that lack the protection afforded by vegetation to the soil, notably in arid and semi-arid zones (Figure 9.12), as well as some humid areas that experience periodic droughts. The redistribution and resorting of particles by wind erosion may have profound effects on the soils affected, their related micro-topography and any agricultural activity

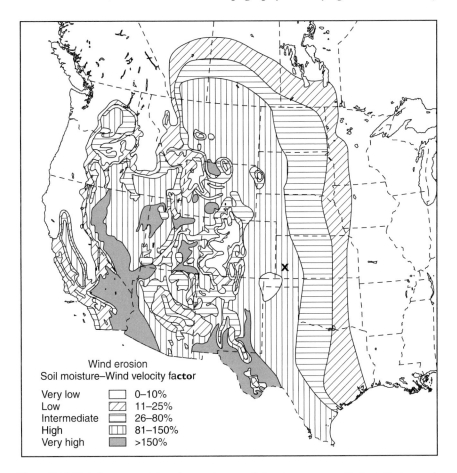

Figure 9.12 Relative potential soil loss by wind in the western United States and southern Canada as a percentage of that in the vicinity of Garden City, Kansas, marked by **X** (after Chepil *et al.*, 1962).

associated with them. Uncultivated fields, in particular, are susceptible to erosion by wind, especially when the soil contains appreciable silt-sized material. The consequences (e.g. soil degradation, crop damage – see Table 9.8) are most serious in agricultural areas that experience low, variable, and unpredictable rainfall, high temperatures and rates of evaporation, and high wind velocity. In such areas, the natural process of wind erosion may be accelerated by imprudent agricultural practices.

Surface winds that are capable of initiating soil particle movement are turbulent. As far as wind erosion is concerned the most important characteristics of soil particles are their size and density. In other words, there is a maximum size for a wind of a given velocity. In terms of the wind itself, the force of the wind at the ground surface is the main factor affecting erosivity. Once particles have started moving, they exert a drag effect on the airflow, which then alters the velocity profile of the wind. The velocity of wind blowing over a surface increases with height above the surface, the roughness of the surface retarding the speed of the wind immediately above it. As more material is picked up, the surface wind speed is reduced. Hence, the more erodible a soil, the greater the proportion of grains involved in saltation and so the greater is the reduction in wind speed (Chepil and Milne, 1941). The trajectories of saltating particles are influenced by the forward force of the wind, the mass of the grains and the force of gravity. When first picked up, particles may rise almost vertically, but as they enter more quickly moving air their trajectories are flattened. As the upward force involved in lifting particles is dissipated, so they fall back to the ground under the influence of gravity. When a particle hits the ground it may rebound back into the airflow. The impact of a particle on a surface of loose material may cause another particle to lift into the air. Entrainment as a result of impact can occur at a lower shear velocity than the threshold velocity required to initiate movement. Saltating particles, depending on their size and a lesser extent on bed roughness, may rise to a height of 2 m and the length of the trajectory is frequently around ten times the height. Generally, wind erosion is more effective in drier soils, so the problem is most serious when wind velocities and evaporation rates are high, and precipitation is low.

Soil particles are also moved in suspension and by traction. Suspension occurs when the terminal velocity of fall is lower than the mean upward velocity in a turbulent airflow. Bagnold (1941) explained traction or creep in terms of surface disruption by particles being impacted from behind and so moved on. Grains may also be rolled across the surface. Obviously, the proportions of soil particles moved in saltation, suspension and traction varies primarily with wind speed and particle size. However, Chepil (1945) maintained that most particles are moved by saltation, and others have agreed. Chepil suggested that between 50 and 75% of soil particles are transported by saltation, with 3 to 40% moving in suspension and 5 to 25% in traction. He also found that very few particles with diameters above 0.84 mm are eroded.

Bagnold (1941) noted that sand flow can increase downwind until it becomes

Table 9.8 Some physical and economic effects of wind erosion (after Cooke and Doornkamp, 1990).

Physical effects	Economic consequences
Soil damage (1) Fine material, including organic matter, may be removed by sorting, leaving a coarse lag. (2) Soil structures may be degraded (3) Fertilizers and herbicides may be lost or redistributed	*Soil damage* (1, 2, 3) Long-term losses of fertility give lower returns per hectare (3) Replacement costs of fertilizers and herbicides.
Crop damage (1) The crop may be covered by deposited material. (2) Sand-blasting may cut down plants or damage the foliage. (3) Seeds and seedlings may be blown away and deposited in hedges or other fields. (4) Fertilizer redistributed into large concentrations can be harmful (5) Soil-borne disease may be spread to other fields. (6) Rabbits and other pests may inhabit dunes trapped in hedges and feed on the crops.	*Crop damage* (1–6) Yield losses give lower returns. (1–3) Replacement costs, and yield losses due to lost growing season. (5) Increased herbicide costs.
Other damage (1) Soil is deposited in ditches and hedges, and on roads. (2) Fine material is deposited in houses, on washing and cars, etc. (3) Farm machinery, windscreens etc. may be abraded, and machinery 'clogged'. (4) Farm work may be held up by the unpleasant conditions during a 'blow'.	*Other damage* (1) Costs of removal and redistribution. (2, 3) Cleaning costs. (4) Loss of working hours and hence productivity declines.

saturated, that is, the amount of material in motion is the maximum sustainable by the wind. Beyond this point no further removal downwind takes place. The downwind increase in quantity of material transported is referred to as avalanching and leads to increased surface abrasion. However, particles that are too large to be removed inhibit wind erosion. Hence, as wind erosion continues the proportion of non-erodible particles increases until eventually they protect the erodible particles from wind erosion. At this point, a wind-stable surface is produced.

Chepil and Woodruff (1963) recognized four types of soil aggregations that produced textures that are more resistant to wind erosion: primary aggregates, secondary aggregates or clods, fine material among clods, and surface crusts. Primary aggregates are bound together by clay and colloids. Clods are cemented together in the dry state by fine particles, and the cohesion between clods is also provided by silt and clay particles. Surface crusts are compacted by raindrop impact. If these aggregations are to be moved by wind, then they should be broken down by weathering, raindrop impact, flowing water or wind abrasion. The resistance of these aggregations to wind abrasion depends upon the soil type and its cohesion.

Vegetation cover, surface roughness, surface length and soil moisture also affect wind erosion. Obviously, vegetation cover affords protection against wind erosion, and the denser and taller it is, the more protection it affords. In addition, plant detritus helps to protect the surface, and organic matter helps to bind soil particles together. Vegetation also interrupts the flow of air and in so doing exerts a drag effect upon the velocity. Figure 9.13 illustrates how wind velocity changes with height within the vegetation zone as a function of leaf area density for a selection of crop types. The root system enhances the strength of the soil, and moving soil is trapped by plants. In general, the rougher the surface, the more effective it is in reducing the velocity of the wind immediately above it and so, in turn, erosion is reduced. The longer the surface over which wind can blow, the more efficient it becomes in terms of erosion. Lastly, soil moisture content imparts a degree of cohesion to soil particles and so helps to impede wind erosion.

In most soils, according to Chepil and Woodruff (1963), the nature and stability of the textural aggregations govern the erodibility of the soil. Hence, in a well-textured soil the number of soil particles that are small enough to be moved may be very low, so the abrasivity of the wind is also low. The stability of the textural aggregations depends upon the soil moisture content, organic content, cation exchange capacity, soil suction and cement bonds on the one hand, and those processes responsible for breakdown on the other.

The Wind Erosion Equation is similar to the Universal Soil Loss Equation and takes into account the major factors involved in wind erosion (Woodruff and Siddoway, 1965). It is as follows:

$$E = F(I, K, C, L, V) \qquad (9.5)$$

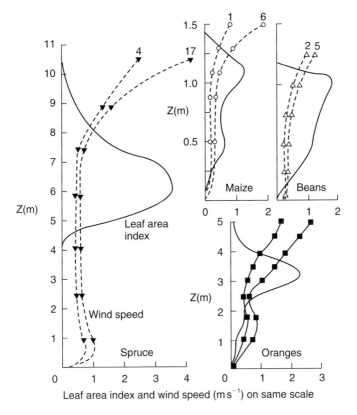

Figure 9.13 Wind velocity as a function of height and leaf area index for four crop types (after Landsberg and James, 1971).

where E is the potential erosion measured in tonnes per hectare per year; I is a soil erodibility index; K is a soil ridge roughness factor; C is a local wind erosion climatic factor (i.e. wind speed and duration); L is the median unsheltered field length along the prevailing wind erosion direction; and V is the equivalent vegetation cover. The wind erodibility, I, is a function of soil aggregates with diameters that exceed 0.84 mm and refers to the soil loss from a wide unsheltered field which has a bare smooth surface. Table 9.9a summarizes soil erodibility values for different textures of soil. However, increased wind erosion around hills (knolls) means that I must be multiplied by adjustment factors (Table 9.9b). Erodibility can be decreased by increasing the proportion of non-erodible clods (>0.84 mm diameter) at the surface of the soil. This can be brought about in some soils by tillage. The roughness factor, K, is a measure of the effect of ridges made by tillage and planting machinery on the erosion rate. Ridges absorb and deflect wind energy, and trap moving particles of soil. On the other hand, too much roughness causes turbulence, which may accelerate the movement of particles. The climatic factor, C, is based on the principle of erosion varying directly as the cube of wind speed and inversely as the square of soil moisture (Chepil *et al.*, 1962). The unsheltered distance is taken as the length from a sheltered edge of a field, parallel to the direction of the prevailing wind, to the end of the unsheltered field. The effect of vegetative cover in the Wind Erosion Equation is expressed in terms of the type, density and orientation of vegetative matter to its equivalent of small grain residue (i.e. flattened wheat straw defined as 254 mm tall stalks lying flat on the soil in rows normal to the wind direction). Since the relationships between the variables in the equation are complex, wind erosion cannot be estimated just by multiplying the variables, and either nomograms and tables or computer programs are needed to obtain answers (Skidmore *et al.*, 1970; Hagen, 1991). The Wind Erosion Equation tends to overestimate wind erosion and to give higher

Table 9.9 Wind erodibility.

(a) Wind erodibility indices for different soil textures*.

Predominant soil texture	Erodibility group	Soil erodibility index I^a $(Mg\,ha^{-1}\,y^{-1})$
Loamy sands and sapric organic material	1	360–700
Loamy sands	2	300
Sandy loams	3	200
Clays and clay loams	4	200
Calcareous loams	4L	200
Noncalcaerous loams, silt loam <20% clay, and humic organic soils	5	125
Noncalcareous and silt loams >20% clay	6	100
Silt, noncalcareous silty loam and fibric organic soils	7	85
Wet or rocky soils not susceptible to erosion	8	

[a] The I factors for Group 1 vary from 360 for coarse sands to 700 for very fine sands. Use 500 for an average.

(b) Knoll erodibility adjustment factors*

Slope change in prevailing wind erosion direction (%)	Knoll adjustment to I (factor)	Increase at crest area where erosion is most severe (factor)
3	1.3	1.5
4	1.6	1.9
5	1.9	2.5
6	2.3	3.2
8	3.0	4.8
10	3.6	6.8

* From US Soil Conservation Service (1988) Wind erosion. In *National Agronomy Manual*, American Society Agronomists, Madison, Wisconsin, 502/1–502/159.

values of soil loss tolerance levels. Obviously, this must be taken into account if the equation is to be used for planning of erosion control.

9.6 Erosion control and conservation practices

Satisfactory erosion control, according to Morgan (1980), implies obtaining the maximum sustained level of production from a given area of land and at the same time keeping soil loss below a given threshold level. In other words, this implies maintaining equilibrium in the soil system so that the rate of loss does not exceed the rate at which new soil is formed. However, this state of equilibrium is difficult to predict and therefore to attain in practice. Accordingly, less rigorous targets are generally pursued, so that the rate of erosion is controlled to maintain soil fertility over a period of 20 to 25 years. Nonetheless, there are doubts regarding whether this target can maintain soil resources adequately.

Tolerable rates of soil loss obviously vary from region to region depending upon type of climate, vegetation cover, soil type and depth, slope, and farming practices. Arnoldus (1977) proposed a range of values for tolerable mean yearly soil loss from $0.2 \text{ kg m}^{-2} \text{ y}^{-1}$ for soils where the rooting depth is up to 250 mm to $1.1 \text{ kg m}^{-2} \text{ y}^{-1}$ where the rooting depth exceeds 1.5 m (cf. Mitchell and Bubenzer, 1980; and Kirby, 1980; Table 9.10). Nevertheless, where erosion rates are very high, higher threshold values may have to be adopted.

As Morgan (1995) indicated, conservation measures must protect the soil from raindrop impact, must increase the infiltration capacity of the soil to reduce the volume of run-off, must improve the aggregate stability of the soil to increase its resistance to erosion, and lastly, must increase the roughness of the ground surface to reduce run-off and wind velocities. In fact, soil conservation practices can be grouped into three categories, namely, erosion prevention practices, which include agronomic measures, and mechanical measures

Table 9.10 Recommended values for maximum permissible soil loss ($\text{kg m}^{-2} \text{ y}^{-1}$) (from Morgan, 1980).

Meso-scale (e.g. field level)	
Deep fertile loamy soils; values used in mid-west of USA	0.6–1.1
Thin, highly erodible soils	0.2–0.5
Very deep loamy soils derived from volcanic deposits, e.g. in Kenya	1.3–1.5
Soil depths: 0–250 mm	0.2
250–500 mm	0.2–0.5
500–1 m	0.5–0.7
1–1.5 m	0.7–0.9
over 1.5 m	1.1
Probable realistic value for very erodible areas, e.g. mountains in the tropics	2.5
Macro-scale (e.g. drainage basins)	0.2
Micro-scale (e.g. construction sites)	2.5

Table 9.11 Soil conservation practices (after Morgan, 1980).

Practice	Control over					
	Rain-splash		Run-off		Wind	
	D	T	D	T	D	T
Agronomic measures						
Covering soil surface	*	*	*	*	*	*
Increasing surface roughness	−	−	*	*	*	*
Increasing surface depression storage	+	+	*	*	−	−
Increasing infiltration	−	−	+	*	−	−
Mechanical measures						
Contouring ridging	−	+	+	*	+	*
Terraces	−	+	+	*	−	−
Shelter belts	−	−	−	−	*	*
Waterways	−	−	+	*	−	−
Soil management						
Fertilizing manures	+	+	+	*	+	*
Increasing infiltration	−	−	+	*	−	−

− no control; + moderate control; * strong control. D = detachment; T = transport

and soil management practices (Table 9.11). Sound soil management is vital to the success of any soil conservation scheme, since it helps to maintain the texture and fertility of the soil. Agronomic measures and soil management have an influence upon the removal of soil particles and their transportation, but mechanical measures tend to affect transportation rather than particle separation. According to Morgan, agronomic measures have frequently proved successful when used on their own and more successful when combined with good soil management practice. On the other hand, mechanical measures are usually ineffective if not used together with agronomic measures. Furthermore, agronomic measures are incorporated more easily into existing farming systems than are mechanical practices. Both agronomic and mechanical measures should be used in conjunction with good soil management in areas where the erosion problem is severe.

Land capability classification is normally used for evaluation of land in soil conservation planning. The classification may be based upon or a modified version of that developed by the United States Soil Conservation Service (Klingebiel and Montgomery, 1966), the rationale of the classification system being that the correct use of land represents the best method of controlling soil erosion. The system makes an assessment of the limitations of agricultural land use, placing special emphasis on erosion risk, soil depth and climate. The land is divided into capability classes according to the severity of these limiting factors.

9.6.1 Conservation measures for water erosion

As far as water erosion is concerned, conservation is directed mainly at the control of overland flow, since rill and gully erosion will be reduced effectively if overland flow is prevented. There is, as pointed out above, a critical slope length at which erosion by overland flow is initiated. Hence, if slope length is reduced by, for example, terracing, then overland flow should be controlled. In other words, terraces break the original slope into shorter units and have been used for agricultural purposes in some parts of the world for centuries. Terraces may be regarded as small detention reservoirs that contain water long enough to help it to infiltrate into the ground. The terraces follow the contours and are raised on the outer side to form a channel. Terrace width is related primarily to ground slope (Anon, 1989) but may also be influenced by climate (i.e. intensity and duration of rainfall) and soil type (Table 9.12). Three principal types of terrace are used for erosion control, namely, diversion, retention and bench terraces (Table 9.13). Terraces help to conserve soil moisture and should remove run-off in a controlled manner. Therefore, in order to remove excess water, drainage ditches must be incorporated into the terrace system. The build-up of excess pore water pressures in a slope is one of the principal reasons for slope failure. Consequently, it is most important that terraced slopes are adequately drained. Morgan (1995) referred to the use of diversion ditches, terrace channels and grass waterways as ways of conveying water from a terraced slope (Table 9.14). Tile drains or plastic pipe drains placed in trenches in the soil and then backfilled are used instead of grass waterways in the more developed countries. These drains need much less maintenance and reduce soil loss. The terrace channel is graded towards a soak-away, which leads into the drain.

Contour farming involves following the contours to plant rows of crops. As with terraces, this interrupts overland flow and thereby reduces its velocity and

Table 9.12 Design criteria for terraces (after Hudson, 1981).

Maximum length:	normal	250 m (sandy soils) to 400 m (clay soils)
	absolute	400 m (sandy soils) to 450 m (clay soils)
Maximum grade:	first 100 m	1:1000
	second 100 m	1:500
	third 100 m	1:330
	fourth 100 m	1:250
	where a constant grade is used	1:250 is recommended
Ground slope:	diversion terraces	usable on slopes up to 7°, on steeper slopes the cost of construction is too great and the spacing too close to allow mechanized farming
	retention terraces	recommended only on slopes up to 4.5°
	bench terraces	recommended on slopes of 7 to 30°

Table 9.13 Types of terrace (from Morgan, 1995).

Diversion terraces	Used to intercept overland flow on a hillside and channel it across slope to a suitable outlet, e.g. grass waterway or soak-away to tile drain; built at slight downslope grade from contour.
Mangum type	Formed by taking soil from both sides of embankment.
Nichols type	Formed by taking soil from upslope side of embankment only.
Broad-based type	Bank and channel occupy width of 15 m
Narrow-based type	Bank and channel occupy width of 3–4 m
Retention terraces	Level terraces; used where water must be conserved by storage on the hillside.
Bench terraces	Alternating series of shelves and risers used to cultivate steep slopes. Riser often faced with stones or concrete. Various modifications to permit inward sloping shelves for greater water storage or protection on very steep slopes or to allow cultivation of tree crops and market garden crops.
Fanyajun terraces	Terraces formed by digging a ditch on the contour and throwing the soil on the upslope side to form a bank.

Table 9.14 Types of waterway (after Morgan, 1995).

Diversion ditches	Placed upslope of areas where protection is required to intercept water from top of hillside; built across slope at slight grade so as to convey the intercepted run-off to a suitable outlet.
Terrace channels	Placed upslope of terrace bank to collect run-off from inter-terraced area; built across slope at slight grade so as to convey the run-off to a suitable outlet.
Grass waterways	Used as the outlet for diversions and terrace channels; run downslope, at grade of the sloping surface; empty into river system or other outlet; located in natural depressions on hillside.

again can help to conserve soil moisture. Contour farming is probably most successful in deep, well-drained soils where the slopes are not too steep. Alternating strips of crops and grass, planted around the contours of slopes, form the basis of strip cropping. Not only do strips reduce the velocity of overland flow, they also protect the soil from raindrop impact and aid infiltration.

Strip cropping may be used on slopes that are too steep for terracing. According to Morgan (1995), the steepest slope on which strip cropping can be practised is around 8.5°. The width of the strip is reduced with increasing slope; for example, on a slope of 3° the width could be 30 m, reducing to half that on a 10° slope (Table 9.15). The widths also vary with the erodibility of the soil. The crop in each strip is normally rotated so that fertility is preserved. However, grass strips may remain in place where the risk of erosion is high.

The use of ground cover is frequently employed to control erosion in plantations of tree crops such as rubber, coffee and palm oil (Morgan and Rickson, 1995). Strips with high erosion risk species alternate with crops associated with low erosion rates or grass. The denser and more uniform the growth of grass, the more effective is the strip at retarding water flow and filtering sediment from flowing water. Any soil removed from the higher erosion strips is intercepted by the low erosion or grass strip. In alley cropping, food crops are grown in rows across the slope between rows of shrubs or trees. The trees reduce the velocity of run-off, so that Lal (1988) was able to show that a hedge spacing of 2 m reduced soil erosion by around 92%. Additional conservation measures such as mulching are required with wider spacings.

Multiple cropping refers to either two or more crops being grown in sequence in the same year on the same plot of land or growing two or more crops simultaneously. In these ways, maximum canopy cover is available throughout the year. It is important that crops with different growth rates are involved in order to take maximum advantage of vegetation cover over as long a period as possible.

Soil or crop management practices include crop rotation, where a different crop is grown on the same area of land in successive years in a four- or five-year cycle. This avoids exhausting the soil and can improve the texture of the soil. The use of mulches, that is, covering the ground with plant residues, affords protection to the soil against raindrop impact and reduces the effectiveness of overland flow. When mulches are ploughed into the soil they increase the organic content and thereby improve the texture and fertility of the soil.

Reafforestation of slopes, where possible, also helps to control the flow of

Table 9.15 Recommended strip widths for strip-cropping (from Morgan, 1995).

Water erosion (soils with fairly high water intake)	
2–5% slope	30 m
6–9%	25 m
10–24%	20 m
15–20%	15 m
Wind erosion (strips perpendicular to wind direction)	
Sandy soil	6 m
Loamy sand	7 m
Sandy loam	30 m
Loam	75 m
Silt loam	85 m
Clay loam	105 m

water on slopes and reduce the impact of rainfall. The tree roots also help to improve the infiltration capacity of the soil.

Gullies can be dealt with in a number of ways. Most frequently, some attempt is made at revegetating them by grassing or planting trees. Small gullies can be filled or partially filled, or traps or small dams can be constructed to collect and control the flow of sediment. A gully can be converted into an artificial channel with appropriate dimensions to convey water away.

Geotextiles can be used to simulate a vegetation cover so as to inhibit erosion processes and thereby reduce soil loss (Rickson, 1995). They can also help to maintain soil moisture and promote the growth of seeds. In fact, some geotextile mats contain seeds. Geotextiles used for erosion prevention can be made from either natural or synthetic materials as an erosion blanket or mat. In the former case, the geotextile is biodegradable, so its role is temporary. Synthetic geotextiles remain in place to provide reinforcement for a soil. If when fully established the vegetation will control erosion, then temporary geotextiles can be used. However, the establishment of vegetation may not coincide with the time at which the geotextile biodegrades. Temporary geotextiles are made from woven fibres of jute, coir, cotton or sisal. Others consist of loose mulch materials (e.g. paper strips, wood shavings or chips, straw, cotton waste, or coconut fibres) held within a lightweight degradable polypropylene mesh. The synthetic geotextiles are normally made of polyethylene and are either three-dimensional meshes or geowebs. Both types of geotextile are used for slope protection, being rolled downslope and pegged into place. Seeding takes place prior to the positioning of temporary geotextiles, whereas synthetic geotextiles tend to be placed first, then seeded and covered with topsoil.

In terms of soil erosion control, geotextiles affect soil erosion processes through soil protection, modifications to the volume of run-off and its velocity, and reinforcing the shear strength of the soil. For example, geotextiles are generally effective in controlling the impact of raindrops on soil if laid on the surface of the soil. On the other hand, buried geotextiles offer little protection against splash erosion. Natural geotextiles have a high water-retention capacity. By intercepting and absorbing rainfall, the weight of the geotextile is increased, so contact with the soil is improved. This, in turn, slows the flow at the geotextile–soil interface, thereby reducing the potential for erosion. Buried geotextiles should reduce run-off volume, as the infiltration rate will be high due to the loose nature of the soil backfill. But, as mentioned, this backfill will be suceptible to raindrop impact and perhaps surface crusting, depending on soil type. Moreover, Rickson (*ibid.*) concluded that, unlike vegetation, geotextiles do not simulate vegetation in controlling run-off and that there is no relationship between percentage cover and reduction in volume of run-off. Obviously, any retarding of the run-off velocity by roughness due to geotextiles will reduce the ability of run-off to erode soil particles. Rickson suggested that geotextiles are, in fact, more successful in controlling run-off velocity than amount. Surface-laid natural geotextiles contribute nothing to soil strength, but once they degrade and the fibres are incorporated into the soil, they may then reinforce the soil. Buried geotextiles reinforce the soil.

9.6.2 Conservation measures for wind erosion

The problem of wind erosion can be mitigated by reducing the velocity of the wind and/or altering the ground conditions so that particle movement becomes more difficult or particles are trapped. Windbreaks or shelter belts are normally placed at right angles to the prevailing wind. On the other hand, if winds blow from several directions, then a herringbone or grid pattern of windbreaks may be more effective in reducing wind speed. Windbreaks interrupt the length over which wind blows unimpeded and thereby reduce the velocity of the wind at the ground surface. The spacing, height, width, length and shape of windbreaks influence their effectiveness. When trees or bushes act as windbreaks, because they are fairly permanent, they should be planned with care. For example, their spacing should be related to the amount of shelter that each windbreak offers. As a rule of thumb, Cooke and Doornkamp (1990) suggested that the velocity of the wind is affected for about 5–10 times the height of the windbreak on its windward side and for approximately 10–30 times on the leeward side. As wind velocities may increase at the ends of windbreaks as a consequence of funnelling, they should be longer rather than shorter. The recommended length of a windbreak is about twelve times its height for prevailing winds that are at right angles to it. However, in order to allow for deviations in the direction of the wind this figure may be doubled. Furthermore, windbreaks should be permeable to avoid creating erosive vortices on their lee side. Permeability can be influenced by the density of vegetation and the width of the windbreak. Slat fence barriers reduce the drag coefficient linearly with increasing permeability until the windbreak is 40% open; from 40 to 60% open the coefficient decreases sharply. The shape of a vegetative windbreak can be influenced by careful selection of the bushes or trees involved. They should preferably be triangular in cross-section. A windbreak consisting of trees and shrubs may need to be up to 9 m in width, comprising two rows of trees and three rows of bushes.

Geotextiles can interrupt the flow of wind over the ground surface and so reduce its velocity, particularly if coarse-fibred products are used. By reducing the initial erosion of particles, further erosion is reduced since there are fewer particles in saltation, suspension and traction. In addition, because of the moisture-retention characteristics, especially of natural geotextiles, the soil remains relatively moist, so it is less susceptible to wind erosion.

Field cropping practices may be simpler and cheaper than windbreaks, and sometimes more effective. The effectiveness of the protection against wind erosion offered by plant cover is affected by the degree of ground cover provided, and therefore by the height and density of the plant cover. It has been suggested that plants should cover over 70% of the ground surface to afford adequate protection against erosion (Evans, 1980). Cover crops can be grown to protect the soil during the period when it would usually be exposed, or strip cropping may be practised. Vegetative cover helps to trap soil particles and is particularly important in terms of reducing the quantity of saltating particles. Strip cropping helps to inhibit soil avalanching. Widths vary according to soil type (see Table 9.15). Although the choice of crop may be limited by the availability of water, it would appear that small grain crops, legumes and grasses are reasonably effective in reducing the effects of wind erosion.

Mulches can be used to cover the soil and should be applied at a minimum rate of 0.5 kg m^{-2}. As well as providing organic matter for the soil, mulches also trap soil particles and conserve soil moisture. Stubble can be left between strips of ploughed soil; the strip should be wide enough to prevent saltating particles from leaping across it. Because wind erosion is reduced by surface roughness, the furrows produced by ploughing should ideally be oriented at right angles to the prevailing wind.

However, the ridges and furrows produced by ploughing are subject to erosion and obviously can be reduced in size and therefore effectiveness in terms of reducing wind velocity and particle transport at the ground surface. Furthermore, ridges and furrows represent a larger surface area from which moisture can be lost.

The sustained use of land in arid and semi-arid regions poses it own problems. Crop yields are generally low near desert margins, so it is rarely worth investing large sums of money on conservation. Consequently, good management becomes essential in these regions. On low slopes it involves controlling yields at low sustainable levels and varying production between the wetter and drier years. On steep slopes or where there are migrating dunes, vegetation must be established and grazing discouraged. However, it goes without saying that such policies are difficult to put into effect.

9.7 Assessment of soil erosion

Morgan (1995) referred to the assessment of soil erosion hazard as a special form of land resource evaluation, the purpose of which was to identify areas of land where the maximum sustained productivity from a particular use of land is threatened by excessive loss of soil. Such an assessment subdivides a region into zones in which the type and degree of erosion hazard are similar. This can be represented in map form and used as the basis for planning and conservation within the region.

Maps of erosivity using the rainfall erosion index, R, have been produced for the United States (see Figure 9.8), and Stocking and Elwell (1976) provided a map of Southern Rhodesia (now Zimbabwe) showing the mean annual erosivity (Figure 9.14). Stocking and Elwell (1973) also devised a rating system for evaluating erosion risk. Each of the parameters in the rating system, namely, erosivity, erodibility, slope, ground cover and human occupation, has five

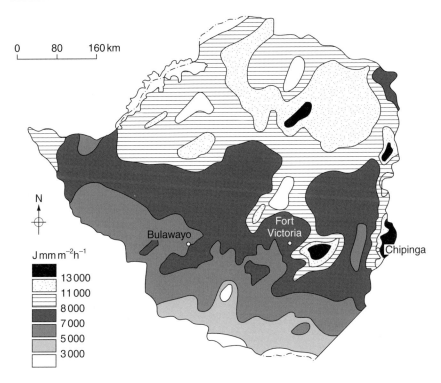

0   80   160 km

N

J mm m^{-2}h^{-1}

13 000
11 000
8 000
7 000
5 000
3 000

Bulawayo

Fort
Victoria

Chipinga

Figure 9.14 Mean annual erosivity in Southern Rhodesia (now Zimbabwe) (after Stocking and Elwell, 1976, by permission of the Royal Geographical Society).

grades. The latter parameter includes the type and density of settlement. Grades range from 1 for low risk of erosion to 5 for high risk, and the ratings for each of the grades is summed to provide a group rating (Figure 9.15). The group rating enables an area to be classified as low, average or high risk. Although such a system of erosion risk classification has shortcomings, as pointed out by Morgan (1995), the method is relatively simple to use.

Land capability classification, as referred to above, was originally developed by the United States Soil Conservation Service for farm planning (Klingebeil and Montgomery, 1966). In other words, it provides a means of assessing the extent to which limitations such as erosion risk soil depth, wetness and climate hinder the agricultural use of land (Table 9.16). Three categories are recognized, namely, capability classes, subclasses and units. A capability class is a group of capability subclasses that have the same relative degree of limitation or hazard. Classes I to IV can be used for cultivation, whereas classes V to VIII cannot because of permanent limitations such as wetness or steeply sloping land. The risk of erosion increases through the first four classes progressively,

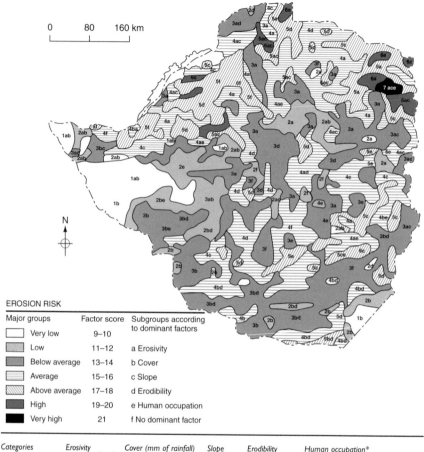

0   80   160 km

EROSION RISK

Major groups		Factor score	Subgroups according to dominant factors
	Very low	9–10	
	Low	11–12	a Erosivity
	Below average	13–14	b Cover
	Average	15–16	c Slope
	Above average	17–18	d Erodibility
	High	19–20	e Human occupation
	Very high	21	f No dominant factor

Categories		Erosivity ($J\,mm\,m^{-2}\,h^{-1}$)	Cover (mm of rainfall) and basal cover est. (%)	Slope (degrees)	Erodibility	Human occupation*
Low	I	below 5000	above 1000 7–10	0–2	orthoferralitic regosols	Extensive large-scale commercial ranching National Parks or Unreserved
Below average	II	5000–7000	800–1000 5–8	2–4	paraferralitic	Large-scale commercial farms
Average	III	7000–9000	600–800 3–6	4–6	fersiallitic	Low-density CLs (below 5 p.p. km²) and SCCF
Above average	IV	9000–11 000	400–600 1–4	6–8	siallitic vertisols lithosols	Moderately settled CLs (5–30 p.p. km²)
High	V	above 11 000	below 400 0–2	above 8	non-calcic hydromorphic sodic	Densely settled CLs (above 30 p.p. km²)

Notes: Cover, erodibility and human occupation are only tentative and cannot as yet be expressed on a firm quantitative basis.
* p.p. km² = persons per square kilometre. CL = Communal Lands. SCCF = Small-scale Commercial Farms.

Figure 9.15 Erosion survey of Southern Rhodesia (now Zimbabwe) (after Stocking and Elwell, 1973, from data compiled and drawn in the Department of Geography, University of Rhodesia, 1972).

Table 9.16 Land-use alternatives of capability classes. Based on the definitions in the original US version of the system (after Kingebiel and Montgomery, 1996).

| Capability class | Limitations | Management under cultivation | | Capability | | | | | | Characteristics and recommended land use |
		Choice of crops	Conservation practices	Cultivation	Pasture (improved pasture)	Range (unimproved)	Woodland	Wildlife food and cover	Recreation water supply aesthetic	
I	few	any	none	√	√	√	√	√	√	Deep, productive soils easily worked, on nearly level land; not subject to overland flow; no or slight risk of damage when cultivated; use of fertilizers and lime, cover crops, crop rotations required to maintain soil fertility and soil structure
II	some	reduced or	moderate	√	√	√	√	√	√	Productive soils on gentle slopes; moderate depth; subject to occasional overland flow; may require drainage; moderate risk of damage when cultivated; use of crop rotations, water-control systems or special tillage practices to control erosion.
III	severe	reduced and/or	special	√	√	√	√	√	√	Soils of moderate fertility on moderately steep slopes, subject to more severe erosion; subject to severe risk of damage but can be used for crops provided plant cover is maintained; hay or other sod crops should be grown instead of row crops.
IV	very severe	restricted and/or	very careful management	√	√	√	√	√	√	Good soils on steep slopes, subject to severe erosion; very severe risk of damage but may be cultivated if handled with great care; keep in hay or pasture but a grain crop may be grown once in 5 or 6 years.

Table 9.16 continued

Class	Limitation					Notes
V	other than erosion	✓	✓	✓	✓	Land is too wet or stony for cultivation but of nearly level slope; subject to only slight erosion if properly managed; should be used for pasture or forestry but grazing should be regulated to prevent plant cover from being destroyed.
VI	severe	✓	✓	✓		Shallow soils on steep slopes; use for grazing and forestry; grazing should be regulated to preserve plant cover, if plant is destroyed, use should be restricted until cover is re-established.
VII	very severe	✓	✓	✓		Steep, rough, eroded land with shallow soils; also includes droughty or swampy land; severe risk of damage even when used for pasture or forestry; strict grazing or forest management may be applied.
VIII	very severe			✓		Very rough land; not suitable even for woodland or grazing; reserve for wildlife, recreation or watershed conservation

Classes I–IV denote soils suitable for cultivation. Classes V–VIII denote soils unsuitable for cultivation.

reducing the choice of crops that can be grown and requiring more expensive conservation practices. A capability subclass is a group of capability units that possess the same major conservation problem or limitation (e.g. erosion hazard; excess water; limited depth of soil or excessive stoniness). Lastly, a capability unit is a group of soil mapping units that have the same limitations and management responses. Soils within a capability unit can be used for the same crops and require similar conservation treatment.

The system developed by the Soil Conservation Service has an obvious agricultural bias and does not consider non-agricultural activities. Hence, it is limited in its application, and this particular system should not be used for purposes for which it was never intended. Nevertheless, the system can be adapted to any environment and to any level of farming technology. Data contained in land capability surveys can be combined with that of erosivity to provide a more detailed assessment of erosion risk. Hannan and Hicks (1980) proposed a land capability classification for urban land-use planning with special reference to erosion control. Information on topographic features, slopes and soils is related to hazards, and this in turn to suitable land use (Figure 9.16).

Terrain evaluation is concerned with analysis, classification and appraisal of a region. The initial interpretation of landform can be made from large-scale maps and aerial photographs. Observation of relief should give particular attention to direction and angle of maximum gradient, maximum relief amplitude and the proportion of the total area occupied by bare rock or slopes. An assessment of the risk of erosion (especially the location of slopes that appear potentially unstable) and the risk of excess sedimentation of water-borne or windblown material should be made. The land system, land facet and land element, in decreasing order of size, are the principal units used. A land systems map shows the subdivision of a region into areas with common physical attributes that differ from those of adjacent areas. Many of the factors considered in land systems analysis are of importance in relation to soil erosion, so terrain evaluation is of value in the evaluation of the risk of soil erosion.

The first types of soil erosion survey consisted of mapping sheet wash, rills and gullies within an area, frequently from aerial photographs. Indices such as gully density were used to estimate erosion hazard. Subsequently, sequential surveys of the same area were undertaken in which mapping, again generally from aerial photographs, was done at regular time intervals. Hence, changes in those factors responsible for erosion could be evaluated. They could also be evaluated in relation to other factors such as changing agricultural practices or land use. Any data obtained from imagery should be checked and supplemented with extra data from the field. It may be possible to rate the degree of severity of soil erosion as seen in the field. A pro forma for recording data is illustrated in Figure 9.17, and a coding system suggested by Morgan (1995) is given in Table 9.17.

Landform

Physical criteria and urban land use

Slope class	Terrain component	Potential hazards related to topographic location and slope and which will affect urban land use	Suitable urban land use
0–5%	Drainage plain	Flooding, seasonally high water tables, high shrink–swell soils, high erosion hazard	Drainage reserves/storm water disposal
	Floodplain	Flooding, seasonally high water tables, high shrink–swell soils, saline soils, gravelly soils	Open space areas, playing fields
	Hillcrests Sideslopes	Shallow soils, stony/gravelly soils; Overland flow, poor surface drainage and profile damage	Residential: all types of recreation; large-scale industrial, commercial and institutional development
	Footslopes	Impedence in lower terrain positions, deep soils; Others – swelling soils, erodible soils, dispersible soils	
5–10%	Hillcrests Sideslopes Footslopes	Shallow soils; Overland flow; Deep soils, poor drainage; Others – swelling soils, erodible soils, dispersible soils	Residential subdivisions, detached housing, medium-density housing/ unit complexes, modular industrial, active recreational pursuits
10–15%	Sideslopes	Overland flow; Geological constraints – possibility of mass movement; Swelling soils; Erodible soils	Residential subdivisions, detached housing, medium-density housing/ unit complexes, modular industrial, passive recreational
15–20%	Sideslopes	Overland flow; Geological constraints – possibility of mass movement; Swelling soils; Erodible soils	Residential subdivisions, detached housing, medium-density housing/ unit complexes, modular industrial, passive recreational
20–25%	Sideslopes	Geological constraints; Mass movement; High to very high erosion hazard	Residential subdivision, passive recreational
25–30%	Sideslopes	Geological constraints; Possible mass movement; High to very high erosion hazard	Upper limit for selective residential use, low-density housing on lots greater than 1 ha, passive recreation
> 30%	Sideslopes	Geological constraints; Mass movement; Severe erosion hazard	Recommended against any disturbance for urban development

LANDFORM

Slope
0–1% — 1
1–5% — 2
5–10% — 3
10–15% — 4
15–20% — 5
20–25% — 6
25–30% — 7

Terrain
Hillcrest — 1
Sideslope — 2
Footslope — 3
Drainage plan — 5
Swamp — 6

From Batemans Bay — George Bass Drive — Road — Beach — Burrewater Pt Road

N

0 200 metres

Figure 9.16 Urban capability classification for soil erosion control (after Hannan and Hicks, 1980).

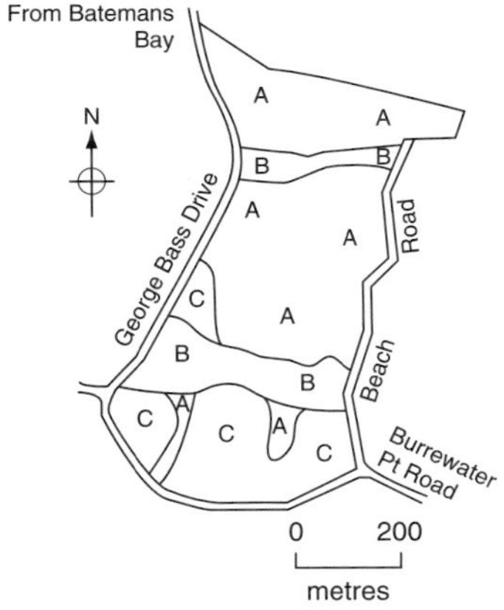

SOILS

Summary of properties of soils

Map unit	Dominant soils	Lithology and physiography	Erosion hazard	Limitations
A	Shallow gravelly soils	Metasediments; cherts, phyllites, etc. with quartz veination Ridges, sideslopes and some footslopes	High	Impeded soil drainage, shallow soil depth, high stone and gravel contents
B	Swamp alluvial soils	Metasediments; cherts, phyllites, etc. with quartz veination Alluvial parent materials, swamp and drainage plains	Very high to extreme	Seasonally high water tables, poor to impeded soil daranage
C	Yelloe duplex soils	Metasediments; cherts, phyllites, etc. Crests, sideslopes and some footslopes	High to very high	Low to moderate shrink–swell potential, poor soil drainage
D	Drainage plain alluvial soils	Metasediments; cherts, phyllites, etc. Alluvia/colluvial parent materials and surface materials	Very high	Seasonally high water tables, poor to impeded soil drainage

Figure 9.16 Continued.

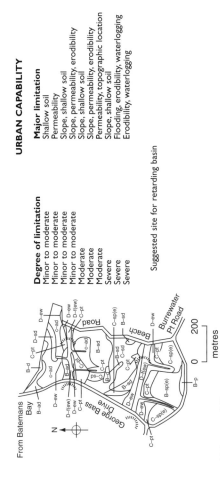

URBAN CAPABILITY

Degree of limitation	Major limitation	Class
Minor to moderate	Shallow soil	B–d
Minor to moderate	Permeability	B–p
Minor to moderate	Slope, shallow soil	B–sd
Minor to moderate	Slope, permeability, erodibility	B–sp(e)
Moderate	Slope, shallow soil	C–sd
Moderate	Slope, permeability, erodibility	C–ap(e)
Moderate	Permeability, topographic location	C–pt
Severe	Slope, shallow soil	D–sd
Severe	Flooding, erodibility, waterlogging	D–f(ew)
Severe	Erodibility, waterlogging	D–ew

///// Suggested site for retarding basin

Urban capability

Definitions of the classes used in urban capability assessment are:

Class A Areas with little or no physical limitations to urban development.

Class B Areas with minor to moderate physical limitations to urban development. These limitations may influence design and impose certain management requirements on development to ensure a stable land surface is maintained both during and after development.

Class C Areas with moderate physical limitations to urban development. These limitations can be overcome by careful design and by adoption of site management techniques to ensure the maintenance of a stable land surface.

Class D Areas with severe physical limitations to urban development which will be difficult to overcome, requiring detailed site investigation and engineering design.

Class E Areas where no form of urban development is recommended because of very severe physical limitations to such development that are very difficult to overcome.

Figure 9.16 Continued.

Recorder: Date: Altitude:		Area: Air photo no: Grid Reference:													FACET NO.
Present land use															
Climate	Month	J	F	M	A	M	J	J	A	S	O	N	D		Erosivity
	Rainfall (mm)														
	Mean temp (C)														
	Maximum intensity														
Vegetation	Type			% Ground cover					% Tree and shrub cover						
Slope	Position	Degree					Distance from crest				Shape				
Soil	Depth		Surface texture									Erodibility			
			Permeability			Clay fracftion									
Erosion															
REMARKS															
EROSION CODE		0	$^1/_2$	1		2		3		4		5			

Figure 9.17 Pro forma for recording soil erosion in the field (from Morgan, 1995).

Table 9.17 Coding system for soil erosiom appraisal in the field (after Morgan, 1995).

Code	Indicators
0	No exposure of tree roots; no surface crusting; no splash pedestals; over 70% plant cover (ground and canopy).
½	Slight exposure of tree roots; slight crusting of the surface; no splash pedestals; soil slightly higher on upslope or windward sides of plants and boulders; 30–70% plant cover.
1	Exposure of tree roots, formation of splash pedestals, soil mounds protected by vegetation, all to depths of 1–10 mm; slight surface crusting; 30–70% plant cover.
2	Tree root exposure, splash pedestals and soil mounds to depths of 10–50 mm; crusting of the surface; 30–70% plant cover.
3	Tree root exposure, splash pedestals and soil mounds to depths of 50–100 mm; 2–5 mm thickness of surface crust; grass muddied by wash and turned downslope; splays of coarse material due to wash and wind; less than 30% plant cover.
4	Tree root exposure, splash pedestals and soil mounds to depths of 50–100 mm; splays of coarse material; rills up to 80 mm deep; bare soil.
5	Gullies; rills over 80 mm deep; blow-outs and dunes; bare soil.

9.8 Desertification

Desertification is a process of environmental degradation that occurs mainly in arid and semi-arid regions, and causes a reduction in the productive capacity of soil. The prime cause of such degradation is excessive human activity and demand in those regions with fragile ecosystems. The net result is that the productivity of agricultural land declines significantly, grasslands no longer produce sufficient pasture, dry farming fails and irrigated areas may be abandoned. Deserts are encroaching into semi-arid regions largely as a consequence of poor farming practices, which include overstocking. The improper use of water resources leading to inefficient use and even to streams drying up aggravates the problem still further. Excessive abstraction of water from wells lowers the water table, which adversely affects plant growth. Desertification can occur within a short time, that is, in 5 to 10 years. Whereas droughts come and go, desertification can be permanent if in order to reverse the situation substantial capital and resources are not available. Nonetheless, when prolonged periods of drought are coupled with environmental mismanagement, they may end in permanent degradation of the land.

Desertification brings with it associated problems such as removal of soil, as well as reduction in its fertility; deposition of windblown sand and silt, which can bury young plants and block irrigation canals and rivers; and moving sand dunes. Also, when rain does fall, a greater proportion of it contributes towards run-off, so erosion becomes more aggressive. This, in turn, means that the amount of sediment carried by streams and rivers increases. It also means that less water is infiltrating into the ground for plant growth.

Initially, desertification may go unrecognized and it may not be until significant changes have occurred in the fertility of the soil that it is identified; hence it can be looked upon as a creeping disaster (Biswas and Biswas, 1978). Some indicators of desertification are given in Table 9.18. It has been suggested that the world loses 20 million hectares of land to desertification each year, and surveys have classified 18% of desertification as slight, 4% as moderate and 28% as severe. But as remarked, desertification is not always readily identified with accuracy. There are a number of reasons for this. First, deterioration of land can take numerous forms, and it may be irregular in pattern rather than advancing as a recognizable front. In addition, the deterioration of soils affects areas of greater size than those that strictly can be regarded as turned to desert.

The expansion of desertification is not estimated easily, because several years may need to elapse before it can be distinguished from an area that has been subjected to a prolonged drought. Be that as it may, the annual loss of land is thought to be about 60 000 km^2. It has been estimated that 14% of the world's population live in drylands that are under threat, and that over 60 million people are affected by desertification. About one-third of the land surface (Figure 9.18) and one-seventh of the population of the world are affected directly, but the rest of the population has to face the indirect effects of lower agricultural production in the form of international aid. Furthermore, in many desert margin regions there have been dramatic increases in population during the twentieth century, and many formerly nomadic peoples in Africa have become settled. Hence, the increased population pressure on soil resources has led to environmental degradation and a decline in productivity.

Table 9.18 Form and severity of desertification (after Warren and Maizels, 1977).

Form	Severity			
	Slight	Moderate	Severe	Very severe
Water erosion	Rills, shallow runnels	Soil hummocks, silt, accumulations	Piping, coarse washout deposits, gullying	Rapid reservoir siltation, landslides, extensive gullying
Wind erosion	Rippled surfaces fluting and small-scale erosion	Wind mounts, wind sheeting	Pavements	Extensive active dunes
Water and wind erosion			Scalding	Extensive scalding
Irrigated land	Crop yield reduced less than 10%	Minor white patches, crop yield reduced 10–50%	Extensive white saline patches, crop yield reduced more than 50%	Land unusable through excessive salinization, soils nearly impermeable, encrusted with salt
Plant cover	Excellent–good range condition	Fair range condition	Poor range condition	Virtually no vegetation

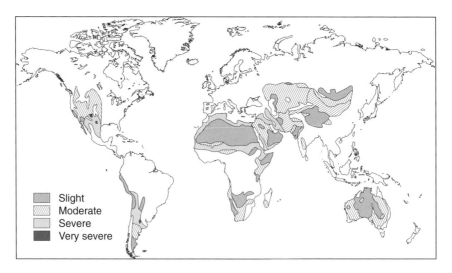

Figure 9.18 World distribution of desertification (after Dregne, 1983).

The loss of vegetation at the margins of deserts leads to diminishing rainfall, increasing dust content in the air and an accelerated rate of desertification. An increased amount of dust in the atmosphere may adversely affect the radiation balance and deplete food production even further. It may also adversely affect people with respiratory problems. In the marginal and degraded areas, there will be a tendency for the number of plant and animal species to decline. For the people involved, desertification can mean loss of income and eventual starvation, which in turn may lead to population migrations, with severe social and economic repercussions. In other words, desertification can be viewed as the interaction between socio-economic pressures and ecosystem fragility (Nnoli, 1990).

Deserts may expand or contract naturally as precipitation varies over time. This process can be charted by various means. Dendrochronology can be used to identify seasons when moisture availability was restricted and growth reduced. Limnology can indicate the expansion and contraction of lakes over time, while archaeology can indicate changes in early settlement patterns in relation to water availability. Nevertheless, most desertification would appear to be attributable to the unwise use of agricultural land by man (Cloudsley-Thompson, 1978).

One of the common causes of desertification is overgrazing of animals on a limited supply of forage and in many cases, especially in the developing world, to a complete absence of range management. Even in the western United States overgrazing in 70% of the rangelands has led to a decline in the original forage potential of about 50%. Drought puts a severe strain on such regions.

Recovery from drought may be a slow process, and in some cases irreversible damage may be done to an ecosystem that is already undergoing serious degradation. It is therefore necessary to attempt to determine the carrying capacity of such grasslands, that is, the number of animals they can support without the vegetation being adversely affected.

As desertification is a process of gradual degradation, it is important that it is monitored effectively so that the changes it brings about are detected as they occur. In this context, remote sensing imagery or aerial photographs taken at successive time intervals can prove useful in that they can record changes in the vegetation cover. Agricultural production might also be of some value.

The cost of reversing desertification is very high, so it is obviously better to prevent it in the first place. In very degraded drylands, retrenchment may be the only solution, and recovery may be out of the question in the short term. In areas that have suffered less deterioration, rotational cropping, reduction in livestock numbers, use of special equipment for ploughing and sowing, use of specially adapted crops or the establishment of irrigation may be appropriate strategies. Over and above this, the pressure on soil resources must be reduced, which in turn must inevitably mean a reduction in population in the region.

9.9 Irrigation

Irrigation has been used in semi-arid and arid regions for centuries in order to increase agricultural production. At present, large-scale irrigation schemes depend upon large sources of water, which can be supplied by large reservoirs or abstracted by deep wells from groundwater sources. However, increasing demands for water, limited availability and concerns about water quality mean that water has to be used effectively, so that an irrigation system must be properly planned and designed, and operated efficiently. In order to make the best use of the water available, the seasonal water requirements of the crops grown must be known as plant requirements vary with the season. Furthermore, weather conditions (i.e. temperature, wind, humidity) affect water demand. In particular, estimates of evapotranspiration must be determined when planning an irrigation system. The duration of periods of insufficient rainfall in humid and subhumid regions influences the economic feasibility of irrigation. Table 9.19 illustrates the water requirements and evapotranspiration of some crops grown in New Mexico. Note that the expected effective rainfall is used in determining the field irrigation requirement. Not all rainfall is effective; only that proportion which contributes towards evapotranspiration is effective.

Easily the more common method of applying irrigation water, particularly in arid regions, is by flooding the ground surface (Hansen *et al.*, 1980). For example, wild flooding is where water is allowed to flow uncontrolled over the soil. Water can also be conveyed to the soil in a more regular manner via canals, ditches and furrows or basins. In the latter cases, the soil should be prepared prior to the irrigation water being applied. Efficient surface irrigation requires

Table 9.19 Seasonal evapotranspiration and irrigation requirements for crops near Deming, New Mexico* (after Jensen, 1973).

Crop	Length of growing season (days)	Evapotranspiration depth (mm)	Effective rainfall depth (mm)	Evapotranspiration less rainfall (mm)	Water application efficiency (%)	Irrigation requirement depth (mm)
Alfalfa	197	915	152	763	70	1090
Beans (dry)	92	335	102	233	65	358
Corn	137	587	135	452	65	695
Cotton	197	668	152	516	65	794
Grain (spring)	112	396	33	363	65	558
Sorghum	137	549	135	414	65	637

* Average frost-free period is April 15 to October 29. Irrigation prior to the frost-free period may be necessary for some crops.

that the ground surface is graded in order to control the flow of water. Obviously, the extent of grading is governed by the nature of the topography. Sprinkler irrigation systems provide a fairly uniform method of applying water and are more efficient than surface irrigation. For example, conveying water via canals, ditches and furrows is regarded, at most, as being only 60% efficient, whereas sprinkler systems are about 75% efficient. Sprinkler irrigation does not entail the land surface being graded, and it allows the rate of water application to be easily controlled. Micro-irrigation or drip systems deliver water in small amounts (1–$10 \, 1 \, h^{-1}$) via special porous tubes to the plant roots and are 90% efficient. However, because of the high cost of drip irrigation it tends to be restricted to high-value crops. Alternatively, perforated or porous tubes are often installed 0.1 to 0.3 m below the surface of the soil, supplying 1–$5 \, 1 \, min^{-1}$ per 100 m of tube. A comparison of the different types of irrigation system in relation to site and situation factors is given in Table 9.20.

Irrigation can raise the soil water content to its field capacity. In either ditch or sprinkler irrigation, the infiltration capacity and the permeability of the soil determine how quickly water can be applied. The field capacity is the water content after a soil is wetted and allowed to drain for 1 or 2 days. It represents the upper limit of water available to plants and in terms of soil suction can be defined as a pF value of about 2. At each point where moisture menisci are in contact with soil particles, the forces of surface tension are responsible for the development of capillary or suction pressure (Table 9.21). Soil suction is a negative pressure and indicates the height to which a column of water could rise due to such suction. Since this height or pressure may be very large, a logarithmic scale has been developed to express the relationship between soil suction and moisture content. This scale is referred to as the pF value. The level at which soil moisture is no longer available to plants is termed the permanent wilting point, and this corresponds to a pF value of about 4.2. The moisture characteristic of a soil provides valuable data concerning the moisture content corresponding to the field capacity and the permanent wilting point, as well as the rate at which changes in soil suction take place with variations in moisture content. This enables an assessment to be made of the range of soil suction and moisture content that is likely to occur in the zone of the soil affected by seasonal changes in climate. The difference between field capacity and permanent wilting point is termed available water. Plants can remove only a proportion of available water before growth and yield are affected.

The use of irrigation in semi-arid and arid regions to increase crop production can lead to the deterioration of water quality and salinity problems (Hoffman *et al.*, 1990). As far as the quality of water for irrigation is concerned, the most important factors are, first, the total concentration of salts; second, the concentration of potentially toxic elements; third, the bicarbonate concentration as related to the concentration of calcium and magnesium; and, fourth, the proportion of sodium to other ions. Most water used for irrigation contains dissolved salts, some of which remain in the soil moisture as a result of

Table 9.20 Comparison of irrigation systems in relation to site and situation factors (from Schwab et al., 1993).

Site and situation factors	Improved surface systems		Sprinkler systems			Micro-irrigation systems
	Redesigned surface systems	Level basins	Intermittent mechanical move	Continuous mechanical move	Solid-set and permanent	Emitters and porous tubes
Infiltration rate	Moderate to low	Moderate	All	Medium to high	All	All
Topography	Moderate slopes	Small slopes	Level to rolling	Level to rolling	Level to rolling	All
Crops	All	All	Generally shorter crops	All but trees and vineyards	All	High value required
Water supply	Large streams	Very large streams	Small streams nearly continuous	Small streams nearly continuous	Small streams	Small streams, continuous and clean
Water quality	All but very high salts	All	Salty water may harm plants	Salty water may harm plants	Salty water may harm plants	All – can potentially use high salt waters
Efficiency	Average 60–70%	Average 80%	Average 70–80%	Average 80%	Average 70–80%	Average 80–90%
Labour requirement	High, training required	Low, some training	Moderate, some training	Low, some training	Low to seasonal high, little training	Low to high, some training
Capital requirement	Low to moderate	Moderate	Moderate	Moderate	High	High
Energy requirement	Low	Low	Moderate to high	Moderate to high	Moderate	Low to moderate
Management skill	Moderate	Moderate	Moderate	Moderate to high	Moderate	High
Machinery operations	Medium to long fields	Short fields	Medium field length, small interference	Some interference circular fields	Some interference	May have considerable interference
Duration of use	Short to long	Long	Short to medium	Short to medium	Long term	Long term, but durability unknown
Weather	All	All	Poor in windy conditions	Better in windy conditions than other sprinklers	Windy conditions reduce performance, good for cooling	All
Chemical application	Fair	Good	Good	Good	Good	Very good

Table 9.21 Suction pressure and pF values.

pF value	Equivalent suction	
	(mm water)	(kPa)
0	10	0.1
1	100	1.0
2	1000	1.0
3	10 000	100.0
4	100 000	1000.0
5	1 000 000	10 000.0

evapotranspiration. If crop production levels are to be maintained, then drainage systems are required to remove excess water and associated salts from the plant root zone. Capillary action also brings salts to the surface. The capillary rise for different soil types is given in Table 9.22. Inefficient irrigation leads to salinization of the soil, that is, the accumulation of salts near the ground surface. Dissolved nitrates from the application of fertilizers is another source of contamination. In addition, the presence of trace elements derived from the rocks in the area, in amounts greater than their threshold values, can prove toxic to plants. Such trace elements include arsenic, boron, cadmium, chromium, lead, molybdenum and selenium. High concentrations of bicarbonate ions can lead to the precipitation of calcium and magnesium bicarbonate from the soil water content, thereby increasing the relative proportions of sodium.

The suitability of water for irrigation depends on the effects that the salt concentration contained therein has on plants (Table 9.23) and soil. One of the principal effects of salinity is to reduce the availability of water to plants. In semi-arid and arid regions the presence of soluble salts in the root zone can pose a serious problem, unlike subhumid and humid regions, where irrigation acts as a supplement to rainfall, and salinity is usually of little concern since rainfall tends to leach salts out of the soil. Salts may harm plant growth in that

Table 9.22 Capillary rise and capillary pressures in soil.

Soil	Capillary rise (mm)	Capillary pressure (kPa)
Fine gravel	Up to 100	Up to 1.0
Coarse sand	100–150	1.0–1.5
Medium sand	150–300	1.5–3.0
Fine sand	300–1000	3.0–10.0
Silt	1000–10 000	10.0–100.0
Clay	Over 10 000	Over 100.0

Table 9.23 Relative tolerance of crops to salt concentrations expressed in terms of specific electrical conductance (after Ayers, 1975).

Crop division	Low salt tolerance	Medium salt tolerance	High salt tolerance
Fruit crops	Avocado Lemon Strawberry Peach Apricot Almond Plum Prune Grapefruit Orange Apple Pear	Cantaloupe Date Olive Fig Pomegranate	Date palm
Vegetable crops*	300 μS cm^{-1} Green beans Celery Radish 4000 μS cm^{-1}	400 μS cm^{-1} Cucumber Squash Peas Onion Carrot Potato Sweet corn Lettuce Cauliflower Bell pepper Cabbage Broccoli Tomato 10 000 μS cm^{-1}	10 000 μS cm^{-1} Spinach Asparagus Kale Gardet beet 12 000 μS cm^{-1}
Forage crops*	2000 μS cm^{-1} Burnet Ladino clover Red clover Alsike clover Meadow foxtail White Dutch clover 4000 μS cm^{-1}	4000 μS cm^{-1} Sickle milkvetch Sour clover Cicer milkvetch Tall meadow grass Smooth brome Big trefoil Reed canary Meadow fescue Blue grama Orchard grass Oats (hay) Wheat (hay) Rye (hay) Tall fescue Alfalfa Hubam clover Sudan grass Dallis grass Strawberry clover Mountain brome Perennial rye grass Yellow sweet clover White sweet clover 12 000 μS cm^{-1}	12 000 μS cm^{-1} Bird's-foot trefoil Barley (hay) Western wheat grass Canada wild rye Rescue grass Rhodes grass Bermuda grass Nuttall alkali grass Salt grass Alkali sacaton 18 000 μS cm^{-1}

Field crops*	4000 µS cm^{-1}	6000 µS cm^{-1}	10 000 µS cm^{-1}
	Field beans	Castor beans	Cotton
		Sunflower	Rape
		Flax	Sugar beet
		Corn (field)	Barley (grain)
		Sorghum (grain)	16 000 µS cm^{-1}
		Rice	
		Oats (grain)	
		Wheat (grain)	
		Rye (grain)	
		10 000 µS cm^{-1}	

* Specific electrical conductance values represent salinity levels at which a 50 percent decrease in yield may be expected compared with yield on non-saline soil under comparable growing conditions. Concentrations refer to soil water. Specific electrical conductance is measured in microsiemens/cm (µS cm^{-1}), which is equivalent to micromhos/cm.

they reduce the uptake of water either by modifying osmotic processes or by metabolic reactions such as those caused by toxic constituents. The effects that salts have on some soils, notably the changes brought about in soil fabric, which in turn affect permeability and aeration, also influence plant growth. In other words, cations can cause deflocculation of clay minerals in a soil, which damages its crumb structure and reduces its infiltration capacity. The quality of water used for irrigation varies according to the climate, the type and drainage characteristics of the soil and the type of crop grown. Plants growing in adverse climatic conditions are susceptible to injury if poor-quality water is used for irrigation. Also, in hot, dry climates plants abstract more moisture from the soil and so tend to concentrate dissolved solids in the soil moisture quickly. Clayey soils may cause problems when poor-quality water is used for irrigation, since they are poorly drained, which means that the capacity for removing excess salts by leaching is reduced. On the other hand, if a soil is well drained, then crops can be grown on it with the application of generous amounts of saline water.

In addition to the potential dangers due to high salinity, a sodium hazard sometimes exists in that sodium in irrigation water can bring about a reduction in soil permeability and cause the soil to harden. Both effects are attributable to cation exchange of calcium and magnesium ions by sodium ions on clay minerals and colloids. The sodium content can be expressed in terms of percent sodium as follows:

$$\%\text{Na} = \frac{\text{Na}}{\text{Ca} + \text{Mg} + \text{Na} + \text{K}} \times 100 \qquad (9.6)$$

where all ionic concentrations are expressed in milliequivalents per litre. The extent of replacement of calcium and magnesium ions by sodium ions, that is, the amount of sodium adsorbed by a soil, can be estimated from the sodium adsorption ratio (SAR), which is defined as:

$$SAR = \frac{Na}{\sqrt{(Ca + Mg)/2}} \tag{9.7}$$

where, again, the concentrations are expressed in milliequivalents per litre. The sodium hazard as defined in terms of the SAR is shown in Figure 9.19.

Generally, soil with more than 0.1% soluble salts within 0.2 m of the surface is regarded as salinized. Heavily salinized land has been abandoned to agriculture in many parts of the world (e.g. in parts of China, Pakistan and the United States). To avoid salinization an adequate system of drainage must be installed prior to the commencement of irrigation. This lowers the water table as well as conveying water away more quickly. Subsurface drainage can be used together with controlled surface drainage. Excess water can be applied to fields during the non-growing season to flush salts from the soil. Usually, the quantity of soil removed by crops is so small that it does not make a significant contribution to salt removal. Water use must be managed, and planting trees for windbreaks and rotating crops, where possible, helps. The other problem that can result from inadequate drainage is waterlogging, which is generally brought about by a rising water table caused by irrigation.

Salt-affected soils have been classified as saline, sodic and saline–sodic (Schwab et al., 1993). Saline soils contain enough soluble salt to interfere with the growth of most plants. They can be recognized by the presence of white crusts on the soil, and by stunted and irregular plant growth. Sodium salts occur in relatively low concentrations compared with calcium and magnesium salts. Sodic soils are relatively low in soluble salts but contain enough exchangeable sodium to interfere with plant growth. Unlike saline soils, which generally have a flocculated structure, sodic soils develop a dispersed structure as the amount of exchangeable sodium increases (see Chapter 5). Saline–sodic soils contain sufficient quantities of soluble salts and adsorbed sodium to reduce the yields of most plants. Both sodic and saline–sodic soils can be improved by the replacement of excess adsorbed sodium by calcium.

The only way by which salts that have accumulated in the soil due to irrigation can be removed satisfactorily is by leaching. Hence, sufficient water must be applied to the soil to dissolve and flush out excess salts. Not only is adequate drainage required for conveying away excess water but it is also needed for water moving through the root zone. The traditional concept of leaching involved ponding of water so that uniform salt removal is achieved from the root zone. However, high-frequency watering should be effective as far as salinity control is concerned. The salt content of the soil should be monitored to ensure that the correct amount of water is being used for irrigation.

Irrigation should be scheduled according to water availability and crop need.

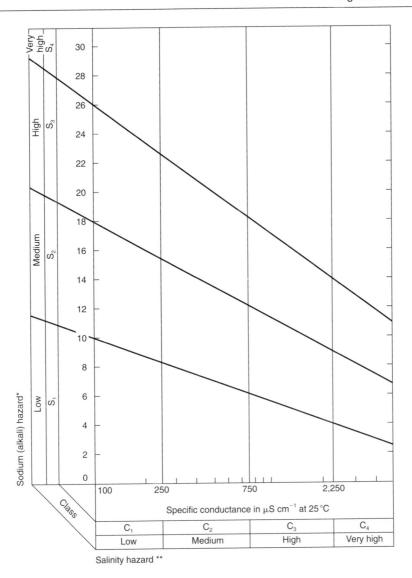

Figure 9.19 Sodium hazard and salinity hazard in relation to agriculture.

S_1 Low-sodium water can be used for irrigation on almost all soils with little danger of the development of harmful levels of exchangeable sodium. However, sodium-sensitive crops such as stonefruit trees and avocado may accumulate injurous concentrations of sodium.

S_2 Medium-sodium water will present an appreciable sodium hazard in fine-textured soils having high cation exchange capacity, especially under low leaching conditions, unless gypsum is present in the soil. This water may be used on coarse-textured or organic soils with good permeability.

S_3 High-sodium water may produce harmful levels of exchangeable sodium in most soils and will require special soil management – good drainage, high leaching and organic matter additions. Gypsiferous soils may not develop harmful levels of exchangeable sodium, except that amendments may not be feasible with waters of very high salinity

Figure 9.19 Continued.

S₄ Very high-sodium water is generally unsatisfactory for irrigation purposes except at low and perhaps medium salinity, where the dissolving of calcium from the soil, or the use of gypsum or other additives, may make the use of these waters feasible.

C₁ Low-salinity water – can be used for irrigation with most crops on most soils with little likelihood that a salinity problem will develop. Some leaching is required, but this occurs under normal irrigation practices except in soils of extremely low permeability.

C₂ Medium-salinity water – can be used if a moderate amount of leaching occurs. Plants with moderate salt tolerance can be grown in most instances without special practices for salinity control.

C₃ High-salinity water – cannot be used on soils with restricted drainage, special management for salinity control may be required and plants with good salt tolerance should be selected.

C₄ Very high-salinity water – is not suitable for irrigation under ordinary conditions but may be used occasionally under very special circumstances. The soil must be permeable; drainage must be adequate; irrigation water must be applied in excess to provide considerable leaching and very salt-tolerant crops should be selected.

Provided that an adequate supply of water is available, then sufficient water is applied to give optimum or maximum yield. If too much water is applied, then this reduces soil aeration, raises the water table and may cause waterlogging, and can flush away fertilizers, thereby decreasing crop yields. On the other hand, when water supply is limited, then the irrigation scheduling strategy is one of maximizing economic return.

As the purpose of irrigation is to provide water for crops, their growth and appearance offers an indication of water need. Various methods can be used to obtain an assessment of soil moisture content. For example, the simplest way to obtain the moisture content is to heat the soil at 105 °C, after weighing, until a constant weight is achieved. The moisture content is then the loss in weight, expressed as a percentage of the dry mass. A tension meter can be used to measure soil water tension, which in turn can indicate the need for irrigation. A neutron probe provides a direct measurement of soil moisture content. Water balance techniques can also be used to obtain the estimated water in the root zone (Schwab *et al.*, 1993).

References

Anon. (1975) *Control of Pollution from Cropland. Volume 1. A Manual for Guideline Development.* Agricultural Research Service, Report. ARS-H-5–1, US Department of Agriculture, Washington, DC.

Anon. (1989) *Design, Layout, Construction and Maintenance of Terrace Systems.* American Society Agricultural Engineers, Standard S268.3, St Joseph, Michigan.

Arnoldus, H.M.S. (1977) Predicting soil losses due to sheet and rill erosion. *Food and Agriculture Organisation Conservation Guide*, Vol. 1, 99–124.

Ayers, R.S. (1975) Quality of water for irrigation. *Proceedings Specialty Conference*, American Society Civil Engineers, Irrigation Drainage Division, Logan, Utah, 24–56.

Bagnold, R.A. (1941) *The Physics of Wind Blown Sand and Desert Dunes*. Methuen, London.

Bates, C.G. (1994) *The Windbreak as a Farm Asset*. US Department of Agriculture, Farmers Bulletin 1045, Washington, DC.

Biswas, M.R. and Biswas, A.R. (1978) Loss of productive soil. *International Journal Development Studies*, 12, 189–197.

Brakensiek, D.L. and Rawls, W.J. (1983) Agricultural management effects on soil water processes. *Transactions American Society Agricultural Engineers*, 26, 1753–1757.

Brune, G.M. (1951) Sediment records in the mid-western United States. *International Association Science Hydrology*, 3, 29–38.

Chepil, W.S. (1945) Dynamics of wind erosion, III. Transport capacity of the wind. *Soil Science*, 60, 475–480.

Chepil, W.S. and Milne, R.A. (1941) Wind erosion of soil in relation to roughness of surface. *Soil Science*, 52, 417–431.

Chepil, W.S., Suddoway, F.H. and Armbrust, D.V. (1962) Climate factor for estimating wind erodibility of farm fields. *Journal Soil Water Conservation*, 17, 162–165.

Chepil, W.S. and Woodruff, N.P. (1963) The physics of wind erosion and its control. *Advances in Agronomy*, 15, 211–302.

Cloudsley-Thompson, J.L. 1978. Human activities and desert expansion. *Geographical Journal*, 144, 416–423.

Cooke, R.U. and Doornkamp, J.C. (1990) *Geomorphology in Environmental Management*. Oxford University Press, Oxford.

Coppin, N.J. and Richards, I.G. (1990) *Use of Vegetation in Civil Engineering*. Construction Industry Research Information Association (CIRIA), Butterworths, London.

Dietrich, W.E., Dunne, T., Humphrey, N.F. and Reid, L.M. (1982) Construction of sediment budgets for drainage basins. In *Sediment Budgets and Routing in Forested Drainage Basins*, US Department of Agriculture Forest Service Technical Report DNW-141, Portland, 5–23.

Dregne, H.E. (1983) *Desertification of Arid Lands*. Harwood, New York.

Ellison, W.D. (1947) Soil erosion studies. *Agricultural Engineering*, 28: 145–146, 197–201, 245–248, 297–300, 402–405, 442–450.

Elwell, H.A. (1984) Soil loss estimation: a modelling technique. In *Erosion and Sediment Yield*, Hadley, R.F. and Walling, D.E. (eds), Geobooks, Norwich, 15–36.

Evans, R. (1980) Mechanics of water erosion and their spatial and temporal controls: an empirical viewpoint. In *Soil Erosion*, Kirby, M.J. and Morgan, R.P.C. (eds), Wiley, Chichester, 109–128.

Forster, G.R. and Wischmeier, W.H. (1974) Evaluating irregular slopes for the soil loss prediction. *Transactions American Society Agricultural Engineers*, 17, 305–309.

Forster, G.R., Lane, L.J., Nowlin, J.D., Lafler, J.M. and Young, R.A. (1981) Estimating erosion and sediment yield on field-sized areas. *Transactions American Society Agricultural Engineers*, 24, 1253–1264.

Fournier, F. (1960) *Limat et erosion: la relation entre l'erosion du sol par l'eau et le precipitations atmospherignes*. Presses Universitaires de France, Paris.

Hagen, L.J. (1991) A wind erosion prediction system to meet user needs. *Journal Soil Water Conservation*, 106–111.

Hannan, I.D. and Hicks, R.W. (1980) Soil conservation and urban land-use planning. *Journal Soil Conservation Service New South Wales*, 36, 134–145.

Hansen, V.E., Israelsen, O.W. and Stringham, G.E. (1980) *Irrigation Principles and Practice.* fourth edition. Wiley, New York.

Hoffman, G.T., Howell, T.A. and Solomon, K.H. (1990) *Management of Farm Irrigation Systems.* American Society Agricultural Engineers Monograph, St Joseph, Michigan.

Hudson, N.W. (1981) *Soil Conservation.* Batsford, London.

Jensen, M.E. (1973) *Consumptive Use of Water and Irrigation Water Requirements.* American Society Civil Engineers, New York.

Kirkby, M.J. (1980) The problem. In *Soil Erosion*, Kirkby, M.J. and Morgan, R.P.C. (eds), Wiley, Chichester, 1–16.

Klingebeil, A.A. and Montgomery, P.H. (1966) *Land Capability Classification.* Soil Conservation Service Agricultural Handbook 210, US Department of Agriculture, Washington, DC.

Lal, R. (1988) Soil erosion control with alley cropping. In *Land Conservation for Future Generations*, Rimwanich, E. (ed.), Department of Land Development, Ministry of Agriculture and Conservation, Bangkok, 237–246.

Landsberg, J.J. and James, G.B. (1971) Wind profiles in plant canopies: studies on an analytical model. *Journal Applied Ecology*, 8, 729–741.

Liebenow, A.M., Elliot, W.J., Laflen, J.M. and Kohl, K.D. (1990) Interrill erodibility: collection and analysis of data from cropland soils. *Transactions American Society Agricultural Engineers*, 33, 1882–1888.

Livesey, R.H. (1974) Corps of Engineers methods for predicting sediment yield. Agriculture Research Service, ARS-5-40, US Department of Agriculture, 16–32.

Mitchell, J.K. and Bubenzer, G.D. (1980) Soil loss estimation. In *Soil Erosion*, Kirkby, M.J. and Morgan, R.C.P. (eds), Wiley, Chichester, 17–62.

Morgan, R.P.C. (1980) Implications. In *Soil Erosion*, Kirkby, M.J. and Morgan, R.P.C. (eds), Wiley, Chichester, 253–301.

Morgan, R.P.C. (1995) *Soil Erosion and Conservation.* second edition, Addison Wesley Longman, Harlow.

Morgan, R.P.C. and Rickson, R.J. (1995) Water erosion control. In *Slope Stabilization and Erosion Control: A Bioengineering Approach*, Morgan, R.P.C. and Rickson, R.J. (eds), E & FN Spon, London, 133–190.

Nnoli, O. (1990) Desertification, refugees and regional conflict in West Africa. *Disasters*, 14, 132–139.

Rausch, D.L. and Heinemann, H.G. (1984) Measurement of reservoir sedimentation. In *Erosion and Sediment Yield*, Hadley, R.F. and Walling, D.E. (eds), Geobooks, Norwich, 179–200.

Rickson, R.J. (1995) Simulated vegetation and geotextiles. In *Slope Stabilization and Erosion Control: A Bioengineering Approach*, Morgan, R.P.C. and Rickson, R.J. (eds), E & FN Spon, London, 95–131.

Schwab, G.O., Fangmeier, D.D., Elliot, W.J. and Frevert, R.K. (1993) *Soil and Water Conservation Engineering*, fourth edition, Wiley, New York.

Skidmore, E.L., Fisher, P.S. and Woodruff, N.P. (1970) Wind erosion equation: computer solution and application. *Proceedings Soil Science Society America*, 34, 931–935.

Smith, D.D. and Wischmeier, W.H. (1962) Rainfall erosion. *Advances in Agronomy*, 14, 109–148.

Stocking, M.A. and Elwell, H.A. (1973) Soil erosion hazard in Rhodesia. *Rhodesian Agricultural Journal*, 70, 93–101.

Stocking, M.A. and Elwell, H.A. (1976) Rainfall erosivity over Rhodesia. *Transactions Institute British Geographers*, New Series, 1, 231–245.

Styczen, M.E. and Morgan, R.P.C. (1995) Engineering properties of vegetation. In *Slope Stabilization and Erosion Control: A Bioengineering Approach*, Morgan, R.P.C. and Rickson, R.J. (eds), E & FN Spon, London.

Thornes, J.B. (1990) The interaction of erosional and vegetational dynamics in land degradation: spatial outcomes. In *Vegetation and Erosion*, Thornes, J.B. (ed.), Wiley, Chichester, 41–53.

Watson, D.A. and Laflen, J.M. (1986) Soil strength, slope and rain intensity effects on interrill erosion. *Transactions America Society Agricultural Engineers*, 29, 98–102.

Warren, A. and Maizels, J. (1977) Ecological change and desertification. In *Desertification: Its Causes and Consequences*, United Nations, Pergamon, Oxford, 171–260.

Williams, J.R. (1985) The physical components of the EPIC model. In *Soil Erosion and Soil Conservation*, El-Swaify, S.A. and Moldenhauer, W.C. (eds), American Society Soil Conservation, Ankeny, Iowa, 272–284.

Wilson, S.J. and Cooke, R.U. (1980) Wind erosion. In *Soil Erosion*, Kirkby, J.M. and Morgan, R.P.C. (eds), Wiley, Chichester, 217–251.

Wischmeier, W.H. (1959) A rainfall erosion index for a universal soil loss equation. *Proceedings Soil Science Society America*, 23, 246–249.

Wischmeier, W.H. (1975) Estimating the soil loss equation's cover and management factor for undisturbed areas. In *Present and Prospective Technology for Predicting Sediment Yields and Sources*, Agricultural Research Service Publication ARS-S-40, US Department of Agriculture, Washington, DC.

Wischmeier, W.H. and Smith, D.D. (1978) *Predicting Rainfall Erosion Losses*. Agricultural Research Service Handbook 537, US Department of Agriculture, Washington, DC.

Wolman, M.G. (1967) A cycle of sedimentation and erosion in urban river channels. *Geografiska Annaler*, 49A, 385–395.

Woodruff, N.P. and Siddaway, F.H. (1965) A wind erosion equation. *Proceedings Soil Science Society America*, 29, 602–608.

Chapter 10

Waste and its disposal

10.1 Introduction

With increasing industrialization, technical development and economic growth, the quantity of waste has increased immensely. In addition, in developed countries the nature and composition of waste has evolved over the decades, reflecting industrial and domestic practices. For example, in Britain domestic waste has changed significantly since the 1950s, from largely ashes and little putrescible content of relatively high density to low-density, highly putrescible waste. Many types of waste material are produced by society, of which domestic waste, commercial waste, industrial waste, mining waste, and radioactive waste are probably the most notable. Over and above this, waste can be regarded as non-hazardous or hazardous. Waste may take the form of solids, sludges, liquids, gases or any combination thereof. Depending on the source of generation, some of these wastes may degrade into harmless products, whereas others may be non-degradable and/or hazardous, thus posing health risks and environmental problems if not managed properly. A further problem is the fact that deposited waste can undergo changes through chemical reaction, resulting in dangerous substances being developed. Indeed, waste disposal is one of the most expensive environmental problems to deal with, and the importance of waste management has increased in terms of the conservation of natural resources. Solving the waste problem is one of the fundamental tasks of environmental protection.

As waste products differ considerably from one another, the storage facilities they require also differ. Despite increased efforts at recycling wastes and avoiding their production, many different kinds of special waste are produced that must be disposed of in special ways. Wastes that do not decompose within a reasonable time, mainly organic and hazardous wastes, and liquids that cannot be otherwise disposed of, should ideally be burned. All the organic materials are removed during burning, converting them to a less hazardous form and leaving an inorganic residue. Solid, unreactive, immobile inorganic wastes can be disposed of at above-ground disposal sites. It is sometimes necessary to treat these wastes prior to disposal. To provide long-term isolation from the environment, high-toxicity, non-degradable wastes should be disposed of underground if they cannot be burned.

The best method of disposal is determined on the basis of the type and amount of waste on the one hand and the geological conditions of the waste disposal site on the other. In terms of locating a site, initially a desk study is undertaken. The primary task of the site exploration that follows is to determine the geological and hydrogeological conditions. Their evaluation provides the basis of models used to test the reaction of the system to engineering activities. Chemical analysis of groundwater, together with mineralogical analysis of rocks, may help to yield information about its origin and hence about the future development of the site. At the same time, the leaching capacity of the water is determined, which allows prediction of reactions between wastes and soil or rock. If groundwater must be protected, or highly mobile toxic or very slowly degradable substances are present in wastes, then impermeable liners may be used to inhibit infiltration of leachate into the surrounding ground.

In terms of waste disposal by landfill, a landfill is environmentally acceptable if it is correctly engineered. Unfortunately, if it is not constructed to sufficiently high standards, a landfill may have an adverse impact on the environment. Surface water or groundwater pollution may result. Consequently, a physical separation between waste on the one hand and ground and surface water on the other, as well as an effective surface water diversion drainage system, are fundamental to design.

The objectives of a landfill classification are to consider waste disposal situations in terms of the type of waste, the amount of waste and the potential for water pollution. The different classes of landfill can then be used as a basis for the establishment of minimum requirements for site selection. Although no waste is completely non-hazardous, non-hazardous waste includes builders', garden, domestic, commercial and general dry industrial waste that can be disposed of in a landfill. Nevertheless, some such landfills may produce significant amounts of leachate and so must incorporate leachate control systems. The size of a non-hazardous landfill operation depends on the daily rate of deposition of waste, which in turn depends upon the size of population served. Hazardous waste has the potential, even in low concentrations, to have a detrimental affect on public health and/or the environment because it may be, for instance, flammable, corrosive or carcinogenic.

In order to carry out a detailed assessment of an existing or potential waste disposal site a large volume of technical information and site-specific data is required (Sara, 1994). The technical information includes:

1   The interaction of wastes in the landfill, their degradation and leaching characteristics and how these change with time.
2   The composition of the leachate, including the solubilities and speciation of toxic elements, organics and major ion pollutants.
3   Changes in the composition and rate of leachate production with time.
4   The generation and behaviour of gases at the site, and their migration through surrounding permeable strata.

5   Groundwater and mixed fluid phase movement away from the site, and how these will change with the geomorphological evolution of the area during the period considered in the assessment.

6   The physical, chemical and biological interactions of the leachate with the rocks and soils of the site, including sorption (desorption reactions, matrix diffusion, dispersion and other diluting/retarding mechanisms).

7   The long-term geotechnical behaviour of the site with respect to engineering structures (including geochemical reactions of leachates and groundwaters with such structures).

One of the most important factors is the requirement for a thorough geological and hydrogeological survey of the site to a standard that will allow predictive leachate transport and gas migration modelling to be carried out. The results of such modelling can then be used, via a sensitivity analysis that takes account of the wastes to be disposed, to design an engineered landfill that meets specific performance criteria. Until a performance assessment is available it is not possible, in advance, and based solely on a survey of the geology, to judge what the engineering requirements or range of acceptable waste types and volumes might be. The form of engineering envisaged for a hazardous waste site depends on whether a safety assessment indicates that containment or predictable dispersion is more appropriate for the wastes concerned.

10.2 Domestic refuse and sanitary landfills

At the present time, increasing concern is being expressed over the disposal of domestic waste products. Although domestic waste is disposed of in a number of ways, quantitatively the most important method is placement in a landfill. For example, in Britain, of approximately 18.6 million tonnes of domestic solid waste produced each year, about 76% is disposed of in landfills.

Domestic refuse is a heterogeneous collection of almost anything (e.g. waste food, garden rubbish, paper, plastic, glass, rubber, cloth, ashes, building waste, metals; Fang, 1995), much of which is capable of reacting with water to give a liquid rich in organic matter, mineral salts and bacteria. This is the leachate and is formed when rainfall infiltrates a landfill and dissolves the soluble fraction of the waste, and from the soluble products formed as a result of the chemical and biochemical processes occurring within decaying wastes. The organic carbon content of waste is especially important, since this influences the growth potential of pathogenic organisms.

Matter exists in the gaseous, liquid and solid states in landfills, and all landfills comprise a delicate and shifting balance between the three states. Any assessment of the state of a landfill and its environment must take into consideration the substances present in the landfill, their mobility now and in the future, the potential pathways along which pollutants can travel, and the targets potentially at risk from the substances involved.

Waste materials disposed of in sanitary landfills have dry densities ranging from 158 to 685 kg m^{-3}, moisture contents of 10 to 35% and load bearing capacities of 19.2 to 33.5 kPa. Baling or shredding the waste materials before deposition improves their *in situ* properties but at greatly increased cost.

As far as the location of landfills in concerned, decisions have to be made on site selection, project extent, finance, construction materials and site rehabilitation. The major requirement in planning a landfill site is to establish exactly, by survey and analysis, the types, nature and quantities of waste involved. A waste survey is undertaken, after which future trends are forecast. These forecasts form the basis of the decision on which potential site for a landfill is chosen. Initially, the potential life of a site is estimated and its distance from the proposed waste catchment area assessed. The design of a landfill is influenced by the character of the material that it has to accommodate.

Modern landfill disposal facilities require detailed investigations to ensure that appropriate design and safety precautions are taken. Furthermore, legislation generally requires those responsible for waste disposal facilities to guarantee that sites are suitably contained to prevent harming the environment. This may require that investigation continues during or after the construction of a landfill. However, sinking boreholes through the base of a leachate-filled landfill to check what is happening is unacceptable without precautions being taken to avoid leakage. Geophysical methods are being used increasingly for this purpose.

Selection of a landfill site for a particular waste or a mixture of wastes involves a consideration of economic and social factors, as well as geological and hydrogeological conditions. The ideal landfill site should be hydrogeologically acceptable, posing no potential threat to water quality when used for waste disposal; be free from running or static water; and have a sufficient store of material suitable for covering each individual layer of waste. It should also be situated at least 200 m away from any residential development. As far as the hydrogeological conditions are concerned, most argillaceous sedimentary, and massive igneous and metamorphic rock formations have low intrinsic permeability and therefore are likely to afford the most protection to water supply. By contrast, the least protection is provided by rocks intersected by open discontinuities or in which solution features are developed. Granular materials may act as filters, leading to dilution and decontamination. Hence sites for disposal of domestic refuse can be chosen where decontamination has the maximum chance of reaching completion and where groundwater sources are located far enough away to enable dilution to be effective. The position of the water table is important as it determines whether wet or dry tipping is involved. Generally, unless waste is inert, wet tipping should be avoided. The position of the water table also determines the location at which flow is discharged. The hydraulic gradient determines the direction and velocity of the flow of leachates when they reach the water table and also influences the amount of dilution that leachates undergo. Aquifers that contain potable supplies of water must be protected. If near a proposed landfill, a thorough hydrogeological investigation is necessary to ensure that site operations will not pollute the aquifer. If pollution is a possibility, the site must be designed to provide some form of artificial protection, otherwise the proposal should be abandoned.

Barber (1982) identified three classes of landfill site based upon hydrogeological criteria (Table 10.1). When assessing the suitability of a site, two of the principal considerations are the ease with which the pollutant can be transmitted through the substrata and the distance it is likely to spread from the site. Consequently, the primary and secondary permeability of the formations underlying a potential landfill area are of major importance. It is unlikely that the first type of site mentioned in Table 10.1 would be considered suitable. There would also be grounds for an objection to a landfill site falling within the second category of Table 10.1 if the site were to be located within the area of diversion to a water supply well. Generally, the third category, in which the leachate is contained within the landfill area, is to be preferred. Since all natural materials possess some degree of permeability, total containment can only be achieved if an artificial impermeable lining is provided over the bottom of the site. However, there is no guarantee that clay, soil cement, asphalt or plastic linings will remain impermeable permanently. Thus the migration of materials from a landfill site into the substrata will occur eventually, although the length of time before this happens may be subject to uncertainty. In some instances the delay will be sufficiently long for the pollution potential of the leachate to be greatly diminished. As mentioned, one of the methods of tackling the problem of pollution associated with landfills is by dilution and dispersal

Table 10.1 Classification of landfill sites based upon their hydrogeology (after Barber, 1982).

Designation	Description	Hydrogeology
Fissured site, or site with rapid subsurface liquid flow	Material with well-developed secondary permeability features	Rapid movement of leachate via fissures, joints, or through coarse sediments. Possibility of little dispersion in the groundwater, or attenuation of pollutants
Natural dilution, dispersion and attenuation of leachate	Permeable materials with little or no significant secondary permeability	Slow movement of leachate into the ground through an unsaturated zone. Dispersion of leachate in the groundwater, attenuation of pollutants (sorption, biodegradation, etc.) probable
Containment of leachate	Impermeable deposits such as clays or shales, or sites lined with impermeable materials or membranes	Little vertical movement of leachate. Saturated conditions exist within the base of the landfill

of the leachate. Otherwise leachate can be collected by internal drains within the landfill and conveyed away for treatment.

One of the difficulties in predicting the effect of leachate on groundwater is the continual changes in the characteristics of leachate as a landfill ages. Leachate may be diluted where it gains access to run-off or groundwater, but this depends on the quantity and chemical characteristics of the leachate, as well as the quantity and quality of the receiving water.

10.2.1 Design considerations for a landfill

The design of landfill sites is influenced by the physical and biochemical properties of the wastes. For instance, the settlement of wastes is attributable to physical mechanisms and decompositon. Post-closure settlement of sites is taken into account at a very early stage in the design of landfills. The need for control of leachate production at a particular landfill is dependent on the extent of the possible pollution problems at that site. Site selection therefore has an important influence on the need for leachate control, so a thorough investigation must be undertaken to determine site suitability. Nonetheless, the use of leachate control and/or treatment methods may permit unsuitable sites to be used for the disposal of solid wastes. Leachate control should be planned during landfill development rather than after the landfill has been constructed, especially if control techniques are to be installed beneath the waste.

The frequency and duration of rainfall for the catchment area of a landfill has to be considered, since this influences the amount of water that can penetrate into a landfill. Data on evapotranspiration and infiltration rates for landfills are also important, since this affects leachate generation. Furthermore, it is necessary to determine how much leachate can be absorbed by the area before run-off occurs. Surface run-off should be considered when the design of a landfill is undertaken. The rate of run-off is affected by the topography of the catchment area, as well as by the nature and type of strata surrounding and below the proposed site.

Various techniques have been developed using water budget methods to estimate the amount of free water in a landfill. These methods consider a mass balance between precipitation, evapotranspiration, surface run-off and waste moisture storage. Some fraction of precipitation will infiltrate the waste, after passing through the cover material. A field capacity or moisture-holding capacity ranging from 50 to 60% by dry weight for compacted refuse has been suggested. Water balance calculations can be used to determine the size of a working cell within a site needed to control the amount of free leachate that will be produced from rainfall or other water inputs. This excess free-draining leachate may then lead to the development of a zone of saturated refuse. The rate of build-up of leachate levels within the site and the amount of leachate held within the saturated zone is directly related to the effective porosity or storativity of the refuse. The ability to manage leachate within a site through leachate drainage schemes and vertical pumping wells is related to both the storativity and permeability of the waste.

The quantity of leachate produced is influenced by the amount of groundwater in a landfill (leachate percolation = net percolation − water absorbed by waste + liquid disposal into landfill). The quantity of water absorbed by waste depends on the age of placement of the waste. Initially, the water-absorbing capacity of the waste exceeds net percolation and the leachate flow is zero. At the other extreme, if the waste is totally saturated, then the leachate flow equals net percolation plus groundwater or subsurface water flow into a landfill plus liquid disposal into the landfill.

The most common means of controlling leachate is to minimize the amount of water infiltrating the site by encapsulating the waste in impermeable material. Hence, well-designed landfills usually possess a cellular structure, as well as a lining and a cover; that is, the waste is contained within a series of cells formed of clay (Figure 10.1). The cells are covered at the end of each working day with a layer of soil and compacted. The dimensions of a cell depend on the volume of waste received and the availability of cover material, but they tend to range from 2.5 to 9.0 m. The thickness of the daily cover is 150 to 300 mm, its purpose being to control flying paper, to minimize gas and moisture percolation, and to provide rodent and fire control, as well as improved appearance. The ability of water to move within a landfill depends on a number of factors. Cell cover may act as an aquiclude. However, uneven settlement of the landfill can cause fractures within the cover, which allow water to percolate

Figure 10.1 Cellular construction of a landfill in Kansas City, United States.

downwards into the fill. In addition, successful landfill design involves the layers of waste being capable of absorbing precipitation during layer formation. Waste consolidation either on site or before tipping also helps to reduce the quantity of leachate produced.

At the present time, there is no standard rule as to how waste should be dumped or compacted. Nonetheless, compaction is important since it reduces settlement and hydraulic conductivity, while increasing shear strength and bearing capacity. Furthermore, the smaller the quantity of air trapped within landfill waste, the lower is the potential for spontaneous combustion. Fang (1995) emphasized that compaction requires planning during the waste disposal process. Wherever possible, waste should be uniformly distributed in thin layers prior to compaction. If non-uniform spreading cannot be achieved, then Fang recommended that heavier items should be placed in the centre of a landfill to help to control its stability. Locally available soil can be mixed with the waste. As far as landfill stability is concerned, the potential for slope failure in a landfill is related to compaction control during disposal, and the heavier the roller that is used for compaction the better. Even so, conventional compaction techniques do not always achieve effective results, especially with highly non-uniform waste. In such instances, dynamic compaction has been used with good results.

Mitchell (1986) maintained that a properly designed and constructed liner and cover offer long-term protection for ground and surface water. Landfill liners are constructed from a wide variety of materials (Figure 10.2). Adequate site preparation is necessary if a lining system is to perform satisfactorily. Nonethless, no liner system, even if perfectly designed and constructed, will prevent all seepage losses. For instance, no liner, no matter how rigid or highly reinforced, can withstand large differential settlement without eventually leaking or possibly failing completely. Jessberger et al. (1995) briefly outlined how to assess the amount of deformation a liner is likely to undergo and whether unacceptable cracking will develop as a result. If extreme concern is warranted, an under-drainage system can be placed beneath the primary liner to collect any leakage, which can either be treated or recirculated back to the containment area. A secondary liner can be placed beneath the under-drainage system.

Clay, bentonite, geomembrane, soil–cement or bitumen cement can be placed beneath a landfill to inhibit movement of leachate into the soil. Clay liners are suitable for the containment of many wastes because of the low hydraulic conductivity of clay and its ability to adsorb some wastes (Daniel, 1993). They are constructed by compaction in lifts of about 150 mm thickness. Care must be taken during construction to ensure that the clay is placed at the specified moisture content and density, and to avoid cracking due to drying out after construction (Quigley et al., 1988). The factors that govern the behaviour of a clay liner include the compaction moisture content, method of compaction, quality control during compaction, and potential increase in the hydraulic conductivity due to interaction with waste liquids (Clarke and Davis, 1996). Clay

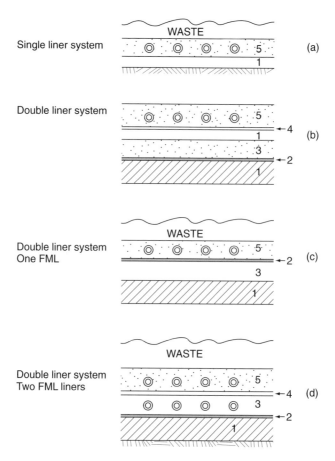

Figure 10.2 Landfill liner systems. 1 compacted, low-permeability clay; 2 flexible membrane liner (FML); 3 leachate collection/detection system; 4 FML; 5 primary
leachate collection system; ◎ collection pipes

liners compacted at wet of optimum moisture content are less permeable than
those compacted dry of optimum.

Certain leachates have been especially troublesome for clay liners. Those consisting of organic solvents and those containing high levels of dissolved salts,
acids or alkalis can give rise to cracking and the development of pipes in clay.
Clays, however, can be treated with polymers that reduce their sensitivity to
potential contaminants. Cation exchange can take place between the ions on
the surfaces of clay particles and those present in the leachate. Nevertheless,
the effectiveness of clay liners in containing many hazardous wastes has been
questioned. Some leachate can seep into a clay liner and could migrate to the
groundwater at some future time.

Between 3 and 6%, by weight, of bentonite can be mixed with soil obtained from borrow pits and this used as a liner. Sands with 20 to 30% fines are usually preferable. Compatibility with the anticipated leachate should be determined, because sodium bentonite is vulnerable to some organic chemicals. The thickness of a bentonite–soil liner is usually 300 mm or more. The material is mixed on site and applied wet of optimum moisture content in two lifts and compacted.

When geomembranes are used as liners the chemical compatibility between waste material and geomembrane should be assessed. The geomembrane must possess sufficient thickness and strength to avoid failure due to physical stresses (Koerner, 1993). The foundation beneath the liner must be able to support the liner and to resist pressure gradients. If the support system settles, compresses or lifts, the liner may rupture. Installation of flexible membrane liners (FMLs) involves seaming the sheets together. The quality of the seams is checked by visual examination, vacuum testing or ultrasonic testing. Underliners and covers may be used to protect the geomembranes from puncture, tearing, abrasion and ultraviolet rays. For example, a bedding course of sand can be placed below the liner and another layer of sand placed on top. Composite liners incorporate both clay blankets and geomembranes. Other measures to protect the liner include a minimum soil cover, elimination of folds, no seaming in cold weather, removal of sharp objects in the subgrade and prohibiting equipment on top of the liner. Table 10.2 outlines the general advantages and disadvantages of some geosynthetic materials.

Table 10.2 Broad advantages and disadvantages of some polymers.

	Advantages	Disadvantages
Polyvinyl chloride	Resistant to inorganics, good tensile strength, elongation, puncture and abrasion resistance, easy to seam	Low resistance to organic chemicals, including hydrocarbons, solvents and oil, poor resistance to exposure
High-density polyethylene	Good, resistance to oils, chemicals and high temperature, available in thick sheets (20–150 mm)	Require more field seams, subject to stress cracking, punctures at low thickness, poor tear propagation
Ethylene propylene rubber	Resistant to dilute concentrations of acids, alkalis, silicates, phosphates and brine, tolerates extreme temperatures, flexible at low temperatures, excellent resistance to exposure	Not recommended for petroleum solvents or halogenated solvents, difficult to seam or repair, low seam strength
Chlorosulphonated polyethylene	Good resistance to ozone, heat, acids and alkalis, easy to seam	Poor resistance to oil, good tensile resistance if reinforced

Shotcrete or gunite may be applied to the bottom and sides of a waste pit. Wire mesh is frequently used with the shotcrete or gunite to reinforce it and to prevent cracking. The technique is not often used except when a landfill is constructed on rock.

Soil–cement liners make use of local soils. The soil should ideally have less than 20% silt/clay fraction and when mixed with 3–12% cement by weight, should perform adequately as a liner. Lining slopes with soil–cement, however, can present difficulties. In addition, soil–cement can be degraded in acidic conditions, which tend to attack the cement. Furthermore, the likelihood of soil–cement shrinking and cracking leads to an increase in its permeability. When used, the minimum design compacted thickness is usually around 300 mm. The mixing is carried out *in situ* in a central plant and the soil–cement is laid by a mechanical spreader in 150–225 mm lifts.

Concrete liners can be reinforced with wire mesh or reinforcement bars, depending on the conditions of the subgrade. Unfortunately, construction and expansion joints, both of which are subject to leaks, must be incorporated in the pour. Waterstops have to be used, but they are expensive and might not be chemically compatible with the leachate. The paving of side slopes with angles greater than 30° to the horizontal presents major construction problems.

Bitumen cement and bituminous concrete have been used for lining landfills. In the latter case, bitumen, cement and high-quality mineral aggregates are hot-mixed. Both are compacted in place and the compacted thickness ranges from 38 to 150 mm. Usually the material is placed in more than one layer. However, a major consideration is the chemical compactibility between the contents of the landfill and the bituminous material.

The double-liner system is intended to prevent leakage. The primary liner of such a system is the upper geomembrane. Leachate should be properly collected above the primary liner at the bottom of the landfill, adequately treated, and disposed of in accordance with accepted environmental principles. A perforated pipe collector system is located below the primary liner and is bedded within a crushed stone or sand drainage blanket. A secondary geomembrane liner is placed beneath the drainage blanket, but it needs to function only if leakage from the primary liner is found in the under-drainage system. If leakage from the primary liner does occur, a downstream well-monitoring system must be deployed to check for possible leakage from the secondary liner.

Leachate drainage systems should be designed to collect the anticipated volume of leachate likely to be produced by the landfill (Jessberger *et al.*, 1995). This will vary during the life of the site and can be estimated by a water balance calculation. One of their functions is to prevent the level of leachate rising to an extent that it can drain from the landfill and pollute nearby watercourses. The system should be robust enough to withstand the load of waste imposed upon it, as well as that of equipment used in the construction of the landfill, and should be able to accommodate any settlement. It should be capable of resisting chemical attack in the corrosive environment of a landfill. Ideally, a leachate drainage collection system should be installed over the whole base of a landfill and, if below ground level, it should extend up the sides. The drainage blanket, consisting of granular material, should be at least 300 mm in thickness, with a minimum hydraulic conductivity of $1–10 \text{ m s}^{-1}$. The perforated collection pipes within the drainage blanket convey leachate to collection points or wells, from which leachate can be removed by pumping.

If lining is considered too expensive, then drainage is a relatively inexpensive alternative. For instance, a drainage ditch can be combined with a layer of free-draining granular material overlying a low-permeability base. The granular material is graded down towards the perimeter of the landfill so that leachate can flow into the drainage ditches. The leachate can be pumped or flow away from the ditches. Because the drainage layer facilitates the flow of leachate to the drains, it avoids the development of a leachate mound. The permeability of the drainage layer is usually at least an order of magnitude higher than that of the waste (typically 10 m s^{-1}) and should have a minimum thickness of 300 mm. Synthetic drainage layers are available and range from 4 to over 40 mm in thickness. Several layers may be used to achieve the drainage capacity. A filter medium may be used between the waste and the drainage layer to allow leachate to pass through but to prevent the migration of fines. A minimum filter thickness of 150 mm is recommended. The use of geotextiles as filter fabrics at the base of a landfill may not ensure trouble-free operation of the system, since they may be prone to clogging. The capability of pipes, filters, drainage and other materials that come into contact with leachate should be assessed by testing.

Any leachate that infiltrates into the ground beneath a landfill may be subject to attenuation as it flows through the surrounding soil and rocks. The slower the flow of leachate, the more effective attenuation is likely to be. The existence of an unsaturated zone of soil beneath a landfill can be particularly beneficial, as the presence of liquid and gas phases delay the movement of leachate towards the water table, thereby promoting gaseous exchange, oxidation, adsorption by clay minerals, and biodegradation.

If a liner beneath a landfill fails, then a leachate plume may develop within the ground beneath. The size of a plume at any particular time depends upon the rate at which the leachate can spread, which in turn is governed by the permeability of the soil or rock and the hydraulic gradient.

The principal function of a cover is to minimize infiltration of precipitation into a landfill; other functions include gas control, future site use, and aesthetics. A properly designed and maintained cover can prevent the entry of liquids into a landfill and hence minimize the formation of leachate. The nature of the soil used for covering waste materials is very important. At many sites, however, the quality of the borrow material is less than the ideal and so some blending with imported soil may be required. The relative suitability of soils for various functions is provided in Table 10.3. Because clays will generally last longer than synthetic materials, a clay cover should usually be chosen.

Table 10.3 Ranking of soil types according to performance of some cover function.

Soil type USCS symbol	Impedence of water percolation	Hydraulic conductivity (approx.) (cm s⁻¹)	Support vegetation	Impedence of gas migration	Resistance to water erosion	Frost resistance	Crack resistance
GW	X	10^{-2}	X	X	I	I	I
GP	XII	10^{-1}	X	IX	I	I	I
GM	VII	5×10^{-4}	VI	VII	IV	IV	III
GC	V	10^{-4}	V	IV	III	VII	V
SW	IX	10^{-3}	IX	VIII	II	II	I
SP	XI	5×10^{-2}	IX	VII	II	II	I
SM	VIII	10^{-3}	II	VI	II	V	I
SC	VI	2×10^{-4}	II	V	IV	VI	IV
ML	IV	10^{-5}	I	III	VI	X	VI
CL	II	3×10^{-8}	III	II	VII	VIII	VIII
OL			VII		VIII	VIII	VII
MH	III	10^{-7}	IV		VII	VIII	IX
CH	I	10^{-9}	VIII	I	IX	IX	X
OH			VIII		X	III	IX
PT			III				

Ranking: I (best) to XIII (poorest).

According to Daniel (1995), most cover systems in the USA consist of a number of components. However, not all the components shown in Figure 10.3 need be present in a cover, and some layers may be combined; for example, the surface layer and protection layer may be combined into a single soil layer. The surface soil layer offers protection to the layers beneath and allows for vegetative growth, thereby helping to stabilize the cover and reduce erosion. Vegetation also helps to reduce infiltration by the process of transpiration. A protection layer not only protects the layers in the cover beneath from excessive wetting and drying, and from freezing, which can cause cracking, but it helps to separate the waste from burrowing animals and plant roots. A sand–gravel drainage layer or bulky needle-punched geotextile can be used to divert surface water from the clay seal to drains along the boundary of the landfill, which convey the water away. Sand–gravel drainage layers must be adequately graded. Where infiltration is heavy, a composite geosynthetic may be required. The drainage layer reduces the pore water pressures in the cover soils, as well as the head on the barrier layer beneath. The barrier layer usually consists of clay and provides a relatively impermeable barrier, helping to prevent surface water entering a landfill. The clay must be compatible with the material in the landfill. It is compacted to a high strength to help to reduce the potential for cover cracking. A gas collection layer may be incorporated in a cover and is composed of coarse to medium gravel. Its purpose is to collect gas for processing or discharge via vents. A filter layer of sand is placed between the clay seal and the underlying gas collection layer. Alternatively, a geotextile filter can be used or a thick geotextile can be used as both drain and filter. A single clay layer cover may be placed on a landfill that is reasonably stable (i.e. it has an unconfined compressive strength of 70 kPa or more). In some circumstances a two-layer cover, consisting of a synthetic membrane overlain by a clay layer, may help to minimize the infiltration of precipitation.

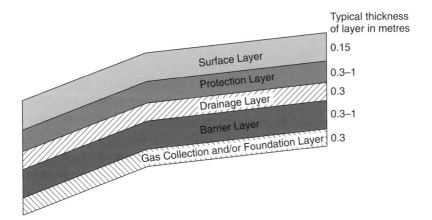

Figure 10.3 Basic components of a final cap.

Geosynthetic clay liners (GCLs) are thin layers of dry bentonite (approximately 5 mm thick) attached to one or more geosynthetic materials. In other words, the bentonite is sandwiched between an upper and lower sheet of geotextile, or the bentonite is mixed with adhesive and glued to a geomembrane. The primary purpose of the geosynthetic component(s) is to hold the bentonite together in a uniform layer. When the bentonite is wetted, it swells and provides a cover with a low permeability. A geosynthetic clay liner can be used in a number of ways in a cover system (Figure 10.4) but should always be overlain by a layer of protective soil. Daniel (1995) compared the performance of GCLs with compacted clay liners (Table 10.4).

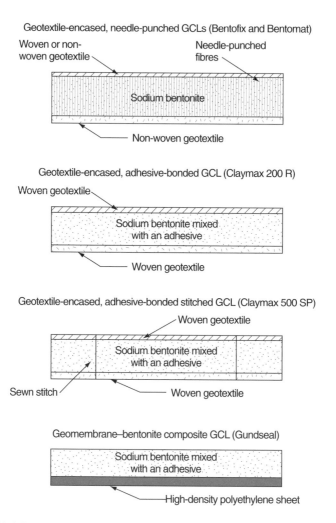

Figure 10.4 Four types of geosynthetic clay liner.

Table 10.4 Differences between GCLs and compacted clay liners (after Daniel, 1995).

Characteristic	Geosynthetic clay liner	Compacted clay liner
Materials	Bentonite, adhesives, geotextiles, and geomembranes	Native soils or blend of soil and bentonite
Thickness	Approximately 12 mm, consumes very little landfill volume	Typically 300 to 600 mm; consumes more landfill volume
Hydraulic conductivity	≤ 1 to 5×10^{-11} m s^{-1}	$\leq 1 \times 10^{-9}$ m s^{-1}
Speed and ease of construction	Rapid, simple installation	Slow, complicated construction
Ease of quality assurance (QA)	Relatively simple, straight forward, common sense procedures	Complex QA procedures requiring highly skilled and knowledgeable people
Vulnerability to damage during construction from desiccation and freeze–thaw	GCLs are essentially dry; GCLs cannot desiccate during construction; not particularly vulnerable to damage from freeze–thaw	Compacted clay liners are nearly saturated; can desiccate during construction; vulnerable to damage from freeze–thaw
Vulnerability to damage from puncture	Thin GCL is vulnerable to puncture	Thick compacted clay liner cannot be punctured
Vulnerability to damage from differential settlement	Can withstand much greater differential settlement than compacted clay liner	Cannot withstand much differential settlement without cracking
Availability of materials	Materials easily shipped to any site	Suitable materials not available at all sites
Cost	Reasonably low, highly predictable cost that does not vary much from project to project	Highly variable – depends greatly on characteristics of locally available soils
Ease of repair	Easy to repair with patch placed over problem area	Very difficult to repair; must mobilize heavy earthmoving equipment if large area requires repair
Experience	Limited due to newness	Has been used for many years
Regulatory approval	Not explicitly allowed in most regulations – owner must gain approval on the basis of equivalence in meeting performance objectives	Compacted clay liners are usually required by regulatory agencies

The key geotechnical factors in cover design are its stability and its resistance to cracking. Cracking of a clay cover may be brought about by desiccation, or by the build-up of gas pressure beneath if a venting system is not functional or not installed. It may also be difficult to maintain the integrity of the cover if large differential settlements occur in the landfill. A hydraulic conductivity of 10^{-9} m s^{-1} is usually specified for clay covers but perhaps is never attained, because of cracking. Reinforcement with high-strength geotextile on the top and bottom of a clay liner may help to reduce the likelihood of cracking. Differential settlement may be reduced by dynamic compaction of refuse. Controlled percolation may not be critical if treatment of leachate is regarded as economically more feasible than an expensive cover.

The slope of the surface of the landfill influences infiltration. Water tends to collect on a flat surface and subsequently infiltrates, whereas water tends to run off steeper slopes. However, when the surface slope exceeds about 8%, the possibility of surface run-off eroding the cover exists. Surface water should be collected in ditches and routed from the site.

As mentioned, ideally a landfill should not be located where wastes are likely to come into contact with the groundwater. This would mean that a leachate collection system would have to be installed. For instance, leachate collection pipes can be installed above the underliner to remove the leachate to a sump (Figure 10.5), from which it is conveyed to a treatment works. Similarly, a landfill should not be located in an area drained by large quantities of surface water. However, streams can be diverted away from a landfill, or can be piped or culverted through a landfill.

Stabilization of soft landfills can be brought about by mixing with soil, fly ash, incinerator residue, lime or cement. When a minimum unconfined compressive strength of 24 kPa occurs due, for example, to irregularities in mixing at the site, geosynthetic material (e.g. a geoweb) should be used to span such areas and should be securely anchored in a trench at the perimeter of the site. A geomembrane should be placed over the geoweb to provide a secondary seal protecting against the infiltration of surface liquids. Clay soil above the geomembrane acts as a seal, but it must be as thin as possible so as not to overload the geoweb–geomembrane.

The major function of all barrier systems is to provide isolation of wastes from the surrounding environment, so offering protection to soil and groundwater from contamination. Hence, the hydraulic and gas conductivities of cut-off systems are of paramount importance, and generally the hydraulic conductivity of an earth or earth-treated cut-off should be less than 10^{-9} m s^{-1}. However, the overall effectiveness of a cut-off system depends on its thickness as well as its hydraulic conductivity. Mitchell (1994) noted that the hydraulic conductivity divided by the thickness, that is, the permittivity, provided a basis for the comparison of the relative effectiveness of such cut-off systems.

Vertical cut-off walls may surround a site but up-gradient and down-gradient walls can also be used. Where an impermeable horizon exists beneath

Figure 10.5 Leachate sump for a landfill in Durban, South Africa.

the site, the cut-off wall should be keyed into it. A very deep cut-off is required where no impermeable stratum exists. If leachate flows from an existing land-fill, then the construction of a seepage cut-off system provides a solution.

Steel sheet piles can be used to form a continuous cut-off wall, the integrity of the wall depending on the interlock between individual piles. Sealable joints have been developed to reduce the possibility of leakage. Geosynthetic sheet piles with sealable joints are now available. Watertight cut-off walls can also be formed by using secant piles. In this system, alternating bentonite–cement piles are constructed first, then concrete piles are constructed between them, parts of the bentonite–cement piles being bored out in order to form the concrete piles and to achieve interlock. A grout curtain can be used as a cut-off wall. However, cut-off walls are more likely to be constructed by jet-grouting than by injection. In its simplest form, jet-grouting involves inserting a triple-fluid phase drill-pipe into the soil to the required depth. The soil is then subjected to a horizontally rotating jet of water from the base of the pipe. At the same time, the soil is mixed with cement or cement–bentonite grout to form plastic soil–cement. The pipe is raised gradually. Alternatively, replacement jet-grouting involves removal of the soil by a high-energy erosive jet of water and air, with grout being emplaced simultaneously. The soil that is removed is brought to the surface by air lift pressure. A grout curtain can be formed by two or three rows of grout holes approximately 1.5 to 2.5 m apart. Particular concerns with grout curtains are the difficulty of ensuring sufficient overlap of the grout columns and the assessment of the overall integrity of the curtain. The latter cannot be determined until the post-construction monitoring data are available.

Compacted clay barriers can be constructed above the water table in trenches and may be suitable for shallow depths. However, most seepage cut-off systems consist of slurry trenches, which may be 1 m thick. The trench is excavated under bentonite slurry in order to stabilize the sides and prevent them collapsing. Once completed, the trench is generally backfilled with soil–bentonite. Usually the addition of 4–5%, by weight, of bentonite to soil is sufficient to reduce the permeability of the soil to less than 10^{-9} m s^{-1}. Cement–bentonite cut-off walls are formed by mixing cement into bentonite slurry so that the slurry sets in place to form a barrier. However, the permeability of such trenches is usually somewhat higher than that of soil–bentonite trenches. Plastic concrete cut-off walls are similar to cement–bentonite cut-offs except that they contain aggregate. Moreover, they are generally constructed in panels rather than continuously, as is done with soil–bentonite and cement–bentonite walls. Mitchell (1994) pointed out that the greater strength and stiffness of plastic concrete cut-off walls means that they are better suited to situations where ground stability and movement control are important. Concrete diaphragm cut-off walls can also be used where high structural strength is required. They are also constructed in panels, which are excavated under bentonite slurry. Once excavation is complete, the concrete is tremied into the panel from the bottom up (Bell and Mitchell, 1986).

A geomembrane sheeting-enclosed cut-off wall consists of a geomembrane that is fabricated to form a U-shaped envelope, which fits the dimensions of a trench. Ballast is placed within the geomembrane in order to sink it into the slurry trench. After initial submergence into the trench, the envelope is filled with wet sand. A system of wells and piezometers can be installed in the sand to monitor water quality and pore pressure, respectively. If the sheeting is damaged, then the system will detect any infiltration of leachate. This can be abstracted by the wells in the sand (Ressi and Cavalli, 1984).

A cut-off wall can also be constructed by forming a narrow cavity in the soil by repeatedly driving or vibrating an I-beam within the soil. As the I-beam is extracted a mixture of bentonite and cement is pumped into the cavity. Successive penetrations of the I-beam are overlapped to develop a continuous membrane. In the driven pile system, a clutch of H-piles is forced into the soil by a vibrator hammer along the line of advance, the end pile being withdrawn and redriven at the head end. Grouting, by a tube fixed to the web, is carried out during withdrawal. Walls are limited to a maximum of 20 m depth with these systems due to the danger of pile deviation during driving.

Leakage from a landfill where the water table is high and there is no impermeable layer beneath can be treated by jet-grouting. The grout pipe is inserted successively into a series of holes at predetermined centres at the base of the landfill and rotated through $360°$. This allows the formation of interlocking discs of grouted soil to be formed. Alternatively, inclined drilling can be used for the injection of grout or jet-grouting, provided that the overall width of the site is not too great. Inclined drilling from opposite sides of a landfill forms an interlocking V-shaped barrier, which can be used along with vertical cut-off walls.

Adverse interactions of wastes on the materials of which liners, covers and cut-offs are constructed can give rise to increases in their hydraulic conductivity. Of particular concern are those landfills where free-phase organics, very acidic or basic solutions and/or high concentrations of dissolved salts are concentrated. High-plasticity clays with a high water content, such as soil–bentonite mixtures, are more susceptible to shrinkage and cracking than densely compacted soils with a lower water content when exposed to aggressive fluids. Advection and diffusion are involved in the transfer of organics and salts through clay barriers. According to Mitchell and Madsen (1987), diffusion coefficients tend to fall within the range 2×10^{-10} to $2 \times 10^{-9} \, \mathrm{m^2 \, s^{-1}}$ for diffusion through soils and around $3 \times 10^{-14} \, \mathrm{m^2 \, s^{-1}}$ for diffusion through geomembranes. Nonetheless, total chemical transfer through soil or geomembrane barriers by diffusion is low when the hydraulic conductivity of the barrier system is low and the difference in head across the barrier is also low.

10.2.2 Degradation of waste in landfills

As remarked above, a major problem associated with landfills is the production of leachate, the movement of which into the surrounding soil, groundwater or surface water can cause pollution problems. Leachate is formed by the action of liquids, primarily water, within a landfill. The generation of leachate occurs once the absorbent characteristics of the refuse are exceeded.

The waste in a landfill site generally has a variety of origins. Many of the organic components are biodegradable. Initially, decomposition of waste is aerobic. Bacteria flourish in moist conditions and waste contains varying amounts of liquid, which may be increased by infiltration of precipitation. Once decomposition starts the oxygen in the waste rapidly becomes exhausted, and so the waste becomes anaerobic.

There are basically two processes by which anaerobic decomposition of organic waste takes place. Initially, complex organic materials are broken down into simpler organic substances, which are typified by various acids and alcohols. The nitrogen present in the original organic material tends to be converted into ammonium ions, which are readily soluble and may give rise to significant quantities of ammonia in the leachate. The reducing environment converts oxidized ions such as those in ferric salts to the ferrous state. Ferrous salts are more soluble and therefore iron is leached from the landfill. The sulphate in the landfill may be reduced biochemically to sulphide. Although this may lead to the production of small quantities of hydrogen sulphide, the sulphide tends to remain in the landfill as highly insoluble metal sulphides. In a young landfill the dissolved salt content may exceed 10 000 mg l^{-1}, with relatively high concentrations of sodium, calcium, chloride, sulphate and iron, whereas as a landfill ages the concentration of inorganic materials usually decreases. Suspended particles may be present in leachate due to the washout of fine material from the landfill.

The second stage of anaerobic decomposition involves the formation of methane. In other words, methanogenic bacteria use the end-products of the first stage of anaerobic decomposition to produce methane and carbon dioxide. Methanogenic bacteria prefer neutral conditions, so if the acid formation in the first stage is excessive, their activity can be inhibited.

The characteristics of leachate depend on the nature of the waste and the stage of waste stabilization (Table 10.5). Its composition varies due to the material of which it consists, because of interactions between the waste, the age of the fill, the hydrogeology of the site, the climate, the season, and moisture routing through the fill. The height of each cell, the overall depth of the fill, top cover and refuse compaction are also important.

As a landfill ages, because much of the readily biodegradable material has been broken down, the organic content of the leachate decreases. The acids associated with older landfills are not readily biodegradable. Consequently, the biochemical oxygen demand (BOD) in the leachate changes with time. The BOD rises to a peak as microbial activity increases. This peak is reached between six months and 2.5 years after tipping. Thereafter, the BOD concentration decreases until the landfill is stabilized after 6 to 15 years.

The microbial activity in a landfill generates heat. Hence the temperature in a fill rises during biodegradation to between 24 and 45 °C, although temperatures up to 70 °C have been recorded.

10.2.3 Attenuation of leachate

Physical and chemical processes are involved in the attenuation of leachate in soils. These include precipitation, ion exchange, adsorption and filtration. In the reducing zone, where organic pollution is greatest, insoluble heavy metal sulphides and soluble iron sulphide are formed. In the area between the reducing and oxidizing zones, ferric and inorganic hydroxides are precipitated. Other compounds may be precipitated, especially with ferric hydroxide. The reduction of nitrate may yield nitrite, nitrogen gas or possibly ammonia, although

Table 10.5 Leachate composition.

(a) Typical composition of leachates from recent and old domestic wastes at various stages of decomposition (after Anon., 1986).

Determinant	Leachate from recent wastes	Leachate from old wastes
pH	6.2	7.5
Chemical oxygen (mg l^{-1})	23 800	1160
Biochemical oxygen (mg l^{-1})	11 900	260
Total organic carbon (mg l^{-1})	8000	465
Fatty acids (mg l^{-1})	5688	5
Ammoniacal-N (mg l^{-1})	790	370
Oxidized (mg l^{-1})	3	1
o-Phosphate (mg l^{-1})	0.73	1.4
Chloride (mg l^{-1})	1315	2080
Sodium (Na) (mg l^{-1})	960	1300
Magnesium (Mg) (mg l^{-1})	252	185
Potassium (K) (mg l^{-1})	780	590
Calcium (Ca) (mg l^{-1})	1820	250
Manganese (Mn) (mg l^{-1})	27	2.1
Iron (Fe) (mg l^{-1})	540	23
Nickel (Ni) (mg l^{-1})	0.6	0.1
Copper (Cu) (mg l^{-1})	0.12	0.3
Zinc (Zn) (mg l^{-1})	21.5	0.4
Lead (Pb) (mg l^{-1})	8.4	0.14

(b) Chemical analyses of leachate from sump at a landfill in Durban, South Africa, taken over a five-year period from 1986 to 1991 (After Bell et al., 1996).

	pH	Suspended solids (mg l^{-1})	Total dissolved solids (mg l^{-1})	Conductivity (mS m^{-1})	Oxygen absorption (mg l^{-1})	Chemical oxygen demand (mg l^{-1})
Max	8.9	6044	45 041	8080	6200	70 900
Min	6.6	200	900	450	145	11.2
Mean	7.7	965	17 029	1174	649	13 546
MPL	5.5–9.5		500	300	10	75

	Sodium (mg l^{-1})	Potassium (mg l^{-1})	Calcium (mg l^{-1})	Magnesium (mg l^{-1})	Sulphate (mg l^{-1})	Ammonium (mg l^{-1})
Max	8249	2646	1236	759	2237	3530
Min	80	355	60	70	6.5	92
Mean	1393	613	146	109	837	1093
MPL	400	400	200	100	600	2

MPL, South African maximum permissible limit for domestic water insignificant risk (Anon., 1993).

ammonia is usually present due to the biodegradation of nitrogen-bearing organic material. The ion-exchange and adsorption properties of a soil or rock are primarily attributable to the presence of clay minerals. Consequently, a soil with a high clay content has a high ion-exchange capacity. The humic material in soil also has a high ion-exchange capacity. Adsorption is susceptible to changes in pH, as the pH affects the surface charge on a colloid particle or molecule. Hence, at low pH values the removal rates due to adsorption can be reduced significantly. However, the removal of pollutants by ion exchange and adsorption can be reversible. For example, after high-level pollution from a landfill subsides and more dilute leachate is produced, the soil or rock may release the pollutants back into the leachate. However, some ions will be irreversibly adsorbed or precipitated.

The degree of filtration brought about by a soil or rock depends on the size of the pores. In many porous soils or rocks, filtration of suspended matter occurs within a short distance. On the other hand, a rock mass containing open discontinuities may transmit leachate for several kilometres (Hagerty and Pavori, 1973). Pathogenic micro-organisms, which may be found in a landfill, do not usually travel far in the soil, because of the changed environmental conditions. In fact, pathogenic bacteria are not normally present within a few tens of metres of a landfill.

10.2.4 Surface and groundwater pollution

Leachates contain many contaminants that can have a deleterious effect on surface water. If, for example, leachate enters a river, then oxygen is removed from the river by bacteria as they break down the organic compounds in the leachate. In cases of severe organic pollution, the river may be completely depleted of oxygen, with disastrous effects on aquatic life. The net effect of oxygen depletion in a river is that its ecology changes and at dissolved oxygen levels below 2 mg l^{-1} most fish cannot survive. The oxygen balance is affected by several factors. For instance, the rate of chemical and biochemical reactions increases with temperature, whereas maximum dissolved oxygen concentration decreases as temperature rises. At high river flows, bottom mud may be re-suspended and exert an extra oxygen demand, but the extra turbulence also increases aeration. At low flows, organic solids may settle out and reduce the oxygen demand of the river.

The principal inorganic pollutants that may cause problems with leachate are ammonia, iron, heavy metals and, to a lesser extent, chloride, sulphate, phosphate and calcium. Ammonia can be present in landfill leachates up to several hundred milligrams per litre, whereas unpolluted rivers have a very low content of ammonia. Discharge of leachate high in ammonia into a river exerts an oxygen demand on the receiving water, ammonia is toxic to fish (lethal concentrations range from 2.5 to 25 mg l^{-1}), and as ammonia is a fertilizer, it may alter the ecology of the river. Leachate that contains ferrous iron is particularly objectionable in a river, since ochrous deposits are formed by chemical or bio-chemical oxidation of the ferrous compounds to ferric compounds. The tur-bidity caused by the oxidation of ferrous iron can lower the amount of light and so reduce the number of flora and fauna. Heavy metals can be toxic to fish at relatively low concentrations. They can also affect the organisms on which fish feed.

Physically, leachate affects river quality in terms of suspended solids, colour, turbidity and temperature. Suspended solids, colour and turbidity reduce light intensity in a river. This can affect the food chain, and the lack of photosyn-thetic activity by plants reduces oxygen replacement in a river. The suspended solids may settle on the bed of a river in significant qualities, and this can destroy plant and animal life.

If a leachate enters the phreatic zone, it mixes and moves with the ground-water (Bell *et al.*, 1996a). The organic carbon content in the leachate leads to an increase in the BOD in the groundwater, which may increase the potential for reproduction of pathogenic organisms. Organic matter is stabilized slowly, because the oxygen demand may deoxygenate the water rapidly, and usually no replacement oxygen is available. If anaerobic conditions develop, then metals such as iron and manganese may dissolve in the water, causing further prob-lems. If the groundwater has a high buffering capacity, then the effects of mix-ing acidic or alkaline leachate with the groundwater are reduced. The worst situation occurs in discontinuous rock masses, where groundwater movement is dominated by fissure flow, or where groundwater movement is slow in a shal-low water table so that little dilution occurs. The velocity of groundwater flow is important in that a high velocity gives rise to more dispersion in the direc-tion of flow and less laterally, so the leachate plume forms a narrow cone in the direction of groundwater flow. Low groundwater velocity leads to a wider plume.

Ideally, a groundwater monitoring programme should be established prior to tipping and continue for anything up to 20 years after completion of the site. The number, location and depth of monitoring wells depend on the par-ticular site (see Chapter 11). Monitoring should include sampling from the wells and analysis for pH, chlorides, dissolved solids, organic carbon and the concentration of any particular hazardous waste that has been tipped in the landfill.

The most serious effect that a leachate can have on groundwater is mineral-ization. Much of the organic carbon and organic nitrogen is biodegraded as it moves through the soil. Heavy metals may be attenuated due to ion exchange on to clay minerals. Mineralization is brought about by inorganic ions, such as chloride. The concentrations can only be reduced by dilution. Continuing input of inorganic ions into potable groundwater eventually means that the ground-water becomes undrinkable.

10.2.5 Landfill and gas formation

The biochemical decomposition of domestic and other putrescible refuse in a landfill produces gas consisting primarily of methane, with smaller amounts of carbon dioxide and volatile organic acids. In the initial weeks or months after placement, the landfill is aerobic and gas production is mainly CO_2, but it also contains O_2 and N_2. As the landfill becomes anaerobic, the evolution of O_2 declines to almost zero and N_2 to less than 1%. The principal gases produced during the anaerobic stage are CO_2 and CH_4, with CH_4 production increasing slowly as methanogenic bacteria establish themselves.

The factors that influence the rate at which gas is produced include the character of the waste, and the moisture content, temperature and pH of the landfill. Concentration of salts, such as sulphate and nitrate, may also be important.

If the refuse is pulverized, then microbial activity in the landfill is higher, which in turn may give rise to a higher production rate of gases. However, the period of time over which gas is produced may be reduced. Compaction or baling of refuse may decrease the rate of water infiltration into the landfill, retarding bacterial degradation of the waste, with gas being produced at a lower rate over a longer period. If toxic chemicals are present in a landfill, then bacterial activity, and especially methanogenesis, may be inhibited. A moisture content of 40% or higher is desirable for optimum gas production. Generally, the rate of gas production increases with temperature. The pH of a landfill should be around 7.0 for optimum production of gas, methanogenesis tending to cease below a pH of 6.2. The amount of gas produced by domestic refuse varies appreciably, and a site investigation is required to determine the amount if such information is required. Nonetheless, it has been suggested that between 2.2 and 250 litres per kilogram dry weight may be produced (Oweis and Khera, 1990).

Methane production can constitute a dangerous hazard because methane is combustible, and in certain concentrations explosive (5–15% by volume in air), as well as asphyxiating. Appropriate safety precautions must be taken during site operation. In many instances, landfill gas is able to disperse safely into the atmosphere from the surface of a landfill. However, when a landfill is completely covered with a soil capping of low permeability in order to minimize leachate generation, the potential for gas to migrate along unknown pathways increases, and there are cases of hazard arising from methane migration. Furthermore, there are unfortunately numerous cases on record of explosions occurring in buildings due to the ignition of accumulated methane derived from landfills near to or on which they were built (Williams and Aitkenhead, 1991). The source of the gas should be identified so that remedial action can be taken. Identification of the source can involve a drilling programme and analysis of the gas recovered. For instance, in a case referred to by Raybould and Anderson (1987), distinction had to be made between household gas and gas from sewers, old coal mines and a landfill before remedial action could be taken to eliminate the gas hazard affecting several houses. Accordingly, planners of residential developments should avoid landfill sites. Proper closure of a landfill site can require gas management to control methane gas by passive venting, power-operated venting or the use of an impermeable barrier. The identification of possible migration paths in the assessment of a landfill site is highly important, especially where old mine workings may be present or where there is residential property nearby. Raybould and Anderson described grouting of old mine workings that had acted as a conduit for the migration of methane from a landfill to residential properties. Although methane is not toxic to plant life, the generation of significant quantities can displace oxygen from the root zone and so suffocate plant roots. Large concentrations of carbon dioxide or hydrogen sulphide can produce the same result.

The movement of gas may be upwards and out through the landfill cover. The slow escape of gas from the top of a landfill can present a problem, particularly if it accumulates in pockets on the site. Artificial channels, such as drains, soak-aways, culverts or shafts can act as pathways for gas movement.

Monitoring of landfill gas is an important aspect of safety. Instruments usually monitor methane, as this is the most important component of landfill gas. Gas can be sampled from monitoring wells in the fill or from areas into which the gas has migrated. Leachate monitoring wells can act as gas collectors. Due caution must be taken when sampling.

Measures to prevent migration of the gas include impermeable barriers (clay, bentonite, geomembranes or cement) and gas venting. An impermeable barrier should extend to the base of the fill or the water table, whichever is the higher. However, it is a problem ensuring that the integrity of the barrier is maintained during its installation and subsequent operation.

Venting either wastes the gas into the atmosphere or facilitates its collection for utilization. Passive venting involves venting gases to locations where it can be released to the atmosphere or burned. The vents are placed at locations where gas concentration and/or pressures are high (Figure 10.6). This usually occurs towards the lower sections of a landfill. A vent that penetrates to the leachate is most effective. Atmospheric vents may be spaced at 30–45 m intervals. Care must be taken to ensure that vents do not become blocked or clogged. In addition, the vent must remain relatively dry in order to maintain its permeability to gases. Therefore drainage, either natural or pumped, must be present to allow the escape of water percolating into the vent. Such vents are not effective in controlling the migration of gas laterally. Perimeter wells are used where there are no geological constraints preventing lateral migration of gas. Alternatively, the gas may be intercepted by a trench filled with coarse aggregate. Where atmospheric venting is insufficient to control the discharge of gas, an active or forced venting system is used. Active ventilation involves connecting a vacuum pump to the discharge end of the vent. The radius of influence depends primarily on the flow rate and the depth of the well screen (e.g. radii of influence of 6 m at 0.85 $m^2 m^{-1}$ flow rate to 22.5 m at a flow rate of 1.4 $m^3 m^{-1}$).

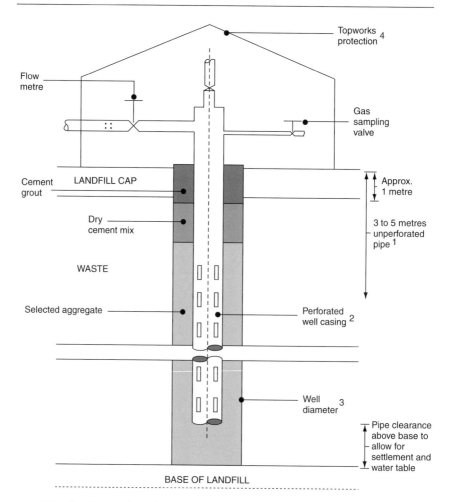

Figure 10.6 Typical landfill gas extraction well (after Attewell, 1993).

If an impermeable cover is placed over an entire landfill, methane will eventually move laterally into any permeable soil and escape. Hence, in such situations methane gas should be properly vented. Each cell cover should be shaped so that a vent is located at the uppermost slope of the bottom of the cover.

A sand–gravel drainage layer located above the liner of a landfill can be used to collect carbon dioxide. Alternatively, a geocomposite of adequate transmissivity, along with a perforated pipe collection system, can be used for gathering the gas.

10.3 Hazardous wastes

A hazardous waste can be regarded as any waste or combination of wastes of inorganic or organic origin that, because of its quantity, concentration, physical, chemical, toxicological or persistency properties, may give rise to acute or chronic impacts on human health and/or the environment when improperly treated, stored, transported or disposed of. Such waste can be generated from a wide range of commercial, industrial, agricultural and domestic activities and can take the form of liquid, sludge or solid. The characteristics of the waste not only influence its degree of hazard but are also of great importance in the choice of a safe and environmentally acceptable method of disposal. Hazardous wastes may involve one or more risks such as explosion, fire, infection, chemical instability or corrosion, or acute toxicity (in particular, is the waste carcinogenic, mutagenic or teratogenic?). An assessment of the risk posed to health and/or the environment by hazardous waste must take into consideration its biodegradability, persistency, bioaccumulation, concentration, volume production, dispersion and potential for leakage into the environment. Hazardous wastes therefore require special treatment and cannot be released into the environment, added to sewage or stored in a situation that is either open to the air or from which aqueous leachate could emanate.

Some of the primary criteria for identification of hazardous wastes include the type of hazard involved (flammability, corrosivity, toxicity and reactivity), the origin of the products, including industrial origins (e.g. medicines, pesticides, solvents, electroplating, oil refining), and the presence of specific substances or groups of substances (e.g. dioxins, lead compounds, PCBs). These criteria and others are used alone or in combination but in different ways in different countries. The compositional characteristics of the waste may or may not be quantified, and where levels of substance concentration are set, they again vary from country to country.

A number of attempts have been made to produce a qualitative system of comparison of wastes in terms of their toxicity or the relative hazards they present. For example, Smith *et al.* (1980) defined a geotoxicity hazard index (GHI) for buried underground toxic materials as:

$$GHI = TI \times P \times A \times C \tag{10.1}$$

where TI is the toxicity index and is equivalent to the volume of water required to dilute the substance to acceptable drinking water levels; P is the persistence factor, which is a function of the decay half-life of the material; C is the build-up correction factor, which takes care of any more toxic daughter products that may be produced during decay; and A is the availability factor. The latter represents an attempt to relate the availability of the substance to man to the amount naturally available in a reference volume of soil. It takes account of the leach rate of a substance, its chemical behaviour and interaction with the soil. It is at this stage where the hazard index concept become impractical, especially for substances with no simple natural analogue (e.g. many organic contaminants found in groundwater).

Assessment of the suitability of a site for hazardous waste disposal is a complex matter that involves the use of models to predict the chemical behaviour of the waste in the ground, and the potential for mobilization and migration in groundwater. The form and rates of release of the waste into the environment, together with the time and place at which releases occur, can be produced with reasonable degrees of confidence. To translate these results into risk assessment requires a parallel prediction of the consequences of a release of waste. Risk associated with the disposal of wastes involves the probability of an event or process that leads to release occurring within a given time period, multiplied by the consequences of that release (or alternatively, the probability that an individual or group will be exposed to a pollutant, multiplied by the probability that this exposure will give rise to a serious health effect). Given adequate epidemiological data on the health effects of toxic substances, the risk can be calculated.

The classification of hazardous wastes and hazard rating facilitates the choice of a cost-effective method of waste management. Obviously, there is no truly non-hazardous waste; some risk always exists when dealing with waste. The quantity of waste involved, the manner and conditions of use, and the susceptibility of man or other living things to a certain waste can be used to determine its degree of hazard. The first step in classifying waste is to identify the hazardous substances present. It can then be placed into a particular group of a classification system, perhaps according to the most dangerous constituent recognized.

Hazard ratings can be categorized as extreme, high, moderate or low. On the one hand, waste that contains significant concentrations of extremely hazardous material, including certain carcinogens, teratogens and infectious substances, is of primary concern. On the other hand, the low category of hazardous waste contains potentially hazardous constituents but in concentrations that represent only a limited threat to health or the environment. If the hazard rating is less than the low category, then the waste can be regarded as non-hazardous and disposed of as general waste.

The minimum requirements for the treatment and disposal of hazardous waste involve ensuring that certain classes of waste are not disposed of without pre-treatment. The objective of treating a waste is to reduce or destroy the toxicity of the harmful components in order to minimize the impact on the environment. In addition, waste treatment can be used to recover materials during waste minimization programmes. The method of treatment chosen is influenced by the physical and chemical characteristics of the waste: that is, is it gaseous, liquid, in solution, sludge or solid; is it inorganic or organic; and what is the concentration of hazardous and non-hazardous components? Physical treatment methods are used to remove, separate and concentrate hazardous and toxic materials. Chemical treatment is used in the application of physical treatment methods and to lower the toxicity of a hazardous waste by changing its chemical nature. This may produce essentially non-hazardous materials. In biological treatments, microbial activity is used to reduce or destroy the toxicity of a waste. For example, microbial action can be used to degrade organic substances or reduce inorganic compounds. The principal objective of such processes as immobilization, solidification and encapsulation is to convert hazardous waste into an inert mass with very low leachability. Macro-encapsulation involves the containment of waste in drums or other approved containers within a reinforced concrete cell that is stored within a landfill. Incineration can be regarded as a means of both treatment and disposal.

Safe disposal of hazardous waste is the ultimate objective of waste management, disposal being in a landfill, by burial, by incineration or by marine disposal. When landfill is chosen as the disposal option, the capacity of the site to accept certain substances without exceeding a specified level of risk has to be considered. The capacity of a site to accept waste is influenced by the geological and hydrogeological conditions, the degree of hazard presented by the waste, the leachability of the waste and the design of the landfill (Figure 10.7). Certain hazardous wastes may be prohibited from disposal in a landfill, such as explosive wastes or flammable gases. Obviously medium- to high-level radioactive waste cannot be disposed of in a landfill.

Protection of groundwater from the disposal of toxic waste in landfills can be brought about by containment. A number of containment systems have been developed that isolate wastes, including compacted clay barriers, slurry trench cut-off walls, geomembrane walls, sheet piling, grout curtains and hydraulic barriers (Mitchell, 1986). A compacted clay barrier consists of a trench that has been backfilled with clay compacted to give a low hydraulic conductivity. Slurry trench cut-off walls are narrow trenches filled with soil–bentonite mixtures which extend downwards into an impermeable layer. Again they have a low hydraulic conductivity. Diaphragm walls are an expensive form of containment. Their use is therefore restricted to situations where high structural stability is required. Grout curtains may be used in certain situations. Extraction wells can be used to form hydraulic barriers and are located so that the contaminant plume flows towards them.

Monitoring of hazardous waste repositories forms an inherent part of the safety requirements governing their operational and post-operational periods. Hence the repository operators are required to conduct monitoring programmes

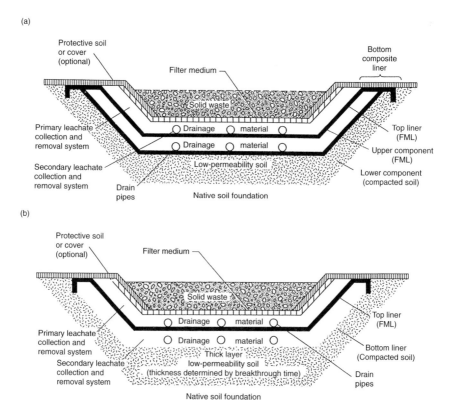

Figure 10.7 Double-liner systems proposed by the 1985 US Environmental Protection Agency guidelines (FML = flexible membrane liner). The leachate collection layer is also considered to function as the geomembrane protection layer. (a) FML/composite double liner; (b) FML/compacted soil double liner (after Mitchell, 1986).

to detect any failure in the waste containment systems so that remedial action can be taken. The nature and duration of a monitoring programme needed to ensure continuing safe isolation of waste depends upon a number of parameters, including the physical condition, composition and nature of the host formation. In addition, the regulatory authorities impose conditions and limitations on the disposal of hazardous wastes, in terms of the type and quantity of waste, and the level of activity for each particular repository.

Much lower-level hazardous waste can be co-disposed, that is, mixed with very much larger quantities of 'non-hazardous' domestic waste and buried in disused quarries, clay pits and other convenient holes in the ground (Chapman and Williams, 1987). The objectives of co-disposal are to absorb, dilute and neutralize liquids, and to provide a source of biodegradable materials to encourage microbial activity. In other words, co-disposal makes use of the attenuation processes in a landfill to minimize the impact of hazardous waste on the environment. The technique helps to bring about changes in the waste so that it is not retained in its original form with the potential to give rise to pollution. Co-disposal is undertaken by discharging waste into a trench located at the working face. The trench is filled with dry domestic waste at the end of the working day or immediately if the waste has an offensive odour.

Disposal of liquid hazardous waste has also been undertaken by injection into deep wells located in rock below freshwater aquifers, thereby ensuring that contamination or pollution of underground water supplies does not occur. In such instances, the waste is generally injected into a permeable bed of rock several hundreds or even thousands of metres below the surface that is confined by relatively impervious formations. However, even where geological conditions are favourable for deep-well disposal, the space for waste disposal is frequently restricted and the potential injection zones are usually occupied by connate water. Accordingly, any potential formation into which waste can be injected must possess sufficient porosity, permeability, volume and confinement to guarantee safe injection. The piezometric pressure in the injection zone influences the rate at which the reservoir can accept liquid waste. A further point to consider is that induced seismic activity has been associated with the disposal of fluids in deep wells. Two important geological factors relating to the cost of construction of a well are its depth and the ease with which it can be drilled.

Monitoring is especially important in deep-well disposal that involves toxic or hazardous materials. A system of observation wells sunk into the subsurface reservoir concerned, in the vicinity of the disposal well, allows the movement of liquid waste to be monitored. In addition, shallow wells sunk into freshwater aquifers permit monitoring of water quality so that any upward migration of the waste can be noted readily. Effective monitoring requires that the geological and hydrogeological conditions are accurately evaluated and mapped before the disposal programme is started.

10.4 Radioactive waste

Radioactive waste may be of low, intermediate or high level. Low-level waste contains small amounts of radioactivity and so does not present a significant environmental hazard if properly dealt with. Intermediate-level waste comes from nuclear plant operations and consists of items such as filters used to purify reactor water, discarded tools and replaced parts. This waste has to be stored for approximately 100 years. When reactors are closed down the decommissioning waste has to be disposed of safely.

Although many would not agree, Chapman and Williams (1987) maintained that low-level radioactive waste can be disposed of safely by burying in carefully controlled and monitored sites where the hydrogeological and geological conditions severely limit the migration of radioactive material. According to Rogers (1994) the most recent trend in low-level radioactive waste disposal in the USA is to provide engineering (concrete) barriers in the disposal facility that prevent the short-term release of contamination. Such disposal facilities include below-ground vaults, above-ground vaults, modular concrete canisters, earth-mounded concrete bunkers, augered holes and mined cavities. A below-ground vault involves placing waste containers in an engineered structure consisting of a reinforced concrete floor, walls and roof, and covering with soil. In this way, infiltration of water into the waste is retricted and gamma radiation at the surface reduced. An above-ground vault is a structure consisting of a reinforced concrete floor, walls and roof, which remains uncovered after closure. The spaces between the waste containers placed in the structure are filled with earth. The modular concrete canisters concept involves placing waste containers in reinforced concrete canisters in trenches in the ground, with the spaces between the containers being filled with grout. When filled, the trench is covered by a layer of soil. The earth-mounded concrete bunker includes a below-ground disposal of waste containers in a concrete bunker and an above-ground disposal of containers with a soil cover, which is usually less than 5 m in thickness. The bunker comprises a number of cells with reinforced concrete floors, walls and roofs. Void spaces between the containers within a cell are filled with concrete. All waste placed above ground is solidified and placed in steel drums or compacted and grouted into concrete canisters. These are stacked on a concrete floor, which separates this part of the structure from the concrete bunker, and the voids between the containers are filled with soil. More active low-level radioactive waste has been disposed of in augered holes, which are either lined or unlined. The linings may be of fibreglass, concrete or steel. A cover (about 3 m thick) is placed over the augered hole. A mined cavity can be either a disused mine or a purpose-built mine and can be used for the disposal of all types of low-level radioactive waste.

High-level radioactive waste unfortunately cannot be made non-radioactive, so disposal has to take account of the continuing emission of radiation. Furthermore, as radioactive decay occurs the resulting daughter product is chemically different from the parent product. The daughter product may also be radioactive, but its decay mechanism may differ from that of the parent. This is of particular importance in the storage and disposal of radioactive material (Krauskopf, 1988). The half-life of a radioactive substance determines the time that a hazardous waste must be stored for its activity to be reduced by half. However, as noted previously, if a radioactive material decays to form another unstable isotope and if the half-life of the daughter product is long compared with that of the parent, then although the activity of the parent will decline with time, that of the daughter will increase since it is decaying more slowly. Consequently, the storage time of radioactive wastes must consider the half-lives of the products that result from decay. The most long-lived radioactive elements need to be stored safely for hundreds of thousands of years.

In general, two types of high-level radioactive waste are being produced, namely, spent fuel rods from nuclear reactors and reprocessed waste. At present, both kinds of waste are adequately isolated from the environment in container systems. However, because much radioactive waste remains hazardous for hundreds or thousands of years, it should be disposed of far from the surface environment and where it will require no monitoring. Baillieul (1987) reviewed various suggestions that have been advanced to meet this end, such as disposal in ice sheets, disposal in the ocean depths, disposal on remote islands and even disposal in space.

Ice sheets have been suggested as a repository for isolating high-level radioactive waste. The presumed advantages are disposal in a cold remote area, in a material that would entomb the wastes for many thousands of years. The high cost, adverse climate and uncertainties of ice dynamics are factors that do not favour such a means of disposal.

Disposal on the deep sea bed involves emplacement in sedimentary deposits at the bottom of the sea (i.e. thousands of metres beneath the surface). Such deposits have a sorptive capacity for many radionuclides that might leach from breached waste packages. In addition, if any radionuclides escaped they would be diluted by dispersal. Currently, however, disposal of radioactive waste beneath the sea floor is prohibited by international convention.

Disposal of radioactive waste on a remote island involves the placement of wastes within deep stable geological formations. It also relies on the unique hydrological system associated with island geology. The remoteness, of course, is an advantage in terms of isolation.

The rock melt concept involves the direct placement of liquids or slurries of high-level wastes or dissolved spent fuel in underground cavities. After evaporation of the slurry water, the heat from radioactive decay would melt the surrounding rock. In about 1000 years the waste–rock mixture would resolidify, trapping the radioactive material in a relatively insoluble matrix deep underground. The rock melt concept, however, is suitable only for certain types of waste. Moreover, since solidification takes about 1000 years, the waste is most mobile during the period of greatest fission product hazard.

The most favoured method, since it will probably give rise to the least problems subsequently, is disposal in chambers excavated deep within the Earth's crust in geologically acceptable conditions (Morfeldt, 1989). Deep disposal of high-level radioactive waste involves the multiple barrier concept, which is based upon the principle that uncertainties in performance can be minimized by conservation in design (Horseman and Volckaert, 1996). In other words, a number of barriers, both natural and man-made, exist between the waste and the surface environment. These can include encapsulation of the waste, waste containers, engineered barriers such as backfills, and the geological host rocks, which are of low permeability.

A deep disposal repository will consist of a large underground system located at least 200 m, and preferably 300 m or more, beneath the ground surface, in which there is a complex of horizontally connected tunnels for transportation, ventilation and emplacement of high-level radioactive waste (Eriksson, 1989).

It will also require a series of inclined tunnels and vertical shafts to connect the repository with the surface. Modern blasting techniques can produce maintenance-free storage chambers and tunnels in rock masses, as can tunnel-boring machines. Ideally, the waste should be so well entombed than none will reappear at the surface, or if it does so, in amounts minute enough to be acceptable.

A monitored interim retrievable storage facility can be integrated into a waste management system. Such a monitored retrievable storage facility would act as a central receiving point for spent fuel and other waste forms. At the facility the waste material would be packaged into standardized disposal canisters and temporarily stored pending transfer to the main repository. It could store waste for 30 to 40 years before final disposal.

The necessary safety of a permanent repository for radioactive wastes has to be demonstrated by a site analysis that takes account of the site geology and the type of waste, and their interrelationship (Langer 1989). The site analysis must assess the thermomechanical load capacity of the host rock so that disposal strategies can be determined. It must determine the safe dimensions of an underground chamber and evaluate the barrier systems to be used. According to the multi-barrier concept, the geological setting for a waste repository must be able to make an appreciable contribution to the isolation of the waste over a long period of time. Hence the geological and tectonic stability (e.g. mass movement or earthquakes), the load bearing capacity (e.g. settlement or cavern stability), and geochemical and hydrogeological development (e.g. groundwater movement and potential for dissolution of rock) are important aspects of safety development. Disposal would be best done in a geological environment with little or no groundwater circulation, as groundwater is the most probable means of moving waste from the repository to the biosphere. The geological system should be complemented by multiple engineered barriers such as the waste form, the waste container, buffer materials and the backfill.

If not already in solid form (e.g. spent fuel rods), then waste should be treated to convert it into solid, ideally non-leachable, material. A variety of different solidification process materials have been proposed, including cement, concrete, plaster, glass and polymers. Currently, borosilicate glass is the most popular agent as it can incorporate wastes of varying composition and has a low solubility in water. The glass would be placed within a metal canister and surrounded by cement or clay. The purpose of the container system is to provide a shield against radiation and so it must be corrosion-resistant. Nonetheless, any container system has to remain intact for a very long time (storage of high-level radioactive waste will have to be for thousands of years), and behaviour over such a period of time cannot be predicted.

These metal canisters would be stored in deep underground caverns excavated in relatively impermeable rock types in geologically stable areas, that is, areas that do not experience volcanic activity, in which there is a minimum risk of seismic disturbance and that are not likely to undergo significant erosion (Figure 10.8). Although earthquakes may represent a potential risk factor

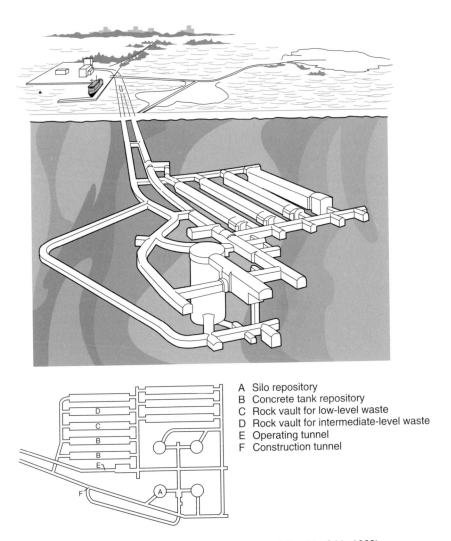

A Silo repository
B Concrete tank repository
C Rock vault for low-level waste
D Rock vault for intermediate-level waste
E Operating tunnel
F Construction tunnel

Figure 10.8 Possible final repository for reactor waste (after Morfeldt, 1989).

for rock chambers and tunnels, experience in mines in earthquake-prone regions have shown that vibrations in rock decrease with depth; for example, even at a depth of 30 m vibrations are only one-seventh of those measured at the surface. Deep structural basins are considered as possible locations (Baillieul, 1987; Eriksson, 1989).

Stress redistribution due to subsurface excavation and possible thermally induced stresses should not endanger the state of equilibrium in the rock mass and should not give rise to any inadmissible convergence or support damage during the operative period. The long-term integrity of the rock formations must be assured. Therefore it is necessary to determine the distribution of stress and deformation in the host rock of the repository. This may involve consideration of the temperature-dependent rheological properties of the rock mass in order to compare them with its load bearing capacity. Obviously, substantial strength is necessary for engineering design of subsurface repository facilities, especially in maintaining the integrity of underground openings.

Completely impermeable rock masses are unlikely to exist, although many rock types may be regarded as practically impermeable, such as large igneous rock massifs, thick sedimentary sequences, metamorphic rocks and rock salt. The permeability of a rock mass depends mainly on the discontinuities present, their surfaces, width, amount of infill and their intersections. Stringent requirements apply to the storage facilities of high-level, long-lived radioactive waste. In particular, the repository needs to be watertight to prevent the transport of radionuclides by groundwater to the surface. Control and test pumping of the groundwater system, and sealing by injection techniques, may be necessary.

Rock types such as thick deposits of salt or shale, or granites or basalts at depths of 300 to 500 m are regarded as the most feasible in which to excavate caverns for disposal of high-level radioactive waste. Once a repository is fully loaded, it can be backfilled, with the shafts being sealed to prevent the intrusion of water. Once sealed, the system can be regarded as isolated from the human environment.

As far as the disposal of high-level radioactive waste in caverns is concerned, thick deposits of salt have certain advantages. Salt has a high thermal conductivity and so will rapidly dissipate any heat associated with high-level nuclear waste; it is 'plastic' at proposed repository depths, so that any fractures that may develop as a result of construction operations will 'self-heal'; it possesses gamma-ray protection similar to concrete; it undergoes only minor changes when subjected to radioactivity; and it tends not to provide paths of escape for fluids or gases (Langer and Wallner, 1988). The attractive feature of deep salt deposits is its lack of water and the inability of water from an external source to move through it. These advantages may be compromised if the salt contains numerous interbedded clay or mudstone horizons, open cavities containing brine or faults cutting the salt beds, so providing conduits for external water. The solubility of salt requires that unsaturated waters are totally isolated from underground openings in beds of salt by watertight linings, isolation seals or cut-offs and/or collection systems. If suitable precautions are not taken in more soluble horizons in salt, any dissolution that occurs can lead to the irregular development of a cavity being excavated. Any water that does accumulate in salt will be a concentrated brine, which, no doubt, will be corrosive to metal canisters. The potential for heavy groundwater inflows during shaft sinking requires the use of grouting or ground freezing. Rock salt is a visco-plastic material that exhibits short- and long-term creep. Hence caverns in rock salt are subject to convergence as a result of plastic deformation of the salt. The rate of convergence increases with increasing temperature and stress in the surrounding rock mass. Since temperature and stress increase with depth, convergence also increases with depth. If creep deformation is not restrained or not compensated for by other means, excessive rock pressures can develop on a lining system that may approach full overburden pressure.

Not all shales are suitable for the excavation of underground caverns in that soft compacted shales would present difficulties in terms of wall and roof stability. Caverns may also be subject to floor heave. Caverns could be excavated in competent cemented shales. Not only do these possess low permeability but they could also adsorb ions that move through it. A possible disadvantage is that if temperatures in the cavern exceed 100 °C, then clay minerals could lose water and therefore shrink. This could lead to the development of fractures. In addition, the adsorption capacity of the shale would be reduced.

Granite is less easy to excavate than rock salt or shale but is less likely to offer problems of cavern support. It provides a more than adequate shield against radiation and will disperse any heat produced by radioactive waste. The quantity of groundwater in granite masses is small, and its composition is generally non-corrosive. However, fissure and shear zones, along which copious quantities of groundwater can flow, do occur within granites. Discontinuities tend to close with depth and faults can be sealed. For location and design of a repository for spent nuclear fuel, the objective is to find 'solid blocks' that are large enough to host the tunnels and caverns of the repository. The Pre-Cambrian shields represent stable granite–gneiss regions.

The large thicknesses of lava flows in basalt plateaux mean that such successions could also be considered for disposal sites. Like granite, basalt can also act as a shield against radiation and can disperse heat. Frequently, the contact between flows is tight and little pyroclastic material is present. Joints may not be well developed at depth and strengthwise basalt should support a cavern. However, the durability of basalts on exposure may be suspect in that they could give rise to spalling from the perimeter of a cavern (Haskins and Bell, 1995). Furthermore, such basalt formations can be interrupted by feeder dykes and sills, which may be associated with groundwater. Groundwater is generally mildly alkaline with a low redox potential.

In the United States, extensive geological studies have been conducted in six states to identify sites for detailed site characteristization studies and ultimately the construction of that country's first high-level nuclear waste disposal facility. The host rock types under consideration included basalts in Washington state, tuffs in Nevada, evaporites (salt) in Utah and Texas, and diapiric salt in Mississippi and Louisiana. The Nuclear Waste Policy Act of 1982 mandated that extensive site characterization activities had to be performed to evaluate the suitability of potential sites for the construction of a high-level nuclear waste repository. The site characterization activities included comprehensive surface and subsurface investigations. Characterization of the subsurface involved geophysical techniques, exploratory drillholes with soil, rock and groundwater sampling, and the construction of an exploratory shaft and underground facilities for *in situ* testing of the target repository geological horizon. The exploratory shaft facility provides a means of verifying the repository design parameters. As of 1 April 1988, the office of Civilian Radioactive Waste Management programme comprises one candidate geological repository site, the Yucca Mountain site in Nevada. It was hoped that this repository would be completed for commencement of operation in January 1998.

Morfeldt (1989) described a tunnel system in Sweden where spent nuclear fuel, contained in copper canisters, has been placed in drillholes in the floor of an individual tunnel and the holes backfilled with highly compacted bentonite. The copper capsules will provide sufficient protection against radiation over a long period of time. The surrounding bentonite swells and seals any fissures in the drillholes. The tunnels will also be backfilled with a mixture of sand and bentonite. Such a design means that the rock mass in the floor of the tunnel can be investigated to locate the best position of the holes for the canisters. If the tunnel should cross an unexpected fracture zone in the rock mass, it could be avoided as far as siting of the canister holes is concerned.

Placement of encapsulated nuclear waste in drillholes as deep as 1000 m in stable rock formations cannot dispose of high volumes of waste. Similarly, injection of liquid waste into porous or fractured strata, at depths from 1000 to 5000 m, can accommodate only limited quantities of waste. Such waste is suitably isolated by relatively impermeable overlying strata and relies on the dispersal and diffusion of the liquid waste through the host rock. The limits of diffusion need to be well defined. Alternatively, thick beds of shale, at depths between 300 and 500 m, can be fractured by high-pressure injection and then waste, mixed with cement or clay grout, can be injected into the fractured shale and allowed to solidify in place. The fractures need to be produced parallel to the bedding planes. This requirement limits the depth of injection. The concept is applicable only to reprocessed wastes or to spent fuel that has been processed in liquid or slurry form. This type of disposal, like that of deep-well injection, can dispose of only limited amounts of waste.

10.5 Waste materials from mining

Mine wastes result from the extraction of metals and non-metals. In the case of metalliferous mining, high volumes of waste are produced because of the low or very low concentrations of metal in the ore. In fact, mine wastes represent the highest proportion of waste produced by industrial activity, billions of tonnes being produced annually. Waste from mines has been deposited on the surface in spoil heaps or tailings lagoons, which disfigure the landscape. Such waste can be inert or contain hazardous constituents but generally is of low toxicity. The chemical characteristics of mine waste and waters arising therefrom depend upon the type of mineral being mined, as well as the chemicals that are used in the extraction or beneficiation processes. Because of its high volume, mine waste historically has been disposed of at the lowest cost, often without regard for safety and often with considerable environmental impacts. Catastrophic failures of spoil heaps and tailings dams, although uncommon, have led to the loss of lives.

The character of waste rock from metalliferous mines reflects that of the rock hosting the metal, as well as the rock surrounding the ore body. The type of waste rock disposal facility depends on the topography and drainage of the site, and the volume of waste. Van Zyl (1993) referred to the disposal of coarse mine waste in valley fills, side-hill dumps and open piles. Valley fills normally commence at the upstream end of a valley and progress downstream, increasing in thickness. Side-hill dumps are constructed by the placement of waste along hillsides or valley slopes, avoiding natural drainage courses. Open piles tend to be constructed in relatively flat-lying areas. Obviously an important factor in the construction of a spoil heap is its slope stability, which includes its long-term stability. Acid mine drainage from spoil heaps is another environmental concern.

10.5.1 Basic properties of coarse discard associated with colliery spoil heaps

Spoil heaps associated with coal mines represent ugly blemishes on the landscape and have a blighting effect on the environment. They consist of coarse discard, that is, run-of-mine material that reflects the various rock types which are extracted during mining operations. As such, coarse discard contains varying amounts of coal that has not been separated by the preparation process. Obviously, the characteristics of coarse colliery discard differ according to the nature of the spoil. The method of tipping also appears to influence the character of coarse discard. In addition, some spoil heaps, particularly those with relatively high coal contents, may be burned, or still be burning, and this affects their mineralogical composition.

In Britain, illites and mixed-layer clays are the principal components of unburnt spoil in English and Welsh tips (Taylor, 1975). Although kaolinite is a common constituent in Northumberland and Durham, it averages only 10.5% in the discard of other areas. Quartz exceeds the organic carbon or coal content, but the latter is significant in that it acts as the major diluent, that is, it behaves in an antipathetic manner towards the clay mineral content. Sulphates, feldspars, calcite, siderite, ankerite, chlorite, pyrite, rutile and phosphates average less than 2%.

The chemical composition of spoil material reflects that of the mineralogical composition. Free silica may be present in concentrations up to 80% and above, and combined silica in the form of clay minerals may range up to 60%. Concentrations of aluminium oxide may be between a few percent and 40% or so. Calcium, magnesium, iron, sodium, potassium and titanium oxides may be present in concentrations of a few percent. Lower amounts of manganese and phosphorus may also be present, with copper, nickel, lead and zinc in trace amounts. The sulphur content of fresh spoils is often less than 1% and occurs as organic sulphur in coal, and in pyrite.

The moisture content of spoil increases with increasing content of fines. It is also influenced by the permeability of the material, the topography and climatic conditions. Generally, it falls within the range 5 to 15% (Table 10.6).

The range of specific gravity depends on the relative proportions of coal, shale, mudstone and sandstone in the waste, and tends to vary between 1.7 and 2.7. The proportion of coal is of particular importance: the higher the coal content, the lower the specific gravity. The bulk density of material in spoil heaps shows a wide variation, most material falling within the range 1.5–2.5 Mg m^{-3}. Of course, the bulk density may vary within an individual deposit of mine waste. Low densities are mainly a function of low specific gravity. Bulk density tends to increase with increasing clay content.

The argillaceous content influences the grading of spoil, although most spoil material would appear to be essentially granular in the mechanical sense. In fact, as far as the particle size distribution of coarse discard is concerned there is a wide variation; often most material may fall within the sand range, but significant proportions of gravel and cobble range may also be present. Indeed, at placement coarse discard very often consists mainly of gravel–cobble size, but subsequent breakdown on weathering reduces the particle size. Once buried within a spoil heap, coarse discard undergoes little further reduction in size. Hence older and surface samples of spoil contain a higher proportion of fines than those obtained from depth.

The liquid and plastic limits can provide a rough guide to the engineering characteristics of a soil. In the case of coarse discard, however, they are only representative of that fraction passing the 425 μm BS sieve, which frequently is less than 40% of the sample concerned. Nevertheless, the results of these

Table 10.6 Examples of soil properties of coarse discard (from Bell, 1996).

	Yorkshire Main*	Brancepath	Wharncliffe
Moisture content, %	8.0–13.6	5.3–11.9	6–13 (7.14)
Bulk density, Mg m^{-3}	1.67–2.19	1.27–1.88	1.58–2.21
Dry density, Mg m^{-3}	1.51–1.94	1.06–1.68	1.39–1.91
Specific gravity	2.04–2.63	1.81–2.54	2.16–2.61
Plastic limit, %	16–25	Non-plastic – 35	14–21
Liquid limit, %	23–44	23–42	25–46
Permeability, m s^{-1}	1.42–9.78 × 10^{-6}		
Size, < 0.002 mm, %	0.0–17.0	} Most material of	2.0–20
Size, > 2.0 mm, %	30.0–57.0	J sand size range	38–67
Shear strength, φ'	31.5–35.0°	27.5–39.5°	29–37°
Shear strength, c'	19.44–21.41 kPa	3.65–39.03 kPa	16–40kPa

* Average value.

consistency tests suggest a low to medium plasticity, while in certain instances spoil has proved to be virtually non-plastic. Plasticity increases with increasing clay content.

The shear strength parameters of coarse discard do not exhibit any systematic variation with depth in a spoil heap so are not related to age, that is, they are not time-dependent. This suggests that coarse discard is not seriously affected by weathering. As far as effective shear strength of coarse discard is concerned, the angle of shearing resistance usually ranges from 25 to 45°. The angle of shearing resistance and therefore the strength increases in spoil that has been burned. With increasing content of fine coal, the angle of shearing resistance is reduced. Also, as the clay mineral content in spoil increases, so its shear strength decreases.

The shear strength of discard within a spoil heap, and therefore its stability, is dependent upon the pore water pressures developed within it. Pore water pressures in spoil heaps may be developed as a result of the increasing weight of material added during construction or by seepage of natural drainage through the heap. High pore water pressures are usually associated with fine-grained materials that have a low permeability and high moisture content. Thus the relationship between permeability and the build-up of pore water pressures is crucial. It has been found that in soils with a coefficient of permeability of less than 5×10^{-9} m s^{-1} there is no dissipation of pore water pressures, while above 5×10^{-7} m s^{-1} they are completely dissipated. The permeability of colliery discard depends primarily upon its grading and its degree of compaction. Coarse discard may have a range of permeability ranging from 1×10^{-4} to 5×10^{-8} m s^{-1}, depending upon the amount of degradation in size that has occurred.

The most significant change in the character of coarse colliery discard brought about by weathering is the reduction of particle size. The extent to which breakdown occurs depends upon the type of parent material involved and the effects of air, water and handling between mining and placing on the spoil heap. After a few months of weathering the debris resulting from sandstones and siltstones is usually greater than cobble size. After that, the degradation to component grains takes place at a very slow rate. Mudstones, shales and seatearth exhibit rapid disintegration to gravel size. Although coarse discard may reach its level of degradation within a matter of months, with the degradation of many mudstones and shales taking place within days, once it has been buried within a spoil heap it suffers little change. When spoil material is burned, it becomes much more stable as far as weathering is concerned.

The presence of expandable mixed-layer clay is thought to be one of the factors promoting the breakdown of argillaceous rocks, particularly if Na^+ is a prominent inter-layer cation. Nevertheless, present evidence suggests that in England, colliery spoils with high exchangeable Na^+ levels are restricted. Sodium may be replaced by Ca^{++} and Mg^{++} cations originating from sulphates and carbonates in the waste, but dilution by more inert rock types is possibly the most usual reason for the generally low levels of sodium. Moreover, after undergoing the preparation process and exposure in the new surface layer of a spoil heap prior to burial, disintegration will have been largely achieved. However, it appears that the low shear strength of some fine-grained English spoils is associated with comparatively high exchangeable Na^+ levels.

Pyrite is a relatively common iron sulphide in some of the coals and argillaceous rocks of the Coal Measures. It is also an unstable mineral, breaking down quickly under the influence of weathering. The primary oxidation products of pyrite are ferrous and ferric sulphates, and sulphuric acid. Oxidation of pyrite within spoil heap waste is governed by the access of air, which in turn depends upon the particle size distribution, amount of water saturation and degree of compaction. However, any highly acidic oxidation products that form may be neutralized by alkaline materials in the waste material.

10.5.2 Spoil heap material and combustion

Spontaneous combustion of carbonaceous material, frequently aggravated by the oxidation of pyrite, is the most common cause of burning spoil. It can be regarded as an atmospheric oxidation (exothermic) process in which self-heating occurs. Coal and carbonaceous materials may be oxidized in the presence of air at ordinary temperatures, below their ignition point. Generally, the lower-rank coals are more reactive and accordingly more susceptible to self-heating than coals of higher rank.

Oxidation of pyrite at ambient temperatures in moist air leads, as mentioned above, to the formation of ferric and ferrous sulphate and sulphuric acid. This reaction is also exothermic. When present in sufficient amounts, and especially when in finely divided form, pyrite associated with coaly material increases the likelihood of spontaneous combustion. When heated, the oxidation of pyrite and organic sulphur in coal gives rise to the generation of sulphur dioxide. If there is not enough air for complete oxidation, then hydrogen sulphide is formed.

The moisture content and grading of spoil are also important factors in spontaneous combustion. At relatively low temperatures an increase in free moisture increases the rate of spontaneous heating. Oxidation generally takes place very slowly at ambient temperatures but as the temperature rises, so oxidation increases rapidly. In material of large size the movement of air can cause heat to be dissipated, while in fine material the air remains stagnant, but this means that burning ceases when the supply of oxygen is consumed. Accordingly, ideal conditions for spontaneous combustion exist when the grading is intermediate between these two extremes, and hot spots may develop under such conditions. These hot spots may have temperatures around 600 °C or occasionally up to 900 °C (Bell, 1996). Furthermore, the rate of oxidation generally increases as the specific surface of particles increases.

Spontaneous combustion can give rise to subsurface cavities in spoil heaps, the roofs of which may be incapable of supporting a person. Burnt ashes may also cover zones that are red-hot to appreciable depths. It has been recommended (Anon., 1973) that during restoration of a spoil heap a probe or crane with a drop-weight could be used to prove areas of doubtful safety that are burning. Badly fissured areas should be avoided and workmen should wear lifelines if they walk over areas not proved safe. Any area that is suspected of having cavities should be excavated by drag line or drag scraper rather than allowing plant to move over suspect ground.

When steam comes in contact with red-hot carbonaceous material, water gas is formed, and when the latter is mixed with air, over a wide range of concentrations, it becomes potentially explosive. If a cloud of coal dust is formed near burning spoil when reworking a heap, then this can also ignite and explode. Damping with a spray may prove useful in the latter case.

Noxious gases are emitted from burning spoil. These include carbon monoxide, carbon dioxide, sulphur dioxide and, less frequently, hydrogen sulphide. Each may be dangerous if breathed in certain concentrations, which may be present at fires on spoil heaps (Table 10.7). The rate of evolution of these gases may be accelerated by disturbing burning spoil by excavating into or reshaping it. Carbon monoxide is the most dangerous since it cannot be detected by taste, smell or irritation and may be present in potentially lethal concentrations. By contrast, sulphur gases are readily detectable in the aforementioned ways and are usually not present in high concentrations. Even so, when diluted they still may cause distress to persons with respiratory ailments. Nonetheless, the sulphur gases are mainly a nuisance rather than a threat to life. In certain situations a gas monitoring programme may have to be carried out. Where danger areas are identified, personnel should wear breathing apparatus.

The problem of combustion in spoil material sometimes has to be faced when reclaiming old tips. Spontaneous combustion of coal in colliery spoil can be averted if the coal occurs in an oxygen-deficient atmosphere that is humid enough with excess moisture to dissipate any heating that develops. Cook (1990), for example, described shrouding a burning spoil heap with a cover of compacted discard to smother the existing burning and prevent further spontaneous combustion. Anon. (1973) also recommended blanketing and compaction, as well as digging out, trenching, injection with non-combustible material and water, and water spraying as methods by which spontaneous combustion in spoil material can be controlled. Bell (1996) described the injection of a curtain wall of pulverized fuel ash, extending to original ground level, around hot spots as a means of dealing with hot spots.

10.5.3 Restoration of spoil heaps

The configuration of a spoil heap depends upon the type of equipment used in its construction and the sequence of tipping the waste. The shape, aspect and height of a spoil heap affects the intensity of exposure, the amount of surface erosion that occurs, the moisture content in its surface layers and its stability.

Table 10.7 Effects of noxious gases (after Anon., 1973).

Gas	Concentration by volume in air		Effect
	%	p.p.m.	
Carbon monoxide	0.01	100	Threshold limit value (TLV)*.
	0.02	200	Headache after about 7 hours if resting or after 2 hours if working.
	0.04	400	Headache and discomfort, with possibility of collapse, after 2 hours at rest or 45 minutes exertion.
	0.12	1200	Palpitation after 30 minutes at rest or 10 minutes exertion.
	0.20	2000	Unconsciousness after 30 minutes at rest or 10 minutes exertion.
Carbon dioxide	0.5	5000	T.L.V. Lung ventilation slightly increased.
	5.0		Breathing is laboured.
	9.0		Depression of breathing commences.
Hydrogen sulphide	0.001	10	T.L.V.
	0.01	100	Irritation to eyes and throat; headache.
	0.02	200	Maximum concentration tolerable for 1 hour.
	0.1	1000	Immediate unconsciousness.
Sulphur dioxide	0.0001	1–5	Can be detected by taste at the lower level and by smell at the upper level.
	0.0005		
	0.0005	5	T.L.V. Onset of irritation to the nose and throat.
	0.002	20	Irritation to the eyes.
	0.04	400	Immediately dangerous to life.

*The Threshold Limit Value is the value below which it is believed nearly all workers may be repeatedly exposed day after day without adverse effect.

Notes:

1 Some gases have a synergic effect, that is, they augment the effects of others and cause a lowering of the concentration at which the symptoms shown in the above table occur. Furthermore, a gas that is not itself toxic may increase the toxicity of one of the toxic gases, for example, by increasing the rate of respiration; strenuous work will have a similar effect.

2 Of the gases listed, carbon monoxide is the only one likely to prove a danger to life, as it is the commonest and the others become intolerably unpleasant at concentrations far below the danger level.

The mineralogical composition of coarse discard from different mines obviously varies, but pyrite frequently occurs in the shales and coaly material present in spoil heaps. When pyrite weathers it results in the formation of sulphuric acid, along with ferrous and ferric sulphates and ferric hydroxide, which gives rise to acidic conditions in the weathered material. Such conditions do not promote the growth of vegetation. Indeed, some spoils may contain elements that are toxic to plant life. To support vegetation, a spoil heap should have a stable surface in which roots can become established, be non-toxic, and contain an adequate and available supply of nutrients. The uppermost slopes of a spoil heap are frequently devoid of near-surface moisture. Hence spoil heaps are often barren of vegetation.

Restoration of a spoil heap represents an exercise in large-scale earthmoving. Since it will invariably involve spreading the waste over a larger area, this may mean that additional land beyond the site boundaries has to be purchased. Where a spoil heap is very close to the disused colliery, spoil may be spread over the latter area. This will involve the burial or removal of derelict colliery buildings and maybe the treatment of old mine shafts or shallow old workings (Johnson and James, 1990). Watercourses may have to be diverted, as may services, notably roads.

Landscaping of spoil heaps is frequently done to allow them to be used for agriculture or forestry. This type of restoration is generally less critical than when structures are to be erected on the site, since bearing capacities are not so important and steeper surface gradients are acceptable. Most spoil heaps offer no special handling problems other than the cost of regrading and possibly the provision of adjacent land so that the gradients on the existing site can be reduced by the transfer of spoil to the adjacent land. Some spoil heaps, however, present problems due to spontaneous combustion of the coaly material they contain. A spoil heap that is burning or that is so acidic that polluted waters are being discharged requires special treatment. Sealing layers have been used to control spontaneous combustion and polluted drainage, the sealing layers being well compacted.

Surface treatments of spoil heaps vary according to the chemical and physical nature of the spoil and the climatic conditions. Preparation of the surface of a spoil heap also depends upon the use to which it is to be put subsequently. Where it is intended to sow grass or plant trees, the surface layer should not be compacted. Drainage plays an important part in the restoration of a spoil heap and should take account of erosion control during landscaping. If the spoil is acid, then this can be neutralized by liming. The chemical composition will influence the choice of fertilizer used and in some instances spoil can be seeded without the addition of topsoil, if suitably fertilized. More commonly, topsoil is added prior to seeding or planting.

10.5.4 Waste disposal in tailings dams

Tailings are fine-grained residues that result from crushing rock that contains ore or are produced by the washeries at collieries. Mineral extraction usually takes place by wet processes in which the ore is separated from the parent rock by flotation, dissolution by cyanide, washing, etc. Consequently, most tailings originate as slurries, in which form they are transported for hydraulic deposition. The water in tailings may contain certain chemicals associated with the metal recovery process, such as cyanide in tailings from gold mines, and heavy metals in tailings from copper–lead–zinc mines. Tailings may also contain sulphide minerals, like pyrite, which can give rise to acid mine drainage. The particle size distribution, permeability and resistance to weathering of tailings affect the process of acid generation (Bell and Bullock, 1996). Acid drainage may also contain elevated levels of dissolved heavy metals. Accordingly, contaminants carried in the tailings represent a source of pollution for both ground and surface water, as well as soil.

Tailings are generally deposited as slurry in specially constructed tailings dams (Figure 10.9). Embankment dams can be constructed to their full height before the tailings are discharged. They are constructed in a similar manner to the earth fill dams used to impound reservoirs. They may be zoned with a clay core and have filter drains. Such dams are best suited to tailings impoundments with high water storage requirements. However, tailings dams usually consist of raised embankments, that is, the construction of the dams is staged over the life of the impoundment. Raised embankments consist initially of a starter dyke, which is normally constructed of earth fill from a borrow pit. This dyke may be large enough to accommodate the first two or three years of tailings production. A variety of materials can be used subsequently to complete such embankments, including earth fill, mine waste or tailings themselves. Tailings dams consisting of tailings can be constructed using the upstream, centreline or downstream method of construction (Figure 10.10). In the upstream method of construction, tailings are discharged from spigots or small pipes to form the impoundment. This allows separation of particles according to size, with the coarsest particles accumulating in the

Figure 10.9 Tailings lagoon with embankment for waste from china clay workings, Cornwall, England.

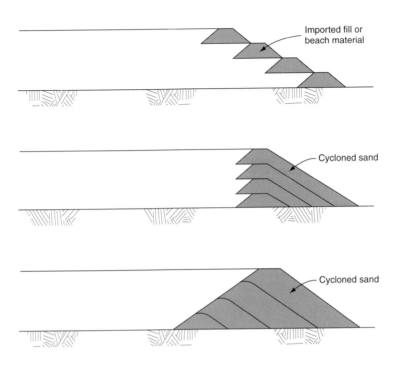

Figure 10.10 Upstream, centreline and downstream methods of embankment construction for tailings dams (after Vick, 1983).

centre of the embankment beneath the spigots and the finer particles being transported down the beach. Alternatively, cycloning may be undertaken to remove coarser particles from tailings so that they can be used in embankment construction. Centreline and downstream construction of embankments uses coarse particles separated by cyclones for the dam. For equivalent embankment heights, water retention embankment dams and downstream impoundments require approximately three times more fill than an upstream embankment. A centreline embankment would require about twice as much fill as an upstream embankment of similar height (Vick, 1983). The design of tailings dams must pay due attention to their stability in terms of both static and dynamic loading. The failure of a dam can have catastrophic consequences. For example, failure of the tailings dam at Buffalo Creek in West Virginia after heavy rain in February 1972 destroyed over 1500 houses and cost 118 lives. In addition, erosion of the outer slopes of tailings dams should be minimized by erosion control measures such as berms, surface drainage and rapid establishment of vegetation.

The tailings slurry can be discharged underwater in the impoundment or it can be disposed of by subaerial deposition. In subaerial disposal, the slurry is discharged from one or more points around the perimeter of the impoundment, with the slurry spreading over the floor to form beaches or deltas. The free water drains from the slurry to form a pond over the lowest area of the impoundment. The discharge points can be relocated around the dam to allow the exposed solids to dry and thereby increase in density. Alternatively, discharge can take place from one fixed point.

The deposition of tailings in a dam may lead to the formation of a beach or mudflat above the water level. When discharged, the coarser particles in tailings settle closer to the discharge point(s), with the finer particles being deposited further away (Blight, 1994). The amount of sorting that takes place is influenced by the way in which the tailings are discharged; for instance, high-volume discharge from one point produces little sorting of tailings. On the other hand, discharge from multiple points at moderate rates gives rise to good sorting. After being deposited, the geotechnical properties of tailings such as moisture content, density, strength and permeability are governed initially by the amount of sorting that occurs, and subsequently by the amount and rate of consolidation that takes place. For example, if sorting has led to material becoming progressively finer as the pool is approached, then the permeability of the tailings is likely to decrease in the same direction. The moisture content of deposited tailings can range from 20 to over 60% and dry densities from 1 to 1.3 Mg m^{-3}. At the end of deposition the density of the tailings generally increases with depth due to the increasing self-weight on the lower material. The coarser particles, which settle out first, also drain more quickly than the finer material, which accumulates further down the beach, and so they develop shear strength more quickly than the latter. These variations in sorting also affect the permeability of the material deposited.

The quantity of tailings that can be stored in a dam of a given volume is dependent upon the density that can be achieved. The latter is influenced by the type of tailings, the method by which they are deposited, whether they are deposited in water or subaerially, the drainage conditions within the dam and whether or not they are subjected to desiccation. Ideally, the pool should be kept as small as possible, and the penstock inlet should be located in the centre of the pool. This depresses the phreatic surface, thus helping to consolidate the tailings above this level. Blight and Steffen (1979) referred to the semi-dry or subaerial method of tailings deposition whereby in semi-arid and arid climates a layer of tailings is deposited in a dam and allowed to dry before the next layer is placed. Since this action reduces the volume of the tailings, it allows more storage to take place within the dam. However, drying out can give rise to the formation of desiccation cracks in the fine discard, which in turn represent locations where piping can be initiated. In fact, Blight (1988) pointed out that tailings dams have failed as a result of desiccation cracks and horizontal layering of fine-grained particles leading to piping failure. The most dangerous situation occurs when the ponded water on the discard increases in size and thereby erodes the cracks to form pipes, which may emerge on the outer slopes of the dam.

The rate at which seepage occurs from a tailings dam is governed by the permeability of the tailings and the ground beneath the impoundment. Climate and the way in which the tailings dam is managed will also have some influence on seepage losses. In many instances, because of the relatively low permeability of the tailings compared with the ground beneath, a partially saturated flow condition will occur in the foundation. Nonetheless, the permeability of tailings can vary significantly within a dam, depending on the nearness to discharge points, the degree of sorting, the amount of consolidation that has taken place, the stratification of coarser and finer layers that has developed, and the amount of desiccation that the discard has undergone.

Seepage losses from tailings dams that contain toxic materials can have an adverse effect on the environment. Hence it is usually necessary to operate the dam as a closed or controlled system from which water can be released at specific times. It is therefore necessary to carry out a water balance determination for a dam, and any associated catchments such as a return water reservoir, to assess whether and/or when it may be necessary to release water from or make up water within the system. Because of variations in precipitation and evaporation, water balance calculations should be undertaken frequently. If provision has been made for draining the pool of the dam, then after tailings deposition has ceased it should be possible to maintain a permanent water deficit, which will reduce seepage of polluted water from the dam.

According to Fell *et al.* (1993), one of the most cost-effective methods of controlling seepage loss from a tailings dam is to cover the whole floor of the impoundment with tailings from the start of the operation. This cover of tailings, provided it is of low permeability, will form a liner. Tailings normally have permeabilities between 10^{-7} and 10^{-9} m s^{-1} or less. Nevertheless, Fell *et al.* pointed out that a problem could arise when using tailings to line an impoundment and that if a sandy zone develops near the point(s) of discharge, then localized higher seepage rates will occur if water covers this zone. This can be avoided by moving the points of discharge or by placing fines. Alternatively, a seepage collection system can be placed beneath the sandy zone prior to its development. Clay liners also represent an effective method of reducing seepage from a tailings dam. The permeability of properly compacted clay soils usually varies between 10^{-8} and 10^{-9} m s^{-1}. However, clay liners may have to be protected from drying out, with attendant development of cracks, by placement of a sand layer on top. The sand on slopes may need to be kept in place by using geotextiles. Geomembranes have tended not to be used for lining tailings dams, primarily because of their cost. Filter drains may be placed at the base of tailings, with or without a clay liner. They convey water to collection dams. A toe drain may be incorporated into the embankment. This will intercept seepage which emerges at that location. Where a tailings dam has to be constructed on sand or sand and gravel, a slurry trench may be used to inter-cept seepage water; but slurry trenches are expensive.

The objectives of rehabilitation of tailings impoundments include their long-term mass stability, long-term stability against erosion, prevention of envi-ronmental contamination and return of the area to productive use. Normally, when the discharge of tailings comes to an end the level of the phreatic sur-face in the embankment falls as water replenishment ceases. This results in an enhancement of the stability of embankment slopes. However, where tailings impoundments are located on slopes, excess run-off into the impoundment may reduce embankment stability, while overtopping may lead to failure by erosion of the downstream slope. The minimization of inflow due to run-off by judi-cious siting is called for when locating an impoundment that may be so affected. Diversion ditches can cater for some run-off but have to be maintained, as do abandonment spillways and culverts. Accumulation of water may be prevented by capping the impoundment, the capping sloping towards the boundaries. Erosion by water or wind can be impeded by placing rip-rap on slopes and by the establishment of vegetation on the waste. Vegetation will also help to return the impoundment to some form of productive use. Where long-term potential for environmental contamination exists, particular precautions need to be taken. For example, as the water level in the impoundment declines, the rate of oxi-dation of any pyrite present in the tailings increases, reducing the pH and increasing the potential for heavy metal contamination. In the case of tailings from uranium mining, radioactive decay of radium gives rise to radon gas. Diffusion of radon gas does not occur in saturated tailings, but after abandon-ment radon reduction measures may be necessary. In both these cases a clay cover can be placed over the tailings impoundment to prevent leaching of con-taminants or to reduce emission of radon gas.

Once the discharge of tailings ceases, the surface of the impoundment is allowed to dry. Drying of the decant pond may take place by evaporation and/or by drainage to an effluent plant. Desiccation and consolidation of the slimes may take a considerable time. Stabilization can begin once the surface is firm enough to support equipment. As mentioned, this will normally involve the establishment of a vegetative cover.

10.6 Contaminated land

In many of the industrialized countries of the world, one of the legacies of the past two centuries is that land has been contaminated. The reason for this is that industry and society tended to dispose of their waste with little regard for future consequences. Hence, when such sites are cleared for redevelopment of urban areas they can pose problems.

Contamination can take many forms and can be variable in nature across a site, and each site has its own characteristics. In addition, the increasing scarcity of acceptable land for development in many Western countries means that poorer quality sites are considered. However, the increasing awareness of environmental issues has meant that within the last 20 years or so contaminated land has become a matter of public concern.

Contaminated land is by no means easy to define. The definition of contaminated land given by Nato (1981, cited in Cairney, 1993) is 'land which contains substances that, when present in sufficient qualities or concentrations, are likely to cause harm, directly or indirectly, to man, to the environment or to other targets.' The British Standards Institution Draft Document (DD175: 1988) defines contaminated land as 'land, that because of its nature or former uses, may contain substances that could give rise to hazards likely to affect a proposed form of development.' Such a definition poses another problem and that is: what is meant by 'hazard'? Hazard implies a degree of risk, but the degree of risk varies according to what is being risked. It depends, for example, upon the mobility of the contaminant(s) within the ground, and different types of soil have different degrees of reactivity to compounds that are introduced. It is also influenced by the future use of a site. However, in a matter-of-fact discussion of the subject of contaminated land, Wood (1994) pointed out that there is no identifiable causal relationship between the concentration of any contaminative substance in the ground and the incidence of ill-health arising from this. Similar statements were made by Cairney (1993), who remarked that demonstrable cases of hazards from contaminated soil are almost totally absent in the technical literature. Consequently, the concept of risk analysis for contaminated land assessment is still being developed and needs a great deal more investigation and assessment of data before it can be used with effect. One of the latest draft documents released by the Department of the Environment (Anon., 1995) indicates that at present there is no consensus as to what is an acceptable risk.

This draft document has defined contaminated land as 'any land which appears to be in such a condition, by reason of substances in, on or under the land that significant harm is being caused or there is a significant possibility of such harm being caused; or pollution of controlled waters is being, or is likely to be caused.' Harm is defined as 'harm to the health of living organisms or other interference with the ecological systems of which they form part, and in the case of man includes harm to his property.' Severe, moderate, mild and minimal degrees of harm are recognized (*ibid.*), as well as high, medium, low and very low degrees of possibility of harm being caused.

The British government maintains a commitment to the 'suitable for use' approach to the control and treatment of existing contaminated land. This supports sustainable development by reducing damage from past activities and by permitting contaminated land to be kept in, or returned to, beneficial use wherever possible. Such an approach only requires remedial action where the contamination poses unacceptable actual or potential risks to health or the environment and where appropriate and cost-effective means are available to do so. The Department of the Environment suggested that guideline values for concentrations of contaminants in soil may be used to indicate that a possibility of significant harm exists, but only in the absence of more detailed information on risks that may be presented by contaminants. So far, however, it has not issued its own guidelines. Furthermore, the Department of the Environment recommends that when using guideline values, it must be demonstrated that the values or assumptions underlying them are appropriate for use, and characteristic of the land and the ecosystems or property in question. The Department is in the process of preparing a guide to risk assessment and management for environmental protection.

No simple definition of hazard can incorporate the variety of circumstances that may arise. Therefore, Beckett (1993) suggested that contamination should be regarded as a concept rather than something that is capable of exact definition. In any given circumstances, the hazards that arise from contaminated land will be peculiar to the site and will differ in significance. Not only does this mean that no acceptable definition of contamination is available, but it also means that generally applicable criteria for the assessment of a contaminated site have not yet been developed.

Nevertheless, like other environmental issues, contaminated land has become a subject that arouses emotions. The emotional response is frequently that all contaminated land poses a risk and that no risk is acceptable. Although this is technically untenable, governments, which have to develop legislation to deal with the problem of contaminated land, can be influenced by the voting public.

In Europe, by comparison, the Dutch attitude towards contaminated land is more stringent. The Dutch recognize the possibility of a regular change in land use and have insisted that when contaminated land is redeveloped, the clean-up involved has to return the land to a standard that will allow any future use of the site in question. However, the Dutch have found that it has been impossible to organize, fund and execute all the clean-ups that have been deemed necessary (Cairney, 1993). Indeed, the concept of total clean-up is a standard of excellence that in practice is usually cost-prohibitive. Hence, standards of relevance become a necessary prerequisite in order to avoid negative land values, which would mean that remedial action would not take place.

In Britain, a contaminated site will undergo clean-up only if it is to be redeveloped, and then only when the contaminants exceed certain threshold levels as, for example, proposed by the Interdepartmental Committee on the Redevelopment of Contaminated Land (ICRCL, 1987; see Table 10.8). These are not the only guideline reference values available, although they are the most commonly used. Therefore, prior to any set of values being used, the original publication(s) should be checked, as their legal standing and applicability differ. In addition, there are no reference limits for certain organic contaminants in the 1987 ICRCL guidelines, and they contain no standards on groundwater quality. In the latter case, the practitioner has to fall back on drinking water abstraction standards. However, in many situations such water quality standards are restrictive, and reliance has been placed on Dutch standards (Table 10.9). Even so, the

Table 10.8 Guidance on the assessment and redevelopment of contaminated land (after ICRCL, 1987).

Contaminants	Use code	Reference value trigger concentrations (mg kg^{-1} air-dried soil)	
		Threshold	Action
Group A: Selected inorganic contaminants that may pose hazards to health			
Arsenic	1	10	NS
	2	40	NS
Cadmium	1	3	NS
	2	15	NS
Chromium total	1	600	NS
	2	1000	NS
Chromium (hexavalent)[1]	1, 2	25	NS
Lead	1	500	NS
	2	2000	NS
Mercury	1	1	NS
	2	20	NS
Selenium	1	3	NS
	2	6	NS
Group B: Contaminants that are phytotoxic, but not normally hazardous to health			
Boron (water-soluble)[2]	4	3	NS
Copper[3,4]	4	130	NS
Nickel[3,4]	4	70	NS
Zinc[3,4]	4	300	NS
Contaminants associated with former coal carbonization sites[5,6,7]			
Polyaromatic hydrocarbons[5,6,7]	1	50	500
	3, 5, 6	1000	10 000
Phenols[5]	1	5	200
	3, 5, 6	5	1000
Free cyanide[5]	1, 3	5	200
	5, 6	5	1000
Complex cyanides[5]	1	250	1000
	3	250	5000
	5, 6	250	NL

Table 10.8 (cont.)

	Use codes		
Thiocyanate[5,7]	All	50	NL
Sulphate[5]	1, 3	2000	10 000
	5	2000[8]	50 000[8]
Sulphide	All	250	1000
Sulphur	All	500	20 000
Acidity (pH less than)	1, 3	pH 5	pH 3
	5, 6	NL	NL

Table 10.9a Dutch standards for concentration values of contamination.

The Dutch use three values of ascending levels of contaminant concentration, namely, A, B, C. These are differentiated according to the nature of the pollution.

- Level A acts as a reference value. This level may be regarded as an indicative level above which there is demonstrable pollution and below which there is no demonstrable pollution.
- Level B is an assessment value. Pollutants above the B level should be investigated more thoroughly. The question asked is: to what extent are the nature, location, and concentration of the pollutant(s) of such a nature that it is possible to speak of a risk of exposure to man or the environment?
- Level C is to be regarded as the assessment value above which the pollutant(s) should generally be treated.

Present in:	Soil ($\mu g\,l^{-1}$ dry matter)			Groundwater ($\mu g\,l^{-1}$)		
component/concentration	A	B	C	A	B	C
I. Metals						
Cr	*	250	800	20	50	200
Co	20	50	300	*	50	200
Ni	*	100	500	*	50	200
Cu	*	100	500	*	50	200
Zn	*	500	3000	*	200	800
As	*	30	50	*	30	100
Mo	10	40	200	5	20	100
Cd	*	5	20	*	2.5	10
Sn	20	50	300	10	30	150
Ba	200	400	2000	50	100	500
Hg	*	2	10	*	0.5	2
Pb	*	150	600	*	50	200
II. Inorganic compounds						
NH₄ (as N)	–	–	–	*	1000	3000
F (as total)	*	400	2000	*	1200	4000
CN (total, free)	1	10	100	5	30	100
CN (total, combined)	5	50	500	10	50	200
S (total, sulphide)	2	20	200	10	100	300
Br (total)	20	50	300	*	500	2000
PO₄ (as P)	–	–	–	*	200	700
III. Aromatic compounds						
Benzene	0.05	0.5	5	0.2	1	5
Ethylbenzene	0.05	5	50	0.2	20	60
Toluene	0.05	3	30	0.2	15	50
Xylenes	0.05	5	50	0.2	20	60
Phenols	0.05	1	10	0.2	15	50
Aromatics (total)	–	7	70	–	30	100
IV. Polycyclic hydrocarbons						
Napthalene	*	5	50	0.2	7	30
Anthracene	*	10	100	0.005	2	30
Phenanthrene	*	10	100	0.005	2	10
Fluoranthene	*	10	100	0.005	2	5

Table 10.9a (continued)

Present in:	Soil ($\mu g\,l^{-1}$ dry matter)			Groundwater ($\mu g\,l^{-1}$)		
component/concentration	A	B	C	A	B	C

IV. Polycyclic hydrocarbons (continued)						
Chrycene	*	5	50	0.005	0.2	2
Benzo (a) anthracene	*	5	50	0.005	0.5	2
Benzo (a) pyrene	*	1	10	0.005	0.2	1
Benzo (k) fluoranthene	*	5	50	0.005	0.2	2
Indeno (1, 2, 3cd) phyrene	*	5	50	0.005	0.5	2
Benzo (ghy) perylene	*	10	100	0.005	1	5
Total polycyclic	1	20	200	0.2	10	40
V. Chlorinated hydrocarbons						
Aliphatic (indiv.)	*	5	50	0.01	10	50
Aliphatic (total)	–	7	70	–	15	70
Chlorobenzenes (indiv.)	*	1	10	0.01	0.5	2
Chlorobenzenes (total)	–	2	20	–	1	5
Chlorophenols (indiv.)	*	0.5	5	0.01	0.3	1.5
Chlorophenols (total)	–	1	10	–	0.5	2
Chlor. polycyclic (total)	*	1	10	–	0.2	1
PCBs (total)	*	1	10	0.01	0.2	1
EOCl (total)	0.1	8	80	1	15	70
VI. Pesticides						
Chlor. organics (indiv.)	*	0.5	5	1/0.1	0.2	1
Chlor. organics (total)	–	1	10	–	0.5	2
Non-chlor. (indiv.)	*	1	10	1/0.1	0.5	2
Non-chlor. (total)	–	2	20	–	1	5
VII. Other pollutants						
Tetrahydrothurane	0.1	4	40	0.5	20	60
Pyridine	0.1	2	20	0.5	10	30
Tetrahydrothiopene	0.1	5	50	0.5	20	60
Cyclohexanes	0.1	5	60	0.5	15	50
Styrene	0.1	5	50	0.5	20	60
Phthalates (total)	0.1	50	500	0.5	10	50
Total polycyclic hydrocarbons oxidized	1	200	2000	0.2	100	400
Mineral oil	*	1000	5000	50	200	600

* Reference value (level A) for soil quality. (For I and II, see Tables 10.9b and 10.9c)

use of Dutch standards is also somewhat restrictive, particularly when applied to former industrial sites, which are unlikely to be used for agricultural land or residential development. Accordingly, the lack of quantitative data on reference limits of some potentially hazardous substances found in soils and groundwater in areas likely to be contaminated at times presents both practitioners

Table 10.9b Reference values (level A) for heavy metals, arsenic and fluorine.

Component	Soil (mg kg^{-1} dry matter)		Groundwater ($\mu g\, l^{-1}$)
	Method of calculation	Standard soil (H = 10/L = 25)	
Cr (chromium)	50 + 2L	100	1
Ni (nickel)	10 + L	35	15
Cu (copper)	15 + 0.6 (L + H)	36	15
Zn (zinc)	50 + 1.5 (2L + H)	140	150
As (arsenic)	15 + 0.4 (L + H)	29	10
Cd (cadmium)	0.4 + 0.007 (L + H)	0.8	1.5
Hg (mercury)	0.2 + 0.0017 (2L + H)	0.3	0.05
Pb (lead)	50 + L + H	85	15
F (fluorine)	175 + 13L	500	–

Notes:
Reference values (level A) for heavy metals, arsenic and fluorine in all types of soil can be calculated by means of the formula given for each element. This formula expresses the reference value in terms of the clay content (L) and/or the organic materials content (H). The clay content is taken to be the weight of mineral particles smaller than 2 μm as a percentage of the total dry weight of the soil. The organic materials content is taken to be the loss in weight due to burning as a percentage of the total dry weight of the soil. As examples, the reference values (level A) are given for an assumed standard soil containing 25% clay (L) and 10% organic material (H). For groundwater in the saturated zone, the reference values are considered independently of the type of soil.

Table 10.9c Reference values (level A) for other inorganic compounds.

Component	Groundwater	Remarks
Nitrate	5.6 mg N/l	Lower values may be
Phosphate	0.4 mg P/l sandy soils	specified for the protection
(Total phosphate)	3.0 mg P/l clay and peaty soils	of soils low in nutrients
Sulphate	150 mg l^{-1}	In maritime areas higher
Bromides	0.3 mg l^{-1}	values occur naturally
Chlorides	100 mg l^{-1}	(saline and brackish
Fluorides	0.5 mg l^{-1}	groundwater)
Ammonium compounds	2 mg N^{-1} l^{-1} 10 mg N^{-1} l^{-1} clay and peaty soils	

and developers with problems. Nonetheless, after an investigation of a site, if the results are found to be below the threshold levels, then the site can be regarded as uncontaminated. Furthermore, the ICRCL guidelines relate hazards to land uses, thereby differing from assessment systems based on concentration limits. In other words, they recognized lower thresholds for certain uses such as residential developments with gardens than for hard cover areas (Table 10.8).

The presence of potentially harmful substances at a site does not necessarily mean that remedial action is required, if it can be demonstrated that they are inaccessible to living things or materials that may be detrimentally affected. However, consideration must always be given to the migration of soluble substances. The migration of soil-borne contaminants is primarily associated with groundwater movement, and the effectiveness of groundwater to transport contaminants is largely dependent upon their solubility. The quality of water can provide an indication of the mobility of contamination and the rate of dispersal. In an alkaline environment, the solubility of heavy metals becomes mainly neutral due to the formation of insoluble hydroxides. Providing groundwater conditions remain substantially unchanged during the development of a site, then the principal agent likely to bring about migration will be percolating surface water. On many sites, the risk of migration off site is of a very low order because the compounds have low solubility, and frequently most of their potential for leaching has been exhausted. Liquid and gas contaminants, of course, may be mobile. Obviously, care must be taken on site during working operations to avoid the release of contained contaminants (e.g. liquors in buried tanks) into the soil. Where methane has been produced in significant quantities, it can be oxidized by bacteria as it migrates through the ground with the production of carbon dioxide. However, this process does not necessarily continue.

If any investigation of a site that is suspected of being contaminated is going to achieve its purpose, it must define its objectives and determine the level of data required, the investigation being designed to meet the specific needs of the project concerned. This is of greater importance when related to potentially contaminated land than an ordinary site. According to Johnson (1994), all investigations of potentially contaminated sites should be approached in a staged manner. This allows for communication between interested parties, and helps to minimize costs and delays by facilitating planning and progress of the investigation. After the completion of each stage, an assessment should be made of the degree of uncertainty and of acceptable risk in relation to the proposed new development. Such an assessment should be used to determine the necessity for, and type of, further investigation.

The first stage in any investigation of a site suspected of being contaminated is a desk study, which provides data for the design of the subsequent ground investigation. The desk study should be supplemented by a site inspection. This is often referred to as a land quality appraisal. The desk study should identify past and present uses of the site, and the surrounding area, and the potential for and likely forms of contamination. The objectives of the desk study are to identify any hazards and the primary targets likely to be at risk; to provide data for health and safety precautions for the site investigation; and to identify any other factors that may act as constraints on development. Hence, the desk study should attempt to provide information on the layout of the site, including structures below ground, its physical features, the geology and hydrogeology of the site, the previous history of the site, the nature and quantities of materials handled and the processing involved, health and safety records, and methods of waste disposal. It should allow a preliminary risk assessment to be made and the need for further investigation to be established.

The preliminary investigation should formulate objectives so that the work is cost-effective. Such an investigation provides the data for the planning of the main field investigation, including the personal and equipment needs, the sampling and analytical requirements, and the health and safety requirements. In other words, it is a fact-finding stage that should confirm the chief hazard and identify any additional ones so that the main investigation can be carried out effectively. The preliminary investigation should also refer to any short-term or emergency measures that are required on the site before the commencement of full-scale operations.

Just as in a normal site investigation, one that is involved with the exploration for contamination needs to determine the nature of the ground. In addition, it needs to assess the ability of the ground to transmit any contaminants either laterally, or upward by capillary action. Permeability testing is therefore required. Investigation of contaminated sites frequently requires the use of a team of specialists, and without expert interpretation many of the benefits of site investigation may be lost. Most sites require careful interpretation of the ground investigation data. Sampling procedures are of particular importance, and the value of the data obtained therefrom is related to how representative the samples are. Some materials can change as a result of being disturbed when they are obtained or during handling. Hence, sampling procedure should take account of the areas of the site which require sampling; the pattern, depth, types and numbers of samples to be collected; their handling, transport and storage; and sample preparation and analytical methods. An assessment should be made as to whether protective clothing should be worn by site operatives.

The exploratory methods used in the main site investigation could include manual excavation, trenching and the use of trial pits, light cable percussion boring, power auger drilling, rotary drilling, and water and gas surveys. Excavation of pits and trenches are perhaps the most widely used techniques for investigating contaminated land (Bell *et al.*, 1996b). Visually different materials should be placed in different piles as a trench or pit is dug to facilitate sampling. The investigation must establish the location of perched water tables and aquifers, and any linkages between them, as well as determining the chemistry of the water on site.

Guidelines on how to conduct a sampling and testing programme are provided by the United States Environmental Protection Agency (USEPA, 1988 and Anon., 1988). A sampling plan should specify the objectives of the investigation, the history of the site, analyses of any existing data, types of sample to be used, sample locations and frequency, analytical procedures, and operational plan. Historical data should be used to ensure that any potential hot spots are sampled satisfactorily. A sampling grid should be used in such areas (Anon., 1988; Sara, 1994).

Volatile contaminants or gas-producing material can be determined by sampling the soil atmosphere by using a hollow gas probe, inserted to the required depth. The probe is connected to a small vacuum pump and a flow of soil gas induced. A sample is recovered using a syringe. Care must be taken to determine the presence of gas at different horizons by the use of sealed response zones. Analysis is usually conducted on site by portable gas chromatography or photo-ionization. Stand pipes can be used to monitor gases during the exploration work.

One factor that should be avoided is cross-contamination. This is the transfer of materials by the exploratory technique from one depth into a sample taken at a different depth. Consequently, cleaning requirements should be considered, and ideally a high specification of cleaning operation should be carried out on equipment between both sampling and borehole locations.

To ensure that the site investigation is conducted in the manner intended and the correct data recorded, the work should be carefully specified in advance. As the investigation proceeds, it may become apparent that the distribution of material about the site is not as predicted by the desk study or preliminary investigation. Hence the site investigation strategy will need to be adjusted. The data obtained during the investigation must be accurately recorded in a manner whereby it can be understood subsequently. The testing programme should identify the types, distribution and concentration (severity) of contaminants, and any significant variations or local anomalies. Comparisons with uncontaminated surrounding areas can be made.

Once completed, the site characterization process, when considered in conjunction with the development proposals, will enable the constraints on development to be identified. These constraints, however, cannot be based solely on the data obtained from the site investigation but must take account of financial and legal considerations. If hazard potential and associated risk are regarded as too high, then the development proposals will need to be reviewed. When the physical constraints and hazards have been assessed, then a remediation programme can be designed that allows the site to be economically and safely developed. It is at this stage that clean-up standards are specified in conjunction with the assessment of the contaminative regime of the surrounding area.

10.7 Remediation of contaminated land

Remedial planning and implementation is a complex process that not only involves geotechnical methodology but may also be influenced by statutory and regulatory compliance. Indeed, it is becoming increasingly common for regulatory standards and guidance to provide very prescriptive procedural and technical requirements (Attewell, 1993; Holm, 1993).

A wide range of technologies are available for the remediation of contaminated sites, and the applicability of a particular method will depend partly upon the site conditions, the type and extent of contamination, and the extent of the remediation required. When the acceptance criteria are demanding and the degree of contamination complex, it is important that the feasibility of the remediation technology be tested in order to ensure that the design objectives can be satisfied (Swane *et al.*, 1993). This can necessitate a thorough laboratory testing programme being undertaken, along with field trials. In some cases, the remediation operation requires the employment of more than one method.

It is obviously important that the remedial works do not give rise to unacceptable levels of pollution either on site or in the immediate surroundings. Hence the design of the remedial works should include measures to control pollution during the operation. The effectiveness of the pollution control measures needs to be monitored throughout the remediation programme.

In order to verify that the remediation operation has complied with the clean-up acceptance criteria for the site, a further sampling and testing programme is required (USEPA, 1989). Swane *et al.* (1993) recommended that it is advisable to check that the clean-up standards are being attained as the site is being rehabilitated. In this way, any parts of the site that fail to meet the criteria can be dealt with there and then, so improving the construction schedule.

The nature of the remedial action depends upon the nature of the contaminants present on site. In some situations it may be possible to rely upon natural decay or dispersion of the contaminants. Removal of contaminants from a site for disposal in an approved disposal facility has frequently been used. However, the costs involved in removal are increasing and can be extremely high. On-site burial can be carried out by the provision of an acceptable surface barrier or an approved depth of clean surface filling over the contaminated material. This may require considerable earthworks. Clean covers are most appropriate for sites with various previous uses and contamination. They should not be used to contain oily or gaseous contamination. Another method is to isolate the contaminants by *in situ* containment by using, for example, cut-off barriers. Encapsulation involves immobilization of contaminants, for instance, by the injection of grout. Low-grade contaminated materials can be diluted below threshold levels by mixing with clean soil. However, there are possible problems associated with dilution. The need for quality assurance is high (unacceptable materials cannot be reincorporated into the site). The dilution process must not make the contaminants more leachable. Other less frequently used methods include bioremediation of organic material, soil flushing and soil washing, incineration, vacuum extraction and venting of volatile constituents, and removal of harmful substances to be buried in containers at a purpose-built facility.

10.7.1 Soil remediation

Most of the remediation technologies differ in their applicability to treat particular contaminants in the soil. Landfill disposal and containment are the exceptions in that they are capable of dealing with most soil contamination problems. However, the removal of contaminated soil from a site for disposal in a special landfill facility transfers the problem from one location to another. Hence the *in situ* treatment of sites, where feasible, is the better course of action. Containment is used to isolate contaminated sites from the environment by the installation of a barrier system such as a cut-off wall (see Subsection 10.1.1). In addition, containment may include a cover placed over the contaminated zone(s) to reduce infiltration of surface water and to act as a separation layer between land users and the contaminated ground.

Soil washing involves using particle size fractionation; aqueous-based systems employing some type of mechanical and/or chemical process; or counter-current decantation with solvents for organic contaminants, and acids/bases or chelating agents for inorganic contaminants, to remove contaminants from excavated soils (Trost, 1993). Particle size fractionation is based on the premise that contaminants are generally more likely to be associated with finer particle sizes. Hence, in particle size fractionation the finer material is subjected to high-pressure water washing, which can include the use of additives. Aqueous-based soil washing systems generally make use of the froth flotation process to separate contaminants. Basically, counter-current decantation uses a series of thickeners in tanks, into which the contaminated soil is introduced and allowed to settle. The contaminated soil is pumped from one tank to the next and the solvent, acid/base or chelating agent is in the last tank. Variation in the type of soil and contaminants at a site mean that soil washing is more difficult and that a testing programme needs to be carried out to determine effectiveness. Steam injection and stripping can be used to treat soils in the vadose zone contaminated with volatile compounds. The steam provides the heat that volatizes the contaminants, which are then extracted (i.e. stripped). The efficiency of the process depends upon the ease with which steam can be injected and recovered. As steam is used in near-surface soils, and since much of it is lost due to gas pressure fracturing the soil, an impermeable cover may be used to impede the escape of steam. One of the disadvantages is that some steam turns to water on cooling, which means that some contaminated water remains in the soil.

Solvents can be used to remove contaminants from the soil. For example, soil flushing makes use of water, water–surfactant mixtures, acids, bases, chelating agents, oxidizing agents and reducing agents to extract semi-volatile organics, heavy metals and cyanide salts from the vadose zone of the soil. The technique is used in soils that are sufficiently permeable (not less than 10^{-7} m s^{-1}) to allow the solvent to permeate, and the more homogeneous the soil is, the better. The solvents are injected into the soil and the contaminated extractant removed by pumping. Every effort should be made to prevent the contaminated extractant from invading the groundwater. In fact, due to the possibility of soil flushing having an adverse impact on the environment, the USEPA (1990) recommended that the technique should be used only when others with lower impacts on the environment are not applicable. Solvents can also be used to treat soils that have been excavated.

Vacuum extraction involves the removal of contaminants by the use of vacuum extraction wells. It can be applied to volatile organic compounds residing in the unsaturated soil or to volatile light non-aqueous phase liquids (LNAPLs) resting on the water table. The method can be used in most types of soil, although its efficiency declines in heterogeneous and high-permeability soils. The vapour from the exhaust pipes can be treated when necessary.

Fixation or solidification processes reduce the availability of contaminants to the environment by chemical reactions with additives (fixation) or by changing the phase (e.g. liquid to solid) of the contaminant by mixing with another medium. Such processes are usually applied to concentrated sources of hazardous wastes. Various cementing materials such as Portland cement and quicklime can be used to immobilize heavy metals.

Some contaminants can be removed from the soil by heating. For instance, soil can be heated to between 400 and 600 °C to drive off or decompose organic contaminants such as hydrocarbons, solvents, volatile organic compounds and pesticides. Mobile units can be used on site, the soil being removed, treated and then returned as backfill.

Incineration, whereby wastes are heated to between 1500 and 2000 °C, is used for dealing with hazardous wastes containing halogenated organic compounds such as polychlorinated biphenyls (PCBs) and pesticides, which are difficult to remove by other techniques. Incineration involves removal of the soil, and it is then usually crushed and screened to provide fine material for firing. The ash that remains may require additional treatment, since heavy metal contamination may not have been removed by incineration. It is then disposed of in a landfill.

In situ vitrification transforms contaminated soil into a glassy mass. It involves electrodes being inserted around the contaminated area and sufficient electric current being applied to melt the soil (the required temperatures can range from 1600 to 2000 °C). The volatile contaminants are either driven off or destroyed, while the non-volatile contaminants are encapsulated in the glassy mass when it solidifies. It may be used in soils that contain heavy metals.

Bioremediation involves the use of micro-organisms to bring about the degradation or transformation of contaminants so that they become harmless (Loehr, 1993). The micro-organisms involved in the process either occur naturally or are artificially introduced. In the case of the former, the microbial action is stimulated by optimizing the conditions necessary for growth (Singleton and Burke, 1994). The principal use of bioremediation is in the degradation and destruction of organic contaminants, although it has also been used to convert some heavy metal compounds into less toxic states. Bioremediation can be carried out on ground *in situ* or ground can be removed for treatment. *In situ* bioremediation depends upon the amenability of the organic compounds to biodegradation, the ease with which oxygen and nutrients can reach the contaminated area, and the permeability, temperature and pH of the soil. Air circulation brought about by soil vacuum extraction enhances the supply of oxygen to micro-organisms in the vadose zone of the soil. Other systems used for *in situ* bioremediation in the vadoze zone include infiltration galleries or injection wells for the delivery of water carrying oxygen and nutrients. In the phreatic zone, water can be extracted from the contaminated area, treated at the surface with oxygen and nutrients, and subsequently injected back into the contaminated ground. This allows groundwater to be treated. However, the *in situ* method of treatment when applied to soil or water alone may prove unsatisfactory due to recontamination by the remaining untreated soil or untreated groundwater. For example, contaminants in the groundwater, notably oils, solvents and creosote, which are immiscible, recontaminate soil as groundwater levels fluctuate. On the other hand, contamination in the soil may be washed out continuously over many years, so contaminating groundwater. Therefore, as Adams and Holroyd (1992) recommended, it is frequently worthwhile to clean both soil and groundwater simultaneously. *Ex situ* bioremediation involves the excavation of contaminated ground and placing it in beds, where it is treated, and then returning it to where it was removed as cleaned backfill. There are a number of different types of *ex situ* bioremediation, including land farming, composting, purpose-built treatment facilities and biological reactors.

10.7.2 Groundwater remediation

Contaminated groundwater can either be treated *in situ* or be abstracted and treated. The solubility in water and volatility of contaminants influence the selection of the remedial technique used. Some organic liquids are only slightly soluble in water and are immiscible. These are known as non-aqueous phase liquids (NAPLs); when dense they are referred to as DNAPLs or 'sinkers' and when light as LNAPLs or 'floaters'. Examples of the former include many chlorinated hydrocarbons such as trichloroethylene and trichloroethane, while petrol, diesel oil and paraffin provide examples of the latter. The permeability of the ground influences the rate at which contaminated groundwater moves and therefore the ease and rate at which it can be extracted.

The pump-and-treat method is the most widely used means of remediation of contaminated groundwater (Haley *et al.*, 1991). The groundwater is abstracted from the aquifer concerned by wells, trenches or pits and treated at the surface. It is then injected back into the aquifer. The pump-and-treat method proves most successful when the contaminants are highly soluble and are not readily adsorbed by clay minerals in the ground. In particular, LNAPLs can be separated from the groundwater either by using a skimming pump in a well or at the surface using oil–water separators. It usually does not prove possible to remove all light oil in this way, so other techniques may be required to treat the residual hydrocarbons. Oily substances and synthetic organic compounds are normally much more difficult to remove from an aquifer. In fact, the successful removal of DNAPLs is impossible at the present. As such, they can be dealt with by containment. Methods of treatment that are used to remove contaminants dissolved in water, once it has been abstracted, include standard water treatment techniques, air stripping of volatiles, carbon adsorption, micro-filtration and bioremediation (Swane *et al.*, 1993).

Active containment refers to the isolation or hydrodynamic control of contaminated groundwater. The process makes use of pumping and recharge systems to develop 'zones of stagnation' or to alter the flow pattern of the groundwater. Cut-off walls are used in passive containment to isolate the contaminated groundwater.

Air sparging is a type of *in situ* air stripping in which air is forced under pressure through an aquifer in order to remove volatile organic contaminants. It also enhances desorption and bioremediation of contaminants in saturated soils. The air is removed from the ground by soil venting systems. The injection points, especially where contamination occurs at shallow depth, are located beneath the area affected. According to Waters (1994), the key to effective air sparging is the contact between the injected air and the contaminated soil and groundwater. Because air permeability is higher than that of water, greater volumes of air have to be used to treat sites. Nevertheless, air sparging can be used to treat sites when pump-and-treat methods are ineffective. The air that is vented may have to be collected for further treatment as it could be hazardous.

In situ bioremediation makes use of microbial activity to degrade organic contaminants in the groundwater so that they become non-toxic. Oxygen and nutrients are introduced into an aquifer to stimulate activity in aerobic bioremediation, whereas methane and nutrients may be introduced in anaerobic bioremediation.

References

Adams, D. and Holroyd, M. (1992) *In situ* soil bioremediation. *Proceedings Third International Conference on Construction on Polluted and Marginal Land*. London, Forde, M.C. (ed.), Engineering Technics Press, Edinburgh, 291–294.

Anon. (1973) *Spoil Heaps and Lagoons: Technical Handbook*. National Coal Board, London.

Anon. (1988) *Draft for Development, DD175: 1988, Code of Practice for the Identification of Potentially Contaminated Land and its Investigation*. British Standards Institution, London.

Anon. (1995) *A Guide to Risk Assessment and Risk Management for Environmental Protection*. Department of the Environment, HMSO, London.

Attewell, P.B. (1993) *Ground Pollution; Environmental Geology, Engineering and Law*. E & FN Spon, London.

Baillieul, T.A. (1987) Disposal of high level nuclear waste in America. *Bulletin Association Engineering Geologists*, 24, 207–216.

Barber, C. (1982) *Domestic Waste and Leachate*. Notes on Water Research No. 31, Water Research Centre, Medmenham.

Beckett, M.J. (1993) Land contamination. In *Contaminated Land, Problems and Solutions*, Cairney, T. (ed.), Blackie, Glasgow, 8–28.

Bell, F.G. (1996) Dereliction: colliery spoil heaps and their rehabilitation. *Environmental and Engineering Geoscience*, 2, 85–96.

Bell, F.G. and Bullock, S.E.T. (1996) The problem of acid mine drainage, with an illustrative case history. *Environmental and Engineering Geoscience*, 2, 369–392.

Bell, F.G. and Mitchell, J.K. (1986) Control of groundwater by exclusion. In *Groundwater in Engineering Geology*, Engineering Geology Special Publication No. 3, Cripps, J.C., Bell, F.G. and Culshaw, M.G. (eds), The Geological Society, London, 429–443.

Bell, F.G., Sillito, A.J. and Jermy, C.A. (1996a) Landfills and associated leachate in the greater Durban area: two case histories. In *Engineering Geology of Waste Disposal*, Engineering Geology Special Publication No. 11, Bentley, S.F. (ed.), The Geological Society, London, 15–35.

Bell, F.G., Bell, A.W., Duane, M.J. and Hytiris, N. (1996b) Contaminated land: the British position and some case histories. *Environmental and Engineering Geoscience*, 2, 355–368.

Blight, G.E. (1988) Some less familiar aspects of hydraulic fill structures. In *Hydraulic Fill Structures*, Van Zyl, D.J.A. and Vick, S.G. (eds), American Society Civil Engineers, Geotechnical Special Publication No. 21, New York, 1000–1064.

Blight, G.E. (1994) Environmentally acceptable tailings dams. *Proceedings First International Congress on Environmental Geotechnics*, Edmonton, Carrier, W.D. (ed.), BiTech Publishers, Richmond, BC, 417–426.

Blight, G.E. and Steffen, O.K.H. (1979) Geotechnics of gold mining waste disposal. In *Current Geotechnical Practice in Mine Waste Disposal*. American Society Civil Engineers, New York, 1–52.

Cairney, T. (1993) International responses. In *Contaminated Land, Problems and Solutions*, Cairney, T. (ed.), Blackie, Glasgow, 1–6.

Chapman, N.A. and Williams, G.M. (1987) Hazardous and radioactive waste management: a case of dual standards? In *Planning and Engineering Geology*, Engineering Geology Special Publication No. 4, Culshaw, M.G., Bell, F.G., Cripps, J.C. and O'Hara, M. (eds), The Geological Society, London, 489–493.

Clark, R.G. and Davis, G. (1996) The construction of clay liners for landfills. In *Engineering Geology of Waste Disposal*, Engineering Geology Special Publication No. 11, Bentley, S.P. (ed.), The Geological Society, London, 171–176.

Cook, B.J. (1990) Coal discard–rehabilitation of a burning heap. In *Reclamation, Treatment and Utilization of Coal Mining Wastes*, Rainbow, A.K.M. (ed.), Balkema, Rotterdam, 223–230.

Daniel, D.E. (1993) Clay cores. In *Geotechnical Practice for Waste Disposal*, Daniel, D.E. (ed.), Chapman & Hall, London, 137–162.

Daniel, D.E. (1995) Pollution prevention in landfills using engineered final covers. *Proceedings Symposium on Waste Disposal, Green '93 – Geotechnics Related to the Environment*, Bolton, Sarsby, R.W. (ed.), Balkema, Rotterdam, 73–92.

Eriksson, L.G. (1989) Underground disposal of high level nuclear waste in the United States of America. *Bulletin International Association Engineering Geology*, 39, 35–52.

Fang, H.Y. (1995) Engineering behaviour of urban refuse, compaction control and slope stability analysis of landfill. *Proceedings Symposium on Waste Disposal by Landfill, Green '93 – Geotechnics Related to the Environment*, Bolton, Sarsby, R.W. (ed.), Balkema, Rotterdam, 47–62.

Fell, R., Miller, S. and de Ambrosio, L. (1993) Seepage and contamination from mine waste. *Proceedings Conference on Geotechnical Management of Waste and Contamination*, Sydney, Fell, R., Phillips, A. and Gerrard, C. (eds), Balkema, Rotterdam, 253–311.

Hagerty, D.J. and Pavori, J.L. (1973) Geologic aspects of landfill refuse disposal. *Engineering Geology*, 7, 219–230.

Haley, J.L., Hanson, B., Enfield, C. and Glass, J. (1991) Evaluating the effectiveness of groundwater extraction systems. *Ground Water Monitoring Review*, 12, 119–124.

Haskins, D.R. and Bell, F.G. (1995) Drakensberg basalts: their alteration, breakdown and durability. *Quarterly Journal Engineering Geology*, 28, 287–302.

Holm, L.A. (1993) Strategies for remediation. In *Geotechnical Practice for Waste Disposal*, Daniel, D.E. (ed.), Chapman & Hall, London, 289–310.

Horseman, S.T. and Volckaert, G. (1996) Disposal of radioactive wastes in argillaceous formations. In *Engineering Geology of Waste Disposal*, Engineering Geology Special Publication No. 11, Bentley, S.P. (ed.), The Geological Society, London, 179–191.

ICRCL (1987) *Guidelines on the Assessment and Redevelopment of Contaminated Land: Guidance Note 59/83*, second edition. Interdepartmental Committee on the Redevelopment of Contaminated Land, Department of the Environment, HMSO, London.

Jessberger, H.L., Manassero, M., Sayez, B. and Street, A. (1995) Engineering waste disposal (Geotechnics of landfill design and remedial works). *Proceedings Symposium on Waste Disposal by Landfill, Green '93 – Geotechnics Related to the Environment*, Bolton, Sarsby, R.W. (ed.), Balkema, Rotterdam, 21–33.

Johnson, A.C. (1994) Site investigation for development on contaminated sites – how, why and when? *Proceedings Third International Conference on Reuse of Contaminated Land and Landfills*, London, Forde, M.C. (ed.), Engineering Technics Press, Edinburgh, 3–7.

Johnson, A.C. and James, E.J. (1990) Granville colliery land reclamation/coal recovery scheme. In *Reclamation and Treatment of Coal Mining Wastes*, Rainbow, A.K.M. (ed.), Balkema, Rotterdam, 193–202.

Koerner, R.M. (1993) Geomembrane liners. In *Geotechnical Practice for Waste Disposal*, Daniel, D.E. (ed.), Chapman & Hall, London, 164–186.

Krauskopf, K.B. (1988) *Radioactive Waste Disposal and Geology*. Chapman & Hall, London.

Langer, M. (1989) Waste disposal in the Federal Republic of Germany: concepts, criteria, scientific investigations. *Bulletin International Association Engineering Geology*, 39, 53–58.

Langer, M. and Wallner, M. (1988) Solution-mined salt caverns for the disposal of hazardous chemical wastes. *Bulletin International Association Engineering Geology*, 37, 61–70.

Loehr, R.C. (1993) Bioremediation of soils. In *Geotechnical Practice for Waste Disposal*, Daniel, D.E. (ed.), Chapman & Hall, London, 520–550.

Mitchell, J.K. (1986) Hazardous waste containment. In *Groundwater in Engineering Geology*, Engineering Geology Special Publication No. 3, Cripps, J.C., Bell, F.G. and Culshaw, M.G. (eds), The Geological Society, London, 145–157.

Mitchell, J.K. (1994) Physical barriers for waste containment. *Proceedings First International Congress on Environmental Geotechnics*, Edmonton, Carrier, W.D. (ed.), BiTech Publishers, Richmond, BC, 951–962.

Mitchell, J.K. and Madsen, F.T. (1987) Chemical effects on clay hydraulic conductivity. In *Geotechnical Practice for Waste Disposal '87*, American Society Civil Engineers,

Geotechnical Special Publication No. 13, 87–116.

Morfeldt, C.O. (1989) Different subsurface facilities for the geological disposal of radioactive waste (storage cycle) in Sweden. *Bulletin International Association Engineering Geology*, 39, 25–34.

Oweis, I.S and Khera, R.P. (1990) *Geotechnology of Waste Management*. Butterworths, London.

Quigley, R.M., Fernandez, F. and Crooks, V.E. (1988) Engineered clay liners: a short review. In *Proceedings Symposium on Environmental Geotechnics and Problematic Soils and Rocks*, Bangkok, Balasubramanian, A.S., Chandra, S., Bergado, D.T. and Natalaya, P. (eds), Balkema, Rotterdam, 63–74.

Raybould, J.G. and Anderson, J.G. (1987) Migration of landfill gas and its control by grouting – a case history. *Quarterly Journal Engineering Geology*, 20, 78–83.

Ressi, A. and Cavalli, N. (1984) Bentonite slurry trenches. *Engineering Geology*, 21, 333–339.

Rogers, V. (1994) Present trends in nuclear waste disposal. *Proceedings First International Congress on Environmental Geotechnics*, Edmonton, Carrier, W.D. (ed.), BiTech Publishers, Richmond, BC, 837–845.

Sara, M.N. (1994) *Standard Handbook for Solid and Hazardous Waste Facilities Assessment*. Lewis Publishers, Boca Raton, Florida.

Singleton, M. and Burke, G.K. (1994) Treatment of contaminated soil through multiple bioremediation technologies and geotechnical engineering. *Proceedings Third International Conference on Reuse of Contaminated Land and Landfills*, London, Forde, M.C. (ed.), Engineering Technics Press, Edinburgh, 97–107.

Smith, C.F., Cohen, J.J. and McKone, T.E. (1980) *Hazard Index for Underground Toxic Material*. Lawrence Livermore Laboratory, Report No. UCRL-52889.

Swane, I.C., Dunbavan, M. and Riddell, P. (1993) Remediation of contaminated sites in Australia. *Proceedings Conference on Geotechnical Management of Waste and Contamination*, Sydney, Fell, R., Phillips, A. and Gerrard, C. (eds), Balkema, Rotterdam.

Taylor, R.K. (1975) English and Welsh colliery spoil heaps – mineralogical and mechanical relationships. *Engineering Geology*, 7, 39–52.

Trost, P.B. (1993) Soil washing. In *Geotechnical Practice for Waste Disposal*, Daniel, D.E. (ed.), Chapman & Hall, London, 585–603.

USEPA (1988) *Guidance for Conducting Remedial Investigations and Feasibility Studies under CERCLA*. Office of Emergency and Remedial Response, US Government Printing Office, Washington, DC.

USEPA (1989) *Methods for Evaluating the Attainment of Clean-up Standards, Volume 1: Soils and Solid Media*. Office of Policy, Planning and Evaluation, EPA/540/2–90/011, US Government Printing Office, Washington, DC.

USEPA (1990) *Subsurface Contamination Reference Guide*. Office of Emergency and Remedial Response, EPA/540/5–91/008, US Government Printing Office, Washington, DC.

Van Zyl, D.J.A. (1993) Mine waste disposal. In *Geotechnical Practice for Waste Disposal*, Daniel, D.E. (ed.), Chapman & Hall, London, 269–287.

Vick, S.G. (1983) *Planning, Design and Analysis of Tailings Dams*. Wiley, New York.

Waters, J. (1994) *In situ* remediation using air sparging and soil venting. *Proceedings Third International Conference on Reuse of Contaminated Land and Landfills*, London, Forde, M.C. (ed.), Engineering Technics Press, Edinburgh, 109–112.

Williams, G.M. and Aitkenhead, N. (1991) Lessons from Loscoe: the uncontrolled migration of landfill gas. *Quarterly Journal Engineering Geology*, 24, 191–208.

Wood, A.A. (1994) Contaminated land – an engineer's viewpoint. In *Proceedings Third International Conference on Reuse of Contaminated Land and Landfills*, London, Forde, M.C. (ed.), Engineering Technics Press, Edinburgh, 205–209.

Chapter 11

Groundwater pollution

11.1 Introduction

Brown *et al.* (1972) defined pollution as the 'addition of chemical, physical or biological substances or of heat, which causes deterioration in the natural quality, generally through the action of man or animals or any other kind of activity.' More specifically, Walker (1969) defined pollution as 'an impairment of water quality by chemicals, heat, or bacteria to a degree that does not necessarily create an actual public health hazard, but that does adversely affect such waters for normal, domestic, farm, municipal or industrial use.' On the other hand, Walker used the term contamination to denote 'impairment of water quality by chemical or bacterial pollution to a degree that creates an actual hazard to public health.'

Groundwater pollution is of common occurrence. In fact, most of man's activities have a direct, and usually adverse, effect upon water quality. A comprehensive list of potential contaminants, the characteristics of the leachate or effluent and the typical rate of production of the effluent or waste has been provided by Jackson (1980).

It was estimated that in 1964 more than 900 million people were without a public water service of any kind, while in the developing countries as many as 500 million people per year were affected by water-borne or water-related diseases (Pickford, 1979). By 1980 the situation had deteriorated. The World Health Organisation then estimated that 1320 million people (57%) of the developing world, excluding China, were without a clean water supply, while 1730 million (75%) were without adequate sanitation. At least 30 000 people per day die in the Third World because they have inadequate water and sanitation facilities. The control of water pollution in developing countries is a necessity; some disastrous environmental results have already occurred through lack of attention to the problem. Consequently, groundwater pollution is not just a hydrogeological problem; many political, social, economic and medical factors are also involved (Van Burkalow, 1982).

The greatest danger of groundwater pollution is from surface sources such as farm animals, man, sewers, polluted streams, refuse disposal sites and so on.

Areas with thin soil cover or where the aquifer is exposed, such as the recharge area, are the most critical from the point of view of pollution potential. Any possible source of contamination in these areas should be carefully evaluated, both before and after well construction, and the viability of groundwater protection measures considered (Raucher, 1983). Changes in land use may pose new threats to water quality. Obvious precautions against pollution are to locate the wells as far from any potential source of contamination as possible and to fence off the tops of wells so that animals cannot defecate adjacent to the well. Good well design and construction are also important.

However, it should be appreciated that the slow rate of travel of pollutants in underground strata means that a case of pollution may go undetected for a number of years. During this period, a large part of the aquifer may become polluted and cease to have any potential as a source of water (Selby and Skinner, 1979).

The concept of safe yield has been used to express the quantity of water that can be withdrawn from the ground without impairing the aquifer as a water source. Draft in excess of safe yield is overdraft and this can give rise to pollution or contamination, or cause serious problems due to severely increased pumping lift. Indeed, this may eventually lead to the exhaustion of wells.

Estimation of the safe yield is a complex problem that must take into account the climatic, geological and hydrological conditions. As such, the safe yield is likely to vary appreciably with time. Nonetheless, the recharge–discharge equation, the transmissivity of the aquifer, the potential sources of pollution and the number of wells in operation must all be given consideration if an answer is to be found. The safe yield (G) is often expressed as follows:

$$G = P - Q_s - ET + Q_g - \Delta S_g - \Delta S_s \qquad (11.1)$$

where P is the precipitation on the area supplying the aquifer, Q_s is the surface stream flow over the same area, ET is evapotranspiration, Q_g is the net groundwater inflow to the area, ΔS_g is the change in groundwater storage, and ΔS_s is the change in surface storage. With the exception of precipitation, all the terms in this expression can be subjected to artificial change. The equation cannot be considered an equilibrium equation or solved in terms of mean annual values. It can be solved correctly only on the basis of specified assumptions for a stated period of years.

11.2 A note on groundwater quality

The quality of groundwater often compares favourably with that from surface sources, and the groundwater from deep aquifers, in particular, can be remarkably pure. However, this does not mean that the quality can be relied upon. In fact, groundwater can undergo cyclic changes in quality (Pettyjohn, 1982), and natural variations in groundwater quality can occur with depth and rock type.

Thus, the purity of groundwater should not be taken for granted, and untreated water should not be put directly into supply. Some form of chlorination or disinfection, at least, is always advisable if the water is to be used for domestic purposes. The most likely source of pollution is from animals, man in particular.

Water for human consumption must be free from organisms and chemical substances in concentrations large enough to affect health adversely. In addition, drinking water should be aesthetically acceptable in that it should not possess unpleasant or objectionable taste, odour, colour or turbidity. For example, the maximum concentration of chloride in drinking water is 250 mg l^{-1}, primarily for reasons of taste. Again for reasons of taste, and also to avoid staining, the recommended maximum concentration of iron in drinking water is 0.3 mg l^{-1}. Oxidation of manganese in groundwater can produce black stains, and therefore the maximum permitted concentration of manganese in domestic water is 0.05 mg l^{-1}. The pH of drinking water should be close to 7, but treatment can cope with a range of 5 to 9. Beyond this range, treatment to adjust the pH to 7 becomes less economic. The bacterial quality of drinking water requires that tests reveal no more than one total coliform organism per 100 ml as the arithmetic mean of all water samples examined per month. In addition, drinking water should contain less than one virus unit (one plaque-forming unit in a cell culture) per 400 to 4000 l. Viruses can be eliminated by effective chlorination. Water should be free from suspended solids in which viruses can be harboured and thereby be protected against disinfectation. European and international standards of drinking water, laid down by the World Health Organisation, are given in Table 11.1.

The quality of water required in different industrial processes varies appreciably; indeed, it can differ within the same industry. Nonetheless, salinity, hardness and silica content are three parameters that are usually important in terms of industrial water. Water used in the textile industries should contain a low amount of iron, manganese and other heavy metals likely to cause staining. Hardness, total dissolved solids, colour and turbidity must also be low. The quality of water required by the chemical industry varies widely depending on the process involved. Similarly, water required in the pulp and paper industry is governed by the type of products manufactured.

The suitability of groundwater for irrigation depends on the effects that the salt concentration contained therein has on plants and soil. Salts may harm plant growth in that they reduce the uptake of water either by modifying osmotic processes or by metabolic reactions such as those caused by toxic constituents. The effects that salts have on some soils, notably the changes brought about in soil fabric, which in turn affect permeability and aeration, also influence plant growth. In other words, cations can cause deflocculation of clay minerals in a soil, which can damage its crumb structure and reduce its infiltration capacity. In hot, dry climates plants abstract more moisture from the soil and so tend to concentrate dissolved solids in the soil moisture quickly. Clayey soils

Table 11.1 Standards for drinking water (after Brown et al., 1972).

	European standards	International standards
Biology[*]		
Coliform bacteria	Nil	Nil
Escherichia coli	Nil	Nil
Streptococcus faecalis	Nil	Nil
Clostridium perfringens	Nil	Nil
Virus	Less than 1 plaque-forming unit per litre per examination in 10 l of water	
Microscopic organisms	Nil	
Radioactivity		
Overall α radioactivity	$< 3 pCi\,l^{-1}$	$< 3\,pCi\,l^{-1}$
Overall β radioactivity	$< 30\,pCi\,l^{-1}$	
Chemical elements	$(mg\,l^{-1})$	$(mg\,l^{-1})$
Pb	< 0.1	0.1
As	< 0.05	< 0.005
Se	< 0.01	< 0.01
Hexavalent Cr	< 0.05	< 0.05
Cd	< 0.01	0.01
Cyanides (in CN)	< 0.05	< 0.05
Ba	< 1.00	< 1.00
Cyclic aromatic hydrocarbon	< 0.20	
Total Hg	< 0.01	< 0.01
Phenol compounds (in phenol)	< 0.001	< 0.001–0.002
NO_3 recommended	< 50	
acceptable	50–100	
not recommended	> 100	
Cu	< 0.05	0.05–1.5
Total Fe	< 0.1	0.10–1.0
Mn	< 0.05	0.10–0.5
Zn	< 5	5.00–15.0
Mg if SO_4 > 250 mg l^{-1}	< 30	< 30
if SO_4 < 250 mg l^{-1}	< 125	< 125
SO_4	< 250	250–400
H_2S	0.05	
Cl recommended	< 200	
acceptable	< 600	
NH_4	< 0.05	
Total hardness	2–10 meq l^{-1}	2–100 meq l^{-1}
Ca	75–200 mg l^{-1}	75–200 mg l^{-1}
F	In the case of fluorine the limits depend upon air temperature:	

Mean annual maximum day time temperature (°C)	Lower limit (mg l^{-1})	Optimum (mg l^{-1})	Upper limit (mg l^{-1})	Unsuitable (mg l^{-1})
10–12	0.9	1.2	1.7	2.4
12.1–14.6	0.8	1.1	1.5	2.2
14.7–17.6	0.8	1.3	1.3	2.0
17.7–21.4	0.7	0.9	1.2	1.8
21.5–26.2	0.7	0.8	1.0	1.6
26.3–32.6	0.6	0.7	0.8	1.4

[*] No 100 ml sample to contain *E. coli* or more than 10 coliform.

may cause problems when poor quality water is used for irrigation, since they are poorly drained, which means that the capacity for removing excess salts by leaching is reduced.

11.3 Rock type, pollution potential and attenuation of a pollutant

The attenuation of a pollutant as it enters and moves through the ground occurs as a result of four major processes (Fried, 1975):

- *Biological processes*   Soil has an enormous purifying power as a result of the communities of bacteria and fungi that live in the soil. These organisms are capable of attacking pathogenic bacteria and can also react with certain other harmful substances.
- *Physical processes*   As water passes through a relatively fine-grained porous medium such as soil and rock, suspended impurities are removed by filtration.
- *Chemical processes*   Some substances react with minerals in the soil/rock, and some are oxidized and precipitated from solution. Adsorption may also occur in argillaceous or organic material.
- *Dilution and dispersion*   The traditional method of disposing of effluent has generally been to dilute it with a large quantity of relatively pure water so that the concentration of the pollutants eventually becomes negligible at some distance from the source, partly as a result of the above processes. The wisdom of this practice has been questioned in recent years, but it still retains support.

From a consideration of the above processes it is apparent that the self-cleansing capacity of an aquifer system will depend upon three principal factors, namely, the physical and chemical form of the pollutant, the nature of the soil/rock comprising the aquifer, and the way in which the pollutant enters the ground.

The form of the pollutant is clearly an important factor with regard to its susceptibility to the various purifying processes. For instance, pollutants that are soluble, such as fertilizers and some industrial wastes, are not affected by filtration. Metal solutions may not be susceptible to biological action. Solids, on the other hand, are amenable to filtration provided that the transmission medium is not coarse-grained, fractured or cavernous. Karst or cavernous limestone areas pose particular problems in this respect. Non-soluble liquids such as hydrocarbons are generally transmitted through a porous medium, although some fraction may be retained in the host material. Usually, however, the most dangerous forms of groundwater pollution are those that are miscible with the water of the aquifer.

In general, the concentration of a pollutant decreases as the distance it has travelled through the ground increases. Thus, the greatest pollution potential exists for wells tapping shallow aquifers that intersect or lie near ground level. An aquifer that is exposed or overlain by a relatively thin formation in the recharge area is also at risk, particularly when the overlying material consists of sand or gravel. Conversely, deeply buried aquifers overlain by relatively impermeable shale or clay beds can generally be considered to have a low pollution potential and are less prone to severe contamination. One approach to groundwater quality management is to mark areas with a high pollution potential on a map and to pay particular attention to activities within these vulnerable areas.

When assessing pollution potential, an additional consideration is the way in which the pollutant enters the ground. If it is evenly distributed over a large area, its probable effect will be less than that of the same amount of pollutant concentrated at one point. Concentrated sources are most undesirable because the self-cleansing ability of the soil/rock in the area concerned is likely to be exceeded. As a result, the 'raw' pollutant may be able to enter an aquifer and travel some considerable distance from the source before being reduced to a negligible concentration.

A much greater hazard exists when the pollutant is introduced into an aquifer beneath the soil. In this case, the powerful purifying processes that take place within the soil are bypassed and attenuation of the pollutant is reduced. This is most critical when the pollutant is added directly to the zone of saturation, because the horizontal component of permeability is usually much greater than the vertical component. For instance, intergranular seepage in the unsaturated zone may have a typical velocity of less than 1 m y^{-1}, whereas lateral flow beneath the water table may be as much as 2 m day^{-1} under favourable conditions. Consequently, a pollutant beneath the water table can travel a much greater distance before significant attenuation occurs. This type of hazard often arises from poorly maintained domestic septic tanks and soak-aways, from the discharge of quarry wastes, farm effluents and sewage to streams, and from the disposal of refuse and commercial wastes. Leaking pipelines and underground storage tanks have resulted in the abandonment of several million domestic and other water wells in the United States.

For non-karst terrains, the minimum recommended safe distance between a domestic well and a source of pollution is shown in Table 11.2. In karst or weathered limestone areas, however, pollutants may be able to travel quickly over large distances. For instance, Hagerty and Pavoni (1973) observed the spread of contaminated groundwater over a distance of 30 km through limestone in approximately three months. They also noted that the degradation, dilution and dispersion of harmful constituents was less effective than in surface waters.

Biological pollution in the form of micro-organisms, viruses and other pathogens is quite common. When considering biological contamination, the first important point that must be appreciated is that not all bacteria are harmful. On the contrary, many are beneficial and perform valuable functions, such as attacking and biodegrading pollutants as they migrate through the soil.

Table 11.2 Recommended safe distance between domestic
water wells and sources of pollution (after Romero, 1970).

Source of pollution	Distance (m)
Septic tank	15
Cess pit	45
Sewage farm	30
Infiltration ditches	30
Percolation zones	30
Pipes with watertight joints	3
Other pipes	15
Dry wells	15

The bacteria that normally inhabit the soil thrive at temperatures of around 20 °C. Bacteria that are of animal origin prefer temperatures of around 37 °C and so generally die quite quickly outside the host body. Consequently, it is sometimes erroneously assumed that pathogenic bacteria cannot survive long underground. However, Brown *et al.* (1972) pointed out that under these conditions some bacteria may have a life span of up to four years. It is generally assumed that bacteria move at a maximum rate of about two-thirds the water velocity. Since most groundwaters move only a few metres per year, the distances travelled are usually quite small and, in general, it is unusual for bacteria to spread more than 33 m from the source of the pollution. Of course, in openwork gravel, cavernous limestone or fissured rock, the bacteria may spread over distances of many kilometres (Romero, 1970).

The coliform group of bacteria, and *Escherichia coli* (E. coli) in particular, are one of the most frequently used indicators of bacterial contamination but are themselves harmless or non-pathogenic. The reason *E. coli* has gained this distinction is because it is easy to detect in the laboratory while being present in large numbers in the intestines and faeces of animals. Since the whole coliform group is foreign to water, a positive *E. coli* test indicates the possibility of bacterial contamination. If *E. coli* is present, then it is possible that the less numerous or harmful pathogenic bacteria, which are much more difficult to detect, are also present. On the other hand, if there is no *E. coli* in a sample of water, then the chances of faecal contamination and of pathogens being present are generally regarded as negligible. However, Morris and Waite (1981) indicated that some viruses are more resistant to disinfection than *E. coli*, so under some circumstances this generalization may not be valid.

Viruses in groundwater are more critical than bacteria in that they tend to survive longer and in addition may cause an infection when ingested. By contrast, ingestion of thousands of pathogenic bacteria may be required before clinical symptoms are developed.

Many outbreaks of water-borne disease have been attributable to viruses originating from human faeces (Gerba *et al.*, 1975). Viruses are parasites that require a host organism before they can reproduce, but they are capable of retaining all their characteristics for fifty days or more in other environments. Because of their relatively small size, perhaps 20×10^{-6} mm, viruses are not greatly affected by filtration but are prone to adsorption, particularly when the pH is around 7 (Jackson, 1980). Brown *et al.* (1972) suggested that viruses are capable of spreading over distances exceeding 250 m, although 20 to 30 m may be a more typical figure.

11.4 Faulty well design and construction

Perhaps the greatest risk of groundwater contamination arises when pollutants can be transferred from the ground surface to an aquifer. Therefore, it is not surprising that one of the most common causes of groundwater contamination is poor well design, construction and maintenance. Indeed, a faulty well can ruin a high quality groundwater resource.

During the construction of a well there is an open hole that affords a direct route from the surface to the aquifer. Apart from the possibility of surface run-off entering the hole during periods of rainfall, various unsanitary materials such as non-potable quality water, drilling fluids, chemicals, casings, screens and so on are deliberately placed in the hole. This provides an ideal opportunity for chemical and bacteriological pollution to occur, but lasting damage can be avoided if the well is completed, disinfected and pumped within a short space of time. Under these circumstances, most of the potentially harmful substances are discharged from the well. However, if there is a lengthy delay in completing the well, the possibility of pollution increases. Similarly, if the well is constructed in a cavernous or highly permeable formation, the chances of recovering all the harmful materials introduced during construction are decreased, while the possibility of pollution at nearby wells is increased.

As far as the well structure itself is concerned, Campbell and Lehr (1973) noted the following ways by which pollution can take place:

- Via an opening in the surface cap or seal, through seams or welds in the casing, or between the casing and the base of a surface-mounted pump.
- As a result of reverse flow through the discharge system.
- Through the disturbed zone immediately surrounding the casing.
- Via an improperly constructed and sealed gravel pack.
- As a result of settlement due to the inability of the basal formation to support the weight of the well structure, sand pumping, or seepage resulting from reduced effectiveness of the surface cap or seal.
- As a result of the grout or cementing material forming the seals failing through cracking, shrinking, etc.
- Through breaks or leaks in the discharge pipes leading to scour and failure of the cement–grout seals.

When a drillhole penetrates only one aquifer the principal concern is the transfer of pollution from the ground surface. However, with multiple aquifers there is the additional possibility of inter-aquifer flow. In such cases, each aquifer (or potential source of pollution) must be isolated using cement–grout seals in the intervening strata. Failure to do this, particularly when the well incorporates a gravel pack, provides a conduit through which water can be transferred from the ground surface and from one aquifer to another. This could be potentially disastrous where a shallow contaminated aquifer overlies a deeper unpolluted aquifer. In such a situation, pollution could also occur as a result of leakage under the bottom of the surface casing, or through the casing itself if it has seams or welded joints, or is severely corroded.

Even after a well has been abandoned, it still provides a means of entry to an aquifer for pollutants and may even be a greater hazard than it was during its operational life. Abandoned wells make suitable receptacles for all kinds of wastes and refuges for vermin. In addition, as the well becomes progressively older, the strength and effectiveness of the casing and sanitary seals deteriorate, thereby increasing the possibility of structural failure and pollution. If a well has been abandoned, any pollution that does happen might go unnoticed until detected at a nearby well.

11.4 Induced infiltration

Induced infiltration occurs where a stream is hydraulically connected to an aquifer and lies within the area of influence of a well. When water is abstracted from a well the water table in the immediate vicinity is lowered and assumes the shape of an inverted cone, which is referred to as the cone of depression. The steepness of the cone is governed by the soil/rock type, being flatter in highly permeable materials like openwork gravels than in less permeable chalk. The size of the cone depends on the rate of pumping, equilibrium being achieved when the rate of abstraction balances the rate of recharge. But if abstraction exceeds recharge, then the cone of depression increases in size. As the cone of depression spreads, water is withdrawn from storage over a progressively increasing area of influence, while water levels are lowered beneath the surface source adjacent to a well. Some of the larger wells in the chalk in England have cones with radii of up to 1.5 km, and this figure may be exceeded by some of those in the Sherwood Sandstone. Eventually, the aquifer is recharged by influent seepage of surface water, so that some proportion of the pumpage from the well is now obtained from a surface source. As a result, the cone of depression becomes distorted, with the hydraulic gradient between the source and the well being steep in comparison with that on the side away from the source. In effect, the stream constitutes a recharge boundary. Assuming relatively uniform conditions, the flow in the distorted cone generally conforms to Darcy's law, so that the flow towards the well is greatest on the side nearest the source, where the gradients are steepest. As pumping continues, the proportion of water entering the cone of depression that is derived from the stream increases progressively.

If influent seepage of surface water is less than the amount required to balance the discharge from a well, the cone of depression spreads up- and downstream until the drawdown and the area of the stream bed intercepted are sufficient to achieve the required rate of infiltration. If the stream bed has a high permeability, the cone of depression may extend over only part of the width of the stream; if it has a low permeability the cone may expand across and beyond the stream. If pumping is continued over a prolonged period, a new condition of equilibrium is established with essentially steady flows. Most of the abstracted water is then derived from the surface source. Since stream bed infiltration occurs as a result of groundwater abstraction and would not occur otherwise, this is termed induced infiltration.

Induced infiltration is significant from the point of view of groundwater pollution in two respects. First, the new condition of equilibrium that is established may involve the reversal of the non-pumping hydraulic gradients, particularly when groundwater levels have been significantly lowered by groundwater abstraction (Figure 11.1). This may result in pollutants travelling in the opposite direction from that normally expected. Second, surface water resources are often less pure than the underlying groundwater, so induced infiltration introduces the danger of pollution. However, induced infiltration does not automatically equate with pollution. On the contrary, induced infiltration is a traditional method of augmenting groundwater supplies in Europe, America and elsewhere. The experience gained from the operation of more than 200 Ranney-collector wells throughout the world has shown that, under normal conditions, the water produced by induced infiltration is generally free from turbidity, pathogenic bacteria, organic matter, tastes and odours. Consequently, the evaluation of situations involving induced infiltration is not as straightforward as might be expected.

Very few materials in nature are completely impermeable. Even relatively impermeable formations such as clay vary significantly in character and frequently contain pockets of permeable material or fissures that allow the passage of water. When a clay deposit is thin, surface water recharge of an aquifer can occur. If the clay is quite thick, surface–groundwater interconnection still cannot be ruled out completely. Consequently, the safest policy is to assume that surface water infiltration or induced infiltration can occur unless it is proved otherwise.

The question of induced infiltration giving rise to pollution depends upon the quality of the surface water source, the nature of the aquifer, the quantity of infiltration involved and the intended use of the abstracted groundwater. Induced infiltration at one extreme can have potentially disastrous consequences or at the other can provide a valuable addition to the overall water resources of an area. What applies at a given location to some extent depends upon whether or not the possibility of induced infiltration was anticipated when the well concerned was designed.

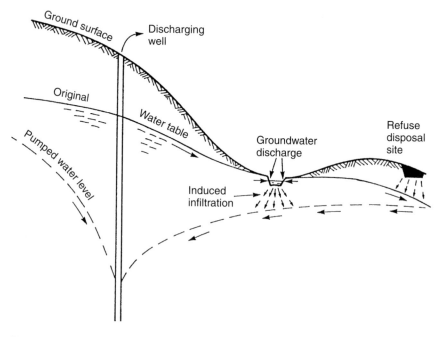

Figure 11.1 Example of induced filtration as a result of pumping. Over much of the area, the original hydraulic gradient has been reversed, so pollutants could travel in the opposite direction to that expected. Additionally, the aquifer has become influent instead of effluent, as it was originally.

As far as estimating the proportion of well discharge that originates from a surface source is concerned, there are basically two approaches that can be adopted. The first uses field data to form a correlation between groundwater level and river leakage, determined from the analysis of gauging station records. In most situations this approach will not be feasible, so recourse must be made to one of the many theoretical techniques that are available. One of the first useful methods was outlined by Theis (1941), who considered a stream as an idealized straight line of infinite extent. By using an image well technique and assuming that the groundwater level under the stream did not change, Theis was able to show that:

- Over half the well discharge derived directly from a stream (or derived indirectly by preventing or diverting groundwater flow to a stream) always originates between points at a distance, x, upstream and downstream of the well, where x is the distance between the well and the stream. The proportion originating from beyond these two points diminishes rapidly with distance.

- If there is no significant increase in recharge or decrease in evapotranspiration (which would significantly reduce the effect of the well on the stream) the proportion of water taken from stream flow varies widely and depends mainly upon the coefficient of transmissivity of the aquifer and the distance of the well from the stream.

Theis (1941) also presented a graph that enabled the proportion of well discharge derived from a surface source to be calculated, based upon values of aquifer characteristics and the duration of pumping (Table 11.3). There is a tendency to underestimate both the area of influence of a well and the amount of water likely to originate from the surface source. Glover and Balmer (1954) also produced a graph for determining the proportion of well discharge derived from a nearby stream (Figure 11.2). Once the quantity and quality of the water derived from the surface source has been established, it is possible to assess whether or not this constitutes a significant threat to groundwater quality.

Table 11.3 Proportion of well discharge derived from a surface source.

Aquifer transmissivity	Distance from well to stream (km)	Duration of pumping (year)	Proportion from stream (%)
$620\ \mathrm{m^2\,d^{-1}}$	1.6	1	30
(i.e. near the lower limit of the		20	90
coefficient for a productive	8	5	2
non-artesian aquifer*)		20	20
$5000\ \mathrm{m^2\,d^{-1}}$	1.6	1	70
(i.e. near the upper limit of the		3	90
coefficient)		1	5
	8	20	75

* The Table is applicable to an unconfined aquifer with a specific yield of 20%.

11.5 Leachate from landfill sites

At the present day, a problem of increasing concern arises from the disposal of domestic and commercial wastes in landfill sites. The decomposition of such waste produces liquids and gases that may present a health hazard by contamination of groundwater supply. They can also threaten the water resource potential of a region and therefore can represent an economic problem of considerable magnitude. Widespread deterioration of groundwater quality means increases in the cost of treatment prior to use. When treatment costs become prohibitive, the volume of unusable water within the system increases and occupies storage space at the expense of potable supplies. Most domestic waste is disposed of in landfills. Typically, this waste comprises 60% degradable solids (e.g. paper, metal, vegetable and putrescible matter), 16% inert solids (e.g. glass

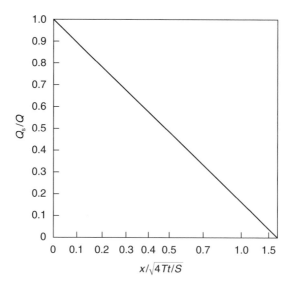

Figure 11.2 Graph for determining the proportion of well discharge derived from a nearby stream. x = perpendicular distance from stream to well; T = coefficient of transmissivity; S = coefficient of storage; t = duration of pumping; Q_s/Q = proportion derived from stream (after Glover and Balmer, 1954).

and plastic) and 24% other varied and unclassified material. Landfills are likely to remain the major method of disposal for domestic refuse in the foreseeable future, and so represent a continuing threat to water quality.

Leachate is formed when rainfall infiltrates a landfill and dissolves the soluble fraction of the waste and those soluble products formed as a result of the chemical and biochemical processes occurring within the decaying waste. If the rate of percolation is very slow, then degradation by natural processes may render these liquids innocuous. In addition, certain substances are removed from suspension or solution by the physico-chemical attraction of the mineral constituents of the substrata beneath a fill. This is largely governed by the ion-exchange capacity of the minerals concerned.

Leachates from landfill sites may be particularly prone to variations in viscosity, which in turn influence their rate of flow. In the zone of aeration of granular materials the storage capacity for leachate may be considerable, but the rate of downward percolation may be slow. Only when the retention capacity is reached does the maximum vertical permeability become effective and leachates then move at their maximum rate. But these rates in granular formations are orders of magnitude lower than movement through fissured materials. Conditions are different below the water table, where leachates may move much more rapidly (see Section 11.3). The hydraulic gradient, which controls the direction and rate of groundwater flow, is generally in a subhorizontal plane, and the horizontal component of permeability is commonly significantly higher than the vertical component.

Generally, the conditions within a landfill are anaerobic, so leachates often contain high concentrations of dissolved organic substances resulting from the decomposition of organic material such as vegetable matter and paper. Recently emplaced wastes may have a chemical oxygen demand (COD) of around 11 600 mg l^{-1} and a biochemical oxygen demand (BOD) in the region of 7250 mg l^{-1}. The concentration of chemically reduced inorganic substances like ammonia, iron and manganese varies according to the hydrology of the site and the chemical and physical conditions within the site. Barber (1982) listed the concentration of ammoniacal nitrogen in a recently emplaced waste as 340 mg l^{-1}, chloride as 2100 mg l^{-1}, sulphate as 460 mg l^{-1}, sodium as 2500 mg l^{-1}, magnesium as 390 mg l^{-1}, iron as 160 mg l^{-1} and calcium as 1150 mg l^{-1}. Previously, Brown et al. (1972) had calculated that 1000 m^3 of waste can yield 1.25 tonnes of potassium and sodium, 0.8 tonnes of calcium and magnesium, 0.7 tonnes of chloride, 0.19 tonnes of sulphate, and 3.2 tonnes of bicarbonate. Furthermore, Barber estimated that a small landfill site with an area of 1 hectare located in southern England could produce up to 8 m^3 of leachate per day, mainly between November and April, from a rainfall of 900 mm y^{-1}, assuming that evaporation is close to the average for the region and run-off is minimal. A site with an area ten times as large would produce a volume of effluent with approximately the same BOD load per year as that received by a small rural sewage treatment works. Accordingly, it can be appreciated that the disposal of domestic wastes in landfill sites can produce large volumes of effluent with a high pollution potential. For this reason, the location and management of these sites must be carefully controlled.

Gray et al. (1974) considered that the major criterion for the assessment of a landfill site as a serious risk was the presence of a toxic or oily liquid waste, although sites on impermeable substrata often merited a lower assessment of risk, depending on local conditions. In this instance, serious risk meant that there was a serious possibility of an aquifer being polluted and not necessarily that there was a danger to life. As such it may involve restricting the type of material that can be tipped at the site concerned. The range of toxic wastes varies from industrial effluents on the one hand to chemical and biological wastes from farms on the other. Unfortunately, the effects of depositing several types of waste together are unknown. Thus site selection for waste disposal must first of all take into account the character of the material that is likely to be tipped. Will this cause groundwater pollution that, because of its toxic nature, will give rise to a health hazard? Although not toxic, will the material increase the concentrations of certain organic and inorganic substances to such an extent that the groundwater becomes unusable? Will the wastes involved be inert and therefore give no risk of pollution?

Selection of a landfill site for a particular waste or a mixture of wastes involves a consideration of economic and social factors as well as the hydrogeological conditions. As far as the latter are concerned, then argillaceous sedimentary, and massive igneous and metamorphic rocks have low permeabilities and therefore afford the most protection to water supply. By contrast, the least protection is offered by rocks intersected by open discontinuities or in which solution features are developed. In this respect, limestones and some sequences of volcanic rocks may be suspect. Granular material may act as a filter. The position of the water table is important as it determines whether wet or dry tipping can take place, as is the thickness of unsaturated material underlying a potential site. Unless waste is inert wet tipping should be avoided. The hydraulic gradient determines the direction and velocity of leachates when they reach the water table and is related to both the dilution that the leachates undergo and to the points at which flow is discharged.

There are two ways in which pollution by leachate can be tackled: first, by concentrating and containing; and second, by diluting and dispersing. Domestic refuse is a heterogeneous collection of almost anything, much of which is capable of being extracted by water to give a liquid rich in organic matter, mineral salts and bacteria. The organic carbon content is especially important since this influences the growth potential of pathogenic organisms. Domestic refuse therefore has an extremely high pollution potential. However, infiltration through granular ground of liquids from a landfill may lead to their decontamination and dilution. Hence sites for disposal of domestic refuse can be chosen where decontamination has the maximum chance of reaching completion and where groundwater sources are located far enough away to enable dilution to be effective. Consequently, waste can be tipped, according to Gray et al. (1974), at dry sites on granular material that has a thickness of at least 15 m. Water supply sources should be located at least 0.8 km away from the landfill site. They should not be located on discontinuous rocks unless overlain by 15 m of superficial deposits. The fill should be completely covered with clay periodically.

As far as potentially toxic waste is concerned it is perhaps best to contain it. Gray et al. (1974) recommended that such sites should be underlain by at least 15 m of impermeable strata and that any source abstracting groundwater for domestic use and confined by such impermeable strata should be at least 2 km away. Furthermore, the topography of the site should be such that run-off can be diverted from the landfill so that it can be disposed of without causing pollution of surface waters.

Since all natural materials possess some degree of permeability, total containment can only be achieved if an artificial impermeable lining is provided over the bottom of the site. However, there is no guarantee that clay, soil-cement, asphalt or plastic linings will remain impermeable; for example, they may be ruptured by settling. Thus migration of leachate from a landfill into the substrata will occur eventually, only the length of time before this happens being in doubt. In some instances, this will be sufficiently long for the problem of pollution to be greatly diminished.

In order to reduce the amount of leachate emanating from a landfill, it is

advisable to construct cut-off drains around the site to prevent the flow of surface water into the landfill area. The leachate that originates from direct precipitation, or other sources, should be collected in a sump and either pumped
to a sewer, transported away by tanker, or treated on site (Frazer and Sims, 1983;
Robinson, 1984). Under no circumstances should untreated leachate be allowed
to enter a surface watercourse, or be allowed to percolate to the water table if this
is likely to pose a significant threat to quality. An account of controlled tipping
was given by Bevan (1971), and Gass (1980) described certain design criteria and
operational methods that can further reduce the danger of pollution.

It has been argued that pollution plumes around landfill sites are often quite
limited in extent, so there is no need to be over-cautious (Anon., 1978). This
view has been disputed by Selby and Skinner (1979), who argued that the
apparent absence of groundwater pollution in the vicinity of landfills is due to
the fact that in the past, landfill has not been allowed in close proximity to
water supply boreholes. In any other situation there are insufficient facilities to
enable groundwater pollution to be detected. Selby and Skinner also recorded
an instance of groundwater pollution at a borehole in the Sherwood Sandstone
in Nottinghamshire, which was located 1.5 km from a waste disposal facility
that was underlain by 5 m of unsaturated aquifer. Therefore, there is evidence
that persistent contaminants can travel substantial distances through aquifers
(see also Hagerty and Pavoni, 1973).

Almost any waste disposal site represents a potential source of groundwater
pollution, but the risk increases when toxic chemicals are involved. There is,
however, an added danger in the form of the illegal dumping of highly toxic
wastes. Although this activity is difficult to stop, it may be detected by a suitable groundwater monitoring scheme (see Section 11.9), and certain precautions
should always be carried out. These include:

- Locating water wells up the hydraulic gradient, if possible, from any waste
  disposal site, or at least ensuring that the well is not directly down-gradient
  of the source of pollution. A flow net may provide a suitable means of study
  initially, followed by tracer surveys if a serious risk is involved.
- Locating a well so that the cone of depression or area of diversion to the
  well does not reach the source of pollution. In fact, the greater the distance
  between a well and the source of pollution, the better.
- Monitoring groundwater quality at points around the periphery of a disposal site and between the site and any water supply wells (see Section
  11.9). Over 20 boreholes may be required to identify the hydrogeology
  and plume migration at a particular site satisfactorily (Cherry, 1983).
  Typical indicators of deteriorating groundwater quality include increases
  in water hardness and in concentration of sulphates and chlorides.
  Concentrations of free CO_2 greater than $20 \, \text{mg} \, \text{l}^{-1}$ create corrosive
  conditions, nitrate in excess of $50 \, \text{mg} \, \text{l}^{-1}$ is considered dangerous, while
  chloride levels of $200 \, \text{mg} \, \text{l}^{-1}$ may cause taste problems in potable water.

By the judicious use of routine water quality surveys it may be possible to detect groundwater deterioration at an early stage. This should ensure that any water wells in the affected area are closed down before the supply becomes contaminated.

- Maintaining a careful check on land use, or changes in land use, in critical areas, such as the recharge zone of the aquifer and in the vicinity of shallow wells. This may detect illegal dumping and other harmful practices.

- Obtaining an estimate of the probable rainfall at the landfill site and thus the volume of leachate (Holmes, 1984). The dilution with water within an aquifer can then be calculated. Flow lines can be used to estimate the proportion of leachate that will arrive at supply wells and hence the quality of the abstracted water. While it is difficult to assess the effects of attenuation, dilution and dispersion on a pollutant, nevertheless such calculations are better than nothing (Naylor *et al.*, 1978).

- Having an emergency plan that can be implemented quickly if polluted water appears in monitoring or supply wells.

11.6 Saline intrusion

Although saline groundwater, originating typically as connate water or from evaporitic deposits, may be encountered, the problem of saline intrusion is specific to coastal aquifers. Near a coast an interface exists between the overlying fresh groundwater and the underlying salt groundwater (Figure 11.3). Excessive lowering of the water table along a coast leads to saline intrusion, the salt water entering the aquifer via submarine outcrops thereby displacing fresh water. However, the fresh water still overlies the saline water and continues to flow from the aquifer to the sea. In the past, the two groundwater bodies have usually been regarded as immiscible, from which it was assumed that a sharp interface exists between them. In fact, there is a transition zone, which ranges from 0.5 m or so to over 100 m in width, although the latter figure may be atypically high.

The shape of the interface is governed by the hydrodynamic relationship between the flowing fresh and saline groundwater. However, if it is assumed that hydrostatic equilibrium exists between the immiscible fresh and salt water, then the depth of the interface can be approximated by the Ghyben–Herzberg formula:

$$Z = \frac{\rho_w}{\rho_{sw} - \rho_w} \times h_w \qquad (11.2)$$

where ρ_{sw} is the density of sea water, ρ_w is the density of fresh water and h_w is the head of fresh water above sea level at the point on the interface (Figure 11.3a). If ρ_{sw} is taken as 1025 kg m^3 and ρ_w as 1000 kg m^3, then $Z = 40/h_w$.

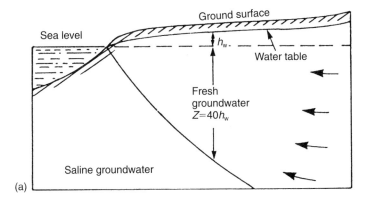

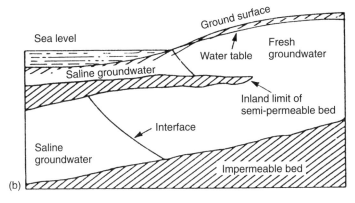

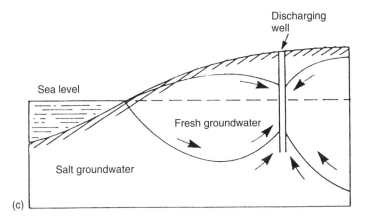

Figure 11.3 Diagrams illustrating, in a simplified manner, the Ghyben–Herzberg hydrostatic relationship in (a) a homogeneous coastal aquifer; (b) a layered coastal aquifer; and (c) a pumped coastal aquifer.

Thus at any point in an unconfined aquifer there is approximately 40 times as much fresh water below mean sea level as there is above it. However, the above expression implies no flow, but groundwater is invariably moving in coastal areas. Where flow is moving upwards, near the coast, the relationship gives too small a depth to salt water but further inland, where the flow lines are nearly horizontal, the error is negligible. This relationship can also be applied to confined aquifers if the height of the water table is replaced by the elevation of the piezometric surface above mean sea level. If the aquifer overlies an impermeable stratum, this formation will intercept the interface and prevent any further saline intrusion (Figure 11.3b).

The Ghyben–Herzberg relationship applies quite accurately to two-dimensional flow at right angles to the shoreline. When flow is three-dimensional, such as when a well is discharging near the coast, most formulae for the position of the interface and the radius of influence of the well are inaccurate. However, the problem is amenable to mathematical modelling (Kishi *et al.*, 1982; Volker and Rushton, 1982).

The problem of saline intrusion usually starts with the abstraction of groundwater from a coastal aquifer, which leads to the disruption of the Ghyben–Herzberg equilibrium condition. Generally, saline water is drawn up towards the well, and this is sometimes termed 'upconing' (Figure 11.3c). This is a dangerous condition that can occur even if the aquifer is not over-pumped and a significant proportion of the fresh water flow still reaches the sea. A well may be ruined by an increase in salt content even before the actual 'cone' reaches the bottom of the well. This is due to 'leaching' of the interface by fresh water. Again this situation is best studied using modelling techniques (Nutbrown, 1976), although Linsley *et al.* (1982) offered a rule of thumb that a drawdown at a well of 1 m will result in a salt water rise of approximately 40 m. This is a particular problem on small oceanic islands, where the fresh water lens is shallow and everywhere rests on salt water. In such cases, shallow wells or horizontal infiltration galleries may be adopted.

The encroachment of salt water may extend for several kilometres inland, leading to the abandonment of wells. The first sign of saline intrusion is likely to be a progressively upward trend in the chloride concentration of water obtained from the affected wells. Chloride levels may increase from a normal value of around 25 mg l^{-1} to something approaching 19 000 mg l^{-1}, which is the concentration in sea water. In the Gravesend area of Kent, chloride ion concentrations of up to 5000 mg l^{-1} have been recorded, which is consistent with sea water encroachment. The recommended limit for chloride concentration in drinking water in Europe is 200 mg l^{-1}, after which the water has a salty taste. Encroaching sea water, however, may be difficult to recognize as a result of chemical changes, so Revelle (1941) recommended the use of the chloride–bicarbonate ratio as an indicator. An additional complication is that there are likely to be frequent fluctuations in chloride content as a result of tides, varying rates of fresh water flow through the aquifer, meteorological phenomena and so on. This also means that the saline–fresh water interface moves seawards or landwards according to the prevailing conditions.

Once saline intrusion develops in a coastal aquifer it is not easy to control. The slow rates of groundwater flow, the density differences between fresh and salt waters and the flushing required usually mean that contamination, once established, may take years to remove under natural conditions. Over-pumping is not the only cause of salt water encroachment; continuous pumping or inappropriate location and design of wells may also be contributory factors. In other words, the saline–fresh water interface is the result of a hydrodynamic balance, hence if the natural flow of fresh water to the sea is interrupted or significantly reduced by abstraction, then saline intrusion is almost certain to occur. Once salt water encroachment is detected, pumping should be stopped, whereupon the denser saline water returns to lower levels of the aquifer.

Although it is difficult to control saline intrusion and to effect its reversal, the encroachment of salt water can be checked by maintaining a fresh water hydraulic gradient towards the sea (according to Darcy's law this means that there must also be fresh water flow towards the sea). This gradient can be maintained naturally, or by some artificial means such as:

- Artificial recharge, which involves either spreading water on the ground surface or injecting water into the aquifer via wells so as to form a groundwater mound between the coast and the area where abstraction is taking place (Brown and Signor, 1973). However, the technique requires an additional supply of clean water. There are also problems associated with the operation of recharge wells (Wood and Bassett, 1975).
- An extraction barrier that abstracts encroaching salt water before it reaches the protected inland wellfield. An extraction barrier consists of a line of wells parallel to the coast. The abstracted water will be brackish and is generally pumped back into the sea. There will probably be a progressive increase in salinity at the extraction wells. These wells must be pumped continuously if an effective barrier is to be created and, apart from being expensive, this can cause logistical problems.

Neither of these two methods of artificially controlling salt water intrusion offers a cheap or foolproof solution. Consequently, when there is a possibility of saline intrusion the best policy is probably to locate wells as far from the coast as possible, select the design discharge with care and use an intermittent seasonal pumping regime (Ineson, 1970). Ideally, at the first sign of progressively increasing salinity, pumping should be stopped. Ineson described a system of selective pumping from the Chalk aquifer in Sussex that represented an attempt to control groundwater deterioration due to saline intrusion. In this area of the south coast of England, groundwater development after 1945 had been limited due to increasing salinity. Ineson recorded that the chloride ion concentration at one point reached 1000 mg l^{-1}, compared with a normal value not exceeding $40 \mathrm{~mg~l}^{-1}$. The scheme adopted involved pumping from the wells near the coast during the winter months, while output was reduced from the inland stations. This led to the interception of fresh groundwater at the coastal boundary that would otherwise flow into the sea. As a result, the storage of groundwater in the landward part of the aquifer increased. During the summer months the pumping scheme was reversed. This method of operation not only controlled saline intrusion but also achieved an overall increase in abstraction.

11.7 Nitrate pollution

There are at least two ways in which nitrate pollution of groundwater is known or suspected to be a threat to health. First, the build-up of stable nitrate compounds in the bloodstream reduces its oxygen-carrying capacity. Infants under one year old are most at risk and excessive amounts of nitrate can cause methaemoglobinaemia, commonly called 'blue-baby syndrome'. Consequently, if the limit of $50 \mathrm{~mg~l}^{-1}$ of NO_3^- recommended by the World Health Organisation (WHO) for European countries is exceeded frequently, or if the concentration lies within the minimum acceptable range of 50–$100 \mathrm{~mg~l}^{-1}$, bottled low-nitrate water should be provided for infants. Second, there is a possibility that a combination of nitrates and amines through the action of bacteria in the digestive tract results in the formation of nitrosamines, which are potentially carcinogenic (Lijinsky and Epstein, 1970). For example, Jensen (1982) reported that the water supply of Aarlborg in Denmark had a relatively high nitrate content of approximately $30 \mathrm{~mg~l}^{-1}$, and there was a slightly greater frequency of stomach cancer in Aarlborg than in other towns during the period 1943 to 1972. This appeared to be due to the nitrate content of the drinking water.

Nitrate pollution is basically the result of intensive cultivation. The major source of nitrate is the large quantity of synthetic nitrogenous fertilizer that has been used since around 1959, although over-manuring with natural organic fertilizer can have the same result. Foster and Crease (1974) estimated that in the 11 years from 1956 to 1967 the application of nitrogen fertilizer to all cropped land in some areas of east Yorkshire increased by about a factor of four (from around 20 to $80 \mathrm{~kgN~ha}^{-1} \mathrm{~y}^{-1}$). Regardless of the form in which the nitrogen fertilizer is applied, within a few weeks it will have been transformed to NO_3^-. This ion is neither adsorbed nor precipitated in the soil and is therefore easily leached by heavy rainfall and infiltrating water. However, the nitrate does not have an immediate affect on groundwater quality, possibly because most of the leachate that percolates through the unsaturated zone as intergranular seepage has a typical velocity of about $1 \mathrm{~m~y}^{-1}$ (Smith et al., 1970). Thus there may be a considerable delay between the application of the fertilizer and the subsequent increase in the concentration of nitrate in the groundwater. So although the use of nitrogenous fertilizer increased sharply in east

Yorkshire after 1959, a corresponding increase in groundwater nitrate was not apparent until after 1970, when the level rose from around $14 \, \text{mg} \, l^{-1} \, \text{NO}_3^-$ (about $3 \, \text{mg} \, l^{-1} \, \text{NO}_3^- - \text{N}$) to between 26 and $52 \, \text{mg} \, l^{-1} \, \text{NO}_3^-$ (6 to $11.5 \, \text{mg} \, l^{-1} \, \text{NO}_3^- - \text{N}$). This just exceeded the EEC's and WHO's recommended limit of $50 \, \text{mg} \, l^{-1} \, \text{NO}_3^-$ ($11.3 \, \text{mg} \, l^{-1} \, \text{NO}_3^- - \text{N}$).

The effect of the time lag, which is frequently of the order of 10 years or more, is to make it very difficult to correlate fertilizer application with nitrate concentration in groundwater. In this respect one of the greatest concerns is that if nitrate levels are unacceptably high now, they may be even worse in the future because the quantity of nitrogenous fertilizer being used has continued to increase steadily. Foster and Young (1980) studied various sites in England and found that in the unsaturated zone the nitrate concentration of the interstitial water was closely related to the history of agricultural practice on the overlying land. Sites subjected to long-term, essentially continuous, arable farming were found to yield the highest concentrations, while lower figures were generally associated with natural grassland. In many of the worst locations, nitrate concentrations exceeded $100 \, \text{mg} \, l^{-1} \, \text{NO}_3^-$ ($22.6 \, \text{mg} \, l^{-1} \, \text{NO}_3^- - \text{N}$), while individual wells with values in excess of $440 \, \text{mg} \, l^{-1} \, \text{NO}_3^-$ ($100 \, \text{mg} \, l^{-1} \, \text{NO}_3^- - \text{N}$) were encountered.

In an update of the work by Foster and Crease (1974), Lawrence et al. (1983) reported on nitrate pollution of groundwater in the Chalk of east Yorkshire a decade after the original study. It appeared that there was a rising trend of nitrate concentration in most of the public groundwater supplies obtained by abstraction from the Chalk. There was also a marked seasonal variation in groundwater quality, which showed a strong correlation with groundwater level. This was tentatively attributed to either rapid recharge directly from agricultural soils or elusion of pore water solutes from within the zone of groundwater level fluctuation. Since nitrate concentrations in the pore water above the water table beneath arable land were found to be two to four times higher than in groundwater supplies, the long-term trend must be towards substantially higher levels of nitrate in groundwater sources.

Measures that can be taken to alleviate nitrate pollution include better management of land use, mixing of water from various sources, or the treatment of high-nitrate water before it is put into supply (Cook, 1991). In general, the ion-exchange process has been recommended as the preferred means of treating groundwaters, although this may not be considered cost-effective at all sources.

11.8 The problem of acid mine drainage

The term 'acid mine drainage' is used to describe drainage resulting from natural oxidation of sulphide minerals that occur in mine rock or waste that is exposed to air and water. This is a consequence of the oxidation of sulphur in the mineral to a higher oxidation state and, if aqueous iron is present and unstable, the precipitation of ferric iron with hydroxide occurs. It can be associated with underground workings, with open-pit workings, with spoil heaps, with tailings ponds or with mineral stockpiles (Brodie *et al.,* 1989).

Acid mine drainage is responsible for problems of water pollution in major coal and metal mining areas around the world. However, it will not occur if the sulphide minerals are non-reactive or if the rock contains sufficient alkaline material to neutralize the acidity. In the latter instance, the pH of the water may be near neutral but it may carry elevated salt loads, especially of calcium sulphate. The character and rate of release of acid mine drainage is influenced by various chemical and biological reactions at the source of acid generation. If acid mine drainage is not controlled, it can pose a serious threat to the environment since acid generation can lead to elevated levels of heavy metals and sulphate in the water, which obviously has a detrimental effect on its quality (Table 11.4). This can have a notable impact on the aquatic environment, as well as vegetation (Figure 11.4). The development of acid mine drainage is time-dependent and at some mines may evolve over a period of years.

Generally, acid drainage from underground mines occurs as point discharges. A major source of acid drainage may result from the closure of a mine. When a mine is abandoned and dewatering by pumping ceases, the water level rebounds. However, the workings often act as drainage systems, so the water does not rise to its former level. Consequently, a residual dewatered zone remains that is subject to continuing oxidation. Groundwater may drain to the surface from old drainage adits, river bank mine mouths, faults, springs and shafts that intercept rock in which water is under artesian pressure. Nonetheless, it may take a number of years before this happens. The mine water quality is determined by the hydrogeological system and the geochemistry of the upper mine levels. Hence, in terms of a working mine it is important that groundwater levels are monitored to estimate rebound potential. In addition, records should be kept of the hydrochemistry of the water throughout the work-

Table 11.4 Examples of acid mine drainage groundwater.

pH	EC	NO$_3$	Ca	Mg	Na	K	SO$_4$	Cl	TDS
6.30	444	11.8	401	289	418	18.5	2937	26.7	4114
5.9	421	4.9	539	373	202	23.8	3275	73.5	4404
3.0	3130	29.4	493	3376	228	12.7	47 720	86.9	51 946
2.9	2370	64.8	627	621	102	6.9	14 580	33.4	16 039
2.8	1680	19.6	509	561	70	0.7	27 070	53.5	28 284
3.2	857	7.4	614	309	103	12.5	2857	173.8	4078
1.9	471	0.1	174	84	247	7	3250	310	4844
2.4	430	0.1	114	48	326	9	1610	431	2968
2.9	377	0.1		61	278	6	1256	324	2490
2.3	404	0.1	113	49	267	4	2124	353	3364

Except for pH and EC (mS m^{-1}), all units are expressed in mg l^{-1}.

Figure 11.4 Vegetation killed by acid mine drainage emanating from a shallow abandoned mine in the Witbank coalfield, South Africa.

ings, drives and adits so that the potential for acid generation on closure can be assessed.

The large areas of fractured rock exposed in opencast mines or open pits can give rise to large volumes of acid mine drainage. Even when abandoned, slope deterioration and failure can lead to fresh rock being exposed, allowing the process of acid generation to continue. Where the workings extend beneath the surrounding topography, the pit drainage system leads to the water table being lowered. Increased oxidation can occur in the dewatered zone.

Spoil heaps represent waste generated by the mining operation. As such they range from waste produced by subsurface mining on the one hand to waste produced by any associated smelting or beneficiation on the other. Consequently, the sulphide content of the waste can vary significantly. Acid generation tends to occur in the surface layers of spoil heaps, where air and water have access to sulphide minerals (Bullock and Bell, 1995).

Tailings deposits that have a high content of sulphide represent another potential source of acid generation. However, the low permeability of many tailings deposits, together with the fact that they are commonly flooded, means that the rate of acid generation and release is limited. Consequently, the generation of acid mine drainage can continue to take place long after a tailings deposit has been abandoned.

Mineral stockpiles may represent a concentrated source of acid mine drainage.

Major acid flushes commonly occur during periods of heavy rainfall after long periods of dry weather. Heap-leach operations at metalliferous mines include, for example, cyanide leach for gold recovery and acid leach for base metal recovery. Spent leach heaps can represent sources of acid mine drainage, especially those associated with low pH leachates.

Certain conditions, including the right combination of mineralogy, water and oxygen, are necessary for the development of acid mine drainage. Such conditions do not always exist. Consequently, acid mine drainage is not found at all mines with sulphide-bearing minerals. The ability of a particular mine rock or waste to generate net acidity depends on the relative content of acid-generating minerals and acid-consuming or neutralizing minerals. Acid waters produced by sulphide oxidation of mine rock or waste can be neutralized by contact with acid-consuming minerals. As a result, water draining from the parent material may have a neutral pH and negligible acidity despite ongoing sulphate oxidation. If the acid-consuming minerals are dissolved, washed out or surrounded by other minerals, then acid generation continues. Where neutralizing carbonate minerals are present, metal hydroxide sludges, such as iron hydroxides and oxyhydroxides, are formed. Sulphate concentration is generally not affected by neutralization unless mineral saturation with respect to gypsum is attained. Hence sulphate may sometimes be used as an overall indicator of the extent of acid generation after neutralization by acid-consuming minerals.

Oxidation of sulphide minerals may give rise to the formation of secondary minerals after some degree of pH neutralization if the pH is maintained near neutral during oxidation. However, other minerals may form instead of or in addition to those listed in Table 11.5. This will depend upon the extent of oxidation, water chemistry and the presence of other minerals such as aluminosilicates. The secondary minerals that are developed may surround the sulphide minerals and in this way reduce the reaction rate.

The primary chemical factors that determine the rate of acid generation include pH, temperature, oxygen content of the gas phase if saturation is less than 100%, concentration of oxygen in the water phase, degree of saturation with water, chemical activity of Fe^{3+}, surface area of exposed metal sulphide, and the chemical activation energy required to initiate acid generation. In addition, the biotic micro-organism *Thiobacillus ferrooxidans* may accelerate reaction by its enhancement of the rate of ferrous iron oxidation. It may also accelerate reaction through its enhancement of the rate of reduced sulphur oxidation. *T. ferrooxidans* is most active in waters with a pH of around 3.2. If conditions are not favourable, the bacterial influence on acid generation will be minimal.

As remarked, acid mine drainage occurs as a result of the oxidation of sulphide minerals, notably pyrite, contained in either mine rock or waste when this is exposed to air and water. In the case of pyrite, the initial reaction for direct oxidation, either abiotically or by bacterial action is:

Table 11.5 Summary of common sulphide minerals and their oxidation products (after Brodie et al., 1989).

Mineral	Composition	Aqueous end-products of complete oxidation*	Possible secondary minerals formed at neutral pH after complete oxidation and neutralization°
Pyrite	FeS_3	Fe_3, SO_4^{2-}, H	Ferric hydroxides and sulphates; gypsum
Marcasite	FeS_3	Fe_3, SO_4^{2-}, H	Ferric hydroxides and sulphates; gypsum
Pyrrhotite	Fe_3S	Fe_3, SO_4^{2-}, H	Ferric hydroxides and sulphates; gypsum
Smythite, Greigite	Fe_3S_4	Fe_3, SO_4^{2-}, H	Ferric hydroxides and sulphates; gypsum
Mackinawite	FeS	Fe_3, SO_4^{2-}, H	Ferric hydroxides and sulphates; gypsum
Chalcopyrite	$CuFeS_3$	Cu_2, Fe_3, SO_4^{2-}, H	Ferric hydroxides and sulphates; copper hydroxides and carbonates; gypsum
Chalcocite	Cu_2S	Cu^{2+}, SO_4^{2-}, H	Copper hydroxides and carbonates; gypsum
Bornite	Cu_3FeS_4	$Cu^{2+}, Fe_3, SO_4^{2-}, H$	Ferric hydroxides and sulphates; copper hydroxides and carbonates; gypsum
Arsenopyrite	$FeAsS$	$Fe_3, AsO_4^{2-}, SO_4^{2-}, H$	Ferric hydroxides and sulphates; ferric and calcium arsenates; gypsum
Realgar	AsS	AsO_4^{2-}, SO_4^{2-}, H	Ferric and calcium arsenates; gypsum
Orpiment	As_2S_3	AsO_4^{2-}, SO_4^{2-}, H	Ferric and calcium arsenates; gypsum
Tennantite and Tennenite	$Cu_{12}(Sb, As)_4S_{10}$	$Cu^{2+}, SbO^{3-}, AsO_4^{2-}, SO_4^{2-}, H$	Copper hydroxides and carbonates; calcium and ferric arsenates; antimony materials; gypsum
Molybdenite	MoS_2	MoO_4^{2-}, SO_4^{2-}, H	Ferric hydroxides; sulphates; molybdenum oxides; gypsum
Sphalerite	ZnS	Zn^{2+}, SO_4^{2-}, H	Zinc hydroxides and carbonates; gypsum
Galena	PbS	Pb^{2+}, SO_4^{2-}, H	Lead hydroxides, carbonates, and sulphates; gypsum
Cinnabar	HgS	Hg^{2+}, SO_4^{2-}, H	Mercuric hydroxide; gypsum
Cobalite	$CoAsS$	Co^{2+}, AsO_4^{2-}, H	Cobalt hydroxides and carbonates; ferric and calcium arsenates; gypsum
Niccolite	$NiAs$	$Ni_2, AsO_4^{2-}, SO_4^{2-}, H$	Nickel hydroxides and carbonates; ferric, nickel and calcium arsenates; gypsum
Pentlandite	$(Fe, Ni)_9S_1$	$Fe_3Ni_2S_4, SO_4^{2-}, H$	Ferric and nickel hydroxides; gypsum

* Intermediate species such as ferrous iron (Fe^{2+}) and $S_2O_3^{2-}$ may be important

° Depending on overall water chemistry other minerals may form with or instead of the minerals listed above.

$$2FeS_2 + 2H_2O + 7O_2 \rightarrow 2FeSO_4 + 2H_2SO_4 \qquad (11.3)$$

Subsequent biotic and abiotic reactions, which lead to the final oxidation of pyrite by ferric ions (indirect oxidation mechanism), can be represented as follows:

$$4FeSO_4 + O_2 + 2H_2SO_4 \rightarrow 2Fe_2(SO_4)^3 + 2H_2O \qquad (11.4)$$
$$Fe_2(SO_4)_3 + 6H_2O \rightarrow 2Fe(OH)_3 + 3H_2SO_4 \qquad (11.5)$$
$$4Fe^{2+} + O_2 + 4H \rightarrow 4Fe^{3+} + 2H_2O \qquad (11.6)$$
$$FeS_2 + 14Fe^{3+} + 8H_2O \rightarrow 15Fe^{2+} + 2SO_4 + 16H \qquad (11.7)$$

Reaction 1 shows the initiation of pyrite oxidation either abiotically (auto-oxidation) or biotically. *Thiobacillus ferrooxidans* converts the ferrous iron of pyrite to its ferric form. The formation of sulphuric acid in the initial oxidation reaction and concomitant decrease in the pH make conditions more favourable for the biotic oxidation of the pyrite by *T. ferrooxidans*. The biotic oxidation of pyrite is four times faster than the abiotic reaction at a pH of 3.0. The presence of *T. ferrooxidans* may also accelerate the oxidation of sulphides of antimony, arsenic, cadmium, cobalt, copper, gallium, lead, molybdenum, nickel and zinc.

Hence, the development of acid mine drainage is a complex combination of inorganic and sometimes organic processes and reactions. In order to generate severe acid mine drainage (pH < 3), sulphide minerals must create an optimum micro-environment for rapid oxidation and must continue to oxidize long enough to exhaust the neutralization potential of the rock.

Accurate prediction of acid mine drainage is required in order to determine how to bring it under control. The objective of acid mine drainage control is to satisfy environmental requirements using the most cost-effective techniques. The options available for the control of polluted drainage are greater at proposed rather than existing operations, as control measures at working mines are limited by site-specific and waste disposal conditions. For instance, the control of acid mine drainage that develops as a consequence of mine dewatering is helped by the approach taken towards the site water balance. In other words, the water resource management strategy developed during mine planning will enable mine water discharge to be controlled and treated prior to release, or to be reused. The length of time that the control measures are required to be effective is a factor that needs to be determined prior to the design of a system to control acid mine drainage.

Prediction of the potential for acid generation involves the collection of available data and carrying out static tests and kinetic tests. A static test determines the balance between potentially acid-generating and acid-neutralizing minerals in representative samples. One of the frequently used static tests is acid–base accounting. Acid–base accounting allows determination of the proportions of acid-generating and acid-neutralizing minerals present. However, static tests

cannot be used to predict the quality of drainage waters or when acid generation will occur. If potential problems are indicated, the more complex kinetic
tests should be used to obtain a better insight into the rate of acid generation.
Kinetic tests involve weathering of samples under laboratory or on-site conditions in order to confirm the potential to generate net acidity, to determine the
rate of acid formation, sulphide oxidation, neutralization and metal dissolution, and to test control and treatment techniques. The static and kinetic tests
provide data that can be used in various models to predict the effect of acid
generation and control processes beyond the time frame of kinetic tests.

There are three key strategies in acid mine drainage management, namely,
control of the acid generation process, control of acid migration, and collection
and treatment of acid mine drainage (Connelly et al., 1995). Control of acid
mine drainage may require different approaches, depending on the severity of
potential acid generation, the longevity of the source of exposure and the sensitivity of the receiving waters. Mine water treatment systems installed during
operation may be adequate to cope with both operational and long-term post-
closure treatment with little maintenance. On the other hand, in many mineral operations, especially those associated with abandoned workings, the
long-term method of treatment may be different from that used while a mine
is operational. Hence there may have to be two stages in the design of a system for treatment of acid mine drainage, one for during mine operation and
another for after closure.

Obviously, the best solution is to control acid generation, if possible. Source
control of acid mine drainage involves measures to prevent or inhibit oxidation, acid generation or containment leaching. If acid generation is prevented,
then there is no risk of the resultant contaminants entering the environment.
Such control methods involve the removal or isolation of sulphide material, or
the exclusion of water or air. The latter is much more practical and can be
achieved by the placement of a cover over acid-generating material such as
waste, or air-sealing adits in mines.

Migration control is considered when acid generation is occurring and cannot be inhibited. Since water is the transport medium, control relies on the
prevention of water entry to the source of acid mine drainage. Water entry may
be controlled by diversion of surface water flowing towards the source by
drainage ditches; prevention of groundwater flow into the source by interception and isolation of groundwater (this is very difficult to maintain over the
long-term); and prevention of infiltration of precipitation into the source by
the placement of cover materials − but again their long-term integrity is
difficulty to ensure.

Release control is based on measures to collect and treat acid mine drainage.
In some cases, especially at working mines, this is the only practical option
available. Collection requires the collection of both ground and surface
water polluted by acid mine drainage, and involves the installation of
drainage ditches, and collection trenches and wells. Treatment processes have

concentrated on neutralization to raise the pH and precipitate metals. Lime or limestone is commonly used, although offering only a partial solution to the problem. Other alkalis such as sodium hydroxide can be used but are more expensive and do not produce such dense sludges as does lime. The sludges recovered from alkali neutralization followed by sedimentation and consolidation are of relatively low density (2–5% dry solids), but drain fairly well. The drained sludge obtains a density of 20–30% dry solids after a few weeks exposure. The iron content of the solids is usually in the range 10–20%, and there are considerable impurities such as calcium sulphate and carbonate. The oxidation of pyrite can be eliminated by flooding a mine with water. However, it is obviously not possible to flood an active mine. It is also not possible to flood abandoned mine workings that are above the water table and are drained by gravity. More sophisticated processes (active treatment methods) involve osmosis (waste removal through membranes), electrodialysis (selective ion removal through membranes), ion exchange (ion removal using resin), electrolysis (metal recovery with electrodes) and solvent extraction (removal of specific ions with solvents).

Cambridge (1995) pointed out that conventional active treatment of mine water requires the installation of a treatment plant, continuous operation and maintenance. Hence the capital and operational costs of active treatment are high. Alternatively, passive systems try to minimize the input of energy, materials and manpower, and so reduce operational costs. Acid mine water treated with active systems tends to produce a solid residue, which has to be disposed of in tailings lagoons. This sludge contains metal hydroxides. However, according to Cambridge, the long-term disposal of sludge in tailings lagoons is not appropriate. Alternatively, sludges could be placed in hazardous waste landfills, but such sites are limited.

Passive treatment involves engineering a combination of low-maintenance biochemical systems (e.g. anoxic limestone drains, aerobic and anaerobic wetlands, and rock filters). Such treatment does not produce large volumes of sludge, and the metals are precipitated as oxides or sulphides in the substrate materials.

Due to the impact on the environment of acid mine drainage, regular monitoring is required. The major objectives of a monitoring programme developed for acid mine drainage are, first, to detect the onset of acid generation before acid mine drainage develops to the stage where environmental impact occurs. If required, control measures should be put in place as quickly as possible. Second, it is necessary to monitor the effectiveness of the prevention–control–treatment techniques and to detect whether the techniques are unsuccessful as early as possible.

11.8 Waste waters and effluents from coal mines

Different types of waste water and process effluent are produced as a result of coal mining. These may arise due to the extraction process, by the subsequent preparation of coal, from the disposal of colliery spoil or from coal stockpiles. The strata from which the groundwater involved is derived, the mineralogical character of the coal and the colliery spoil, and the washing processes employed all affect the type of effluent produced. Problems also result from past mining operations.

Generally, the major pollutants associated with coal mining are suspended solids, dissolved salts (especially chlorides), acidity and iron compounds (Bell and Kerr, 1993). However, the character of drainage from coal mines varies from area to area and from coal seam to coal seam. Hence, mine drainage waters are liable to vary in both quality and quantity, sometimes unpredictably, as the mine workings develop. Nonetheless, drainage water from, for example, coal mines in Britain can be classified as hard, alkaline, moderately saline, highly saline, alkaline and ferruginous, or acidic and ferruginous (Best and Aikman, 1983; Table 11.6). Colliery discharges have little oxygen demand, the BOD normally being very low. Elevated levels of suspended matter are associated with most coal mining effluents, with occasional high values being recorded. Although not all mine waters are highly mineralized, a high level of mineralization is typical of many coal mining discharges and is reflected in the high values of electrical conductivity. Highly mineralized mine waters usually contain high concentrations of sodium and potassium salts, and mine waters that do not contain sulphate may contain high levels of strontium and barium. Similarly, not all mine waters are ferruginous, and in fact some are of the highest quality and can be used for potable supply. Nonetheless, they are commonly high in iron and sulphates. The low pH of many mine waters is commonly associated with highly ferruginous discharges. In some nearby streams the pH is less than 4.0, the iron concentration is greater than several hundred milligrams per litre and the sulphates exceed 1000 mg l^{-1}.

Table 11.6 Composition of mine drainage waters (from Bell and Kerr, 1993).

Column number	1	2	3	4	5	6
Approximate percentage of waters in each class	55%	25%	10%	7%	7%	2%
Quality	hard alkaline		moderately saline	alkaline and ferruginous	acidic and ferruginous	highly saline
pH value	7.8	6.8	8.2	6.9	2.9	7.5
Alkalinity, mg l^{-1} CaCO$_3$	260	850	240	340	nil	190
Calcium, mg l^{-1}	75	28	90	190	125	2560
Magnesium, mg l^{-1}	90	17	40	130	90	720
Dissolved iron, mg l^{-1}	0.1	0.5	0.1	25	122	0.6
Suspended iron, mg l^{-1}	0.1	2	0.1	21	0.1	0.2
Manganese, mg l^{-1}	0.1	0.1	0.1	6	7	0.9
Chloride, mg l^{-1}	180	200	3400	42	50	30 800
Sulphate, mg l^{-1}	170	210	250	1720	1250	350

The high level of dissolved salts that is often present in mine waters represents the most intractable water pollution problem connected with coal mining, because dissolved salts are not readily susceptible to treatment or removal. In some situations this option is not available. The range of dissolved salts encountered in mine water is variable, with electrical conductivity values up to $335\,000\ \mu S\ cm^{-1}$ and chloride levels of $60\,000\ mg\ l^{-1}$ being recorded (Woodward and Selby, 1981). Some average values for various coal mining effluents associated with the Nottinghamshire coalfield, England, are given in Table 11.7.

The principal groups of salts in mine discharge waters are chlorides and sulphates. The former occur in the waters lying in the confined aquifers between coal seams in most coalfields in Britain, South Wales being a notable exception. These salts are released into the workings by mining operations. In general, the salinity increases with depth below the surface and with distance from the outcrop or incrop. The concentration of ions in mine waters conforms to established ratios that are remarkably consistent throughout British coalfields. The more saline waters contain significant concentrations of barium, strontium, ammonium and manganese ions. Dissolved sulphates occur only in trace concentrations in waters of confined aquifers.

Mine drainage waters at the point of discharge almost invariably contain sulphates, which are either present in the waters lying in the more shallow unconfined aquifers or are generated in the workings by the action of atmospheric oxygen on pyrite. The physical changes such as delamination, bedding plane separation and fissuring of rock masses caused by mining permit air to pene-

Table 11.7 Average quality characteristics of coal mining effluents in the Nottinghamshire coalfield (from Bell and Kerr, 1993).

Type of effluent	BOD (ATU) $(mg\ l^{-1})$	Suspended solids $(mg\ l^{-1})$	Chloride $(mg\ l^{-1})$	Electrical conductivity $(\mu S\ cm^{-1})$	Minimum pH	Other potential contaminants
Mine waters	2.1	57	4900	14 000	3.5	Iron barium nickel aluminium sodium sulphate
Drainage from coal stocking sites	2.6	128	600	2200	2.2	Iron, zinc
Spoil tip drainage	3.1	317	1600	4100	2.7	Iron, zinc
Coal preparation plant discharges	2.1	39	1500	4200	3.2	Oil from flotation chemicals
Slurry lagoon discharges	2.4	493	2000	6100	3.8	

trate a much larger surface area than the immediate boundaries of the working faces and associated roadways. These changes also alter the hydrogeological conditions within a coalfield and allow wider movement of groundwater through the rock masses than existed prior to mining. These two factors mean that groundwater comes into contact with a large surface area of rock that is exposed to atmospheric oxidation and hence the groundwater becomes contaminated. Pyrite may occur in high concentrations, particularly in the upper part of a coal seam or in associated black shales. As noted above, primary oxidation products of pyrite are ferrous and ferric sulphates, and sulphuric acid. Sulphates and sulphuric acid react with clay and carbonate minerals to form secondary products, including manganese and aluminium sulphates. Further reactions with these minerals and incoming waters give rise to tertiary products such as calcium and magnesium sulphates. Generally, the stratal waters are sufficiently alkaline to ensure that only the tertiary products appear in the discharge at the surface (columns 1, 2, 3 and 6 of Table 11.6). Exceptionally, both primary and secondary products may appear in waters from intermediate depths (column 4 of Table 11.6). The primary oxidation products tend to predominate in very shallow workings liable to leaching by meteoric water (column 5 of Table 11.6).

Precipitation of metal hydroxides occurs as acid mine water is neutralized by water into which it flows. Acidic ferruginous mine water may also contain high concentrations of aluminium, which precipitates as hydroxide as the pH rises on entering a receiving body of water, giving a milky appearance to the water. Concentrations of heavy metals may be high in some acid waters and may exert a toxic effect.

The prevention of pollution of groundwater by coal mining effluent is of particular importance. Movement of pollutants through strata is often very slow and is difficult to detect. Hence effective remedial action is often either impractical or prohibitively expensive. Because of this there are few successful recoveries of polluted aquifers. In the coalfields of the east Midlands and south Staffordshire, England, colliery spoil is often tipped on top of the Sherwood Sandstone, which is the second most important aquifer in Britain. Although modern tipping techniques may render spoil impervious, surface water run-off can leach out soluble salts, especially chloride. This may result in the loss of up to 1 tonne of chloride per hectare of exposed spoil heap per annum under average rainfall conditions (spoil heaps may be many hundreds of hectares in extent). The run-off from these spoil heaps may discharge directly into drainage ditches or to land around the periphery of the heap and infiltrate into the aquifer. Figure 11.5 shows the isopleths for chloride ion concentration in the groundwater of the Sherwood Sandstone in the concealed part of the Nottinghamshire coalfield. It also indicates the locations of the collieries and so demonstrates the relationship between elevated chloride ion level in the groundwater and mining activity. Such pollution of an aquifer can be alleviated by lining the beds of influent streams that flow across the aquifer or by

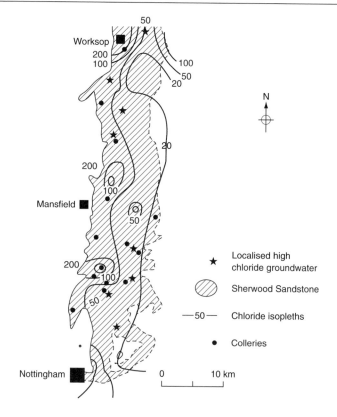

Figure 11.5 Chloride isopleths (in mg l^{-1}) for groundwater in the Sherwood Sandstone of the concealed Nottinghamshire coalfield, England (from Bell and Kerr, 1993).

providing pipelines to convey mine discharge to less sensitive watercourses that do not flow across the aquifer.

Old adits are often unmapped and unknown, and even currently discharging ones are not always immediately evident. An examination of the catchment data is often the only way that such discharges come to light. Discharges from old adits and mine mouths are usually gravity flows.

11.9 Other causes of pollution

The list of potential groundwater pollutants is almost endless. Some of the most common sources of pollution have already been described, but there are others that merit close attention, such as sewage sludge disposal. The sludge arises from the separation and concentration of most of the waste materials found in sewage. Since the sludge contains nitrogen and phosphorus it has a value as a fertilizer. As a result, about 50% of the 1.24 million dry tonnes per year of sludge produced in the United Kingdom is used on agricultural land. While this does not necessarily lead to groundwater pollution, the presence in the sludge of contaminants such as metals, nitrates, persistent organic compounds and pathogens does mean that the practice must be carefully controlled (Anon., 1977; Davis, 1980). In particular cadmium, which can also originate as metal waste from mine workings, may give cause for concern. Too much cadmium can cause kidney damage in humans, and this metal was implicated in *itai-itai* disease in Japan. Food is the usual source of the cadmium found in humans, although small amounts are also present in water. The European standard for drinking water recommends a cadmium concentration of less than 0.01 mg l^{-1}.

The widespread use of chemical and organic pesticides and herbicides is another possible source of groundwater pollution. For instance, Zaki *et al.* (1982) recorded a groundwater pollution incident in Suffolk County, New York, involving the pesticide aldicarb. Previous laboratory and field studies had suggested that the use of this pesticide would not result in groundwater pollution. Not surprisingly, Zaki *et al.* called for more comprehensive testing of pesticides and regular monitoring of groundwater quality in sensitive areas.

In Britain, there is also concern about the number of organic contaminants in both raw and treated waters. For instance, Fielding *et al.* (1981) conducted a survey of treated water (some from groundwater sources) and identified 324 organic compounds. Some of the substances found, if present in higher concentrations, would have had grave medical implications. The source of this contamination was described as sewage effluent containing treated domestic and industrial wastes, surface run-off from roads and agricultural land, and atmospheric fallout and rain.

According to Mackay (1998), volatile organic chemicals (VOCs) are the most frequently detected organic contaminants in water supply wells in the United States. Of the VOCs, by far the most commonly found are chlorinated hydrocarbon compounds. Conversely, petroleum hydrocarbons are rarely present in supply wells. This may be due to their *in situ* biodegradation.

Many of the VOCs are liquids and are usually referred to as non-aqueous phase liquids (NAPLs), which are sparingly soluble in water. Those that are lighter than water, such as the petroleum hydrocarbons, are termed LNAPLs, whereas those that are denser than water, such as the chlorinated solvents, are called DNAPLs. Of the VOCs, the DNAPLs are the least amenable to remediation. If they penetrate the ground in large enough quantities, depending on the hydrogeological conditions, DNAPLs may percolate downwards into the saturated zone. This can occur in granular soils or discontinuous rock masses. Plumes of dissolved VOCs develop from the source of pollution. Although dissolved VOCs migrate more slowly than the average velocity of groundwater, Mackay and Cherry (1989) indicated that there are many examples of chlorinated VOC plumes several kilometres in length in the United States occurring in sand and gravel aquifers. Such plumes contain billions of litres of contaminated water. Because VOCs are sparingly soluble in water, the time taken for complete dissolution, especially of DNAPLs by groundwater flow in granular soils, is estimated to be decades or even centuries.

In some parts of the world such as Canada and Scandinavia, 'acid rain', originating from the combination of industrial gases with atmospheric water, is acidifying not only surface water resources but groundwater as well. The results from several sites in Norway indicate that in affected areas the groundwater has a pH lower than normal (Henriksen and Kirdhusmo, 1982).

The potential of the run-off from roads to cause pollution is often overlooked. This water can contain chemicals from many sources, including those that have been dropped, spilled or deliberately spread on the road. For instance, hydrocarbons from petroleum products and urea and chlorides from de-icing agents are all potential pollutants and have caused groundwater contamination (Bellinger et al., 1982). There is also the possibility of accidents involving vehicles carrying large quantities of chemicals. The run-off from roads can cause bacteriological contamination as well.

Cemeteries and graveyards form another possible health hazard. According to Bouwer (1978), the minimum distance between a potable water well and a cemetery required by law in England is 91.4 m (100 yards). However, it was suggested that a distance of around 2500 m is better because the purifying processes in the soil can sometimes break down. Decomposing bodies produce fluids that can leak to the water table if a non-leakproof coffin is used. Typically, the leachate produced from a single grave is of the order of $0.4 \text{ m}^3 \text{ y}^{-1}$, and this may constitute a threat for about ten years. It is recommended that the water table in cemeteries should be at least 2.5 m deep and that an unsaturated depth of 0.7 m should exist below the bottom of the grave.

While most of the cases of groundwater pollution so far considered have been the result of chemical or inorganic compounds, it is worth remembering that over-pumping an aquifer can also be regarded as a potential hazard. The danger lies in the natural variations of water quality with depth within the aquifer, the reversal of hydraulic gradients so that pollutants travel in the opposite direction to that normally expected, induced infiltration, saline intrusion, and the exhaustion of supplies. All these can have serious consequences, so careful consideration must be given to the selection of the pumping rate and the duration of the abstraction. Over-pumping or 'groundwater mining' during periods of water shortage, on the assumption that groundwater levels will recover at a later date, should be avoided if possible. The groundwater level may recover, but the quality may not. Essentially, this is a question of good aquifer management, part of which includes the routine analysis and monitoring of groundwater quality.

11.10 Groundwater monitoring and artificial recharge

The routine monitoring of groundwater level and water quality is a fundamental part of aquifer management. This provides an early warning system for pollution incidents and phenomena such as over-abstraction, induced infiltration and saline intrusion.

In almost all situations where groundwater monitoring is undertaken, it is important that adequate background samples be obtained before a groundwater abstraction scheme is inaugurated. Without this background data it will be impossible to assess the effects of new development. It should also be remembered that groundwater can undergo cyclic changes in quality, so any apparent changes must be interpreted with caution. Hence routine all-year-round monitoring is essential.

A mass of fluid introduced into an aquifer tends to remain intact rather than mix with, or be diluted by, the groundwater. In many instances of groundwater pollution, the contaminating fluid is discharged into the aquifer as a continuous or nearly continuous flow. Hence the polluted groundwater often takes the form of a plume or plumes. An important part of an investigation of groundwater pollution is to locate and define the extent of the contaminated body of groundwater in order to establish the magnitude of the problem and, if necessary, to design an efficient abatement system.

Pfannkuch (1982) pointed out that the first important step in designing an efficient groundwater monitoring system is the proper understanding of the mechanics and dynamics of contaminant propagation (e.g. soluble or multiphase flow), the nature of the controlling flow mechanism (e.g. vadose or saturated flow), and the aquifer characteristics (e.g. permeability, porosity). He listed the objectives of a monitoring programme as follows:

1   Determination of the extent, nature and degree of contamination (source monitoring).
2   Determination of the propagation mechanism and hydrological parameters so that the appropriate countermeasures can be initiated.
3   Detection and warning of movement into critical areas.
4   Assessment of the effectiveness of the immediate countermeasures undertaken to offset the effects of contamination.
5   Recording of data for long-term evaluation and compliance with standards.
6   Initiation of research monitoring to validate and verify the models and assumptions upon which the immediate countermeasures were based.

Obviously, these objectives may have to be changed to suit the physical, political or other conditions prevailing at the site of a particular pollution incident. Frequently it is not practical to initiate countermeasures to combat groundwater pollution once it has occurred, the eventual attenuation of the pollutant being the result of time, degradation, dilution and dispersion. In the case of a recent accidental spill, however, it may be possible to excavate affected soil and use scavenger wells to intercept and recover some of the pollutant before it has dispersed significantly (Tester and Harker, 1982). Alternatively, in a few situations, artificial recharge may be undertaken in an attempt to dilute the pollutant, or to form a hydraulic pressure barrier that will divert the pollutant away from abstraction wells. Unfortunately, none of these options provides a reliable method of dealing with groundwater pollution and they should be considered only as a last resort.

The design of a water quality monitoring well and its method of construction must be related to the geology of the site. The depth and diameter of the well should be as small as possible so as to reduce the cost, but not so small that the well becomes difficult to use or ineffective. However, an additional requirement where water quality is concerned is that the well structure should not react with the groundwater. Clearly, if the groundwater does react with the well casing or screen, any subsequent analysis of samples taken from the well will be affected. Thus monitoring wells are frequently constructed using plastic casing and screens, partly for economy and partly because plastic is relatively inert. Plastic does react with some types of pollutant, so well materials must be selected to suit the anticipated conditions. The well screen should be provided with a gravel or sand pack to prevent the migration of fine material into the well. The selection of a suitable screen slot width and particle size distribution for the pack is accomplished using the same procedures as for a water supply well, although the design entrance velocity should be lower for a monitoring well than for a water well. Generally, a sample of the aquifer material is obtained, its grain size distribution determined and the appropriate particle size for the pack decided. The pack should extend at least 0.3 m above and below the screened zone.

When deciding the diameter of a monitoring well, some consideration must be given to the method that is to be used to obtain the water sample. If a bailer or some form of sampler is to be inserted down the casing, then obviously the well must be of a sufficiently large diameter to permit this. There are, however, several alternatives. Water samples can be obtained using a specially designed submersible pump that is small enough to fit into a 50 mm diameter well.

After a well has been completed it should be developed, by pumping or bailing, until the water becomes clear. Background samples should then be collected over a lengthy period prior to the commencement of whatever it is that the well has been constructed to monitor.

When designing a monitoring network the problem is to ensure that there are sufficient wells to allow the extent, configuration and concentration of the pollution plume to be determined, without incurring the unnecessary expense of constructing more wells than are actually required. The network of monitoring wells must be designed to suit a particular location and modified, as necessary, as new information is obtained or in response to changing conditions.

Diefendorf and Ausburn (1977) recommended that at least three monitoring wells should be used to observe the effects of a new development such as a landfill site. However, as a result of aquifer heterogeneity, non-uniform flow, variations in quality with depth and so on, it was suggested that three well clusters (rather than individual wells) may be required. Each well cluster should contain two or more wells located at various depths within the aquifer, or within different aquifers in the case of a multiple aquifer system. One cluster should be located close to the source of the pollution, for early warning purposes, with another installation some suitable distance down-gradient to assess the propagation of the plume. A third installation should be located up-gradient of the monitored site to detect changes in background quality attributable to other causes.

Williams (1980) described several instances where over 20 boreholes were needed to identify fully the hydrogeology and plume development originating from a landfill site. In some situations it may be possible to install multiple piezometers and groundwater sampling devices in a single well, so reducing the number of boreholes required (Berk and Yare, 1977). Pfannkuch (1982) quoted an example where 26 holes were drilled to monitor an unanticipated spill, 15 of the holes being equipped as monitoring wells. The procedure for establishing a suitable monitoring network for an accidental spill, according to Pfannkuch, is first to undertake a rapid preliminary survey. While background water samples are being taken for comparison purposes, the direction of movement of the contaminant is established. As a first approximation, it can be assumed that the water table is a subdued replica of the ground surface. To obtain more detailed information, three wells should be sunk in a triangular pattern that encloses the site of the spill. One well must be up-gradient from the source of pollution so that uncontaminated background samples can be obtained. Using three-point interpolation it is possible to draw the groundwater level contours and to construct flow lines, so that a more exact estimate of groundwater movement can be obtained. The flow line(s) that pass beneath the spill site can be identified and at least two monitoring wells should be sunk on this line. The first should be located as close as possible to, but not actually in, the extending plume and the second at some suitable distance downstream. This enables the time of travel of the contaminant between the two wells to be measured, and its propagation velocity assessed. Finally, two wells, or two sets of wells, should be constructed perpendicular to the central flow line so that the lateral spread of the contamination plume can be investigated. Again, these wells should be close to the advancing plume but not actually in it, so that time of travel data can be obtained. This completes the rapid preliminary survey, and the monitoring phase begins with samples being taken at frequent intervals initially. As time progresses and the contaminant begins to disperse and degrade, the monitoring frequency can be reduced. After a number of months or years, only routine sampling need be undertaken to assess the long-term effects of the spill.

Under favourable conditions, the resistivity method can be used to determine the boundaries of a plume of polluted groundwater. Vertical electrical soundings are made in areas of known pollution in an attempt to define the top and bottom of the plume. Borehole logs are used to establish geological control. Next, resistivity profiling is carried out to determine the lateral extent of the polluted groundwater. In this way, a quantitative assessment can be made of groundwater pollution (Kelly, 1976). The method is based on the fact that formation resistivity depends on the conductivity of the pore fluid as well as the properties of the porous medium. Generally the resistivity of a rock is proportional to that of the water with which it is saturated. The groundwater resistivity decreases if its salinity increases due to leachate pollution from a landfill, and hence the resistivity of the rock concerned decreases. Obviously, there must be a contrast in resistivity between the contaminant and groundwater in order to obtain useful results. Contrasts in resistivity may be attributed to mineralized groundwater with a higher than normal specific conductance due to pollution.

However, the resistivity of a saturated porous sandstone or limestone aquifer depends not only on the salinity of the saturated groundwater but also on the aquifer's porosity and the amount of conductive minerals, notably clay, in the rock matrix. The accurate determination of water quality can take place only if the effects of porosity and clay content are insignificant, or at least understood (Anon., 1988). Fortunately, it is often the case, especially in coastal aquifers, that extreme variations in groundwater salinity preclude the necessity of considering porosity and matrix conductivity effects. As an example, Oteri (1983) used the results of an electrical resistivity survey, with a Wenner configuration, to delineate the extent of saline intrusion into the shingle aquifer at Dungeness, Kent.

In simple situations, ground conductivity profiling can be used to detect plumes of polluted water. For example, a ground conductivity survey can be carried out if saline groundwater is near the surface. If the depth to water is too great, then the thickness of overlying unsaturated sediments can mask any contrasts between polluted and natural groundwater. In addition, the geology of the area has to be relatively uniform so that the conductivity values and profiles can be compared with others. Reported uses of this technique include tracing polluted groundwater from landfills, septic tanks, oil-field brine-disposal pits, acid mine drainage, sewage treatment effluent, industrial process waters and spent sulphur liquor.

The ionic content of the groundwater in a borehole can be monitored by measuring the resistivity of the fluid at a short electrode spacing (Brown, 1971). The quality of groundwater can also be estimated from the spontaneous potential deflection on an electric log, provided that the specific conductance of the drilling mud is lower than the specific conductance of the pore water. The method tends to give better results as the salinity of the pore water increases.

Artificial recharge may be undertaken to prevent large reductions in groundwater levels as a result of over-abstraction, to improve the standard of water in a poor-quality aquifer or to check or reverse saline intrusion. The source of water may be storm run-off, river or lake water, water used for cooling purposes, industrial waste water, or sewage effluent. Many of these sources require some form of treatment before use, not just because of the risk of polluting the aquifer, but because of the possibility of an interaction between the recharged water and the groundwater. Interaction may lead to precipitation of, for example, calcium carbonate or iron and manganese salts. Bacterial action may lead to the development of sludges. These two processes tend to clog well screens and reduce the permeability of aquifers, resulting in lower rates of recharge. Nitrification or denitrification and possibly sulphate reduction may occur during the early stages of recharge.

Before embarking upon an artificial recharge scheme, the suitability of an aquifer for this type of operation must be investigated. In particular, the aquifer must have an adequate storage potential and the bulk of the water recharged should be recoverable and should not be lost rapidly by discharge to a nearby river. Attention must also be paid to the compatibility of the recharge water with respect to the groundwater, otherwise chemical interaction may lead to operational problems.

There are several ways in which artificial recharge may be accomplished. These include recharge basins or ditches, spray irrigation, and recharge wells. Recharge basins are one of the most widely adopted means of artificially recharging groundwater. To be effective, however, the ground must have a high infiltration capacity and the aquifer must be unconfined and preferably within 2–3 m of the ground surface. The bottom of the basin or trench may penetrate the aquifer provided that a filter is employed, although in some countries it is considered safer to have a natural unsaturated zone between the water table and the bottom of the recharge works.

The quality of the water used in basin recharge is not critical. In fact, basin recharge can constitute an important element in water treatment, acting rather like a slow sand filter, leading to the elimination of suspended solids, ammonia, bacteria and viruses. Under such circumstances the basin requires periodic cleaning, the top 10–30 mm of the floor being replaced by clean material. Recharge basins may yield quality water despite some pollution of the original water supply to the basin.

Recharge basins are often of permanent construction, but temporary basins or ditches can be constructed by building dykes across a river at the beginning of each dry season, or by digging ditches leading from the river on to the flood plain. In these cases, the intention is simply to spread the river water over as large an area as possible to increase the natural rate of infiltration. Old gravel pits can also be utilized successfully as recharge basins.

Spray irrigation needs to be carried out over large areas of land. Recharge rates are generally quite low, because no head of water can be applied, although enhanced water quality is obtained. Spray irrigation is limited to ground with a high infiltration capacity, areas where the water table is near the ground surface and to periods of active plant growth. Care must be taken not to flush salts and nutrients from the soil into the groundwater. In regions with hot climates, there may also be a danger that excessive evaporation of the recharge water will cause a build-up of salts in the soil. Both of these mechanisms could result in decreased fertility and reduced crop yields.

Recharge wells are most frequently employed where the aquifer to be recharged is deep or confined, or where there is insufficient space for recharge basins. Where the flow is intergranular, recharge wells are typically between 500 and 900 mm in diameter and penetrate a good way below groundwater level, with a screened section extending up to the highest anticipated groundwater level under recharged conditions.

References

Anon. (1970) Ground water pollution. *Water Well Journal,* 24, No. 7, 31–61.

Anon. (1977) *Report of the Working Party on the Disposal of Sewage Sludge to Land.* Department of the Environment/National Water Council Technical Committee Report No. 5, National Water Council, London.

Anon. (1978) *Co-operative Programme of Research on the Behaviour of Hazardous Wastes in Landfill Sites.* Final Report of the Policy Review Committee, Department of Environment, HMSO, London.

Anon. (1988) Engineering geophysics. Engineering Group Working Party Report. *Quarterly Journal Engineering Geology,* 27, 207–273.

Barber, C. (1982) *Domestic Waste and Leachate, Notes on Water Research No. 31.* Water Research Centre, Medmenham, England.

Bell, F. G. and Kerr, A. (1993) Coal mining and water quality with illustrations from Britain. *Proceeding International Conference on Environmental Management, Geowater and Engineering Aspects,* Wollongong, Chowdhury, R. N. and Sivakumar, S. (eds), A.A. Balkema, Rotterdam, 607–614.

Bellinger, E. G., Jones, A. D. and Tinker, J. (1982) The character and dispersal of motorway run-off water. *Water Pollution Control,* 81, 372–390.

Berk, W. J. and Yare, B. S. (1977) An integrated approach to delineating contaminated ground water. *Ground Water,* 15, No. 2, 138–145.

Best, G. T. and Aikman, D. T. (1983) The treatment of ferruginous groundwater from an abandoned colliery. *Water Pollution Control,* 82, 557–566.

Bevan, R. (1971) *Notes on the Science and Practice of the Controlled Tipping of Refuse.* Institute Public Cleansing, London.

Bouwer, H. (1978) *Groundwater Hydrology.* McGraw-Hill, New York.

Brodie, M. J., Broughton, L. M. and Robertson, A. (1989) A conceptional rock classification system for waste management and a laboratory method for ARD prediction from rock piles. In *British Columbia Acid Mine Drainage Task Force,* Draft Technical Guide 1, 130–135.

Brown, D. L. (1971) Techniques for quality of water interpretation from calibrated geophysical logs, Atlantic Coastal Area, *Ground Water,* 9, No. 4, 25–38.

Brown, R. F. and Signor, D. C. (1973) Artificial recharge – state of the art. *Ground Water,* 12, No. 3, 152–160.

Brown, R. H., Konoplyantsev, A. A., Ineson, J. and Kovalevsky, V. S. (eds). (1972) *Groundwater Studies, An International Guide for Research and Practice.* Studies and Reports in Hydrology 7, UNESCO, Paris.

Bullock, S. E. T. and Bell, F. G. (1995) An investigation of surface and groundwater quality at a mine in the north west Transvaal, South Africa. *Transactions Institution Mining and Metallurgy*, 104, Section A, A125–A133.

Cambridge, M. (1995) Use of passive systems for treatment of mine outflows and seepages. *Minerals Industry International Bulletin*, Institution of Mining and Metallurgy, No. 1024, 35–42.

Campbell, M. D. and Lehr, J. H. (1973) *Water Well Technology*, McGraw-Hill, New York.

Cherry, J. A. (ed.) (1983) Migration of contaminants in groundwater at a landfill: a case study. Special issue, *Journal Hydrology*, 1–398.

Connelly, R. J., Harcourt, K. J., Chapman, J. and Williams, D. (1995) Approach of remediation of ferruginous discharge in the South Wales coalfield and its application to closure planning. *Minerals Industrty International Bulletin*, Institution of Mining and Metallurgy, No. 1024, 43–48.

Cook, H. F. (1991) Nitrate protection zones: targeting and land-use over an aquifer. *Land Use Policy*, 8, 16–28.

Davis,. R. D. (1980) *Control of Contamination Problems in the Treatment and Disposal of Sewage Sludge.* Technical Report 156, Water Research Centre, Medmenham, England.

Diefendorf, A. F. and Ausburn, R. (1977) Groundwater monitoring wells. *Public Works*, 108, No. 7, 48–50.

Dohrman, J. A. (1982) Source protection – an ounce of prevention. *Journal New England Water Works Association*, 96, No. 4, 322–328.

Fielding, M. Gibson, T. M., James, H. A., McLoughlin, K. and Steel, C. P. (1981) *Organic Micropollutants in Drinking Water.* Technical Report 159, Water Research Centre, Medmenham, England.

Foster, S. S. D. and Crease, R. I. (1974) Nitrate pollution of chalk groundwater in east Yorkshire – a hydrological appraisal. *Journal Institution Water and Engineers Scientists*, 28, 178–194.

Foster, S. S. D. and Young, C. P. (1980) Groundwater contamination due to agricultural land-use practices in the United Kingdom. In *Aquifer Contamination and Protection,* Jackson, R. E. (ed.), Chapter 11–3, Project 8.3 of the International Hydrological Programme, Studies and Reports in Hydrology, 30, UNESCO, Paris.

Frazer, J. A. L. and Sims, A. F. E. (1983) Tip leachate using hydrogen peroxide. *Water Pollution Control*, 82, 243–255.

Fried, J. J. (1975) *Groundwater Pollution Theory, Methodology, Modelling and Practical Rules.* Elsevier, Amsterdam.

Gass, T. E. (1980) Hydrogeologic considerations of landfill siting and design, *Water Well Journal*, 43–45.

Gerba, C. P., Wallis, C. and Melnick, J. L. (1975) Fate of waste-water bacteria and viruses in soil. *Proceedings American Society Civil Engineers, Journal Irrigation and Drainage Division*, 101, 157–174.

Glover, R. E. and Balmer, C. E. (1954) River depletion resulting from a well pumping near a river. *Transactions American Geophysical Union*, 35, 468–470.

Gray, D. A., Mather, J. D. and Harrison, I. B. (1974) Review of groundwater pollution for waste sites in England and Wales with provisional guidelines for future site selection. *Quarterly Journal Engineering Geology*, 7, 181–196.

Hagerty, D. J. and Pavoni, J. L. (1973) Geologic aspects of landfill refuse disposal. *Engineering Geology*, 7, 219–230.

Henriksen, A. and Kirkhusmo, L. A. (1982) Acidification of groundwater in Norway. *Nordic Hydrology*, 13, 183–192.

Holmes, R. (1984) Comparison of different methods of estimating infiltration at a landfill site in south Essex with implications for leachate management and control. *Quarterly Journal Engineering Geology*, 17, 9–18.

Ineson, J. (1970) Development of groundwater resources in England and Wales. *Journal Institution Water Engineers*, 24, 155–177.

Jackson, R. E. (ed.) (1980) *Aquifer Contamination and Protection*. Project 8.3 of the International Hydrological Programme, Studies and Reports in Hydrology 30, UNESCO, Paris.

Jackson, R. E. (1982) The contamination and concept of aquifers. *Nature and Resources*, 18, No. 3, 26.

Jensen, O. M. (1982) Nitrate in drinking water and cancer in northern Jutland, Denmark, with special reference to stomach cancer. *Ecotoxicology and Environmental Safety*, 6, 258–267.

Kelly, W. E. (1976) Geoelectric sounding for delineating ground-water contamination. *Ground Water*, 14, 6–10.

Kishi, Y., Fukuo, Y., Kakinuma, T. and Ifuku, M. (1982) The regional steady interface between fresh water and salt water in a coastal aquifer. *Journal Hydrology*, 58, 63–82.

Lawrence, A. R., Foster, S. S. D. and Izzard, P. W. (1983) Nitrate pollution of chalk groundwater in east Yorkshire – a decade on. *Journal Institution Water Engineers and Scientists*, 37, 410–420.

Lijinksky, W. and Epstein, S. S. (1970) Nitrosamines as environmental carcinogens. *Nature*, 225, 21–23.

Linsley, R. K., Kohler, M. A. and Paulhus, J. L. H. (1982) *Hydrology for Engineers*, third edition. McGraw-Hill, New York.

Mackay, D. M. (1998) Is clean-up of VOC-contaminated groundwater feasible? In *Contaminated Land and Groundwater – Future Directions*, Engineering Geology Special Publication No. 13. Lerner, D. N and Walton, N. (eds), Geological Society, London (in press).

Mackay, D. M. and Cherry, J. A. (1989) Groundwater contamination: limits of pump-and-treat remediation. *Environmental Science and Technology*, 23, 630–636.

Morris, R. and Waite, W. M. (1981) Environmental virology and its problems. *Journal Institution Water Engineers and Scientists*, 35, 232–244.

Naylor, J. A. Rowland, C. D., Young, C. P. and Barber, C. (1978) *The Investigation of Landfill Sites*. Technical Report 91, Water Research Centre, Medmenham, England.

Nutbrown, D. A. (1976) Optimal pumping regimes in an unconfined coastal aquifer. *Journal Hydrology*, 31, 271–280.

Oteri, A. U. E. (1983) Delineation of saline intrusion in the Dungeness shingle aquifer using surface geophysics. *Quarterly Journal Engineering Geology*, 16, 43–52.

Pettyjohn, W. A. (1982) Cause and effect of cyclic changes in ground water quality. *Ground Water Monitoring Review*, 2, No. 1, 43–49.

Pfannkuch, H. A. (1982) Problems of monitoring network design to detect unanticipated contamination. *Ground Water Monitoring Review*, 2, No. 1, 67–76.

Pickford, J. A. (1979) Control of pollution and disease in developing countries. *Water Pollution Control*, 78, 239–253.

Raucher, R. L. (1983) A conceptual framework for measuring the benefits of ground-water protection. *Water Resources Research*, 19, 320–326.

Revelle, R. (1941) Criteria for recognition of sea water in ground-waters. *Transactions American Geophyical Union*, 22, 593–597.

Robinson, H. (1984) On site treatment of leachate using aerobic biological techniques. *Quarterly Journal Engineering Geology,* 17, 31–37.

Robinson, H. D. and Maris, P. J. (1979) *Leachate from Domestic Waste, Generation, Composition and Treatment, A Review.* Technical Report 108, Water Research Centre, Medmenham, England.

Romero, J. C. (1970) The movement of bacteria and viruses through porous media. *Ground Water*, 8, No. 2, 37–48.

Selby, K. H. and Skinner, A. C. (1979) Aquifer protection in the Severn-Trent region: policy and practice. *Water Pollution Control*, 78, 320–326.

Smith, D. B., Wearn, P. L., Richards, H. J. and Rowe, P. C. (1970) Water movement in the unsaturated zone of high and low permeability strata by measuring natural tritium. *Proceedings Symposium Isotope Hydrology*, International Atomic Energy Association, Vienna, 73–87.

Tester, D. J. and Harker, R. J. (1981) Groundwater pollution investigations in the Great Ouse basin. *Water Pollution Control*, 80, 614–631.

Theis, C. V. (1941) The effect of a well on the flow of a near stream. *Transactions American Geophysical Union*, 22, 734–738.

Van Burkalow, A. (1982) Water resources and human health, the viewpoint of medical geography. *Water Research Bulletin*, 18, 869–874.

Volker, R. E. and Rushton, K. R. (1982) An assessment of the importance of some parameters for seawater intrusion in aquifers and a comparison of dispersive and sharp-interface modelling approaches. *Journal Hydrology*, 56, 239–250.

Walker, W. H. (1969) Illinois groundwater pollution. *Journal American Water Works Association*, 61, No. 1, 31–40.

Williams, G. M. (1980) The Control of Pollution Act 1974, 3 Implications of changes in landfill practice. *Journal Institution Water Engineers and Scientists*, 34, 153–160.

Wood, W. W. and Basset, R. L. (1975) Water quality changes related to the development of anaerobic conditions during artificial recharge. *Water Resources Research*, 11, 553–558.

Woodward, G. M. and Selby, K. (1981) The effect of coal mining on water quality. *Proceedings Symposium on Mining and Water Pollution*, Nottingham, Institution Water Engineers and Scientists, 11–19.

Zaki, M. H., Moran, D. and Harris, D. (1982) Pesticides in groundwater, the aldicarb story in Suffolk County, New York. *American Journal Public Health*, 72, 1391–1395.

Chapter 12

Ground subsidence

12.1 Introduction

Subsidence of the ground surface takes place when mineral deposits, be they gas, fluid or solid, are removed from within the ground. It reflects the movements that occur in the area undergoing extraction or abstraction. Unfortunately, subsidence can have serious effects on surface structures, be responsible for flooding, and lead to the sterilization of land, and can call for special constructional design in site development or extensive remedial measures in developed areas.

In some parts of the world mining has gone on for centuries and methods of mining have changed with time. In addition, mining methods differ according to the type of mineral deposit exploited. One of the problems associated with old mineral workings is that there may be no record of their existence. For example, coal mining has gone on in Britain for several centuries, but the first statutory obligation to keep mine records dates from only 1850, and it was not until 1872 that the production and retention of mine plans became compulsory. Even if old records exist, they may be inaccurate. Old abandoned workings, particularly in coal, occur at shallow depth beneath the surface of many urban areas of Western Europe and North America. Such old workings can represent a hazard during any subsequent redevelopment.

12.2 Pillar and stall workings and their potential failure

Stratified deposits such as coal, limestone, gypsum, salt, sedimentary iron ore, etc. have been and are worked by partial extraction methods whereby pillars of the mineral deposit are left in place to support the roof of the workings. Again taking Britain and coal as an example, the geometry of old workings differed from region to region (Figure 12.1). Also, early workings tended to be rather irregular in their layout, but with the passage of time the arrangement of pillars and voids (variously referred to as stalls, rooms, bords) became more regular.

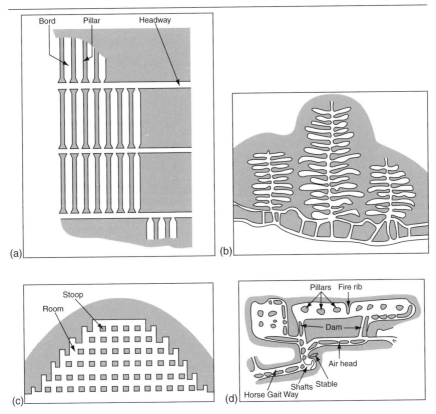

Figure 12.1 (a) Bord and pillar workings, Newcastle upon Tyne, seventeenth century. (b) Post and stall workings, South Wales, seventeenth century. (c) Stoop and room workings, Scotland, seventeenth century. (d) Staffordshire squarework, developed for conditions involving spontaneous combustion – airtight stoppings could be placed in narrow accessways to prevent the spread of fire.

In pillared workings the pillars sustain the redistributed weight of the overburden, which means that they and the rocks immediately above and below are subjected to added compression. Stress concentrations tend to be located at the edges of pillars, and the intervening roof rocks tend to sag. The effects on ground level are normally insignificant. Although the intrinsic strength of a stratified deposit varies, the important factor in the case of pillars is that their ultimate behaviour is a function of bed thickness to pillar width, the depth below ground and the size of the extraction area. The mode of failure also involves the character of the roof and floor rocks. Pillars in the centre of the mined-out area are subjected to greater stress than those at the periphery.

Individual pillars in dipping seams tend to be less stable than those in horizontal seams, since the overburden produces a shear force on the pillar.

Collapse in one pillar can bring about collapse in others in a sort of chain reaction, because increasing loads are placed on those remaining (Bryan *et al.*, 1964). Slow deterioration and failure of pillars may take place after mining operations have long since ceased. The effective pillar width can be reduced by blast damage or local geological weaknesses. Furthermore, the collapse of roof strata around pillars alters their geometry and so can affect their stability. Old pillars at shallow depth have occasionally failed near faults, and they may fail if they are subjected to the effects of subsequent mining. The yielding of a large number of pillars can bring about a broad shallow subsidence over a large surface area, which Marino and Gamble (1986) referred to as a sag. The ground surface in a sag displaces radially inwards towards the area of maximum subsidence. This inward radial movement generates tangential compressive strain, and circumferential tension fractures are frequently developed. Sag movements depend on the mine layout, in particular the extraction ratio, and geology, as well as on the topographic conditions at the surface. They tend to develop rather suddenly, the major initial movements lasting, in some instances, for about a week, with subsequent displacements occurring over varying periods of time. The initial movements can produce a relatively steep-sided bowl-shaped area. Nonetheless, the shape of a sag profile can vary appreciably, and because it varies with mine layout and geological conditions, it can be difficult to predict accurately. Normally, the greater the maximum subsidence, the greater is the likelihood of variation in the profile. Maximum profile slopes and curvatures frequently increase with increasing subsidence. The magnitude of surface tensile and compressive strains can range from slight to severe.

In the past, especially in coal mining, pillars were frequently robbed on retreat. Extraction of pillars during the retreat phase can simulate longwall conditions (see section 12.7), although it can never be assumed that all pillars have been removed. At moderate depths pillars, especially pillar remnants, are probably crushed and the goaf (i.e. the worked-out area) compacted, but at shallow depths lower crushing pressures may mean the closure is variable. This causes foundation problems when large or sensitive structures are erected above. The potential for pillar failure should not be ignored, particular warning signs being strong roofs and floors allied with high extraction ratios and moderately to steeply dipping seams. Prediction of subsidence as a result of pillar failure requires accurate data regarding the layout of the mine. Such information frequently does not exist in the case of abandoned mines. On the other hand, when accurate mine plans are available and in the case of a working mine, the method outlined by Goodman *et al.* (1980) can be used to evaluate collapse potential.

Basically, the method involves calculating vertical stress based upon the tributary area load concept, which assumes that each pillar supports a column of rock with an area bounded by room centres and a height equal to the depth from the surface, and comparing this with the strength of the pillar. When a structure is to be built over an area of old pillared workings the additional load on the pillars can be estimated simply by adding the weight of the appropriate part of the structure to the weight of the column of strata supported by a given pillar. This method is very conservative, except when used for large concentrated loads where old workings are located at shallow depth. Bell (1992a) provided a number of other means of calculating the strength of or stresses on pillars (Table 12.1).

Even if pillars are relatively stable, the surface may be affected by void migration. This can take place within a few months of or a very long period of years after mining. Void migration develops when roof rock falls into the worked-out areas. When this occurs the material involved in the fall bulks, which means that migration is eventually arrested, although the bulked material never completely fills the voids. Nevertheless, the process can, at shallow depth, continue upwards to the ground surface, leading to the sudden appearance of a crown hole (Figure 12.2). The factors that influence whether or not void migration will take place include the width of the unsupported span; the height of the workings; the nature of the cover rocks, particularly their shear strength and the incidence and geometry of discontinuities; the thickness and dip of the seam; the depth of overburden; and the groundwater regime.

Garrard and Taylor (1988) summarized four methods that have been used to predict the collapse of roof strata above rooms. These are clamped beam analysis, which considers the tensile strength of the immediate roof rocks (Wardell and Eynon, 1968; Hoek and Brown, 1980); bulking equations, which consider the maximum height of collapse before a void is choked (Tincelin, 1958; Price et al., 1969; Piggott and Eynon, 1978); arching theories, which estimate the height to which a collapse will occur before a stable arch develops (Terzaghi, 1946; Szechy, 1970); and coefficients based on experience and field observations, which act as multipliers of either seam thickness or span width (Walton and Cobb, 1984).

It is frequently maintained that the maximum height of void migration is directly proportional to the thickness of seam mined, assuming that the total thickness is worked, and inversely proportional to the change in volume of the collapsed material. Garrard and Taylor (1988) also showed that the height of collapse in pillared workings in coal is frequently proportional to the width of the excavation and that the larger the span, the more likely collapse is to occur. The maximum height of migration in exceptional cases might extend to 10 times the height of the original room; however, it is generally 3 to 5 times the room height.

In a sequence of differing rock types, if a competent rock beam is to span an opening, then its thickness should be equal to twice the span width in order to allow for arching to develop. A bed of sandstone will usually arrest a void, especially if it is located some distance above the immediate roof of the working (Van Besian and Rockaway, 1988). Sandstones apart, however, most voids are bridged when the span decreases through corbelling to an acceptable width, rather than when a more competent bed is encountered. Chimney-type

Table 12.1 Some expressions for determining the strength of and the stress on coal pillars (after Bell, 1992a)

Author(s)	Expression	Remarks
1 Greenwald et al. (1941)	Pillar strength $(\tau_p) = \dfrac{2800 \times w^{0.5}}{h^{0.85}}$ psi w = width of pillar in inches h = height of seam in inches	Based on the results of *in situ* tests on square pillars in Pittsburg bed. Pillars varied in width from 1 to 5.3 ft and height from 1.4 to 5.3 ft.
2 Terzaghi (1943)	$\sigma_v = \dfrac{\rho 0.5\beta}{K \tan \phi} \left(-\exp^{-K\tan\sigma/\beta}\right) + q \exp^{-K\tan\tau/\beta z}$ σ_v = vertical pillar loading pressure K = constant representing ratio between horizontal and vertical pressure B = breadth of room ρ = density or unit weight of overburden z = depth ϕ = angle of internal friction q = uniform surcharge carried by soil per unit area	This expression was developed in relation to theories of arching and more particularly to the vertical load above a tunnel. It has been suggested that it could be used to calculate the total vertical load, including any proposed surface structure, above old mine workings.
3 Denkaus (1962)	$\sigma_\gamma = wz \dfrac{1}{1-r}$ r = extraction ratio $\quad = \dfrac{2ab + b^2}{(a+b)^2}$ (see Figure)	This expression is based on two assumptions, namely, that each pillar supports the column of rock over an area that is the sum of the cross-sectional area of the pillar plus a part of the room area; and that the load is vertical only and uniformly distributed across the cross-sectional area. Steart (1954) has pointed out, however, that the pressure is not evenly distributed; it frequently assumes a parabolic distribution with maximum pressure exerted on the central pillars of the mined-out area. He further stated that the pressure gradient gradually increased with increasing area of development to a maximum given by Eq. 3, and that this maximum was reached when the development, if roughly circular, attained a radius equal to the depth, divided by the ratio $(a+b)^2/a^2$ (see Figure below).

Table 12.1 (cont.)

Author(s)	Expression	Remarks

PLAN VIEW

4 Coates (1965)

$$\sigma_v = \frac{1.1z}{1 - \frac{r}{100}} \text{ psi}$$

r = extraction ratio

According to Coates, this expression gives valid results where the depth of the workings is less than half their extent, as is the case with most old mine workings.

5 Wardell and Wood (1965)

$$\sigma_v = \frac{z}{1 - r} \text{ psi}$$

When a structure is to be built over old pillared workings the additional weight on the pillars can be estimated by adding the weight of the appropriate part of the structure to the overburden pressure supported by the pillar. The total load on a pillar determined in this manner is probably greater than the true load. Moreover, the distribution of load over the area of the pillar is not uniform, for stress is concentrated at the pillar edges. The transfer of the weight of a surface structure to residual pillars is normally calculated on the assumption that the additional weight acts vertically downwards and that there is no lateral spreading of the load. This is based on the fact that void migration causes dilation above the seam as well as joint opening, so reducing or preventing such lateral distribution,

Table 12.1 (cont.)

Author(s)	Expression	Remarks
6 Salamon and Munro (1967) and Salamon (1967)	(a) $\tau_p = \dfrac{1320\,w^{0.46}}{h^{0.66}}$ psi (dimensions in ft) (b) $\sigma_v = 1.1z\left(\dfrac{w+\sigma}{w}\right)^2$ psi (c) factor of safety (FS) against pillar failure $FS = \dfrac{\text{strength of pillar}}{\text{pillar loading pressure}}$	These expressions are based upon a statistical analysis of a survey of pillar dimensions in both stable (98 cases) and collapsed (27 cases) mining areas in South Africa. *Summary of data* *Stable* *Collapsed* 1 Depth (m) 20–220 21–192 2 Room height (m) 1.2–4.9 1.5–5.5 3 Pillar width (m) 2.7–21.3 3.4–15.9 4 Extraction ratio 37–89 49–91 5 Pillar width:height ratio 1.2–8.8 0.9–3.6 Salamon and Munro concluded that 99 percent of collapses occurred at safety factors lower than 1.48. Holland (1964), however, had previously suggested that a factor of 1.8 was generally necessary and that in critically important areas it was 2.2.
7 Bieniawski (1967, 1970)	$\tau_p = \dfrac{400 + 200}{h}\,w$ psi	Based on tests carried out on coal pillars in South Africa, the pillars varying in height and width from 2 to 6.7 ft.

8 Wilson (1972)	(a) Wide pillars, $w > 0.003\ mz$ ft
	  (i) Square pillars
	    $\sigma_v = 4\rho z(w^2 - 3wmz \times 10^{-3} + 3m^2z^2 \times 10^{-6})$ tons
	  (ii) Rectangular pillars
	    $\sigma_v = 4\rho z(wl - 1.5(w + l)mz \times 10^{-3} + 3m^2z^2 \times 10^{-6})$ tons
	  (iii) Long pillars
	    $\sigma_v = 4\rho z(w - 1.5mz \times 10^{-3})$ tons/ft run
		Wilson suggested that a pillar was surrounded by an outer yield zone in which the stress distribution varied in linear fashion from zero at the surface to the point of failure at the pillar core. The constraint given to coal in the pillar core can increase its strength appreciably. The six expressions take no account of the strength of the coal and it has been suggested that although they may be admissable for deep workings they could give underestimates of strength of pillars at shallow depths.

Table 12.1 (cont.)

Author(s)	Expression	Remarks
8 Wilson (1972) (cont.)	(b) Narrow pillars, $w < 0.003\ mz$ ft (i) Square pillars $$\sigma_v = 4\rho\frac{w^3}{m}\ \text{tons}$$ (ii) Rectangular pillars $$\sigma_v = 1333\rho\frac{w^2}{m}\left(1 - \frac{w+1}{2} + \frac{w}{3}\right)$$ (iii) Long pillars $$\sigma_v = 667\rho\frac{w^2}{m}\ \text{tons/ft run}$$ where σ_v = load which pillar will carry, in tons ρ = average density of rock in tons/ft^3 l = length of pillar in ft m = height of roadway in ft w = width of pillar in ft z = depth of overburden in ft	
9 Goodman et al. (1980)	(a) $\tau_p = \dfrac{\text{UCS} \times \text{N-shape} \times \text{N-size}}{\text{FS}}$ UCS = unconfined compressive strength N-shape = $(0.875 + 0.250\ w/h_p)$ h_p = height of pillar FS = factor of safety (b) $\sigma_v = \dfrac{\sigma_{vi} \times A_t}{A_p - A_w}$	Allows zones of minimum stability in abandoned pillar and stall workings to be located. Locations and dimensions of pillars plotted on mine plan. Factor of safety for each pillar determined and plotted on plan. Pillars which have factors of safety of less than one are considered potentially unstable and removed from plan. Their tributary areas are reassigned to adjacent pillars and the calculation repeated. In second calculation more pillars may fail. They are removed for further iteration of the calculation, if so required. Although method requires knowledge of exact shape of pillars, as well as strength of rocks involved, it offers practical approach to recognizing

Table 12.1 (cont.)

Author(s)	Expression	Remarks
	σ_{vi} = initial vertical stress at roof level A_t = area tributary to each pillar A_p = cross sectional area of pillar A_w = area of pillar lost from load carrying (c) $FS = \tau_p$	where the most potentially unstable areas exist so that appropriate measures may be taken optimally.
10 MacCourt et al. (1986)	(a) $\tau_p = 7176\,w^{0.459}\,h^{-0.66}$ kPa (b) $\sigma_v = \dfrac{\rho g z_f c^2}{w^2}$ ρ = density of overburden g = acceleration due to gravity z_f = depth to floor of workings c = pillar centre distance (c) $FS = \tau_p \sigma_v$	MacCourt et al. (1986) extended the work of Salamon and Munro (1967). Like Salamon and Munro, MacCourt et al. adopted tributary area theory, assuming that each pillar carried the mass of superincumbent strata immediately above it. This theory is valid when pillars have reasonably uniform geometry and are mined over an area where the width of mining exceeds the depth of the seam. After the survey of recent pillar collapses in South Africa collieries, MacCourt et al. concluded that the majority occurred when the depth was less than 60 m, the pillar width was less than 6 m, the pillar width-to-height ratio was less than 1.75 and the extraction ratio exceeded 70 percent. They found that subsidence as a proportion of mined height decreased from around 0.8 at very shallow depths to between 0.1 and 0.5 at depths exceeding 100 m.

Figure 12.2 Crown hole developed by void migration from abandoned mine workings at shallow depth, Stockton, South Island, New Zealand.

collapses can occur to abnormally high levels of migration in massive strata in which the joints diverge downwards.

Exceptionally, void migrations in excess of 20 times the worked height of a seam have been recorded. The self-choking process may not be fulfilled in dipping seams (Statham *et al.*, 1987), especially if they are affected by copious quantities of water, which can redistribute the fallen material. The redistribution of collapsed material can lead to the formation of super-voids, and their migration to rockhead then produces large-scale subsidence at ground level. Under such circumstances, simple analysis according to bulking factors proves inadequate (Carter, 1985). Depth of cover should not include superficial deposits or made ground, since low bulking factors are characteristic of these materials. Weak superficial deposits may flow into voids that have reached rockhead, thereby forming features that may vary from a gentle dishing of the surface to inverted cone-like depressions of large diameter.

12.3 Investigations in areas of abandoned mine workings

A site investigation for an important structure requires the exploration and sampling of all strata likely to be significantly affected by the structural loading. The location of subsurface voids due to mineral extraction is of prime importance in this context. In other words, an attempt should be made to determine the number and depth of mined horizons, the extraction ratio, the pattern of the layout, and the condition of the old room and pillar workings. The sequence and type of roof rocks may provide some clue as to whether void migration has taken place and if so, its possible extent. Of particular importance is the state of the old workings. Careful note should be taken of whether they are open, partially collapsed or collapsed, and the degree of fracturing and bed separation in the roof rocks should be recorded, if possible. This helps to provide an assessment of past and future collapse, which is obviously very important.

Each investigation should be designed to meet the requirements of the construction operations to be carried out. The first stage of the site investigation involves a desk study and a reconnaissance survey, which are then followed by the necessary field exploration. The desk study includes a survey of appropriate maps, documents, records and literature. The presence on geological maps of mineral deposits that could have been mined suggests the possibility of past mining unless there is evidence to the contrary, and geological and topographic maps may show evidence of past workings such as old shafts, adits and spoil heaps. All the geological and topographic maps of the area in question, going back to the first editions, should be examined. Abandoned mines record offices, when they exist, represent primary sources of information relating to past mining activity. Other sources include public record offices, museums, libraries, specialist contractors and consultants, private collections, and geological surveys.

The use of remote sensing imagery and aerial photography for the detection of surface features caused by subsidence is more or less restricted to rural areas, and scale is a critical factor. The resolution necessary for the detection of relatively small subsidence features (1.5–3 m across) is provided by aerial photographs with scales between 1:25 000 and 1:10 000. Colour photographs may be more useful than black and white ones in the detection of past workings, since they can reveal subtle changes in vegetation related to subsidence and, if there are differences in thermal emission, then infrared (false colour) photographs should show these differences. The detail obtained from aerial photographs should be represented on a site plan at a scale of 1:2500 or larger.

Over the past 25 years, attempts have been made to develop geophysical methods for the location and delineation of abandoned mine workings. Unfortunately, no one geophysical method has yet been developed that resolves all problems of this nature. A variety of surface traversing techniques are available that provide readings at close station intervals for the location of shallow voids where the lateral dimensions of the void are of the same order as the depth of burial.

Considerable care should be exercised at the planning stage of a geophysical survey for the location of subsurface voids because of the variable nature of the target (McCann et al., 1987; Cripps et al., 1988). The selection of the most appropriate technique necessitates consideration of four parameters, namely, penetration, resolution, signal-to-noise ratio and contrast in physical properties. The size and depth of the workings and the character of any infill control the likelihood of the workings being detected as an anomaly. With the information obtained from the desk study and the reconnaissance survey, many of the available geophysical methods can be assessed at the selection stage, using a model study, and accepted, or rejected, without any requirement for field trials. Generally, it is possible to detect a cavity whose depth of burial is less than twice its effective diameter. However, since the presence of a void is likely to affect the physical properties and drainage pattern of the surrounding rock mass, this can give rise to a larger anomalous zone than that produced by the void alone.

The nature of the environment around a site affects the success of geophysical surveys. For instance, traffic vibrations adversely affect the results obtained from seismic surveys, as do power lines and electricity cables in the case of electro-magnetic and magnetic techniques. Of particular importance is that there should be sufficient physical property contrast between the void and the surrounding rock mass so that an anomaly can be detected.

Seismic refraction has not been used particularly often in searching for voids at shallow depth created by previous mining, since such voids are often too small to be detected by this method because of attenuation of seismic waves in the rock mass. Furthermore, if the workings are dry, then high attenuation of the energy from the seismic source occurs, so penetration into the rock mass is poor. McCann *et al.* (1982) concluded that there was little likelihood of a cavity producing a measurable anomaly where it is buried at a depth greater than its diameter. However, the disturbed zone around an anomaly may increase its effective size. Voids in coal seams above a competent sandstone can sometimes be detected by a localized increase in travel time and decrease in amplitude of the seismic event, provided that the anomalous region is comparable in dimensions with the seismic wavelengths. For instance, a seismic wave having a dominant frequency of 100 Hz and a velocity of 2000 m s^{-1} has a wavelength of 20 m. As a consequence, voids less than 20 m in width probably will not be recognized (Anon., 1988). Recent advances in seismic reflection techniques have improved the potential for detection of shallow voids.

Generally, where workings occur at a depth greater than 5 m, it is unlikely that resistivity profiling will detect the presence of dry room and pillar workings. In addition, where the depth of a cavity is equal to its diameter, the maximum disturbance in the resistivity profile is only about 10%, in which case cavities at depths greater than twice their average dimension are usually not recorded. Electrical resistivity depth sounding can be applied to the location of voids where the width–depth ratio is large. Mine workings that produce an air-filled layer can often be identified on the sounding curve as an increase in apparent resistivity.

Terrain conductivity meters have several advantages over conventional electrical resistivity equipment when used for locating small, near-surface anomalous features. For example, in the depth range up to 30 m terrain conductivity surveys are more effective than resistivity traversing. Conductivity values are taken at positions set out on a grid pattern, and the results can be contoured to indicate the presence of any anomalies. Penetration into the ground achieved by electromagnetic radiation can be limited by excessive attenuation in ground of high conductivity. However, problems with detecting an extremely small secondary field in the presence of a strong primary signal can be overcome by the use of pulsed radiation.

Ground-probing radar is capable of detecting small subsurface cavities directly and appears to be one of the most promising methods for the future, especially for surveys in urban areas. The method is based upon the transmission of pulsed electromagnetic waves, the travel time of the waves reflected from subsurface interfaces being recorded as they arrive at the surface, so allowing the depth to an interface to be obtained (White, 1992). The high frequency of the system provides high resolution, and characteristic arcuate traces are produced by air-filled voids. Depths to voids can be determined from the two-way travel times of reflected events if velocity values can be assigned to the strata above the void. The conductivity of the ground imposes the greatest limitation on the use of radar probing in site investigation. In other words, the depth to which radar energy can penetrate depends upon the effective conductivity of the strata being probed. This, in turn, is governed chiefly by the water content and its salinity. Furthermore, the value of effective conductivity is also a function of temperature and density as well as the frequency of the electromagnetic waves being propagated. The least penetration occurs in saturated clayey materials or when the pore water is saline.

Generally speaking, voids in shallow abandoned mine workings are too small and located at depths too great to be detected by normal magnetic or gravity surveys. However, the flux-gate magnetic field gradiometer permits surveys of shallow depths to be carried out. It provides a continuous recording of lateral variations in the vertical gradient of the Earth's magnetic field rather than giving the total field strength. The gradiometer tends to give better definition of shallow anomalies by automatically removing the regional magnetic gradient. Quantitative analysis of the depth, size and shape of an anomaly can generally be made more readily for near-surface features, using the gradiometer, than from total field measurements, obtained with a proton magnetometer. The sensor of gradiometers is small (0.5 or 1 m) in order to record features within 2–3 m of the ground surface. On the other hand, a proton magnetometer can more easily detect larger and deeper features, and it yields results that are more suitable for contouring.

Micro-gravity meters, accurate to 2 μgal (0.01 g.u.),* may be successful when the voids have a significant lateral extent, as in some room and pillar workings (Emsley et al., 1992). A series of traverses can be used to map the

*1 mgal = 10 g.u. (gravity units).

lateral extent of such features, but there are still size/depth constraints and an inherent ambiguity in the interpretation of results.

Cross-hole techniques can be used when the depth of burial of the void is more than two or three times the diameter of the void. In inter-drillhole acoustic scanning an electric sparker, designed for use in a liquid-filled drillhole, produces a highly repetitive pulse. This signal is received by a hydrophone array in an adjacent drillhole similarly occupied by liquid. Generally, the source and receiver are at the same level in the two drillholes and are moved up and down together. Drillholes must be spaced closely enough to achieve the required resolution of detail. The method can be used to detect subsurface cavities if the cavity is directly in line between two drillholes and has at least one-tenth of the drillhole separation as its smallest dimension. Air-filled cavities are more readily detectable than those filled with water. Cross-hole seismic testing has also been used, employing two or more drillholes, to detect near-vertical subsurface anomalies. Acoustic tomography techniques are now being developed to map voids between adjacent drillholes.

The location of old workings generally has been done by exploratory drilling, the locations of drillholes being influenced by data obtained from the desk study or from the data gathered by indirect methods. However, it must be admitted that although frequently successful in locating the presence of old mine workings, exploratory drilling is not necessarily able to establish their layout. Nevertheless, when the results from drilling are combined with a study of old mine plans, if they exist, then it should be possible to obtain a better understanding of methods of working, sizes of voids, and directions of roadways and galleries.

Drilling to prove the existence of old mine workings is usually done by open holes, which allow relatively quick probe drilling (Bell, 1986). Rotary percussion drilling with a cruciform bit may be used. The drillholes should be taken to a depth where any voids present are not likely to influence the performance of the structures to be erected. If a grid pattern of drillholes is used some irregularity should be introduced to avoid holes coinciding with pillar positions. The stratigraphic sequence should be established by taking cores in at least three drillholes. The presence of old voids is indicated by the free-fall of the drill string and the loss of flush.

One of the principal objectives of investigations of abandoned mine workings is to determine their extent and condition. Accordingly, core material may need to be obtained. Double-barrel sampling tubes with inner plastic liners can be used to obtain cores, which can then be photographed and logged, and the rock quality designation (RQD) or fracture spacing index recorded. Drilling penetration rates, water flush returns and *in situ* permeability tests can be used to assess the degree of fracturing. The degree of fracturing is important in that it tends to increase as old workings are approached. Determination of groundwater conditions is necessary, especially where a grouting programme is required to treat the workings.

Detailed mapping of galleries is best done by driving a heading from the outcrop if this is close at hand or by sinking a shaft to the level of the mineral deposit to obtain access to the workings. Sometimes access can be gained via old shafts. Radial holes may be drilled from a shaft to establish the dimensions of rooms and pillars. As old workings may prove dangerous, it is advisable that exploration be undertaken with the advice and aid of experts.

Below-surface workings can be examined by using drillhole cameras or closed-circuit television, information being recorded photographically or on videotape and used to assess the geometry of voids and, possibly, the percentage extraction. However, their use in flooded old workings has not proved very satisfactory. Occasionally, smoke tests or dyes have been used to aid the exploration of subsurface cavities.

It is possible to study the interior of large abandoned mine workings that are flooded by using a rotating ultrasonic scanner. The ultrasonic survey can be carried out within the void, the probe being lowered down a drillhole to mine level. Horizontal scans are made at 1 m intervals over the height of the old workings and then tilting scans at 15° intervals are taken from vertical to horizontal. The echoes received during the horizontal scan are processed by computer to provide a plan view of the mine. Vertical sections are produced from the vertical scan. Hence, it is possible to determine the positions of pillars and the extent of the workings.

12.4 Old mine workings and hazard zoning

Assessment of mining hazards has usually been on a site basis, regional assessments being much less common. Nonetheless, regional assessments can offer planners an overview of the problems involved and could help them to avoid imposing unnecessarily rigorous conditions in areas where they really are not warranted.

Hazard maps of areas where old mine workings are present should ideally represent a source of clear and useful information for planners. In this respect, they should realistically present the degree of risk (i.e. the probability of the occurrence of a hazard event). Descriptive terms such as high, medium and low risk must be defined, and there are social and economic dangers in overstating the degree of risk in the terms used to describe it. Ideally, numerical values should be assigned to the degree of risk. However, this is no easy matter, for numerical values can only be derived from a comprehensive record of events, which unfortunately in the case of abandoned mine workings is available only infrequently. Furthermore, many events in the past may not have been recorded, which throws into question the reliability of any statistical analysis of data. Indeed, subsidence is affected by so many mining and geological variables that many regard it as, to all intents and purposes, a random process. The matter of risk assessment is further complicated in terms of its tolerance; for example, people are less tolerant in relation to loss of life than to loss of property. Hence, the likelihood of an event causing loss of life would be assigned a higher value than loss of property (e.g. the probability of a 1 in 100 loss of life would presumably be regarded as very risky, while in terms of loss of property very risky may be accepted as 1 in 10). Nonetheless, if planners can be provided with some form of numerical assessment of the degree of risk, then they are better able to make sensible financial decisions and to provide more effective solutions to problems arising.

One of the first attempts at hazard zoning in an area of old mine workings was made by Price (1971). After a detailed investigation of a site in Airdrie, Scotland, underlain by shallow abandoned mine workings, he was able to propose safe and unsafe zones (Figure 12.3). In the safe zones the cover rock was regarded as thick enough to preclude subsidence hazards (about 10 m of rock or 15 m of till was regarded as sufficient to ensure that crown holes did not appear at the surface), and normal foundations could be used for the two-storey dwellings that were to be erected. On the boundaries between the safe and unsafe zones, the dwellings were constructed with reinforced foundations, or

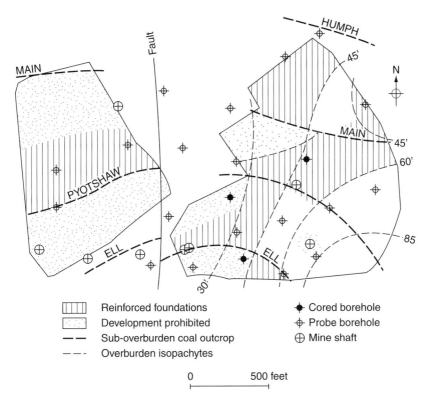

Figure 12.3 Permissible foundation zones at a site in Airdrie, Scotland. Normal foundations are permitted in unshaded areas (i.e. safe areas) (after Price, 1971).

rafts, as an added precaution against unforeseen problems. Development was prohibited in zones designated unsafe. In effect, Price produced a thematic mining information plan of the site to facilitate its development.

A zoning system based upon the depth of cover above old mine workings in the Central Rand area of South Africa has been referred to by Stacey and Bakker (1992). They described the building restrictions that are imposed on surface development related to those zones. Basically, these prohibit development where old mine workings occur at a depth of less than 90 m. They then limit the permissible heights of proposed buildings in accordance with depth of mining below the site. These restrictions are progressivley relaxed to a depth of 240 m, after which no restrictions apply. The restrictions are flexible in that they also take account of the number and separation of the reefs worked, the extent of mining, the stoping widths, the type and amount of any artificial support, and the dip of the reefs.

In recent years, thematic maps have been produced in the United Kingdom of both urban and rural areas with a view to benefiting planners and civil engineers concerned about ground stability and associated land use. Early thematic maps produced by the British Geological Survey depicted areas of undermining assumed to be within 30 m of the surface on the one hand and at depths exceeding 30 m below the surface on the other (McMillan and Browne, 1987). This 30 m depth is based on limited information and is therefore subject to interpretation. It assumes that bulking factors of 10–20% will affect the strata involved in void migration. Initially, known and suspected mining areas were not differentiated on these maps. However, only areas of mining shown on mine plans were represented on the map of the Glasgow district, and no areas of suspected mining were shown (Figure 12.4). No attempt was made to infer the extent of working beyond the limits defined by mine plans other than to plot relevant drillhole data. The recognition of single- and multiple-seam working led to the requirement that areas of shallow working (i.e. less than 30 m below rockhead) should be identified in terms of seams worked. Separate maps were prepared illustrating areas of total known mining, current mining, and known mining within 30 m of rockhead, together with the locations of shafts and drillholes encountering shallow old workings; and mining for minerals other than coal and ironstone. In Fife, known and inferred shallow old mining are differentiated on the same map. Each modification has reflected an attempt to clarify the presentation of known and inferred past mining. An indication of the area in which old mine workings might be expected can be obtained by plotting all drillholes that encounter spoil outside areas of workings known from abandonment plans. In areas where mineral outcrops are reasonably well known, areas of suspected workings can be mapped as a separate category, although it is then necessary to assume that all workable beds have been exploited, at least in the near-surface area.

In an assessment of the degree of risk due to subsidence incidents associated with abandoned mine workings in South Wales carried out by Statham *et al.*

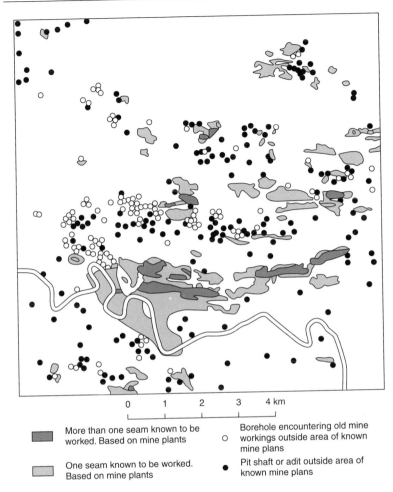

0   1   2   3   4 km

More than one seam known to be
worked. Based on mine plants

One seam known to be worked.
Based on mine plants

○ Borehole encountering old mine
   workings outside area of known
   mine plans

● Pit shaft or adit outside area of
   known mine plans

Figure 12.4 Thematic map of part of Glasgow, Scotland (after Browne *et al.,* 1986).

(1987), they found that 64% of the 388 events occurred in open land and so posed no threat to person or property. Twenty-one percent had occurred when people were nearby or property threatened. The remainder caused damage to highways, buildings or other property, and only one of these resulted in minor injuries. In the context of the South Wales coalfield this represents a low level of hazard. Assuming that a typical incident affects an area of 5 m^2, then the probability of collapse occurring on any 25 m^2 plot is of the order of 10^{-7} per year. Even if the number of subsidence incidents that have remained undiscovered increased the above total figure by a factor of 3, then the overall risk would still be low. Statham *et al.* found that over 90% of the incidents occurred within 100 m of the outcrop of the seam concerned. They produced a development advice map for South Wales that showed two zones inside the outcrops of worked seams corresponding to migration ratio (thickness of rock cover ÷ extracted thickness) values of 6 and 10, which were expected to contain 90 and 100%, respectively, of relevant subsidence incidents. The map makes a contribution towards regional planning by taking account of a possible development constraint at an early stage and offers an early warning on the likely scale of ground investigation required at specific sites.

However, it must be borne in mind that thematic maps that attempt to portray the degree of risk of a hazard event represent generalized interpretations of the data available at the time of compilation. Therefore, they cannot be interpreted too literally, and areas outlined as 'undermined' should not automatically be subject to planning blight. Obviously, there is a tendency to assume that the limits of old mine workings represented on a map indicate the full extent of the workings, but the interpretation of their location is based on scanty information and includes assumptions, some of which may be unfounded. It should be recognized that engineering problems in areas of past mining occur only if buildings are not properly planned, designed and constructed with reference to the state of undermining. Also, zoning based entirely upon depth of cover above workings cannot be relied on completely, since occasionally subsidences have occurred in zones labelled 'safe' (Carter, 1985).

Thematic geological maps of past mining areas have their limitations due to factors such as scale of presentation, and reliability and availability of data. They are transitory and require amendment as new data become available. These maps must not be used as a substitute for site investigation.

12.5 Measures to reduce or avoid subsidence effects due to old mine workings

Where a site that is proposed for development is underlain by shallow old mine workings, there are a number of ways in which the problem can be dealt with (Healy and Head, 1984). However, any decision regarding the most appropriate method should be based upon an investigation that has evaluated the site conditions carefully and examined the economic implications of the alternatives. One of the most difficult assessments to make is related to the possible effects of progressive deterioration of the workings and associated potential subsidence risk. Also, the placement of a new structure might alter a quasi-stable situation.

The first and most obvious method is to locate the proposed structure on sound ground away from old workings or over workings proved to be stable. It is not generally sufficient to locate immediately outside the area undermined, as the area of influence should be considered. In such cases, the angle of influence or draw is usually taken as 25°; in other words, the area of influence is defined by projecting an angle of 25° to the vertical from the periphery and depth of the workings to the ground surface. Such location is, of course, not always possible.

If old mine workings are at very shallow depth, then it might be feasible, by means of bulk excavation, to found on the strata beneath. This is an economic solution at depths of up to 7 m or on sloping sites.

Where the allowable bearing capacity of the foundation materials has been reduced by mining, it may be possible to use a raft. Rafts can consist of massive concrete slabs or stiff slab-and-beam cellular constructions. The latter are suitable for the provision of jacking sockets in the upstand beams to permit the columns or walls to be re-levelled if subsidence distorts the raft. A raft can span weaker and more deformable zones in the foundation, thus spreading the weight of the structure well outside the limits of the building. However, rafts are expensive and therefore tend to be used where no alternative exists. For low buildings, up to four storeys in height, it is occasionally possible to use an external reinforced ring beam with a lightly reinforced central raft as a practical and more economic foundation.

Reinforced bored pile foundations have also been resorted to in areas of abandoned mine workings. In such instances, the piles bear on a competent stratum beneath the workings. They should also be sleeved so that concrete is not lost into voids, and to avoid the development of negative skin friction if overlying strata collapse. However, some authorities have suggested that piling through old mine workings seems inadvisable because, first, their emplacement may precipitate collapse and, second, subsequent collapse at seam level could possibly lead to piles being either buckled or sheared (Price *et al.*, 1969). There may also be a problem with lateral stability of piles passing through collapsed zones above mine workings or through large remnant voids.

Where old mine workings are believed to pose an unacceptable hazard to development and it is impracticable to use adequate measures in design or found below their level, then the ground itself can be treated. Such treatment involves filling the voids in order to prevent void migration and pillar collapse. In exceptional cases, where, for example, the mine workings are readily accessible, barriers can be constructed underground and the workings filled hydraulically with sand or pneumatically with some suitable material. Hydraulic stowing may also take place from the surface via drillholes of sufficient diameter. Pneumatic or gravity stowing is often considered where large subsurface voids have to be filled. If the fill material, once drained, is likely to suffer adversely from any inflow of groundwater, for instance if it may become liquefied and flow from the treated area, or if fines may be washed out so causing consolidation of the fill, then it may be necessary to prevent this by constructing barriers around the area to be treated.

Grouting is brought about by drilling holes from the surface into the mine workings, on a systematic basis, generally on a grid pattern, and filling the remnant voids with an appropriate grout mix. If it has been impossible to obtain accurate details of the layout and extent of the workings, then the zone beneath the intended structure can be subjected to consolidation grouting. The grouts used in these operations commonly consist of cement, fly ash and sand mixes, economy and bulk being their important features. Pea gravel may be used as a bulk filler where a large amount of grout is required for treatment. Alternatively, foam grouts can be used. If the workings are still more or less continuous, then there is a risk that grout will penetrate the bounds of the zone requiring treatment. In such instances, dams can be built by placing pea gravel down large-diameter drillholes around the periphery of the site. When the gravel mound has been formed it is grouted. The area within this barrier is then grouted. If the old workings contain water, then a gap should be left in the dam through which water can drain as the grout is emplaced. This minimizes the risk of trapped water preventing the voids being filled. As bulk grouting results in a significant change in mass permeability, consideration must be given to the implication of such treatment for groundwater flow.

12.6 Old mine shafts

Centuries of mining in many countries has left behind a legacy of old shafts. Unfortunately, many, if not most, are unrecorded or are recorded inaccurately (Dean, 1967). In addition, there can be no guarantee of the effectiveness of their treatment unless it has been carried out in recent years. The location of a shaft is of great importance as far as the safety of a potential structure is concerned, for although shaft collapse is fortunately an infrequent event, its occurrence can prove disastrous. Moreover, from the economic point of view the sterilization of land due to the suspected presence of a mine shaft is unrealistic. Shafts that were used solely for ventilation and pumping were usually smaller in cross-section than winding shafts and the number of shafts per mine varied. Some of the shallower old coal mines in Britain had as many as 10 shafts, affording easy access to all parts of the workings. On the other hand, some deep mines had only one shaft. If the function of a shaft is known, it may be possible to guess its proximity to other shafts. For example, pumping shafts sometimes occur within a metre or so of each other. Again in Britain, in the latter half of the eighteenth century ventilation shafts for coal mines were connected to the main shaft and at the surface could be from 2 to 8 m away from the latter. After 1863, the legal minimum distance between two shafts had to be 10 feet (3 m), and this was increased to 41 feet 6 inches (13.6 m) in 1887.

The ground around a shaft may subside or, worse still, collapse suddenly (Figure 12.5). Collapse of filled shafts can be brought about by deterioration of the fill, which is usually due to adverse changes in groundwater conditions, or to the surrounding ground being subjected to vibration or overloading. Collapse at an unfilled shaft is usually due to the deterioration and failure of the shaft lining. Shaft collapse may manifest itself as a hole roughly equal to the diameter of the shaft if the lining remains intact or if the ground around the shaft consists of solid rock. More frequently, however, shaft linings deteri-

Figure 12.5 Collapse of a shaft, trapping a car, in Cardiff, South Wales.

orate with age to a point at which they are no longer capable of retaining the surrounding material. If superficial deposits surround a shaft that is open at the top, the deposits eventually collapse into the shaft to form a crown hole at the surface. The thicker these superficial deposits are, the greater will be the dimensions of the crown hole. Such collapses may affect adjacent shafts if they are interconnected.

For obvious reasons, the search for old shafts on land that is about to be developed must extend outside the site in question for a sufficient distance to find any shafts that could affect the site itself if a collapse occurred. An investigation should include a survey of maps, literature and aerial photographs (Anon., 1976). The principal sources of information include plans of abandoned mines, geological records of shaft sinking, geological maps, all available editions of topographic maps, aerial photographs, and archival and other official records.

The success of geophysical methods in locating old shafts depends on the existence of a sufficient contrast between the physical properties of the shaft and its surroundings to produce an anomaly. If no contrast exists a shaft cannot be detected. Moreover, a shaft will frequently remain undetected when the top of the shaft is covered by more than 3 m of fill. The size, especially the diameter, of a shaft influences whether or not it is likely to be detected.

A resistivity survey may be used where there is a significant contrast in electrical resistance between a shaft and its surroundings. Anomalies are detectable if the depth of cover around a shaft is less than its radius, and success has generally been achieved where the cover is less than 1 m in thickness. An isoresistivity map is drawn from the results.

Magnetic surveys, especially those using the proton magnetometer and, more recently, the flux-gate magnetic field gradiometer, have had some success in locating old mine shafts. An isomagnetic contour map is produced from the results (Figure 12.6).

An open shaft offers a gravity contrast, but unfortunately most old shafts are too small to produce a difference in density that is sufficient to yield a gravity

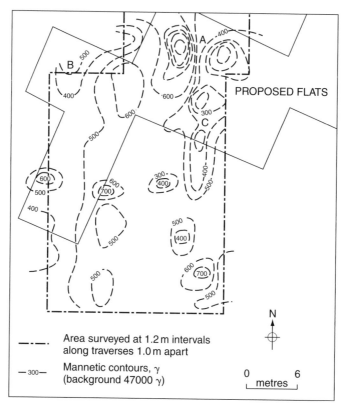

Figure 12.6 Isomagnetic map of a site containing old mine shafts at A, B and C.

anomaly. The maximum depth at which an open shaft can be located by a gravity survey is about 1.5 times the diameter of the shaft. If the shaft has been backfilled the depth is less. The production of a gravity contour map allows any anomalies to be recognized.

Seismic surveys are not generally used to locate old mine shafts. However, cross-hole seismic testing, which involves sending both compressional and shear waves between drillholes, could possibly detect open shafts.

Geochemical exploration depends on identifying chemical changes such as changes in mineral content or the chemical character of the moisture content in the soil associated with old mine workings. In addition, gases such as carbon dioxide, carbon monoxide, methane and nitrogen may accumulate in open or partially filled shafts. Indeed, the most effective geochemical method yet used in the location of abandoned coal mine shafts is that of methane detection. Methane, being a light gas, may escape from old shafts, and methane detectors can record concentrations as low as 1 ppm. Anomalies associated with old coal mine shafts have generally ranged between 10 and 100 ppm. The detector is carried over the site near the ground and should not be used on a windy day. A contour map showing methane concentration is produced.

Confirmation of the existence of a shaft is accomplished by excavation. As this may be a hazardous task, necessary safety precautions should be taken. Gregory (1982) suggested that when a possible position of a shaft has been determined, a mechanical boom-type digger can be used to reveal its presence. The excavator is anchored outside the search area, and harness and lifelines should be used by the operatives. A series of parallel trenches are dug at intervals reflecting the possible diameter of the shaft. If excavation to greater depth than can be achieved by the excavator is required, resort can be made to a dragline. Again, the dragline is safely anchored outside the search area.

Where a site is considered potentially dangerous, in that shaft collapse may occur, or where obstructions prevent the use of earthmoving equipment, a light mobile rig can be used to drill exploratory holes. The rig should be placed on a platform or girders long enough to give protection against shaft collapse. Since many old shafts have diameters around 2 m or even less, drilling should be undertaken on a closely spaced grid. Rotary percussion drilling can be done quickly, and the holes can be angled. Changes in the rate of penetration may indicate the presence of a shaft, or the flush may be lost. Significant differences in the depth of unconsolidated material may indicate the presence of a filled shaft, or the fill may differ from the surrounding superficial material.

Once a shaft has been located, the character of any fill occurring within it needs to be determined. A drillhole alongside the shaft to determine the thickness of the overburden and the stratal succession, especially if the latter is not available from mine plans, proves very valuable. In particular, in old coal mines the positions of the seams may mean that mouthings open into the shaft at those levels. It is also important to record the position of the water table and, if possible, the condition of the shaft lining.

If the depth is not excessive and the shaft is open, it can be filled with suitable granular material, the top of the fill being compacted. If, as is more usual, the shaft is filled with debris in which there are voids, these should be filled with pea gravel and grouted. Dean (1967) suggested that if the exact positions of the mouthings in an abandoned open shaft are known, then these areas should be filled with gravel, the rest of the shaft being filled with mine waste. However, the latter will tend to consolidate much more than will gravel. Anon. (1982) supplied details concerning the concrete cappings needed to seal mine shafts. The concrete capping should take the form of an 'inverted cone'. The zone immediately beneath the capping is grouted.

Obviously, the easiest way to deal with old mine shafts is to avoid locating structures in their immediate vicinity. Because of the possible danger of collapse and cratering it is generally recommended that the minimum distance for siting buildings from open to poorly filled shafts is twice the thickness of the superficial deposits up to a depth of 15 m, unless they are exceptionally weak. A consideration of the long-term stability of shaft linings is required, as well as the effects of increased lateral pressure due to the erection of structures nearby. Protective sheet piles or concrete walls can be constructed around shafts to counteract cratering, but such operations should be carried out with due care.

12.7 Longwall mining and subsidence

In longwall mining, the coal is exposed at a face of up to 220 m between two parallel roadways (Whittacker and Reddish, 1989). The roof is supported only in and near the roadways and at the working face. After the coal has been won and loaded the face supports are advanced, leaving the rocks, in the areas where coal has been removed, to collapse. Subsidence at the surface more or less follows the advance of the working face and may be regarded as immediate (Brauner, 1973a). Trough-shaped subsidence profiles associated with longwall mining develop tilt between adjacent points that have subsided by different amounts, and curvature results from adjacent sections that are tilted by differing amounts. Maximum ground tilts are developed above the limits of the area of extraction and may be cumulative if more than one seam is worked up to a common boundary. Where movements occur, points at the surface subside downwards and are displaced horizontally inwards towards the axis of the excavation (Figure 12.7). Differential horizontal displacements result in a zone of apparent extension on the convex part of the subsidence profile (over the edges of the excavation), while a zone of compression develops on the concave section over the excavation itself. Differential subsidence can cause substantial damage, the tensile strains thereby generated usually being the most effective in this respect.

The surface area affected by ground movement is greater than the area worked in the seam (Figure 12.7). The boundary of the surface area affected is defined

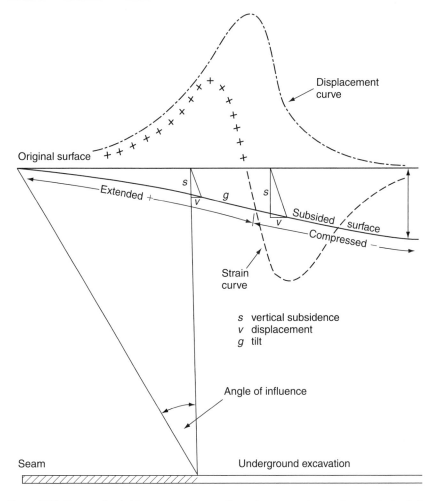

Figure 12.7 Curve of subsidence showing tensile and compressive strains, vertical subsidence and tilt, together with angle of draw (not to scale).

by the limit angle of draw or angle of influence, which varies from 8 to 45°, depending on the coalfield. It would seem that the angle of draw may be influenced by depth, seam thickness and local geology, especially the location of the self-supporting strata above the coal seam.

Comparatively slight deviations in the subsidence profile are accompanied by appreciable variations in strain. In fact, ground movement is three-dimensional, and movements of the vertical and two horizontal components may occur simultaneously.

As mentioned, removal of roof support in longwall mining is followed by collapse of those rocks that were immediately above the coal, since they are subjected to bending and tensile stresses. These broken rocks offer partial support to the super-incumbent roof layers. Nevertheless, stresses in the rock mass remaining in place are increased significantly, and the resultant fracture and associated dilation mean that the rock strength is reduced from a peak to a residual value, with loss in load bearing capacity and redistribution of stress. In addition, laminated rocks may suffer bed separation. The fracture zone defined by the extent of dilation extends at least half the face width above seam level.

There is usually an appreciable difference between the volume of mineral extracted and the amount of subsidence at the surface. For example, Orchard and Allen (1970) showed that where maximum subsidence of 90% of seam thickness occurred, the volume of the subsided ground was only 70% of the coal extracted. This is mainly attributable to bulking.

One of the most important factors influencing the amount of subsidence is the width–depth relationship of the panel removed. In fact, in Britain it has usually been assumed that maximum subsidence generally begins at a width–depth ratio of 1.4:1 (this assumes an angle of draw of 35°; Anon., 1975). This is the critical condition above and below which maximum subsidence is and is not achieved, respectively (Figure 12.8).

However, the width–depth ratio necessary to cause 90% subsidence can usually be achieved only in shallow workings because with deeper workings the critical area of extraction is made up of a number of panels, often with narrow pillars of coal left *in situ* to protect one or other of the roadways. These pillars reduce the subsidence, and as a result some coalfields have no experience of subsidence exceeding 75–80% of maximum.

Orchard (1957) found that ground movements over shallow workings reveal many trends that are not so apparent over deep workings, mainly because of the masking effect of thick overburden. For instance, for a given maximum subsidence the curvature of the ground surface is more marked over shallow workings than over deep workings owing to the smaller distance over which the subsidence curve is spread. The horizontal strains are proportional to subsidence and inversely proportional to the depth of workings. The maximum slope in the ground in a subsidence trough is also proportional to the subsidence at the bottom of the trough and inversely proportional to the maximum horizontal strain.

According to Brauner (1973b), in addition to the rate of advance and size of critical area, the duration of surface subsidence depends on geological conditions, depth of extraction, kinds of packing and previous extraction. For example, it lasts longer for thickly bedded or stronger overburden and for complete caving. Depth, in particular, influences the rate of subsidence: first, because the diameter of the area of influence and therefore the time taken for a working, with a given rate of advance, to traverse it, increases with depth; and, second, because at greater depths several workings may be necessary before

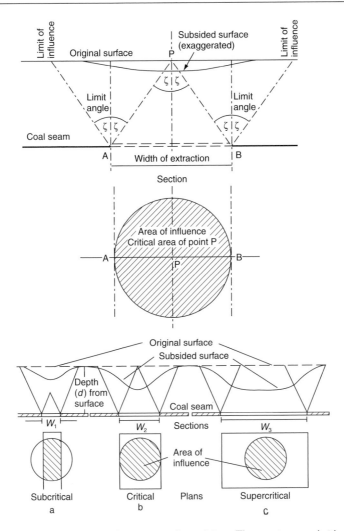

Figure 12.8 Subcritical, critical and supercritical conditions. The maximum subsidence at a point at the surface generally begins when the depth to seam/width of extraction ratio equals 1:1.4. This is the critical condition, above and below which maximum subsidence is not achieved (subcritical condition) and is achieved over a wider area (supercritical condition). The figure also shows that the area affected by subsidence is greater than the area from which coal is extracted.

the area of influence is completely worked out. Consequently, the time that elapses before subsidence is complete varies according to circumstances.

Residual subsidence takes place at the same time as instantaneous subsidence and may continue after the latter events for periods of up to two years. The magnitude of residual subsidence is proportional to the rate of subsidence of the surface and is related to the mechanical properties of the rocks above the seam. For instance, strong rocks produce more residual subsidence than weaker ones. Residual subsidence rarely exceeds 10% of total subsidence if the face is stopped within the critical width, but falls to 2–3% if the face has passed the critical width. Very occasionally, values greater than 10% have been recorded.

12.7.1 Geological factors and subsidence due to longwall mining

Ground movements induced at the surface by mining activities are influenced by variations in the ground conditions, especially by the near-surface rocks and superficial deposits. However, the reactions of surface deposits to ground movements are usually difficult to predict reliably. Indeed, it has been suggested that 25% of all cases of mining subsidence undergo some measure of abnormal ground movement that, at least in part, is attributable to the near-surface strata.

In concealed coalfields, the strata overlying the Coal Measures often influence the basic movements developed by subsidence. In fact, abnormal subsidence behaviour and inconsistent movements are much more common in concealed than exposed coalfields. The occurrence of abnormally thick beds of sandstone can modify stratal movement due to mining. Such beds may resist deflection, in which case stratal separation occurs and the effective movements at the surface are appreciably less than would otherwise be expected. The differences in behaviour disappear when the extraction becomes wide enough for the sandstone to collapse, when subsidence behaviour reverts to normal.

The necessary readjustment to subsidence in weak strata can usually be accommodated by small movements along joints. However, as the strength of the surface rock and the joint spacing increases so the movements tend to become concentrated at fewer points, so that in massive limestones and sandstones movements may be restricted to master joints. Well-developed joints or fissures in such rocks concentrate differential displacement. Tensile and compressive strains many times the basic values have been observed at such discontinuities (Bell and Fox, 1991). For example, joints may gape anything up to a metre in width at the surface. According to Shadbolt (1978) it is quite common for the total lateral movement caused by a given working to concentrate in such a manner. In such instances, no strain is measurable on either side of the discontinuity concerned.

Drift deposits are often sufficiently flexible to obscure the effects of movements at rockhead. In particular, thick deposits of till tend to obscure tensile effects. On the other hand, superficial deposits may allow movements to affect larger areas than otherwise.

Faults also tend to be locations where subsidence movement is concentrated, thereby causing abnormal deformation of the surface. While subsidence damage to structures located close to or on the surface outcrop of a fault can be very severe, in any particular instance the areal extent of such damage is limited, often being confined to within a few metres of the outcrop. Also, many faults have not reacted adversely when subjected to subsidence (Hellewell, 1988). The extent to which faults influence and modify subsidence movements cannot be quantified accurately.

In fact, faults tend to act as boundaries controlling the extent of the subsidence trough. When workings terminate against a fault plane that has an angle of hade larger than the angle of draw, then the subsidence profile extends to the surface with associated permanent strains, probably accompanied by severe differential subsidence. Indeed, a subsidence step may occur at the outcrop of the fault. Where the width of the extraction is large enough, the surface subsidence at a point vertically above the intersection of the fault and the workings is approximately one-half that of the maximum subsidence for that working. When the hade of the fault is less than the angle of draw, the fault again determines the extent of the subsidence trough, which in this case is less than that normally expected.

By accommodating most of the subsidence movement, faults can reduce the amount of damage, which would normally be experienced over a wider area. The presence of a fault does not increase the amount of subsidence on the side of the fault nearer the workings compared with normal conditions, but subsidence on the opposite side is considerably reduced (by upwards of 50%).

Faults are most likely to react adversely when their hade is less than 30°, when they have a simple form, and when the material occupying the fault zone does not offer high frictional resistance to movement. For example, subsidence steps tend to occur when faults represent single, sharp stratal breaks (Lee, 1966). By contrast, if a fault consists of a relatively wide shatter zone, then the surface subsidence effects are usually less pronounced but more predictable in terms of location and amount. The thickness and nature of the unconsolidated deposits above bedrock can influence the magnitude of such features.

The location of the workings in relation to a fault and method of mining determines the position and nature of anomalies. The most notable steps occur when the coal is worked beneath the hade of a fault, faces in other positions being much less likely to cause differential movement. Steps are usually down towards the goaf, but if old workings exist then steps may occasionally occur away from the face. Furthermore, a single working of small width–depth ratio approaching a fault at right angles is less likely to cause a step than a large width–depth working parallel to the fault. A fault step is much more likely to develop when the fault has been affected by previous workings in shallow seams than it is for a single working in a virgin area. Workings on the upthrown side of a fault are more likely to cause stepping than similar workings on the downthrown side.

Once differential movement has occurred, further mining in the area can cause renewed movement, which at times may be out of all proportion to the thickness and extent of the coal extracted. As successive workings increase the effect on a fault plane, so the chance of steps developing is increased.

In concealed coalfields where, for example, a fault passes through block-jointed sandstones or limestones, severe fissures or fractures can occur at the surface. The fissures are generally parallel to the line of the fault but can occur up to 300 m away from it.

12.7.2 Prediction of subsidence due to longwall mining

An important feature of subsidence due to longwall mining is its high degree of predictability. Usually, movements parallel and perpendicular to the direction of face advance are predicted. Although adequate for many purposes, such treatment does not consider the three-dimensional nature of ground movement.

Empirical methods of subsidence prediction, such as those developed by the former National Coal Board, have been developed by continuous study and analysis of survey data from British coalfields (Anon., 1975). Since they do not take into account the topography, the nature of the strata and geological structure involved or how the rock masses are likely to deform, such empirical relationships can be applied only under conditions similar to those in which the original observations were made. Nevertheless, these methods were continually being refined so that they could yield more accurate results. In fact, they allow the amount of subsidence due to longwall mining in Britain to be predicted, usually within $\pm$ 10%.

Burton (1978) developed an empirical three-dimensional model of subsidence prediction, which can be represented graphically. In this way, a series of average subsidence and strain-profile shapes, which are related to width–depth ratios, can be portrayed as a single graph. Burton showed that surface points move inwards towards the panel and also move backwards and forwards in the line of advance. By fixing one point and then calculating the displacements of another point in relation to this, a series of differential movements is obtained. Accordingly, strains can be calculated along the line between the two points under consideration. In other words, strain can be calculated in any direction with respect to face advance. This is important because the value of strain, or change in surface length, is direction-related, that is, a zone of tension or compression, which is related to a line normal to the ribside, has no meaning for a line running in another direction. Hence, a strain contour map must be related to a surface direction.

The theoretical methods of subsidence assessment assume that stratal displacement behaves according to one of the constitutive equations of continuum mechanics over most of its range. The continuum theories have been developed from the analysis of a displacement discontinuity produced by a slit in an infinite elastic half-space. Analytical procedures were subsequently developed for three types of subsurface excavation based on elastic ground conditions, that is, non-closure, partial closure and complete closure. Further work extended the closed-form solution to transversely isotopic ground conditions in both two and three dimensions.

Numerical models permit quantitative analysis of subsidence problems and are not subject to the same restrictive assumptions required for the closed-form analytical solutions. Finite-element modelling has frequently been applied to subsidence problems, since it can accommodate non-homogeneous media, non-linear material behaviour and complicated mine geometries. Although the finite-element method has the advantage of allowing the elastic constants to differ for each element so that appropriate constants can be used for each layer, because of the involvement of considerable depths below the surface and either side of the excavation zone, it is perhaps of questionable value in terms of estimating subsidence.

Alternatively, finite-difference models can be used for the large-strain, non-linear phenomena associated with subsidence development. Other elastic approaches employing numerical techniques include boundary-element methods.

Most viable methods of subsidence prediction fall into the category of semi-empirical methods, since fitting field data to theory often results in good correlations between predicted and actual subsidence. There are two principal methods of semi-empirical prediction, namely, the profile-function and the influence-function methods (Brauner, 1973b; Hood et al., 1983). The profile-function method basically consists of deriving a function that describes a subsidence trough. The equation produced is normally for one half of the subsidence profile, and is expressed in terms of maximum subsidence and the location of the points of the profile. Provided that the input data are available, the profile-function method provides an accurate technique for the prediction of subsidence where the mine geometry is relatively simple, as in longwall mining.

The influence-function approach to prediction of subsidence is based upon the principle of superposition. The subsidence trough is regarded as a combination of many infinitesimal troughs formed by a number of infinitesimal extraction elements. There are two types of influence function, namely, the zone area or circle method and complementary influence functions.

In the zone area or circle method of subsidence prediction, surface subsidence is estimated by constructing a number of concentric zones around a surface point, the radius of the outer zone being equal to the radius of the area of influence. The subsidence at a surface point is obtained from the summation of the proportions of coal that are extracted in each zone, multiplied by its particular subsidence factor. In practice, it has been found that three, five or seven zones are suitable for most estimations. The method permits estimation of subsidence that develops when panels of irregular shape are mined. Ren et al. (1987) compared the results obtained by using the zone area method of prediction with those derived by the National Coal Board method (Anon., 1975) and found that they were similar for both subsidence and displacement.

The fundamental concept of complementary influence functions is that the separate influence functions describing the response of mined and unmined zones act together to produce subsidence (Sutherland and Munson, 1984). Each influence function is defined by the response of a limit element, that is, an unmined element for the coal left in place and a mined element for the void created by extraction. The amount of subsidence is predicted by appropriately summing these elements over the entire seam, it being the sum of the influence of both the mined and unmined response. Hence complementary influence functions can be used to compute subsidence above mines with complex geometry (that is, for room and pillar as well as longwall workings). This method, however, tends to overestimate the subsidence directly over the ribside.

12.8 Damage associated with ground movements

The amount of damage caused by subsidence depends upon the magnitude and type of the ground movements, structural factors and geological factors. The effects on a structure of the rate at which ground movements occur and their magnitude are governed by its location and orientation in relation to the underground workings and the depth at which mining is taking place, as well as panel width and extraction. The different types of ground movement associated with mining subsidence affect different structures in different ways. For instance, vertical subsidence may seriously affect drainage systems, and tilt may cause serious concern as far as railroads and tall chimneys are concerned. Damage to buildings is generally caused by differential horizontal movements and the concavity and convexity of the subsidence profile that gives rise to compression and extension in the structure itself, the latter generally being the more serious. Usually, however, it is not just a simple matter of examining the reaction of a structure to a particular value of tensile or compressive strain. For example, it is quite common for a building to undergo compressive strains in one direction and tensile strains in another direction. A building may also be subjected to alternate phases of tensile and compressive ground movements, so there is a dynamic effect to consider. Thus any acceptable design for a structure situated in an active mining area must have regard for the nature, degree and periodicity of the ground movements likely to be caused by mining. The structural factors that influence the amount of damage caused include the size and shape of the structure, the type of foundation, the method of construction and the type of materials used, along with the existing state of repair of the structure.

Ground movement that adversely affects the safety or function of a structure is unacceptable. However, with many buildings their appearance is also of concern, and therefore significant cracking of architectural features is also unacceptable. Hence an estimation of the amount of subsidence that will adversely affect structural members and/or architectural features is required. This is influenced by many factors, including the type and size of the structure, and the properties of the materials of which it is constructed, as well as the rate and nature of the subsidence. Because of the complexities involved, critical movements have not been determined analytically. Instead, almost all criteria for tolerable subsidence or settlement have been established empirically on the basis of observations of ground movement and damage in existing buildings.

Two parameters have commonly been used for developing correlations between damage and differential settlement, namely, angular distortion and deflection ratio (Figure 12.9). Angular distortion (β) describes the rotation of a straight line joining two reference points. It is often related to the tilt (ω). In this way, the modified value is more representative of the deformed shape of the structure. The deflection ratio (Δ/L) is defined as the maximum displacement (Δ) relative to a straight line between two points divided by the distance (L) separating the points.

Skempton and MacDonald (1956) selected angular distortion as the critical index of ground movement. They concluded that cracking of load bearing walls or panel walls in frame structures is likely when δ/L exceeds $1/150$. Skempton and MacDonald also suggested that a value of $\delta/L = 1/500$ could be used as a design criterion that provides some factor of safety against cracking. Subsequently, Bjerrum (1963) also suggested limits for damage criteria based on angular distortion (Table 12.2).

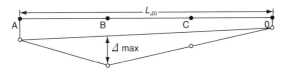

(a) Definitions of settlement ρ, relative settlement $\delta\rho$, rotation θ, and angular strain α

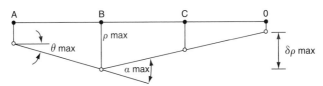

(b) Definitions of relative deflection Δ, and deflection ratio Δ/L

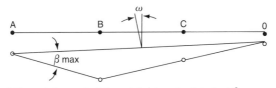

(c) Definitions of tilt ω and relative rotation (angular distortion) β

Figure 12.9 Definitions of ground movement (after Burland *et al.*, 1977).

Table 12.2 Limiting angular distortion (after Bjerrum, 1963).

Category of potential damage	Δ/L
Danger to frames with diagonals	1/600
Safe limit for no cracking of buildings*	1/500
First cracking of panel walls	1/300
Tilting of high rigid buildings becomes visible	1/250
Considerable cracking of panel and brick walls	1/150
Danger of structural damage to general buildings	1/150
Safe limit for flexible brick walls, L/H > 4*	

*Safe limits include a factor of safety.

Polshin and Tokar (1957) defined allowable displacement in terms of deflection ratio (Δ/L). There are a number of differences between these criteria and those presented by Skempton and MacDonald (1956). For instance, frame structures and load bearing walls are treated separately. The allowable displacement for frames is expressed in terms of the slope or the differential displacement between adjacent columns. The limiting values quoted by Polshin and Tokar range from 1/500 for steel and concrete frame infilled structures to 1/200 where there is no infill or danger of damage to cladding. The maximum allowable deflection ratio was assumed to be related to the development of a critical level of tensile strain in a wall. For brick walls, the critical tensile strain was taken as 0.05%. Polshin and Tokar adopted more stringent limits for differential movement of load bearing brick walls than did Skempton and MacDonald. The deflection ratio at which cracking occurs in brick walls was related to the length-to-height ratio (L/H) of the wall. For L/H ratios of less than 3 the maximum deflection ratios quoted varied from 0.3×10^{-3} to 0.4×10^{-3}, while for L/H ratios greater than 5 the values ranged from 0.5×10^{-3} to 0.7×10^{-3}.

Burland and Wroth (1975) and Burland *et al.* (1977) also used the deflection ratio (Δ/L) at which the critical tensile strain (0.075%) is reached as a criterion for allowable ground movement. They proposed that limiting deflection criteria should be developed for at least three different cases. Diagonal strain is critical in the case of frame structures (which are relatively flexible in shear) and for reinforced load bearing walls (which are relatively stiff in direct tension). Bending strain is critical for unreinforced masonry walls and structures that have relatively low tensile resistance. Hence unreinforced load bearing walls, particularly when subjected to hogging, are more susceptible to damage than frame buildings. For unreinforced load bearing walls in the sagging mode, Burland and Wroth gave values of 0.4×10^{-3} (1/2500) for an L/H ratio of 1, and 0.8×10^{-3} (1/1250) for a ratio of 5. However, they pointed out that cracking in the hogging mode occurs at half of these values of deflection ratio.

Investigations carried out by the National Coal Board (Anon., 1975) have revealed that typical mining damage starts to appear in conventional structures when they are subjected to effective strains of 0.5 to 1.0 mm m^{-1}, and damage can be classified as negligible, slight, appreciable, severe or very severe (Figure 12.10 and Table 12.3). However, this relationship between damage and change in length of a structure is valid only when the average ground strain produced by mining subsidence is equalled by the average strain in the structure. In fact, this is commonly not the case, strain in the structure being less than it is in the ground. Hence Figure 12.10 serves only as a crude guide to the likely damaging effects of ground strains induced by mining subsidence. In addition, it takes no account of the design of the structure or of construction materials. Nevertheless, it indicates that the larger a building, the more susceptible it is to differential vertical and horizontal ground movement.

More recent criteria for subsidence damage to buildings has been proposed by Bhattacharya and Singh (1985), who recognized three classes of damage: namely, architectural, which is characterized by small-scale cracking of plaster and doors and windows sticking; functional damage, which is characterized by instability of some structural elements, jammed doors and windows, broken

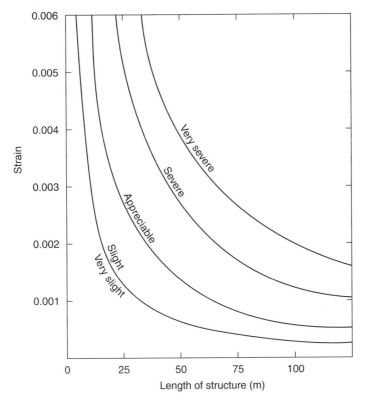

Figure 12.10 Relationship of damage to length of structure and horizontal strain (Anon., 1975).

Table 12.3 National Coal Board classification of subsidence damage (after Anon., 1975).

Change in length of structure (mm)	Class of damage	Description of typical damage
Up to 30	1 Very slight or negligible	Hair cracks in plaster. Perhaps isolated slight fracture in the building not visible on outside.
30–60	2 Slight	Several slight fractures showing inside the building. Doors and windows may stick slightly. Repairs to decoration probably necessary.
60–120	3 Appreciable	Slight fracture showing on outside of building (or main fracture). Doors and windows sticking, service pipes may fracture.
120–180	4 Severe	Service pipes disrupted. Open fractures requiring rebonding and allowing weather into the structure. Window and door frames distorted, floors sloping noticeably. Some loss of bearing in beams. If compressive damage, overlapping of roof joints and lifting of brickwork with open horizontal fractures.
> 180	5 Very severe	As above, but worse and requiring partial or complete rebuilding. Roof and floor beams lose bearing and need shoring up. Windows broken with distortion. Severe slopes on floors. If compressive damage, severe buckling and bulging of the roof and walls.

window panes, and restricted building services; and structural damage, in which primary structural members are impaired, there is a possibility of collapse of members, and complete or large-scale rebuilding is necessary. Their conclusions are summarized in Table 12.4. They observed that basements were

Table 12.4 Recommended damage criteria for buildings (after Bhattacharya and Singh, 1984).

Category	Damage level	Angular distortion (mm/m)	Horizontal strain (mm/m)	Radius of curvature (km)
1	Architectural	1.0	0.5	–
	Functional	2.5–3.0	1.5–2.0	20
	Structural	7.0	3.0	–
2	Architectural	1.3	–	–
	Functional	3.3	–	–
	Structural	–	–	–
3	Architectural	1.5	1.0	–
	Functional	3.3–5.0	–	–
	Structural	–	–	–

the most sensitive parts of houses with regard to subsidence damage, and therefore that basements usually suffered more damage than the rest of a building.

12.9 Measures to mitigate the effects of subsidence due to longwall mining

The contemporaneous nature of subsidence associated with longwall mining sometimes affords the opportunity for planners to phase long-term surface development in relation to the cessation of subsidence (Bell, 1987). However, the relationship between future programming of surface development and that of subsurface working may be difficult to coordinate because of the differences that may arise between the programmed intention and the performance achieved. Another important aspect, as far as planning and development at the surface are concerned, is that this type of subsidence is predictable. Damage attributable to subsidence due to longwall mining can be reduced or controlled by precautionary measures incorporated into new structures in mining areas, and preventive works applied to existing structures (Anon., 1977). Several factors have to be considered when designing buildings for areas of active mining. First, where high ground strains are anticipated, the cost of providing effective rigid foundations may be prohibitive. Second, experience suggests that buildings with deep foundations, on which thrust can be exerted, suffer more damage than those in which the foundations are more or less isolated from the ground. Third, because of the relationship between ground strain and size of structure, very long buildings should be avoided unless their long axes can be oriented normal to the direction of principal ground strain. Lastly, although tall buildings may be more susceptible to tilt rather than to the effects of horizontal ground strain, tilt can be corrected by using jacking devices.

The most common method of mitigating subsidence damage is by the introduction of flexibility into a structure (Bell, 1988a). In flexible design, structural elements deflect according to the subsidence profile. The foundation therefore remains in contact with the ground as subsidence proceeds. One of the most notable examples of a flexible form of construction is the CLASP (Consortium of Local Authorities Special Programme) system, which incorporates a structural framework that is sufficiently flexible to accommodate differential subsidence without cantilevering over the approaching subsidence depression. In other words, the building is constructed on a raft, with a superstructure of lightweight steel, jointed with pins and braced with a spring-loaded system. The frame is clad so that movement can occur without distortion, and special attention is paid to the flexibility of window openings, stairs and services. If a large building is required, it is desirable to separate it into small units and to provide a gap of at least 50 mm between each pair of units, the space extending to foundation level.

Flexibility can also be achieved by using specially designed rafts. Raft foundations should be as shallow as possible, preferably above ground, so that compressive strains can take place beneath them instead of transmitting direct compressive forces to their edges. They should be constructed on a membrane so that they will slide as ground movements occur beneath them. For instance, reinforced concrete rafts laid on granular material reduce friction between the ground and the structure. Where relatively small buildings (up to about 30 m in length) are concerned, they can be erected on a 'sandwich' raft foundation. Cellular rafts have been used for multi-storey buildings.

The use of piled foundations in areas of mining subsidence presents its own problems. The lateral and vertical components of ground movement as mining progresses mean that the pile caps tend to move in a spiral fashion and that each cap moves at a different rate and in a different direction according to its position relative to the mining subsidence. Such differential movements and rotations would normally be transmitted to the structure, with a corresponding readjustment of the loadings on the pile cap. In order to minimize the disturbing influence of these rotational and differential movements it is often necessary to allow the structure to move independently of the piles by the provision of a pin joint or roller bearing at the top of each pile cap. It may be necessary to include some provision for jacking the superstructure where severe dislevelment is likely to occur.

Preventive techniques can frequently be used to reduce the effects of movements on existing structures. The type of technique depends upon the type of structure and its ability to withstand ground movements, as well as on the amount of movement likely to occur. Again, the principal objective is to introduce greater flexibility or reduce the amount of ground movement that is transmitted to the structure. In the case of buildings longer than 18 m, damage can be reduced by cutting them into smaller, structurally independent units. The space produced should be large enough to accommodate deflection. In particular, such items as chimneys, lift shafts, machine beds, etc., can be planned independently of the other, generally lighter, parts of a building and separated from the main structure by joints through foundations, walls, roof and floor that allow freedom of movement. Extensions and outbuildings should similarly be separated from the main structure by such joints.

The excavations of trenches around buildings subjected to compressive ground strains can reduce their effect on the buildings by as much as 50%. The trench is about 1 m from the perimeter wall of a structure and extends down to foundation level, which effectively breaks the continuity of the surface and hence the foundations are isolated from side-thrust. The trenches are backfilled with compressible material and covered with concrete slabs.

Structures that are weak in tensile strength can be afforded support by strapping or tie-bolting together where they are likely to undergo extension. However, some distortion can occur at the points where the ties are fixed.

Flexible joints can be inserted into pipelines to combat the effects of subsidence. Thick-walled plastics, which are more able to withstand ground movements, have been used for service pipes. Where pipes are already laid they can be exposed and flexible joints inserted or freed from contact with the surrounding ground. The latter reduces the ground strains transmitted to the pipeline. The pipe can be backfilled with, for example, pea gravel so that the friction between it and the surrounding soil is lowered. If failure is likely to cause severe problems, then services can be relocated or duplicated.

Damage to surface structures can also be reduced by adopting specially planned layouts of underground workings that take account of the fact that surface damage to structures is primarily caused by ground strains. Thus to minimize the risk of damage, underground extraction must be planned so that surface strain is reduced or eliminated. Harmonic mining reduces the effects of strain but not necessarily the amount of subsidence at the surface. The system involves two or more faces being worked simultaneously but with careful selection of mining dimensions and rates of advance so that the resultant strains from each panel tend to cancel each other out. The stepped-face layout system has been used in Britain, and it has been alleged that it can reduce the tensile strain by up to 50%. Unfortunately, it is seldom possible to arrange the direction and location of underground workings to suit the need to minimize surface movement in the neighbourhood of a particular surface structure.

Pillars of coal can be left in place to protect surface structures above them. In panel and pillar mining, pillars are left between relatively long but narrow panels. Adjacent panels are designed with a pillar of sufficient width in between so that the interaction of the ground movements results in flat subsidence profiles at the surface with low ground strains. The resultant surface subsidence ranges from 3 to 20% of the thickness of the seam. However, subsidence increases with depth of mining due to the greater loading carried by the pillars. Hence pillar widths have to be increased with increasing depth. Narrow shortwall (e.g. panels 40 m wide with intervening pillars about 50 m wide) or single-entry panel methods of extraction that are narrow enough to allow strata to bridge the goaf can be used, so resulting in little collapse and therefore reduced ground movement at the surface.

Maximum subsidence can also be reduced by packing the goaf. This subsidence factor, because it varies with method, packing materials and depth, needs to be determined from actual observations of individual cases in each mining district. Pneumatic stowing can reduce subsidence by up to 50% (Anon., 1975).

12.10 Metalliferous mining

Several underground mining methods are used for the extraction of metalliferous deposits. In partial extraction methods, solid pillars are left unmined to provide support to the underground workings. For example, sill pillars and shaft pillars are designed to protect important underground areas. Pillars can be variable in shape and size and in many cases, since pillars often represent reserves of ore, they are extracted or reduced in size towards the end of the life of the mine. Pillar design takes account of the strength of the pillars, the strength of the roof and floor rocks, the capacity of the rock to span between the pillars, and the stresses acting on the pillars. In modern practice it is possible to design against the occurrence of surface subsidence. However, this procedure was not practised in earlier times. In such situations, subsidence of the surface is a possibility. The collapse of the roof or hanging wall of the workings between pillars can result in localized surface subsidence. The subsidence profile may be very severe, with large differential subsidence, tilts and horizontal strains. Pillar collapse can give rise to localized or substantial areas of subsidence. If variable-sized pillars are present, or if larger barrier pillars are provided at regular spacings, the area of subsidence is likely to be restricted. The resulting subsidence profile is likely to be irregular, with large differential subsidence, tilts and horizontal strains at the perimeter of the subsidence area. The presence of faults can give rise to fault steps (Stacey and Rauch, 1981). Failure of roof and/or floor foundations leads to a similar type of subsidence behaviour. When mining inclined ore bodies, the steeper the dip, the more localized the subsidence or potential subsidence is likely to be.

Mining of tabular stopes with substantial spans is practised in the gold and platinum mines of southern Africa. Extraction of one, two or more reefs has taken place, each typically with a stoping width of 1 to 2 m. On the other hand, when this type of mining occurs at shallow depth, detrimental subsidence has frequently taken place. The surface profile is commonly very irregular and tension cracks have often been observed. The effects are dependent on the dip of the reefs, which vary from as little as 10° to vertical. Closure of stopes is more likely for flatter dips, and hence surface movements are more likely to occur under such conditions. Steeply dipping stopes tend to remain open and present a longer-term hazard. Adverse subsidence is not usually observed when the mining depth exceeds 250 m. Most of the detrimental subsidence that has occurred has been associated with geological features such as dyke contacts and faults (Stacey and Rauch, *ibid.*).

Many metalliferous deposits occur in large disseminated ore bodies, and often the only way in which such ore bodies can be mined economically is by means of very high-volume production. When the rock mass is sufficiently competent, large open stopes can be excavated. The sizes of the open stopes are the subject of design, and some form of pillar usually separates adjacent open stopes. The method is therefore a partial extraction approach, but open stopes may have spans of between 50 and 100 m. In caving, material from above is allowed to cave into the opening created by the mined extraction. Since caving occurs above the ore body, it usually progresses through to the surface to form a subsidence crater. This crater usually has several scarps around its perimeter and contains material that has subsided *en bloc* overlying caved and rotated material, and loosely consolidated ravelled material. Owing to the large volumes that are extracted during mining, the extent and depth of the subsidence crater can be very large.

12.11 Subsidence associated with fluid abstraction

12.11.1 Subsidence associated with groundwater abstraction

Subsidence of the ground surface occurs in areas where there is intensive abstraction of groundwater, that is, where abstraction exceeds natural recharge and the water table is lowered, the subsidence being attributed to the consolidation of the sedimentary deposits as a result of increasing effective stress (Bell, 1988b). The total overburden pressure in partially saturated or unsaturated deposits is borne by their granular structure and the pore water. When groundwater abstraction leads to a reduction in pore water pressure by draining water from the pores, this means that there is a gradual transfer of stress from the pore water to the granular structure. For instance, if the groundwater level is lowered by 1 m, then this gives rise to a corresponding increase in average effective overburden pressure of 10 kN m^{-3}. As a result of having to carry this increased load the fabric of the deposits affected may deform in order to adjust to the new stress conditions. In particular, the void ratio of the deposits concerned undergoes a reduction in volume, the surface manifestation of which is subsidence (Figure 12.11). Surface subsidence does not occur simultaneously with the abstraction of fluid from an underground reservoir; rather it occurs over a longer period of time than that taken for abstraction.

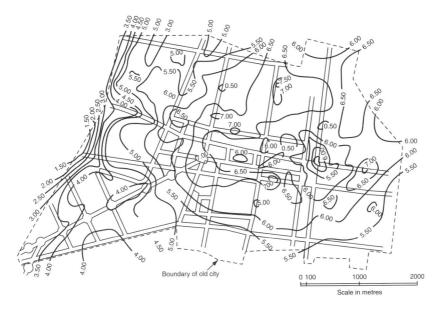

Figure 12.11 Subsidence due to the abstraction of groundwater in Mexico City, 1891–1959. Subsidence contours are given in 0.5 m intervals.

The amount of subsidence that occurs is governed by the increase in effective pressure (that is, the magnitude of the decline in the water table), the thickness and compressibility of the deposits involved, the depth at which they occur, the length of time over which the increased loading is applied, and possibly the rate and type of stress applied (Lofgren, 1968). The rate at which consolidation occurs depends on the thickness of the beds concerned as well as the rate at which pore water can drain from the system, which in turn is governed by its permeability. Thick, slow-draining, fine-grained beds may take years or decades to adjust to an increase in applied stress, whereas coarse-grained deposits adjust rapidly. However, the rate of consolidation of slow-draining aquitards reduces with time and is usually small after a few years of loading.

Consolidation may be elastic or non-elastic, depending on the character of the deposits involved and the range of stresses induced by a decline in the water level. In elastic deformation stress and strain are proportional, and consolidation is independent of time and reversible. Non-elastic consolidation occurs when the grain structure of a deposit is rearranged to give a decrease in volume, that decrease being permanent. Generally, recoverable consolidation represents compression in the pre-consolidation stress range, while irrecoverable consolidation represents compression due to stresses greater than the pre-consolidation stress (Lofgren, 1979).

A formation is generally in equilibrium with the effective pressure prior to the water table being lowered by groundwater abstraction. As mentioned above, as the water table is lowered, the effective weight of the deposits in the dewatered zone increases since the buoyancy effect of the pore water is removed. This increase in load is transmitted to the deposits beneath the newly established level of the water table. If a confined aquifer is overlain by an unconfined aquifer, then the effective pressure is affected by changes in the level of the piezometric surface as well as in the position of the water table.

Subsidence is measured by periodic precision levelling of bench marks referenced to some stable datum outside the area affected. Use can also be made of marker points attached to structures and surface settlement plates. Photogrammetric techniques have also been used to monitor subsidence (Gubellini et al. 1986).

Borehole extensometers can be used to monitor the amount of consolidation suffered by subsurface deposits, the change in thickness of the deposits in the depth interval above some bench mark set in the formation being recorded. Inclinometers can be used to observe horizontal ground movements. Long-term records from standard extensometers, in conjunction with those from piezometers, have made possible the determination of the properties of aquifer systems that control subsidence. Newer, extra high-resolution extensometers permit definition of the compressibility and hydraulic conductivity of thin individual aquitards by means of short-term pumping tests.

Lofgren (1979) recorded that twenty years of precise field measurements in the San Joaquin and Santa Clara valleys of California have indicated a close correlation between hydraulic stresses induced by groundwater pumping and consolidation of the water bearing deposits. He went on to note that the stress–strain characteristics of the producing aquifer systems have been established and that the storage parameters of the systems have been determined from such data. What is more, this has provided a means by which the response of a groundwater system to future pumping stresses can be predicted. According to Lofgren, the storage characteristics of compressible formations change significantly during the first cycle of groundwater abstraction. Such abstraction is responsible for permanent consolidation of any fine-grained interbedded formations, consolidation being brought about by the increase in effective pressure. The water released by consolidation represents a one-time, and sometimes important, source of water to wells. During a second cycle of prolonged pumping overdraft much less water is available to wells and the water table is lowered much more rapidly.

Methods that can be used to arrest or control subsidence caused by groundwater abstraction include reduction of pumping draft, artificial recharge of aquifers from the ground surface and repressurizing the aquifer(s) involved via wells, or any combination thereof. The aim is to manage the rate and quantity of water withdrawal so that its level in wells is either stabilized or raised somewhat so that the effective stress is not further increased.

Reduction of pumping draft can be brought about by importing water from outside the area concerned, by conserving or reducing the use of water, by treating and reusing water, or by decreasing the demand for water. Some of these measures may require legal enforcement. The geological conditions determine whether or not artificial recharge of aquifers from the ground surface is feasible. If confining aquitards or aquicludes inhibit the downward percolation of water, then such treatment is impractical. Repressurizing confined aquifer systems from wells may prove the only viable means of slowing down and eventually halting subsidence. Generally, the results of repressurizing aquifers from wells prove satisfactory when clean water is used; the problems that have occurred have usually involved clogging of well or aquifer. Some of the causes of clogging include air entrainment, micro-organisms and fine particles in the recharge water.

12.11.2 Sinkholes and subsidence

Many, if not most, sinkholes are induced by man's activities, that is, they result from declines in groundwater level, especially those due to excessive abstraction. For example, Jammal (1986) recorded that 70 sinkholes appeared in Orange and Seminole Counties in central Florida over the previous 20 years. Most developed in those months of the year when rainfall was least (that is, April and May) and withdrawal of groundwater was high. Nonetheless, the appearance of a sinkhole at the surface represents a late-stage expression of processes that have been in operation probably for thousands of years.

Most collapses forming sinkholes result from roof failures of cavities in unconsolidated deposits. These cavities are created when the unconsolidated deposits move or are eroded downwards into openings in the top of bedrock. Collapse of bedrock roofs, compared with the migration of unconsolidated deposits into openings in the top of bedrock, is rare.

Some induced sinkholes develop within hours of the effects of man's activity being imposed upon the geological and hydrogeological conditions. Several collapse mechanisms have been assessed and have included the loss of buoyant support to roofs of cavities or caverns in bedrock previously filled with water and to residual clay or other unconsolidated deposits overlying openings in the top of bedrock; increase in the amplitude of water level fluctuations; increases in the velocity of movement of groundwater; and movement of water from the land surface to openings in underlying bedrock where most recharge previously had been rejected since the openings were occupied by water.

Areas underlain by highly cavernous limestones possess most dolines, hence doline density has proven a useful indicator of potential subsidence, as has sinkhole density. As there is preferential development of solution voids along zones of high secondary permeability because these concentrate groundwater flow, data on fracture orientation and density, fracture intersection density and the total length of fractures have been used to model the presence of solution cavities in limestone. Therefore the location of areas of high risk of cavity collapse has been estimated by using the intersection of lineaments formed by fracture traces and lineated depressions (dolines). Aerial photographs have proved particularly useful in this context.

Brook and Alison (1986) described the production of subsidence susceptibility maps of a covered karst terrain in Dougherty County, southwest Georgia. These have been developed using a geographical information system that incorporates much of the data referred to in the previous paragraph. The county was partitioned into 885 cells, each 1.18 km^2 in area. Five cell variables were used in modelling, namely, sinkhole density, sinkhole area, fracture density, fracture length and fracture intersection density. Broadly similar subsidence susceptibility models were developed from cell data by intersection and separately by linear combination. In the intersection technique, cells having specified values for all variables were located and mapped. In the linear combination technique, a map value, $MV = W_1\tau_1 + \dots + W_n\tau_n$, where W is an assigned variable weight and τ an assigned weight value, was calculated for each cell.

Dewatering associated with mining in the gold bearing reefs of the Far West Rand, South Africa, which underlie dolostone and unconsolidated deposits, has led to the formation of sinkholes (Figure 12.12) and produced differential subsidence over large areas (Bezuidenhout and Enslin, 1970). Hence certain areas became unsafe for occupation and were evacuated. It then became a matter of urgency that the areas that were subject to subsidence or to the occurrence of sinkholes be delineated.

Sinkholes formed concurrently with the lowering of the water table in areas

Figure 12.12 Collapse of crushing plant into a sinkhole in 1962 at the Westdriefontein Mine, Transvaal, South Africa (courtesy of Gold Fields of South Africa Ltd).

that formerly, in general, had been free from sinkholes. A sinkhole develops when the self-supporting arch of unconsolidated material collapses into a cavity in the dolostone beneath. Initially, the neck of the cavity was small enough to permit the unconsolidated deposits to bridge across it. Normally the cover of unconsolidated material is less than 15 m, otherwise it will choke the cavity on collapse. Bezuidenhout and Enslin (*ibid.*) divided these areas into three groups. First, there were those areas where the original water table was less than 15 m below, and less frequently within 30 m of, the surface. Second, sinkholes formed in the scarp zones that bordered deep buried valleys. Third, sinkholes occurred in narrow buried valleys where the limited width meant that initially unconsolidated material bridged sinks in the dolostone below. Bezuidenhout and Enslin found that sinkholes of larger dimensions than would normally be expected to occur develop in such valleys. The largest one they recorded had a diameter of 80 m and a depth of 50 m.

Slow subsidence occurred as a consequence of consolidation taking place in unconsolidated material as the water table was lowered. The degree of subsidence that took place reflected the thickness and proportion of unconsolidated deposits consolidated. The thickness of these deposits varies laterally, thereby giving rise to differential subsidence, which in turn caused large fissures to occur at the surface. In fact, the most prominent fissures frequently demarcated the areas of subsidence. The total subsidence varied from several centimetres to over 9 m. The time lag between the lowering of the water table and surface subsidence, where observed, has been fairly short.

12.11.3 Subsidence due to land drainage

Poorly drained areas, especially low-lying marshes and swamplands, generally contain a high proportion of organic matter in their associated soils. When such soils are drained, the ground level subsides. The subsidence is not only due to the consolidation that occurs as a result of the loss of the buoyant force of groundwater but is also attributable to desiccation and shrinkage associated with drying out in the zone of aeration and oxidation of the organic material in that zone. For instance, differential subsidence of up to 1 m has been measured in New Orleans, and its amount is related to the type of sediment as well as to the amount by which the water table has been lowered and the time elapsed since reclamation began. The silts and sands of the natural levees are affected least by subsidence. Backswamp and interdistributary clay deposits are subject to shrinkage on drying. Where the organic matter in these clays is more abundant, their subsidence potential is increased.

Peat deposits have by far the greatest potential for subsidence when drained. In fact, because peat is highly porous, with moisture contents that can range up to 2000%, it is by far the most compressible soil type. In order to predict subsidence due to drainage of peat deposits, it is necessary to know the thickness, type, density and rate of decomposition of the peat; the position of the water table and the rate and amount of groundwater lowering; the nature of the reclamation; and the type of climate.

12.11.4 Subsidence due to the abstraction of oil or gas

Subsidence due to the abstraction of oil occurs for the same reason as does subsidence associated with the abstraction of groundwater. In other words, pore pressures are lowered in the oil-producing zones due to the removal not only of oil but also of gas and water, and the increased effective load causes consolidation in compressible beds. Such an explanation was probably first advanced to account for the spectacular and costly subsidence at Wilmington oilfield, California. This area is underlain by approximately 1800 m of Miocene to recent sediments consisting largely of sands, silts and shales. Subsidence was first noticed in 1940 after oil production had been under way for three years. Gilluly and Grant (1949) were among the first to demonstrate that the area of maximum subsidence showed a remarkable coincidence with the productive area of the oilfield. They also indicated that there was a very close agreement between the relative subsidence of the various parts of the oil field and the pressure decline, the thickness of oil sand affected and the mechanical properties of the oil sands. By 1947, subsidence was occurring at a rate of 0.3 m per year and had reached 0.7 m annually by 1951, when the maximum rate of withdrawal was attained. By 1966, an elliptical area of over 75 km^2 had subsided by more than 8.8 m. Because of the seriousness of the subsidence (and the realization that it was due to declines in fluid pressures), remedial action was taken in 1957 by injecting water into and thereby repressurizing the abstraction zones. By 1962, this had brought subsidence to a halt in most of the field. In fact, Yerkes and Castle (1970) referred to more than 40 known examples of differential subsidence, horizontal displacement and surface fissuring/faulting associated with 27 oil and gas fields in California and Texas. The maximum subsidences recorded in these fields are not comparable with that of Wilmington and are generally less than 1 m.

Yerkes and Castle (*ibid.*) also reported that centripetally directed horizontal displacements commonly accompany differential subsidence. Again the Wilmington oilfield offers an example. There the maximum horizontal displacement exceeded 35% of the maximum subsidence. It was located about halfway along the flanks of the subsidence basin, from which it decreased progressively to zero in both directions, that is, towards the centre and periphery of the basin. A maximum horizontal displacement amounting to 3.66 m was recorded between 1937 and 1966, which gave rise to horizontal strains exceeding 1.2%. Between 1957, when repressurization commenced, and 1967, several survey stations along the eastern area of the subsidence basin (an area where vertical rebound approximating to 17% of total differential subsidence has taken place) recovered as much as 80% of their measured horizontal displacement.

The Groningen gas field in the Netherlands has been in production since 1965. Regular precision levelling surveys have been carried out since abstraction of gas commenced in order to monitor the amount of subsidence occurring at ground surface. Contour maps showing the amount of subsidence have been produced from this data. Another method of subsidence monitoring has involved measuring the consolidation of near-surface sediments by using borehole extensometers. Lastly, movements of reservoir consolidation were taken by monitoring the relative displacement of radioactive bullets shot into the formation at regular intervals down a well (Schoonbeck, 1976).

The initial levelling surveys indicated that subsidence was occurring at an amount considerably less than had been predicted. Furthermore, the consolidation coefficients of sediments determined in the laboratory turned out to be about three times those derived from field measurements. As a consequence, the mathematical model that had been developed to predict subsidence had to be modified (Geertsma and Van Opstal 1973).

The modified model has been used to predict the amount of subsidence up to the year 2050, when it is estimated that between 0.25 and 0.3 m will have occurred. Although not appreciable, this amount of subsidence still means that it is necessary to make provision for maintaining water management in the polder region of Groningen province. It has also been used to predict that the greatest amount of subsidence would have occurred before 1990. This does not include any subsidence attributable to consolidation in the overlying Quaternary sediments. These average about 400 m in thickness and are consolidating at between 0.2 and 2 mm per year.

12.11.5 Subsidence due to the abstraction of brine

Those deposits that readily go into solution, notably salt, can be extracted by solution mining. Salt has been obtained by brine pumping in a number of areas in the United Kingdom, Cheshire being by far the most important. Consequently, subsidence due to salt extraction has been an inhibiting factor, especially in parts of Cheshire, as far as major developments have been concerned (Bell, 1992b). This was chiefly because of the unpredictable nature of such subsidence.

Today, wild brine operations, which were responsible for subsidence, have almost ceased in Cheshire. Ninety-nine percent of the brine is now obtained by controlled solution mining from cavities formed in the salt. This has not given rise to any subsidence. The most successful wild brine pumping in Cheshire was carried out on the major natural brine runs. Such pumping accelerated the formation of solution channels. Active subsidence normally concentrated at the head and sides of a brine run, where fresh water first entered the system. Hence, serious subsidence used to occur at considerable distances, anything up to 8 km, from pumping centres.

Subsidence due to wild brine pumping also gave rise to the formation of flashes, that is, water-filled linear hollows. They formed as a result of collapse above brine runs. Their sides are cambered, which is a surface manifestation of the associated subsidence curves. The flanks may also be interrupted by tension scars, along which movements may occur.

Because the exact area from which salt was extracted was not known, the magnitude of subsidence developed could not be related to the volume of salt worked. Consequently, there was no accurate means of predicting the amount of ground movement or strain. However, the rate at which subsidence occurred appears to have been related to the rate at which brine was extracted. Although it is not easy to tell, it seems that subsidence did not necessarily occur immediately after the removal of salt. For example, surface subsidence in some areas was found to continue for six months after the particular well thought to be responsible ceased to operate. Indeed, residual subsidence often continued for one or two years. The maximum strains developed by wild brine pumping could ruin buildings. What was worse, as far as structural damage was concerned, was the formation of tension scars (small faults) on the convex flanks of subsidence hollows (Figure 12.13). These features developed in the surface tills. The vertical displacement along a scar is usually less than 1 m.

Wassman (1980) described subsidence that resulted from the solution mining of salt in the area around Hengelo in the Netherlands. Cavities have been

Figure 12.13 Tension scar produced by wild brine pumping, Cheshire, England. Note the flash in the top right-hand corner.

produced in the salt, and they have subsequently become interconnected. Wassman quoted 1.6 m of subsidence as having taken place in the centre of one subsidence trough, and at Hengelo vertical subsidence amounting to about 50 mm per year is still occurring. Again, certain similarities with subsidence due to coal mining were observed. However, the duration over which the subsidence occurs differs and the area of influence at the surface in relation to depth of cavity is small. Furthermore, Wassman found that some cavities tended to collapse several years after pumping had ceased. In addition, brecciation has taken place in the overlying Red Beds of Bunter age (this claystone or marl loses its strength when wetted). Unfortunately, brine from the cavities in the underlying salt permeates into the marl, rising as a result of capillary action. Eventually this weakened roof material collapses into the cavity. Void migration then takes place and, depending on the bulking factor of the rocks involved, the void may move into the overlying Tertiary clays. Pronounced subsidence occurs when this happens. The subsidence basin in these latter deposits is demarcated by an angle of influence of 45°, which extends upwards from the contact between the Red Beds and Tertiary clays. This explains the relatively small area of influence in relation to the depth of the original cavity. An additional cause of ground movement is attributable to consolidation, which occurs in the brecciated marls. This is responsible for the slowly decreasing but long-lasting subsidence.

Deere (1961) described differential subsidence (resulting from the development of a concavo-convex subsidence profile), simultaneous horizontal displacements and surface faulting associated with the abstraction of sulphur from numerous wells in the Gulf region of Texas. Shortly after mining operations commenced, vertical and horizontal movements were noticed. During the first 31 months of operation, differential subsidence amounting to 1.75 m occurred over an elliptical area exceeding 5 km^2. The subsidence basin was centred directly above the narrow linear producing zone. A normal fault, around 650 m in length, with a downthrow of 0.1 m on the mining side and peripheral to the subsidence basin, running more or less parallel to the subsidence contours, developed suddenly during the fifth month of production (Lee and Strauss 1970). By the thirty-first month the length of the fault had increased to 800 m and the displacement to 0.3 m. The fault dipped at about 40° directly inwards towards the top of the producing zone, and it formed at a point of maximum surface tension (*cf.* the tension scars associated with salt extraction referred to above). Mining of sulphur and potash in Louisiana has given rise to vertical subsidence ranging up to 9 and 6 m, respectively.

12.11.6 Ground failure and the abstraction of fluids

Ground failure is often associated with subsidence due to the abstraction of groundwater, oil, natural gas or brine (Figure 12.14). Because the withdrawal of groundwater easily exceeds that of all the other fluids, some of the most notable failures are found in such subsidence areas. For instance, ground failures associated with subsidence due to groundwater abstraction occur in at least 14 areas in six states in the United States. Some of the best examples of ground failure attributable to the withdrawal of groundwater are found in southern Arizona, the Houston–Galveston region of Texas and the Fremont valley in California. By contrast, only four ground failures have been reported in the San Joaquin valley in California, one of the most noteworthy areas of subsidence due to groundwater abstraction.

Two types of ground failure, namely fissures and faults, are recognized. Fissures may appear suddenly at the surface, and the appearance of some may be preceded by the occurrence of minor depressions at the surface. Within a matter of a year or so most fissures become inactive. In a few instances new fissures have formed in close proximity to older fissures.

The longest fissure zone in the United States is 3.5 km, and lengths of hundreds of metres are typical (Holzer 1980). Individual fissures are commonly not continuous but consist of a series of segments with the same trend. Exceptionally, segments form an *en echelon* pattern. Segments are not connected at the surface. Occasionally a zone, defined by several closely spaced parallel fissures, occurs, the width of which may be of the order of 30 m. Secondary cracks are frequently associated with fissures. These cracks develop subparallel to the main fissure and are usually only a few metres long. They occur at dis-

Figure 12.14 Broken wall due to fissuring of ground caused by abstraction of water, Xian, China.

tances of up to 15 m from the parent fissure. Fissure traces range from linear to curvilinear in outline. Fissures commonly intersect but do not cut through one another.

Fissure separations tend not to exceed a few centimetres, the maximum reported being 64 mm. However, they are frequently enlarged by erosion to form gullies 1–2 m in width and 2–3 m in depth (Figure 12.15). The greatest measured depth so far is 25 m.

Fissure development has caused an increasing problem in the alluvial basins of southern Arizona over several decades. On average, groundwater levels have declined by some 45 to 130 m, accompanied by more than 0.3 m of subsidence. According to Larson (1986), subsidence generally begins when the decline in water level exceeds 30 m, and with continued subsidence fissures form at points of maximum horizontal tensile stress associated with maximum convex upward

Figure 12.15 Earth fissure in south-central Arizona. The fissure results from erosional
enlargement of tension cracks by differential subsidence. The subsidence is
caused by declining groundwater level (courtesy of Dr Thomas L. Holzer).

curvature of the subsidence profile. Certain conditions are believed to localize
differential subsidence, such as buried bedrock hills and scarps; changes in sedimentary facies; the hinge or zero line of subsidence; the edge of an advancing
subsidence front; man-made changes in vertical loading; and neighbouring
recharge mounds. Geophysical surveys (notably gravity surveys) indicate that
most fissures in southern Arizona are associated with bedrock hills. The most
critical depths for bedrock features would appear to be from 30 to 500 m
because the greatest amount of consolidation of sediments occurs at this interval
as groundwater levels decline. In fact, most irregularities in bedrock surface are
inferred to occur at depths of less than 250 m. The association of fissure occurrence with variable aquifer thickness also suggests that differential consolidation is taking place near these fissures as groundwater levels fall. Theoretical
estimates of horizontal strains thereby generated indicate that this mechanism
is the dominant source of horizontal tension causing fissures (Jachens and
Holzer 1980). Tensile strains at fissures at the time of their formation have
ranged from 0.1 to 0.4%.

However, gravity and magnetic surveys in areas of the Lucerne valley,
California, as well as in some parts of southeastern Arizona, where the fissures
have polygonal patterns, did not indicate special subsurface conditions beneath

the fissures according to Holzer (1986). He suggested that these fissures are probably caused by tension induced by capillary stresses in the zone above the declining water table. Holzer further noted that in the Las Vegas valley, Nevada, and the Fremont and San Jacinto valleys, California, some fissures are coincident with existing faults.

Those faults that are suspected of being related to groundwater withdrawal are much less common than fissures. They frequently have scarps more than 1 km in length and more than 0.2 m high. The longest such fault scarp in the United States is 16.7 km long, and the highest scarp is around 1 m, both being in the Houston–Galveston region which is the most affected by such faulting in the United States. These fault scarps have been found to increase in height by dip-slip creep along normal fault planes. Measured rates of vertical offset range from 4 to 60 mm annually; however, movement tends to vary with time. Although some short-term episodic movement has been reported, seasonal variations of offset that correlate both in magnitude and timing with seasonal fluctuations of water level are remarkably widespread (Holzer, *ibid.*). The land surface near a scarp may tilt, tilting being greatest near a scarp but having been observed to extend as far as 500 m from a scarp.

The relationship between faulting and fluid abstraction (of both oil and gas, as well as groundwater) in the Houston–Galveston region is ambiguous. Many faults in this region are thought to act as hydrological barriers. Consequently, fluid production on one side of a fault may cause the piezometric surface to decline, with attendant consolidation on that side of the fault and not the other. This differential consolidation may be translated to the surface as fault movement. It has also been maintained that fluid abstraction has brought about the reactivation of faults.

12.11.7 Prediction of subsidence due to the abstraction of fluids

Obviously, it is necessary to determine the amount of subsidence that is likely to occur as a result of the withdrawal of fluids from the ground, as well as to estimate the rate at which it may occur. Unfortunately, a large number of prediction methods have been developed, some of which are relatively simple, while others are complex. One of the probable reasons for this is that stratal sequences are different in different areas where subsidence has occurred and consequently different models have been devised. Nonetheless, a number of steps can be taken in order to evaluate the subsidence likely to occur due to the abstraction of fluids from the ground. These include defining the *in situ* hydraulic conditions; computing the reduction in pore pressure due to removal of a given quantity of fluid; conversion of the reduction in pore pressure to an equivalent increase in effective stress; and estimating the amount of consolidation likely to take place in the formation affected from consolidation data and the increased effective load. In addition to depth of burial, the ratio between maximum subsidence and reservoir consolidation should also take account of the lateral extent of the reservoir in that small reservoirs that are deeply buried do not give rise to noticeable subsidence, even if undergoing considerable consolidation, whereas extremely large reservoirs may develop significant subsidence. The problem is therefore three-dimensional rather than one of simple vertical consolidation. Subsidence prediction in relation to the rate of groundwater pumping should therefore involve relating the consolidation model to a two- or three-dimensional hydrogeological model based on the groundwater flow equation. Variations in the hydraulic head in both time and space in response to groundwater abstraction are obtained from the hydrogeological model. These values can then be used to derive the time-dependent consolidation curve at any point in the system. This provides an indication of the amount of subsidence likely to occur.

Subsidence above water, oil or gas reservoirs is frequently treated as a mechanical problem involving the elastic behaviour of the reservoir and is analysed with the aid of a mathematical model. The materials concerned include in their mechanical properties the combined behaviour of their individual components (that is, elasticity and plasticity of solids; viscosity of liquids; compressibility of gases; decay of organic matter; attraction and repulsion of ionic charges, etc.). Hence the mechanical properties are anisotropic, as well as dependent on stress history and time. Accordingly, such material is difficult, if not impossible, to deal with in a theoretical model of subsidence. Resort has to be made to simplifying assumptions in order to develop any type of model of subsidence prediction; in particular, the properties of the ground must be idealized. Other simplifications may include the assumption that strata are horizontal; that flow in aquifers is horizontal, while it is vertical in aquitards; and that subsidence is due primarily to consolidation of aquitards. Obviously, the greater the number of simplifications that are incorporated into a model, the more restricted its use becomes.

References

Anon. (1975) *Subsidence Engineer's Handbook.* National Coal Board, London.

Anon. (1976) *Reclamation of Derelict Land: Procedure for Locating Abandoned Mine Shafts.* Department of the Environment, London.

Anon. (1977) *Ground Subsidence.* Institution of Civil Engineers, London.

Anon. (1982) *Treatment of Disused Mine Shafts and Adits.* National Coal Board, London.

Anon. (1988) Engineering geophysics: Report by the Geological Society Engineering Group Working Party. *Quarterly Journal Engineering Geology*, 21, 207–271.

Bell, F.G. (1986) Location of abandoned workings in coal seams. *Bulletin of the International Association of Engineering Geology*, 30, 123–132.

Bell, F.G. (1987) The influence of subsidence due to present day coal mining on surface development. In *Planning and Engineering Geology*, Engineering Geology Special Publication No. 4, Culshaw, M.G., Bell, F.G., Cripps, J.C. and O'Hara, M., (eds)

The Geological Society, London, 359–368.

Bell, F.G. (1988a) Land development: State-of-the-art in the search for old mine shafts. *Bulletin International Association of Engineering Geology*, 37, 91–98.

Bell, F.G. (1988b) Subsidence associated with the abstraction of fluids. In *Engineering Geology of Underground Movements*, Engineering Geology Special Publication No. 5, Bell, F.G., Culshaw, M.G., Cripps, J.C. and Lovell, M.A. (eds), The Geological Society, London 363–376.

Bell, F.G. (1992a) Ground subsidence: a general review. *Proceedings Symposium on Construction over Mined Areas*, Pretoria, South African Institution of Civil Engineers, Yeoville, 1–20.

Bell, F.G. (1992b) Salt mining and associated subsidence in mid-Cheshire, England, and its influence on planning. *Bulletin Association of Engineering Geologists*, 22, 371–386.

Bell, F.G. and Fox, R.M. (1991) The effects of mining subsidence on discontinuous rock masses and the influence on foundations: the British experience. *The Civil Engineer in South Africa*, 201–210.

Bezuidenhout, C.A. and Enslin, J.F. (1970) Surface subsidence and sinkholes in dolomitic areas of the Far West Rand, Transvaal, Republic of South Africa. *Proceedings of the First International Symposium on Land Subsidence*, Tokyo International Association of Hydrological Sciences, UNESCO Publication No. 88, 2, 482–495.

Bhattacharya, S. and Singh, M.M. (1985) *Development of Subsidence Damage Criteria.* Office of Surface Mining, Department of the Interior, Contract No J5120129, Engineering International Inc. Washington, DC, 226pp.

Bieniawski, Z.T. (1967) *An analysis of the results from underground tests aimed at determining the in situ strength of coal pillars.* CSIR Report MEG 569, Pretoria.

Bieniawski, Z.T. (1970) In situ large scale testing of coal. *Proceedings Conference on In Situ Testing of Rocks and Soils*, British Geotechnical Society, London, 67–74.

Bjerrum, L. (1963) Discussion, Session IV. *Proceedings European Conference on Soil Mechanics and Foundation Engineering*, Wiesbaden, 2, 135–137.

Brauner, G. (1973a) *Subsidence due to Underground Mining: Part II, Ground Movements and Mining Damage.* Bureau of Mines, Department of the Interior, US Government Printing Office, Washington DC, 53pp.

Brauner, G. (1973b) *Subsidence due to Underground Mining. Part I, Theory and Practice in Predicting Surface Deformation.* Bureau of Mines, Department of the Interior, US Government Printing Office, Washington, DC, 56pp.

Brook, C.A. and Alison, T.L. (1986) Fracture mapping and ground subsidence susceptibility modelling in covered karst terrain: the example of Dougherty County, Georgia. *Proceedings Third International Symposium on Land Subsidence*, Venice. International Association of Hydrological Sciences, Publication No. 151, 595–606.

Browne, M.A.E., Forsyth, I.H. and Macmillan, A.A. (1986) Glasgow: a case study in urban geology. *Journal Geological Society*, 143, 509–520.

Bryan, A., Bryan, J.G. and Fouche, J. (1964) Some problems of strata control and support in pillar workings. *Mining Engineer*, 123, 238–266.

Burland, J.B. and Wroth, C.P. (1975) Allowable and differential settlement of structures including damage and soil-structure interaction. In *Settlement of Structures*, Pentech Press, London, 611–654.

Burland, J.B., Broms, B.B. and De Mello, V.F.P. (1977) Behaviour of foundations and structures. *Proceedings Ninth International Conference on Soil Mechanics and Foundation*

Engineering, Tokyo, 2, 495–547.

Burton, D.A. (1978) A three-dimensional system for the prediction of surface movements due to mining. *Proceedings First International Conference on Large Ground Movements and Structures*, Cardiff, Geddes, J.D. (ed.), Pentech Press, London, 209–28.

Carter, P.G. (1985) Case histories which break the rules. In *Mineworkings '84, Proceedings of International Conference on Construction in Areas of Abandoned Mineworkings*, Edinburgh, Forde, M.C., Topping, B.H.V. and Whittington, H.W. (eds), Engineering Technics Press, Edinburgh, 20–29.

Coates, D.F. (1965) *Pillar load. Part I. Literature survey and new hypothesis.* Department of Mining and Technical Surveying, Ottawa, Mines Branch, Research Report R168.

Cripps, J.C., McCann, D.M., Culshaw, M.G. and Bell, F.G. (1988) The use of geophysical methods as an aid to the detection of abandoned shallow mine workings. In *Minescape '88, Proceedings of Symposium on Mineral Extraction, Utilisation and Surface Environment*, Harrogate, Institution of Mining Engineers, Doncaster, 281–289.

Dean, J.W. (1967) Old mine shafts and their hazard. *Mining Engineer*, 127, 368–377.

Deere, D.U. (1961) Subsidence due to mining – a case history from the Gulf region of Texas. *Proceedings Fourth Symposium on Rock Mechanics.* Bulletin of the Mining Industries Experimental Station, Engineering Series, Hartman, H.L. (ed.), Pennsylvania State University, 59–64.

Denkaus, H.G. (1962) Critical review of the present state of Scientific knowledge related to the strength of mine pillars. *Journal South African Institute of Mining and Metallurgy*, 63, 59–75.

Emsley, S.J., Summers, J.W. and Styles, P. (1992) The detection of sub-surface mining related cavities using micro-gravity technique. *Proceedings Symposium on Construction over Mined Areas*, Pretoria, South African Institution of Civil Engineers, Yeoville, 27–35.

Garrard, G.E.G. and Taylor, R.K. (1988) Collapse mechanisms of shallow coal mine workings from field measurements. In *Engineering Geology of Underground Movements*, Engineering Geology Special Publication No. 5, Bell, F.G., Culshaw, M.G., Cripps, J.C. and Lovell, M.A. (eds), The Geological Society, London, 181–192.

Geertsma, J. and Van Opstal, G. (1973) A numerical technique for predicting subsidence above compacting reservoirs, based on the nucleus strain concept. *Geologie en Mijnbouw*, 28, 63–74.

Gilluly, J. and Grant, U.S. (1949) Subsidence in the Long Beach area, California. *Bulletin Geological Society of America*, 60, 461–560.

Goodman, R.E., Korbay, S. and Buchignani, A. (1980) Evaluation of collapse potential over abandoned room and pillar mines. *Bulletin Association of Engineering Geologists*, 17, 27–37.

Greenwald, H.P., Howarth, H.C. and Hartman, I. (1941) *Experiments on strength of small pillars of coal in the Pittsburg bed.* US Bureau of Mines, Technical Paper 605 (1939), Report Investigation 3575, Washington DC.

Gregory, O. (1982) Defining the problem of disused coal mine shafts. *Chartered Land Surveyor/Chartered Mine Surveyor*, 4, No. 2, 4–15.

Gubellini, A., Lombardini, G. and Russo, P. (1986) Application of high precision levelling and photogrammetry to the detection of the movements of an architectonic complex produced by subsidence in the town of Bologna. *Proceedings Third International Symposium on Land Subsidence*, Venice, International Association of Hydrological Sciences, Publication No. 151, 257–267.

Healy, P.R. and Head, J.M. (1984) *Construction over Abandoned Mine Workings.* Construction Industry Research and Information Association, Special Publication 32, London.

Hellewell, F.G. (1988) The influence of faulting on ground movement due to coal mining. The UK and European experience. *Mining Engineer*, 147, 334–337.

Hoek, E. and Brown, E.T. (1980) *Underground Excavations in Rock.* Institution of Mining and Metallurgy, London.

Holland, C.T. (1964) The strength of coal mine pillars. *Proceedings Sixth Symposium Rock Mechanics*, Missouri University, Rolla, Pergamon Press, 450–466.

Holzer, T.L. (1980) Faulting caused by groundwater level declines, San Joaquin Valley, California. *Water Resources Research*, 16, 1065–1070.

Holzer, T.L. (1986) Ground failure caused by groundwater withdrawal from unconsolidated sediments. *Proceedings Third Symposium on Land Subsidence*, Venice, International Association of Hydrological Sciences, Publication No. 151, 747–56.

Hood, M., Ewy, R.T. and Riddle, R.L. (1983) Empirical methods of subsidence prediction – a case study from Illinois. *International Journal Rock Mechanics and Mining Science and Geomechanical Abstracts*, 20, 153–170.

Jachens, R.C. and Holzer, T.L. (1980) Geophysical investigations of ground failure related to groundwater withdrawal, Picacho basin, Arizona. *Ground Water*, 17, No. 6, 574–585.

Jammal, S.E. (1986) The Winter Park sinkhole and Central Florida sinkhole type subsidence. *Proceedings Third International Symposium on Land Subsidence*, Venice, International Association of Hydrological Sciences, Publication No. 151, 585–594.

Larson, M.K. (1986) Potential for fissuring in the Phoenix, Arizona, USA, area. *Proceedings Third International Symposium on Land Subsidence*, Venice, International Association of Hydrological Sciences, Publication No. 151, 291–300.

Lee, A.J. (1966) The effect of faulting on mining subsidence. *Mining Engineer*, 125, 417–427.

Lee, K.L. and Strauss, M.E. (1970) Prediction of horizontal movements due to subsidence over mined areas. *Proceedings First International Symposium on Land Subsidence*, Tokyo, International Association of Hydrological Sciences, UNESCO Publication No. 88, 2, 515–22.

Lofgren, B.N. (1968) *Analysis of stress causing land subsidence.* US Geological Survey, Professional Paper 600–B, 219–225.

Lofgren, B.N. (1979) Changes in aquifer system properties with groundwater depletion. In *Evaluation and Prediction of Subsidence, Proceedings Specialty Conference of the American Society of Civil Engineers*, Gainsville, Saxena, S.K. (ed.). 26–46.

Marino, G. and Gamble, W. (1986) Mine subsidence damage from room and pillar mining in Illinois, *International Journal of Mining and Geological Engineering*, 4, 129–150.

McCann, D.M., Suddaby, D.L. and Hallam, J.R. (1982) *The Use of Geophysical Methods in the Detection of Natural Cavities, Mineshafts and Anomalous Ground Conditions.* Open File Report No. EG82/5, Institute of Geological Sciences, London.

McCann, D.M., Jackson, P.D. and Culshaw, M.G. (1987) The use of geophysical surveying methods in the detection of natural cavities and mineshafts. *Quarterly Journal Engineering Geology*, 20, 59–73.

McMillan, A.A. and Browne, M.A.E. (1987) The use and abuse of thematic information maps. In *Planning and Engineering Geology*, Engineering Geology Special Publication No. 4, Culshaw, M.G., Bell, F.G., Cripps, J.C. and O'Hara, M. (eds), The Geological Society, London, 237–246.

Orchard, R.J. (1957) Prediction of the magnitude of surface movements. *Colliery Engineer*, 34, 455–62.

Orchard, R.J. and Allen, W.S. (1970) Longwall partial extracation systems. Mining Engineer, 127, 523–535.

Piggott, R.J. and Eynon, P. (1978) Ground movements arising from the presence of shallow abandoned mine workings. *Proceedings First International Conference on Large Ground Movements and Structures*, Cardiff, Geddes, J.D. (ed.), Pentech Press, London, 749–780.

Polshin, D.E. and Tokar, R.A. (1957) Maximum allowable non-uniform settlement of structures. *Proceedings Fourth International Conference Soil Mechanics and Foundation Engineering*, London, 1, 402–406.

Price, D.G. (1971) Engineering geology in the urban environment. *Quarterly Journal Engineering Geology*, 4, 191–208.

Price, D.G., Malkin, A.B. and Knill, J.L. (1969) Foundations of multi-storey blocks on Coal Measures with special reference to old mine workings. *Quarterly Journal Engineering Geology*, 1, 271–322.

Ren, G., Reddish, D.J. and Whittacker, B.N. (1987) Mining subsidence and displacement prediction using influence function methods. *Mining Science and Technology*, 5, 89–104.

Salamon, M.D.G. (1967) A method of designing bord and pillar workings. *Journal South African Institute of Mining and Metallurgy*, 68, 68–78.

Salamon, M.D.G. and Munro, A.H. (1967) A study of the strength of coal pillars. *Journal South African Institue of Mining and Metallurgy*, 68, 53–67.

Schoonbeck, J.B. (1976) Land subsidence as a result of natural gas extracation in the province of Groningen. *Journal Society Petroleum Engineers*, American Institute Mining Engineers Paper No. SPE 5751, 1–20.

Shadbolt, C.H. (1978) Mining subsidence. *Proceedings First International Conference on Large Ground Movements and Structures*, Cardiff, Geddes, J.D. (ed.), Pentech Press, London, 705–48.

Skempton, A.W. and MacDonald, D.H. (1956) Allowable settlement of buildings. *Proceedings Institution Civil Engineers*, 5, Part III, 727–768.

Stacey, T.R. and Rauch, H.P. (1981) A case history of subsidence resulting from mining at considerable depth. *Transactions South African Institution of Civil Engineers*, 23, 55–58.

Stacey, T.R. and Bakker D. (1992) The erection or construction of buildings and other structures on undermined ground. *Proceedings Symposium on Construction over Mined Areas*, Pretoria, South African Institution of Civil Engineers, Yeoville, 282–288.

Statham, I., Golightly, C. and Treharne, G. (1987) The thematic mapping of the abandoned mining hazard – a pilot study for the South Wales Coalfield. In *Planning and Engineering Geology*, Engineering Geology Special Publication No. 4, Culshaw, M.G., Bell, F.G., Cripps, J.C. and O'Hara, M. (eds), The Geological Society, London, 255–268.

Steart, F.A. (1954) Strength and stability of pillars in coal mines. *Journal of the Chemical, Metallurgical and Mining Society of South Africa*. 54, 309–325.

Sutherland, H.J. and Munsen, D.E. (1984) Prediction of subsidence using complementary influence functions. *International Journal Rock Mechanics and Mining Science and Geomechanical Abstracts*, 21, 195–202.

Szechy, K. (1970) *The Art of Tunnelling*. Akademia Kiado, Budapest.

Terzaghi, K. (1943) *Theoretical Soil Mechanics*. Wiley, New York.

Terzaghi, K. (1946) Introduction to tunnel geology. In *Rock Tunnelling with Steel Supports*, Proctor, R.V. and White, T. (eds), Commercial Shearing and Stamping Company, Youngstown, Ohio, 17–99.

Tincelin, E. (1958) *Pression et deformations de terrain dans les mines de fer de Lorraine*. Jouve Editeurs, Paris.

Van Besian, A.C. and Rockaway, J.D. (1988) Influence of overburden on subsidence development over room and pillar mines. In *Engineering Geology of Underground Movements*, Engineering Geology Special Publication No. 5, Bell, F.G., Culshaw, M.G., Cripps, J.C. and Lovell, M.A. (Eds), The Geological Society, London, 215–220.

Walton, G. and Cobb, A.E. (1984) Mining subsidence. In *Ground Movements and Their Effects on Structures*, Attewell, P.B. and Taylor, R.K. (eds), Survey University Press, London, 216–242.

Wardell, K. and Eynon, P. (1968) Structural concept of strata control and mine design. *Transactions Institution of Mining and Metallurgy*, 77, Section A, A125–A150.

Wardell, K. and Wood, J.C. (1965) Ground instability arising from old shallow mine workings. *Proceedings of the Midlands Soil Mechanics and Foundations Society*, 7, 5–30.

Wassmann, T.H. (1980) Mining subsidence in Twente, east Netherlands. *Geologie en Mijnbouw*, 59, 225–231.

White, H. (1992) Accurate delineation of shallow subsurface structure using ground penetrating radar. *Proceedings Symposium on Construction over Mined Areas*, Pretoria, South African Institution of Civil Engineers. Yeoville, 23–25.

Whittacker, B.N. and Reddish, D.J. (1989) *Subsidence: Occurrence, Prediction and Control*. Elsevier, Amsterdam.

Wilson, A.H. (1972) An hypothesis concerning pillar stability. *The Mining Engineer*, 131, 409–417.

Yerkes, R.F. and Castle, R.O. (1970) Surface deformation associated with oil and gas field operation in the United States. *Proceedings First International Symposium on Land Subsidence*, Tokyo, International Association of Hydrological Sciences, UNESCO Publication No. 88, 1, 55–66.

Index

Entries in **bold** refer to figures and tables